Advances in Insect Physiology

Volume 38

Advances in Insect Physiology

Insect Integument and Colour

edited by

Jérôme Casas

Université de Tours,
Institut de Recherche en Biologie de l'Insecte
IRBI UMR CNRS, Tours, France

Stephen J. Simpson

School of Biological Sciences,
The University of Sydney,
Sydney, Australia

Volume 38

ELSEVIER

AMSTERDAM • BOSTON • HEIDELBERG • LONDON • NEW YORK • OXFORD
PARIS • SAN DIEGO • SAN FRANCISCO • SINGAPORE • SYDNEY • TOKYO
Academic Press is an imprint of Elsevier

ACADEMIC PRESS

Academic Press is an imprint of Elsevier
32 Jamestown Road, London NW1 7BY, UK
Radarweg 29, PO Box 211, 1000 AE Amsterdam, The Netherlands
Linacre House, Jordan Hill, Oxford OX2 8DP, UK
30 Corporate Drive, Suite 400, Burlington, MA 01803, USA
525 B Street, Suite 1900, San Diego, CA 92101-4495, USA

First edition 2010

ISBN: 978-0-12-381389-3
ISSN: 0065-2806

Printed and bound by CPI Group (UK) Ltd, Croydon, CR0 4YY

Transferred to Digital Print 2012

Contents

Contributors

Jérôme Casas
Institut de Recherche sur la Biologie de l'Insecte, Université de Tours, IRBI UMR CNRS, Tours, France; and INRA, UR Zoologie Forestière, Orléans, France

Ephraim Cohen
Robert H. Smith Faculty of Agriculture, Food and Environment, The Hebrew University of Jerusalem, Rehovot, Israel

Helen Ghiradella
Department of Biology, The University at Albany, Albany, New York, USA

Doris Gomez
Département d'Ecologie et de Gestion de la Biodiversité, CNRS UMR 7179, Muséum National d'Histoire Naturelle, Brunoy, France

Ali Miserez
School of Materials Science and Engineering, Nanyang Technological University, Singapore, Singapore

H. Frederik Nijhout
Department of Biology, Duke University, Durham, North Carolina, USA

Daniel J. Rubin
School of Engineering and Applied Sciences, Harvard University, Cambridge, Massachusetts, USA

Priscilla Simonis
Research Center in Physics of Matter and Radiation (PMR), University of Namur (FUNDP), Namur, Belgium

Stephen J. Simpson
School of Biological Sciences, University of Sydney, Sydney, New South Wales, Australia

Marc Théry
Département d'Ecologie et de Gestion de la Biodiversité, CNRS UMR 7179, Muséum National d'Histoire Naturelle, Brunoy, France

Jean-Pol Vigneron
Research Center in Physics of Matter and Radiation (PMR), University of Namur (FUNDP), Namur, Belgium

J. Herbert Waite
Molecular, Cell and Developmental Biology Department, University of California, Santa Barbara, California, USA

Increasing Demands and Vanishing Expertise in Insect Integrative Biology

Jérôme Casas*,† and Stephen J. Simpson‡

*Institut de Recherche sur la Biologie de l'Insecte, Université de Tours, IRBI UMR CNRS, Tours, France
†INRA, UR Zoologie Forestière, Orléans, France
‡School of Biological Sciences, University of Sydney, Sydney, New South Wales, Australia

We regard *Advances in Insect Physiology* as the journal in which deep, lasting reviews and syntheses can be published, at a level of comprehensiveness which precludes publication in other venues. The planning and production of each volume is therefore an exciting moment for the editors. It is the time when we hope to identify those fields which are setting tomorrow's agenda, and to discover those areas which are losing ground, often silently. We were unexpectedly surprised by several conflicting trends while producing this volume; trends which apply more broadly than to insect integument and colour, but rather are relevant to the entire field of insect integrative biology.

Insects can be astonishingly colourful and often display extremely delicate patterning. This variety is endless, and there is sustained interest in explaining these patterns, in particular iridescence, as exemplified in two recent books (Berthier, 2009; Kinoshita, 2008). Even the interdisciplinary *Journal of the Royal Society Interface*, which puts equal emphasis on physics, chemistry, mathematics and biology, had a thematic issue on iridescence, in which insects have a prominent place as study objects (Meadows *et al.*, 2009), sandwiched between a thematic issue on quantitative fluorescence microscopy and one on Biomaterials Research in Japan. Why such a widespread interest? The potential applications in art, design and industry are worth billions per year, as in more fundamental optics (Vigneron and Simonis, this volume). For example, the paint industry is interested in iridescence. Because it contains fragments of multilayer slabs that orient themselves due to surface tension effects, some paints can change colour with the angle of viewing, as measured from the surface normal. The same effects can be achieved, for instance, in cosmetics. As Vigneron and Simmonis explain (this volume), in nature there are examples of total reflection, thin-film filtering, gratings, photonic crystals, lenses, parabolic mirrors, optical fibres, solid-state light sources and much more.

ADVANCES IN INSECT PHYSIOLOGY VOL. 38
ISBN 978-0-12-381389-3
DOI: 10.1016/S0065-2806(10)38012-X

Natural selection has put two gratings on the same surface, and yielded other effects that we have not yet invented. Chitin, one of the building blocks of these structures, is also used worldwide for numerous purposes. Due to their biodegradability, biocompatibility and non-toxicity, chitin, chitosan and their chemical modification products cover a large range of useful applications. The textile and pharmaceutical industries as well as agriculture, water treatment, cosmetics, food and photography use them (Cohen, this volume) and one can only wonder how this complex material is produced and works. It also represents a high point in the evolution of load-bearing tissues due to the rate of maturation, low density, tunability and robustness of its mechanical properties, as well as its resistance to moisture, among other properties (Rubin *et al.*, this volume).

Biomimetics, which we have already referred to in an earlier volume (Casas and Simpson, 2008), is becoming an increasing force in the development of new technologies; insects, by their sheer variety, are an endless source of inspiration. Behavioural ecologists, too, show increasing interest in colours, as they may correlate with the health status or sexual readiness of the displayer, for example. We human do not necessarily see colours in the same way as our study organism; one should therefore consider the visual system of the receiver to make meaningful interpretations of animal communication. This shift in understanding brought a healthy dose of the sensory physiology of colour vision into behavioural ecology, insects being again very good models (Théry and Gomez, this volume). Thus, the interest in and need for dedicated work on the physiology of insect integument and colours is mounting and the future is bright.

Are we up to the task? It seems not, and the situation is getting worst. Take insect pigments, a notable and regrettable omission in this volume. More precisely, let us consider the different kinds of melanins. Melanins are being studied intensively, by many different groups of scientists, most of them related to human health. Ecologists, too, are proposing theories regarding melanin, including on insects, and carry out studies with melanin bioassays run on a routine basis. The number of excellent recent reviews on the mechanisms of melanin production and melanin physico-chemical properties is large and increasing rapidly (see Meredith and Sarna, 2006, or Simon *et al.*, 2009 and references therein). None is, however, coming from entomological quarters, despite important advances in relation to the control and use of melanins (see, e.g., Hiruma and Riddiford, 2009).

The melanins are not an isolated case, as shown in another example, the ommochromes (Nijhout, this volume). This pigment class is present in nearly every insect, as a light barrier in the eye (latest references in Insausti and Casas, 2008, 2009). Many arthropods use it for yellow, red or brown colouration. Its biochemistry has been partially elucidated from the 1960s to the 1980s, culminating with many reviews and leading to the discovery of a new class of compounds, the papiliochromes. The prospects for profound work on the biochemistry and photoreactions of these pigments were excellent at that time. Indeed, several workers then hinted that there were potential biochemical links between ommochromes, pterines and melanin. Such prospects, which would have partially solved the incredibly complex matrices of pigment production and maybe animal colouration, were considered of sufficient importance to warrant publication in the journal *Science* (Bagnara *et al.*, 1979). Why did this expertise nearly vanish two decades ago, just before the advent of molecular biology and gene silencing? Today, there is little activity on the chemistry, physics and biochemistry of ommochrome pigments, for insects or any other organisms. Many of those who worked on such pigments went into retirement. Those who are still in the science of photochemistry and photobiology of pigments have moved to other, related, substances of more promising applicability (e.g., in one notable case to bioactive-pharmaceutical anti-tumour compounds, which descend from the xanthommatin scaffold). The mainstream journal in the field changed from 'Pigment Cell Research' to 'Pigment, Cell and Melanoma Research' and now focuses nearly exclusively on biomedical issues. Newer techniques, even when adapted specifically for such difficult chemical groups (Vogliardi *et al.*, 2004), are not even picked up by the community for later use—it is plainly too late, the community seems not to exist anymore. Is this work then obsolete, irrelevant? No. As a single example, the latest work on wing pattern development and on the sensory biology of colour vision of *Heliconius*, one of the most important butterfly groups for understanding mimicry, would greatly benefit from such knowledge (Briscoe *et al.*, 2010; Ferguson and Jiggins, 2009; Nijhout, this volume). The same litany can be extended to functional morphology in general. The chapter of Gihradella (this volume) is indeed a reminder that a lot is still to be discovered by describing intriguing structures at a very small scale, on mites for example, and by asking basic questions about how and why such structures function as they do. In an age often described as the age of nanotechnology, functional morphology seems to be of highest potential value, but is in fact on the brink of extinction. France, for example, seems no longer to have a single academic position within its ca. 70 universities for a dedicated insect functional morphologist.

We think that the fate of insect integrative biology will be determined through the relative timing of two opposite and concurrent processes: the disappearance of skilled knowledge on one hand and the increased interest and dedication of sciences at a higher integrative level, most likely behavioural ecology *sensu lato*, on the other hand. A third process, molecular biologists moving up to physiology via systems biology, is also at play but might be too slow to reach such levels of integration. For us editors, the duty in the meantime is to keep the flame alive, by bringing like minds together and making their knowledge available such that it might inspire new directions of research in other adventurous scientists. The same spirit explains why we opted for having a chapter on protein sclerotization in marine invertebrates (Rubin *et al.*, this volume): the findings are obviously highly relevant to insect biology and the expertise too scant to restrict its dissemination on the basis of (shifting) taxonomical boundaries.

We hope that this volume of *Advances in Insect Physiology*, too, contributes to the revival of insect integrative biology. Should you feel moved to suggest a thematic volume for the future that has similar aspirations, please feel free to contact us.

Acknowledgements

We warmly thank the referees who dedicated many hours checking large manuscripts. The work of J. Casas is supported by the Customized Intelligent Life Inspired Arrays (CILIA) project (FP6-IST-016039), funded by the European Community under the Information Society Technologies (IST) Program, Future and Emergent Technologies (FET), Lifelike Perception Systems Action, by the French ANR Project Blanc Physique 'Entomopter: The physics of flapping flight inspired by insects', and by a sabbatical grant from INRA. S. Simpson is funded by the Australian Research Council Federation and Laureate Fellowship schemes.

References

Bagnara, J. T., Matsumoto, J., Ferris, W., Frost, S. K., Turner, W. A. Jr., Tchen, T. and Taylor, J. D. (1979). Common origin of pigment cells. *Science* **203**, 410–415.

Berthier, S. (2009). Iridescence: The Physical Colors of Insects. Springer-Verlag, Paris.

Briscoe, A. D., Bybee, S. M., Bernard, G. D., Yuan, F. R., Sison-Mangus, M. P., Reed, R. D., Warren, A. D., Llorente-Bousquets, J. and Chiao, C. C. (2010). Positive selection of a duplicated UV-sensitive visual pigment coincides with wing pigment evolution in *Heliconius* butterflies. *Proc. Natl. Acad. Sci. USA* **107**, 3628–3633.

Casas, J. and Simpson, S. J. (2008). Insect Mechanics and Control. Academic Press/Elsevier, San Diego, CA.

Ferguson, L. C. and Jiggins, C. D. (2009). Shared and divergent expression domains on mimetic *Heliconius* wings. *Evol. Dev.* **11**, 498–512.

Hiruma, K. and Riddiford, L. (2009). The molecular mechanisms of cuticular melanization: the ecdysone cascade leading to dopa decarboxylase expression in *Manduca sexta*. *Insect Biochem. Mol. Biol.* **39**, 245–253.

Insausti, T. and Casas, J. (2008). The functional morphology of color changing in a spider: development of ommochrome pigment granules. *J. Exp. Biol.* **211**, 780–789.

Insausti, T. and Casas, J. (2009). Turnover of pigment granules: cyclic catabolism and anabolism of ommochromes within epidermal cells. *Tissue Cell* **41**, 421–429.

Kinoshita, S. (2008). Structural Colors in the Realm of Nature. World Scientific Press, Singapore.

Meadows, M. G., Butler, M. W., Morehouse, N. I., Taylor, L. A., Toomey, M. B., McGraw, K. J. and Rutowski, R. (2009). Iridescence: more than meets the eye. *J. R. Soc. Interface* **6**, S107–S265(A theme supplement).

Meredith, P. and Sarna, T. (2006). The physical and chemical properties of eumelanin. *Pigment Cell Res.* **19**, 572–594.

Simon, J. D., Peles, D., Wakamatsu, K. and Ito, S. (2009). Current challenges in understanding melanogenesis: bridging chemistry, biological control, morphology, and function. *Pigment Cell Melanoma Res.* **22**, 563–579.

Vogliardi, S., Bertazzo, A., Comai, S., Costa, C. V. L., Allegri, G., Seraglia, R. and Traldi, P. (2004). An investigation on the role of 3-hydroxykynurenine in pigment formation by matrixassisted laser desorption/ionization mass spectrometry. *Rapid Commun. Mass Spectrom.* **18**, 1413–1420.

Chitin Biochemistry: Synthesis, Hydrolysis and Inhibition

Ephraim Cohen

Robert H. Smith Faculty of Agriculture, Food and Environment, The Hebrew University of Jerusalem, Rehovot, Israel

1 Introduction

The etymological origin of the word chitin (chitine in French; coined in 1836) stems from the word 'chiton' that in Latin stands for mollusc that in turn may be related to the Greek word 'khiton' meaning tunic or shirt. Chitin is described (in

ADVANCES IN INSECT PHYSIOLOGY VOL. 38
ISBN 978-0-12-381389-3
DOI: 10.1016/S0065-2806(10)38005-2

1913 Webster's Revised Unabridged Dictionary) as ''a white amorphous horny substance forming the harder part of the outer integument of insects, crustacea, and various other invertebrates''. Chitin is believed to be a major structural component of animal skeletons since at least the Cambrian Period, more than 550 million years ago, although it probably originated in eukaryotic protozoans sometime in the Proterozoic Eon (Miller, 1991). Chemically detectable remains of chitin have been verified in 25-million-year-old insect fossils (Stankiewicz *et al.*, 1997) and even in crustacean fossils dating back 65 million years (Bierstedt *et al.*, 1998).

Chitin is one among a number of extracellular polymers with fundamentally important physiological functions in prokaryotes and eukaryotes (Table 1). Yet in terms of biomass, chitin is a globally abundant extracellular biopolymer, second only to cellulose and possibly to lignin (Cohen, 1993a). The annual production of cellulose (by plants, algae and phytoplankton) and chitin (by marine zooplankton and invertebrates as well as by terrestrial arthropods) was estimated in the range of 10^{11} and 10^{10} tons, respectively (Duchesne and Larson, 1989; Gooday, 1990c). There is a worldwide market for chitin and for certain derived products such as chitosan (the deacetylated form of chitin) and glucosamine. The source of such products is the fishing industry, which generates large quantities of marine arthropods (crabs, lobsters and shrimps) waste. Due to their biodegradability, biocompatibility and non-toxicity, chitin, chitosan and their chemical modification products cover a large range of useful applications in the textile and the pharmaceutical industries as well as *inter alia* in agriculture, water treatment, cosmetics, food and photography products (Kumar, 2000;

TABLE 1
Distribution of extracellular biopolymers

Extracellular biopolymer[a]	Distribution
Polysaccharides[b]	
Cellulose	Plants, algae, fungi[c]
Hemicellulose	Plants
Glucan, mannan	Plants, fungi
Chitin, chitosan	Invertebrates, fungi, diatom algae
Alginate	Algae, bacteria
Hyaluronan	Vertebrates, bacteria, algae
Proteins, proteoglycans	
Collagen	Invertebrates, vertebrates
Keratin, heparin,	
Chondroitin, dermatan sulphate	Vertebrates
Exoskeletal and PM proteins	Invertebrates (arthropods)
Aromatic polymers	
Lignin	Plants

[a] Composite polymers: lignocellulose, chito-protein, chitokeratan.
[b] Polymerization mediated by glycosyltransferase activity.
[c] The Oomycetes group.

Kurita, 2001; Tharanathan and Kittur, 2003). Still the potential applications and use of the available chitin biomass are underexploited.

Chitin and cellulose are long-chained extracellular polysaccharides formed by progressive additions of their respective glycoside residues by catalytic reactions involving transfers of glycosyl units (mediated by processive glycosyl transferase enzymes), and quantitatively such reactions are the paramount global biotransformation event. Chitin does not accumulate in the oceanic environment due to widespread hydrolytic action of marine micro-organisms, notably bacteria (Bassler *et al.*, 1991). It was shown that chitin is readily and efficiently degradable in open water layers down to the anaerobic compartments (Poulicek and Jeuniaux, 1991). Degradation (up to mineralization) of chitin is efficiently attained by a network of myriad terrestrial and marine organisms and micro-organisms.

The cycles of chitin synthesis and hydrolysis in chitin-forming organisms and micro-organisms comprise a network of strictly integrated spatially and temporally regulated processes. Such processes involve specific genes and their transcripts, translational events and post-translational modifications, intracellular traffic of catalytic units, integration into cell membrane compartments and catalysis that form the chitin biopolymer. The biopolymers are translocated across cellular plasma membrane barriers and subsequently crystallized into microfibrils that associate with other extracellular components forming rigid supportive scaffold structures such as exoskeletons in insects or cell walls in fungi. Such structures, vital for survival, are cyclically degraded by regulated hydrolysis to allow and accommodate necessary modifications in shape or size required for growth and development. The physiologically essential processes of chitin synthesis and degradation have been regarded as targets for interference, and potent inhibitors were subjected to further scrutiny in a hope of developing commercial pharmaceuticals and pest control agents. Various inhibitors of catalytic and post-catalytic processes as well as post-transcriptional gene silencing (using RNAi knockout methodology) were useful tools in gaining insights into and better understanding of the intricate biochemical mechanisms involved in chitin metabolism and the functional analysis of the genes involved. Crystal structures of several chitinolytic enzymes facilitated our perception of the three-dimensional organization, catalytic mechanisms and binding of inhibitors at the active sites. X-ray diffraction data of enzyme–inhibitor complexes and structure–activity relationship (SAR) studies may contribute in designing better compounds for further optimization and developing of effective drugs and biopesticides.

Intricate mechanisms implicated in chitin synthesis, deposition and degradation are based to a large extent on fundamental studies with micro-organisms like yeast and filamentous fungi. The plethora of mutants in such systems helped in elucidating gene function and regulation, and facilitated the mechanistic dissection of complex chitin-related anabolic and catabolic processes. Obviously, caution should be exercised by researchers studying insect chitin metabolism to directly adapt conclusions and theories based on taxonomically distant organisms like yeast and fungi. Moreover, one should be prudent in presenting any generalization based on or derived from research performed with limited dipteran, lepidopteran or coleopteran model insect species.

The scope of the present review covers diverse aspects revolving around chitin as a crucial extracellular biopolymer. Its physicochemical properties and structural functions as chito-protein complexes in integuments and peritrophic membranes (PMs) are defined and discussed. Later, I emphasize the catalytic and post-catalytic processes that are involved in chitin synthesis and deposition focusing on genes of the polymerizing enzymes as well as the packaging, trafficking and incorporation of the translated catalytic proteins into plasma membranes. Also, the mechanism of the catalysis and the intricate event of translocating the catalytic products across the plasma membrane barrier are described. Then, I consider in great detail chitin hydrolysis delineating the variety of genes involved, the three-dimensional structure of the hydrolytic enzymes, and mechanisms of catalysis. Next, the knowledge concerning expression patterns and hormonal control and regulation of chitin synthesis and hydrolysis is reviewed. This chapter is followed by a detailed account of inhibition and inhibitors of chitin metabolism acting on the various anabolic and catabolic enzymes and processes outlined before. Finally, in the concluding remarks, I present the numerous gaps in knowledge regarding chitin metabolism and deposition that requires, and pretty much deserve, future intensive research.

2 Chemical and physical aspects

Chitin is a large, straight-chained, water insoluble amino-sugar homopolymer of β-(1-4)-linked *N*-acetyl-D-glucosamine (Fig. 1) synthesized by a highly conserved membrane-bound glycosyltransferase enzyme (chitin synthase (CS), EC 2.4.1.16). The chemically stable polymer is hydrolytically degradable by strong acids or by concentrated hot alkali. Chitin contributes the necessary rigidity and mechanical strength to complex supramolecular structures of arthropod cuticles

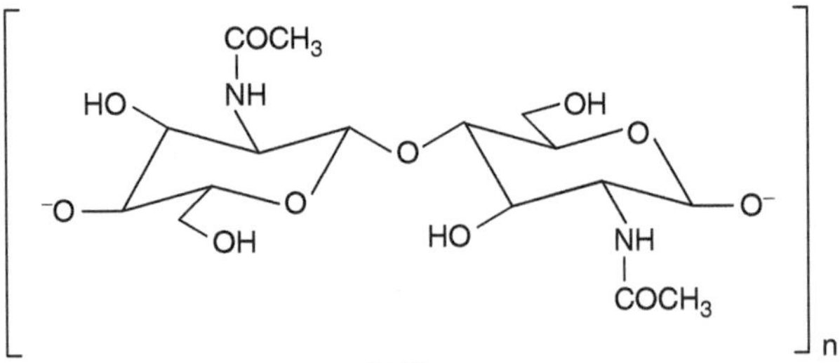

Chitin

FIG. 1 Repetitive unit of chitin.

or fungal cell walls. Such properties, although essential for an external supporting structure, pose problems to the growth and development of the organism beyond the limits of its initial shape or size. To overcome such constraints, the polymer is biologically degraded by a dual enzymatic hydrolysis carried out by a combined action of chitinases and β-*N*-acetyl-glucosaminidases (Kramer and Koga, 1986). This process is required to facilitate budding, mycelial growth, hyphal branching or spore formation in fungi and moulting in arthropods. On the other hand, chitin is subjected to extensive hydrolysis by non-chitinous organisms (plants, invertebrates, vertebrates) and micro-organisms (notably bacteria) for largely nutritional purposes as a source of carbon and nitrogen.

The molecular configuration of a single polymer chain is helicoidal (6 amino-sugar residues per turn) that is stabilized by intramolecular hydrogen bonds. The crystalline state of chitin, in which a number of molecules (~20) coalesce to form microfibrils (~3 nm in diameter), is generated by intermolecular hydrogen bonds. Three highly ordered crystalline structures of chitin polymorphs were described largely by X-ray diffraction studies (Jang *et al.*, 2004; Minke and Blackwell, 1978; Rudall and Kenchington, 1973). The most abundant (in insect cuticles, crustacean shells, and cell walls of fungi) and most thermodynamically stable crystalline polymorph is the α-chitin in which one polymer chain is anitiparallel to the other. It has been speculated that such arrangement is due to either folding back of nascent chitin polymers or by virtue of *in situ* antiparallel biosynthesis at the catalytic sites (Cohen, 1987a). In the less abundant β-chitin polymorph, which is found in squid pens (Fan *et al.*, 2008; Nagahama *et al.*, 2008), PMs in certain beetles (Lehane, 1997; Rudall and Kenchington, 1973), certain diatoms (Blackwell *et al.*, 1967; Herth and Zugenmaier, 1979; Imai *et al.*, 2003) and in setae of brachiopods (Tanaka *et al.*, 1988), the polymer chains are packed in a parallel fashion (Gardner and Blackwell, 1975; Herth and Barthlott, 1979), with reducing ends pointing to the same direction (Imai *et al.*, 2003). The structure of a less common third polymorphic form (γ-chitin), where the orientation of chains alternates so that for every three polymer chains two are parallel, was described in the PM matrix and insect cocoons (Lehane, 1997; Rudall and Kenchington, 1973). The distinct molecular packing patterns of the chitin polymorphs confer different physico-chemical properties to their corresponding crystalline structures. Difference in hydrogen bonding has a marked impact on physical properties of chitin polymorphs. Thus, due to its extensive hydrogen bonds [C3-OH···O-C5 and C6-OH···O=C (intramolecular); NH···O=C and C6-OH···OH-C6 (intermolecular)], α-chitin is a tightly packed crystalline structure (Minke and Blackwell, 1978). It is resistant to swelling in water, moderately withstands hydrolysis by dilute alkali and acids, and is dissolved by concentrated solutions. The β-polymorph with fewer intermolecular hydrogen bonds in the crystallite is less tightly packed, structurally more flexible and easily swells in water. Readers are referred to a recent detailed studies of Jang *et al.* (2004), describing the crystalline structure and the physicochemical characterization of the above three chitin polymorphs, as well as to Kameda *et al.* (2005), who characterized α-chitin using solid-state NMR analysis.

3 Taxonomic occurrence

Chitin as an extracellular structural component, which is generated for largely supportive functions, is found in diverse and unrelated taxonomic groups. Chitin is widespread in invertebrates, particularly in exoskeletons and PMs of arthropods (Andersen, 1979; Neville, 1975). Recently, a chitin-like material was also detected in mosquito egg shells and ovaries (Moreira *et al.*, 2007). Along with its deacetylated form, chitosan, chitin is an important component of cell wall and septum structures as well as in spore formation in yeasts and most filamentous fungi. Chitin is present in calcified layers of mollusc shells and squid pens (Nagahama *et al.*, 2008; Poulicek *et al.*, 1986), in egg shell of nematodes (Brydon *et al.*, 1987; Fuhrman and Piessens, 1985; Zhang *et al.*, 2005) as well as in eukaryotic unicellular organisms like centric diatoms (Herth and Zugemaier, 1977, 1979), certain amoebae cysts (Arroyo-Begovich and Carabez-Trego, 1982; Van Dellen *et al.*, 2006; Ward *et al.*, 1985) and in protozoan parasites (Lanuza *et al.*, 1996). Interestingly, chitin was found to be deposited in the unicellular alga, *Chlorella*, infected by a chlorovirus containing a gene encoding for a chitin polymerizing enzyme (Kawasaki *et al.*, 2002). Chitooligosaccharides are morphogenetic factors implicated in the communication between leguminous plants and *Rhizobium*, and even in vertebrates where they may be important during early stages of embryogenesis (Bakkers *et al.*, 1999; Barny *et al.*, 1996; de Iannino *et al.*, 1995). Chitin and chitosan oligosaccharides act as potent signals eliciting defence responses in plants (Shibuya and Minami, 2001). Chitin is absent in bacteria, algae, plants as well as in vertebrates, although its presence was reported in certain chordates and bony fish (Wagner, 1994; Wagner *et al.*, 1993). Due to its taxonomic distribution, chitin has been considered as a selective and safe target for developing pest control agents (Cohen, 1993b).

4 Structural and functional aspects

Chitin hardly occurs in pure form but rather as an integral element of various extracellular structures such as cuticles and PMs in arthropods or cells' walls in fungi. Chitin crystallites form a basic scaffold for the construction of such structures jointly with either extracellular proteins or polysaccharides in arthropod and fungal systems, respectively. Characteristics of the chito-protein interactions and the organization and length of chitin crystallites determine the architectures and properties of the chitin-containing extracellular formations.

4.1 THE CHITO-PROTEIN COMPLEX OF CUTICLES

The major extracellular chito-protein complex in insects is produced by a monolayer of ectodermal cells lining the body surface, the fore- and midgut, and the tracheal system. Chitin crystallites, embedded in proteinaceous matrix, function akin to long rods in steel-reinforced concrete, giving high tensile strength to supporting elements such as insect exoskeletons. Ultrastructural studies demonstrated that chitin microfibrils may have various organized patterns (helicoidal, unidirectional or pseudo-orthogonal arrangements) that contribute to different mechanical qualities of the cuticle (Neville, 1975). Still the mechanism by which the spatial cuticular patterns are oriented is unknown. One possibility invokes a direct deposition by the underlying epidermal cells (primary orientation according to Neville, 1967), while self-assembly of the chito-protein complexes *per se* was also suggested. It is noteworthy that chitin crystallites in aqueous suspensions coalesce spontaneously and reproduce the helicoidal organization of arthropod cuticles *in vitro* (Revol and Marchessault, 1993). Being efficient in thermodynamic terms, self-assembly was viewed as suitable for extracellular skeletal structures like cuticles (Neville, 1986). In addition, it was proposed that the secondary structure of certain cuticular proteins, which interact with chitin, dictates formation of the helicoidal architecture (Iconomidou *et al.*, 1999).

Electron micrograph transverse sections of pro-cuticles normally depict a lamellar pattern that is parallel to the surface of the epidermis (Fig. 2). Each lamella consists of packed chitin crystallites and a helicoidal pattern is instigated by a gradual progressive rotation of microfibril sheets by a near constant angle. There are different models to explain the helicoidal and unidirectional orientation of the microfibrillar lamellae (Neville, 1986). The pseudo-orthogonal arrangement (plywood type) occurs when both types of orientations pass from one major direction to the next (Neville and Luke, 1969).

4.2 CHITO-PROTEIN COMPLEX OF PERITROPHIC MEMBRANES

The PM is a semi-selective permeable membrane with filtering properties that is restricted to and formed by the midgut of most insects (Hegedus *et al.*, 2009; Lehane, 1997). It protects the epithelial cells of the alimentary canal from mechanical disruption by abrasive ingested food, shields insects from insult by xenobiotic molecules and provides a physical barrier against attack by various pathogens. Lehane (1997) distinguished between type I PM that is secreted by the entire midgut epithelium from type II PM that is formed by a discrete region of the anterior midgut (cardia). In contrast to the highly structurally organized exoskeletal chito-protein complex of cuticles, the PM is a composite amalgam of various proteins, entwined randomly with chitin microfibrils. A recent proteomic study of the mosquito *Anopheles gambiae* PM matrix

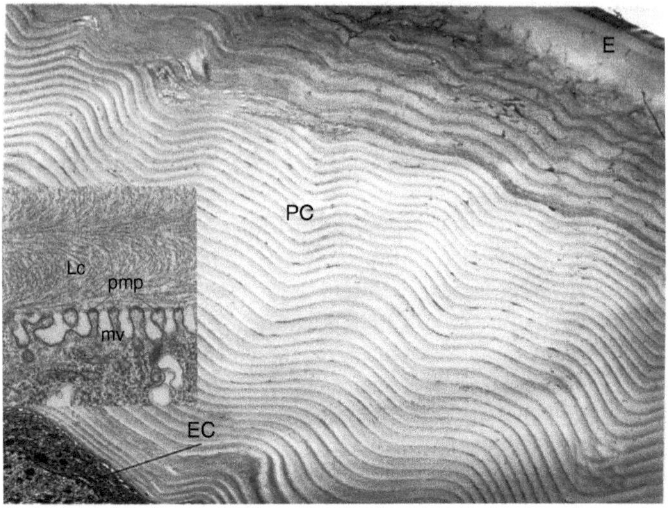

FIG. 2 Cross-section of an insect integument. EC, epithelial cells; PC, procuticle displaying lamellate structure; E, epicuticle. Inset: portraying microvilli at the apical region of epithelial cells, and plasma membrane plaques (pmp) on tips of microvilli (mv), where the putative CS catalytic units are localized. Similar microvilli with apical plaques were depicted for midgut cells producing the PM chito-protein matrix.

identified as many as 209 different proteins (Dinglasan *et al.*, 2009). Chitin content in PM matrices ranges from 3% to 13% with α and γ chitin as the predominant polymorphs (Lehane, 1997). The complex mixture of proteins, glycoproteins and proteoglycans in the PM are divided into four classes according to the ease of their extraction procedure (Tellam *et al.*, 1999). Emphasis was placed on class III known as peritrophins that are glycosylated proteins containing multiple chitin-binding, cysteine-rich domains. For example, the deduced amino acid sequence of a major PM protein (peritrophin-44) of the sheep blowfly fly, *Lucilia cuprina*, larvae revealed five cysteine-rich segments (Elvin *et al.*, 1996). It was also suggested that peritrophins interact with chitin and play an important role in binding chitin microfibrils together, hence contributing tensile strength to PM matrices (Hegedus *et al.*, 2009; Tellam *et al.*, 1999).

5 Synthesis and deposition (from gene to microfibril)

Chitin formation is a cyclical, orderly event that involves a multi-faceted dynamic display of integrated and harmonized intracellular and extracellular biochemical, structural and functional processes and modifications, part of which is poorly understood. Intracellular events involve gene expression of polymerizing enzyme units and their delivery to and integration into plasma membranes. Since chitin is an extracellular material, its formation is topologically restricted to cell areas facing the exterior. Combined ultrastructural and biochemical studies indicated that the catalytic machinery involved in chitin formation is basically composed of clusters of tightly packed membrane-bound polymerizing enzymes with their catalytic units facing the cytoplasm (Cohen, 1991; Duran *et al.*, 1975; Sentandreu *et al.*, 1984). Such an arrangement most likely facilitates spatial closeness of nascent chitin chains that facilitates formation of crystallite microfibrils by accessible intercellular hydrogen bonding. Histological and histochemical investigations in insects like the large cana leafroller *Calpodes ethlius*, the triatomid bug *Rhodnius prolixus* and the tobacco hornworm *Manduca sexta* clearly showed typical structures of distinct plasma membrane plaques on tips of projected microvilli that emerge from neighbouring epithelial cells facing the cuticle or the midgut lumen (Locke, 1991; Merzendorfer and Zinoch, 2003; Fig. 2, inset). Such plaques undergoing periodic turnover were implicated in chitin formation required for cuticle deposition at moulting (Locke and Huie, 1979). Electron microscopy and immunohistochemical studies revealed that enzymes involved in PM chitin synthesis are localized in midgut brush border membranes of the *M. sexta* and the European corn borer, *Ostrinia nubilalis*, is restricted to apical ends of microvilli (Hopkins and Harper, 2001; Zimoch and Merzendorfer, 2002). Also similar microvilli structures with apical plasma membrane plaques associated with PM formation were described in the digestive track of the mite *Acarus siro* (Sobotnik *et al.*, 2008).

5.1 CHITIN SYNTHASE (CS)—GENES

A large number of genes encoding the chitin polymerizing enzyme (CS) in yeast and filamentous fungi have been isolated and sequenced in the past three decades (Bullawa, 1993; Yarden and Yanofsky, 1991). Only in the last decade a number of *CS* genes from invertebrates notably nematodes (Fanelli *et al.*, 2005; Foster *et al.*, 2005; Harris *et al.*, 2000; Veronico *et al.*, 2001; Zhang *et al.*, 2005) have been reported. Also recently, a *CS* gene from a marine bivalve mollusc (*Atrina rigida*) was cloned, sequenced and characterized (Weiss *et al.*, 2006). The first cloned and sequenced *CS* cDNA (5757 bp) from an insect was reported for the blowfly *L. cuprina* (Tellam *et al.*, 2000). Since then, a number of other cloned and sequenced *CS* genes have been available from dipteran, lepidopteran and coleopteran species including the mosquitoes *Aedes aegypti* (Ibrahim *et al.*, 2000; Kato *et al.*, 2006) and *Anopheles quadrimaculatus* (Zhang and Zhu, 2006), the fruit fly *Drosophila melanogaster* (Gagou *et al.*, 2002), the tobacco hornworm *M. sexta* (Hogenkamp *et al.*, 2005), the fall armyworm *Spodoptera frugiperda* (Bolognesi *et al.*, 2005), the beet armyworm *Spodoptera exigua* (Chen *et al.*, 2007; Kumar *et al.*, 2008), the diamondback moth *Plutella xylostella* (Ashfaq *et al.*, 2007), and the red flour beetle *Tribolium castaneum*

(Arakane *et al.*, 2004). *CS* genes of the honey bee *Apis mellifera*, the mosquito *A. gambiae* and the fruit fly *Drosophila pseudo-obscura* were deduced from their respective genomic libraries. As more insect genomes are sequenced, additional *CS* genes are expected to become available in the future.

Multiple *CS* genes (grouped into two divisions and seven classes) encoding for a variety of CS isoenzymes with different biochemical properties and physiological functions were reported. Three and four CS isozymes were cloned in the yeasts *Saccharomyces cerevisiae* and *Candida albicans*, respectively (Lenardon *et al.*, 2007; Lesage and Bussey, 2006), six genes in *Botrytis cinerea* (Choquer *et al.*, 2004), and according to amino acid sequence similarities, up to 10 isozymes were identified in various filamentous fungi (Abramczyk *et al.*, 2009). Similarly, multiple isozymes were described in the cloned (CS analogous) cellulose polymerizing enzyme genes (Pear *et al.*, 1996). At least 10 isoforms of cellulose synthase isozymes were identified in *Arabidopsis thaliana* (Lindeboom *et al.*, 2008), while 12 members of the gene family were detected in maize (Appenzeller *et al.*, 2004).

In contrast to multiple fungal CS or cellulose synthase genes, basically only two tissue specific genes were discovered in insects (Gagou *et al.*, 2002), derived perhaps from a gene duplication episode (Merzendorfer, 2006). It was found that the two genes are positioned adjacent to and on either side in the centromere of *Drosophila* third chromosome (Gagou *et al.*, 2002). One gene (*CS*1) expressed in ectodermal cells encodes the integumental enzymes engaged in the formation of cuticles, while the other (*CS*2), which is relatively smaller in size, is restricted to midgut epithelial cells and is involved in the formation of PM chitinous matrices (Arakane *et al.*, 2005; Hogenkamp *et al.*, 2005; Zhu *et al.*, 2002). A close analysis by Zimoch *et al.* (2005) revealed that although *CS*1 is expressed in *M. sexta* midgut tissue, its expression is due to the presence of tracheal cells developing in between midgut columnar cells. An inverse pattern of *CS* gene expression was demonstrated as *CS*1 mRNA levels are high during moulting and the wandering larval stage, while *CS*2 transcripts are elevated in feeding larvae during the inter-moult periods. Analyses using RNAi methodology, show that *T. castaneum* CS1 and CS2 are the contributors of cuticular and PM chitin, respectively (Arakane *et al.*, 2005). When *CS*2 mRNA transcripts were knocked down in the mosquito *A. aegypti*, the PM was almost completely absent (Kato *et al.*, 2006). The integumental gene (*CS*1) but not the gut gene (*CS*2) is expressed in two alternatively spliced forms referred as A and B. The spliced variants of *M. sexta*, which differ in amino acid sequence (a segment of 59 amino acid residues), are expressed differently during development and between tissues such as epidermis and ectodermal cells in tracheae (Arakane *et al.*, 2004; Hogenkamp *et al.*, 2005; Merzendorfer, 2006). It has been suggested that *CS*1B present in *M. sexta* midgut tracheae provide a potential glycosylation site that confer some unknown enzymatic function different from that of its *CS*1A counterpart

(Hogenkamp *et al.*, 2005). The two alternatively spliced transcripts of *CS*1 of the diamondback moth, *P. xylostella*, are expressed in all its developmental stages (Ashfaq *et al.*, 2007). The *CS*1A variant is expressed during most of the larval and pre-pupal and pupal stages, while the *CS*1B form is expressed mainly during the pupal stage of *T. castaneum* (Arakane *et al.*, 2004, 2005). *S. exigua CS*1 gene is highly expressed in early and late stages of each larval instar and high levels of transcripts throughout the pupal stage were detected (Chen *et al.*, 2007). The *CS*2 gene from *S. exigua* gut was similar to the isolated one from *M. sexta* including intron/exons organization and absence of alternate exons (Kumar *et al.*, 2008). The *CS*2 genes in *M. sexta*, *T. castaneum* and *S. frugiperda* are expressed in the midgut, with a peak of expression in the actively feeding stages (Arakane *et al.*, 2005; Bolognesi *et al.*, 2005; Hogenkamp *et al.*, 2005; Zimoch and Merzendorfer, 2002). The transcripts' pattern in *T. castaneum CS*2 reveals increased level of expression in late larval and adult stages (Arakane *et al.*, 2004). Increased levels of *A. aegypti CS*2 transcripts following blood meal were detected (Ibrahim *et al.*, 2000; Kato *et al.*, 2006). Blood feeding by the mosquito adults also induced transcriptional up-regulation of glutamine: fructose-6-phosphate amidotransferase, a key precursor in the pathway of the CS substrate formation (Kato *et al.*, 2006).

5.2 CHITIN SYNTHASE (CS)—PROTEINS

The membrane-bound CS [UDP-acetamido-2-deoxy-D-glucose: chitin 4-β-acetamidodeoxyglucosyltransferase (EC2.4.1.16)] is the key enzyme in chitin biosynthesis. Like cellulose synthase and hyaluronan synthase, CS catalyses the transfer of its respective glycosyl units at the non-reducing end of the polymer chain (Imai *et al.*, 2003). CS belongs to a large family of processive and non-processive glycosyltransferases that display sequence homology to hyaluronan synthase, cellulose synthase and the bacterial NodC protein (Coutinho *et al.*, 2003; Kamst and Spaink, 1999). The enzymes are involved in biosynthesis of biologically essential oligosaccharides, polysaccharides and glycoconjugates, and belong to glycosyltransferses family 2. Conserved signature motifs that have the following amino acid residue sequences QXXRW, EDR and WGTRE, are essential for glycosyltransferase activity that catalyses β1-4 linkages (Nagahashi *et al.*, 1995; Takeo *et al.*, 2004). These motifs are located in the central cytoplasmic loop domain and site-directed mutagenesis studies confirmed their role in the catalysis carried by analogous enzymes such as cellulose synthase, hyaluronan synthase, and β-1,3 glucan synthase (Campbell *et al.*, 1997; Coutinho *et al.*, 2003). The insect *CS* gene codes for a large membrane-integrated protein of ~1600 deduced amino acid residues, a molecular weight of ~170 kDa and a slightly acidic isoelectric point (Merzendorfer, 2006). Interestingly, an unusual *CS* gene encoding a much shorter protein of 516 amino acid residues was isolated from a chlorovirus infecting *Chlorella* cells (Kawasaki *et al.*, 2001).

Based on putative amino acid sequence alignments of various insect enzymes, the CS is divided into three major domains where the peripheral A and C are essentially hydrophobic transmembrane segments, whereas the central hydrophilic cytoplasmic region, B (containing about 400 amino acid residues), is highly conserved and harbours the catalytic site (Merzendorfer, 2006; Fig. 3). There is a consensus that the CS catalytic sites face the cytoplasm and thus are accessible to a flow of substrate molecules required for the polymerization step. Such a topological state has posed the lingering, yet unresolved, dilemma in chitin deposition since nascent chitin polymers must be extruded across the plasma membrane barrier. On the basis of the deduced amino acid sequences, the number of the peripheral hydrophobic transmembrane segments in insects ranges from 15 to 18, much larger than in their fungal counterparts (Cohen, 2001). The transmembrane domains apparently anchor the enzyme to the plasma membranes, and helices in the C terminal domain, which are close to the central catalytic region, may function in post-catalytic processes such as translocation of chitin polymers across cell membranes (Cohen, 2001; Merzendorfer, 2006).

The catalytic mechanism that results in chitin or cellulose biosynthesis involves glycosyl transfers to acceptor molecules forming β-1,4-linked products. However, the exact catalytic mechanism of glycosyl transfer reaction, where the donor (UDP-GlcNAc) (Fig. 4) and the acceptor molecule is the

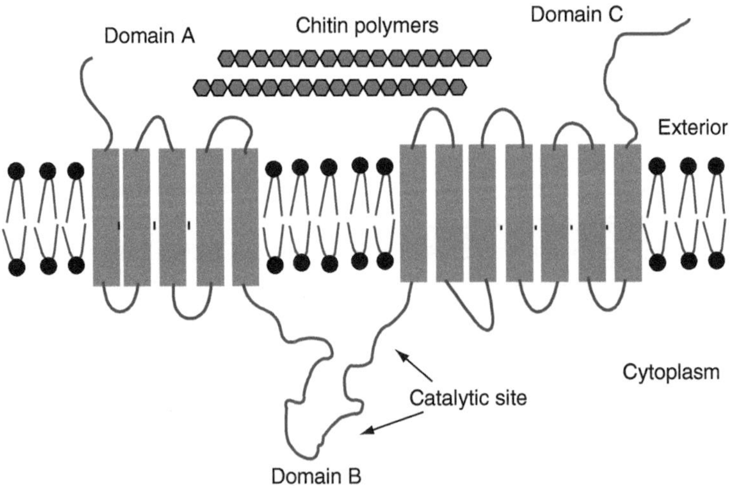

FIG. 3 The schematic tripartite structure of CS. The cytoplasmic domain B harbouring the catalytic site is flanked by transmembrane segments (domains A and C). The translocated chitin polymers are parallel to the cell surface.

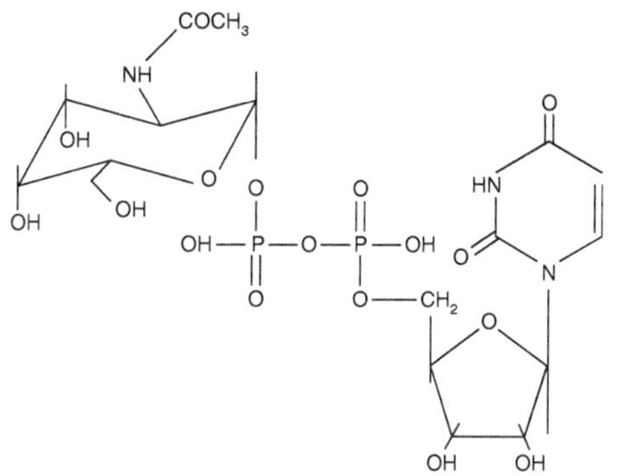

UDP-*N*-acetyl-D-glucosamine (UDP-GlcNAc)

FIG. 4 Chemical structure of the CS substrate.

growing carbohydrate chain or a lipid carrier, is largely unknown. Each sugar residue is rotated 180° relative to the preceding one and it has been hypothesized that three aspartic acid residues, which serve as polar amino acids, are required for the acid–base catalytic reaction in which two monosaccharides are simultaneously added in each catalytic cycle (Saxena *et al.*, 1995; Yabe *et al.*, 1998). The mechanistic design for chitin polymerization must tackle the structural opposed orientation of adjacent glycosyl residues (Yeager and Finney, 2004a). UDP-chitobiose is a poor CS substrate and chitin polymerization as suggested could involve two catalytic sites, one for each orientation of glycosyl unit, acting in close proximity to each other (Chang *et al.*, 2003). The observation that uridine-derived dimeric inhibitors are more potent in comparison to the monomeric counterpart supports the hypothesis concerning the presence of two CS active sites. However, inhibition of CS by dimeric inhibitors is rather low (in the mM range) in comparison to potent monomeric inhibitors like polyoxins and nikkomycin (Yeager and Finney, 2004b). In the case of hyaluronan synthase, which catalyse the alternate transfer of two different monoaminosaccharides, two independent catalytic sites within a single polypeptide were defined and analysed (DeAngelis, 1999; Jing and DeAngelis, 2003). Another alternative mechanistic solution for chitin catalysis that entails two adjacent catalytic sites invokes oligomeric assembly of CS molecules (Merzendorfer, 2006). Such assembly is indicated by the existing conserved coiled-coils oligomerization motifs in CS systems, and also supported by the oligomerization architecture of the analogous cellulose synthase forming the rosette structures. The oligomeric configuration is solidly corroborated by the recent purification of an active CS complex from *M. sexta* midgut tissue (Maue *et al.*, 2009). A comprehensive hypothetical model for insect CS was presented by Merzendorfer (2006). In this elaborate model, concrete and tentative structural features, derived from data of various processive glycosyltransferases, helped to construct the design. Experimental evidence based largely on CS crystal structure analysis could solve long-lasting ambiguities and dilemmas concerning the configuration of the catalytic site and the exact mechanisms of chitin polymerization and extrusion. Still the large membrane-integrated glycosyl transferases like CS or cellulose synthase forestalls attaining protein crystals for X-ray three-dimensional structure analysis.

5.3 PROPERTIES OF CS AND CATALYSIS

An orderly sequence of numerous and complex catalytic events is involved in the process of chitin formation (Fig. 5). Clusters of CS molecules integrated into cell membranes polymerize *N*-acetylglucosamine (GlcNAc) units into chitin chains. Glaser and Brown (1957) using the filamentous fungus *Neurospora crassa*, were the first to demonstrate *in vitro* chitin polymerization in a cell-free system, and later, formation of chitin from UDP-GlcNAc was initially demonstrated in a cell-free system extracted from the southern armyworm *Prodenia eridania* (Jaworsky *et al.*, 1963). The substrate for CS polymerization, 5′-uridine diphospho-*N*-acetyl-D-glucosamine (UDP-GlcNAc; Fig. 4), is a terminal metabolite of an elaborate cascade of biochemical transformations (depicted in Fig. 5) starting with glucose or with the insect typical disaccharide trehalose. A series of seven catalytic steps are required to mediate formation of the substrate from trehalose. This sequence of metabolite formation include hydrolysis of trehalose, phosphorylation of glucose, transmutation to form phosphorylated fructose, a trans-amination step forming glucosamine-6-phosphate, subsequent acetylation forming *N*-acetylamine 6-phosphate followed by its conversion to *N*-glucosamine-1-phosphate, and finally the formation of the nucleotide amino-sugar conjugate UDP-GlcNAc. The amination step, in which fructose-6-phosphare is converted to glucosamine-6-phosphate, is a catalytic 'crossroad' event in the pathway for down of the road incorporation of amino-sugar metabolites into various glycoconjugates and chitin. The above catalytic steps and the various ensuing metabolites related to the generation of CS substrate were identified in yeast and filamentous fungi (Cabib, 1981; Gooday, 1979; Gooday and Trinci, 1980), as well as in insects (Benson and Friedman, 1970; Candy and Kilby, 1962; Fristrom, 1968; Silvert and Fristrom, 1980), and in crayfish (Gwinn and Stevenson, 1973; Pahlic and Stevenson, 1978, Stevenson and Hettick, 1980).

A large number of other subcellular systems from yeasts, filamentous fungi and arthropods yielded substantial information related to the chitin polymerization (Cabib, 1987; Cohen, 1987b; Gooday, 1990b; Kramer and Koga, 1986). It should be emphasized that obtaining enzymatically functional cell-free systems for studying chitin biosynthesis has been intricate as the CS enzyme is tightly membrane-integrated, and due to its basic intrinsic fragility or instability upon tissue disruption (Cohen, 1987a). So far, data are available either from crude

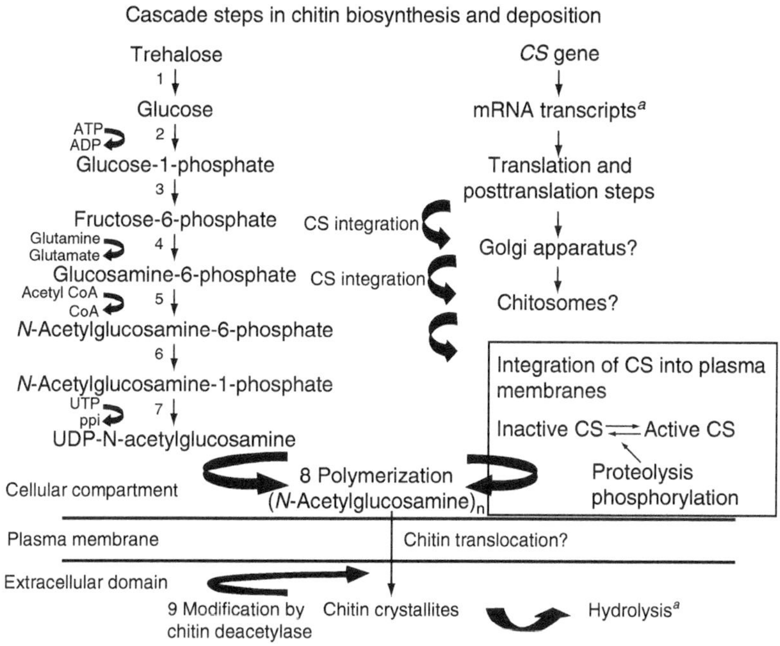

FIG. 5 Cascade steps in chitin biosynthesis and deposition. On the left the various catalytic steps culminating in the synthesis of the CS substrate (UDP-GlcNAc) are listed. 1: trehalase; 2: hexokinase; 3: glucose-6-P isomerase; 4: glutamine fructose-6-aminotransferase; 5: glucosamine-6-N-acyltransferase; 6: phosphoacetylglucosamine mutase; 7: UDP-*N*-acetylglucosamine pyrophosphorylase. Following the polymerization step (8) nascent chitin polymers may undergo a modification step of deacetylation by chitin deacetylase (9). On the right are sequential transcriptional, translational and post-translational steps involved in the formation of CS molecules. Probably the CS enzymes become associated with microvesicular structures (chitosomes) by which the proteins are conveyed to and integrated into apical plasma membrane domains. Before or after integration into the plasma membranes the CS catalytic units might be modified by partial proteolysis and/or phosphorylation that determine their functional status. [a]denotes evidence for hormonal regulation at the transcriptional level of the CS and the chitinolytic enzymes (chitinase and β-*N*-acetylglucosaminidase). ?—uncertain mechanism.

solubilized preparations or, at best, from a partially purified enzyme. Ordinarily, the enzyme is extracted from integumental or gut tissues and its catalytic activities were detected in either the mitochondrial or the microsomal fractions. Attempts to solubilize the crude particulate CS enzymes using various detergents were unsuccessful and high levels of the detergent digitonin, helpful for fungal CS (Duran and Cabib, 1978; Gooday and de Rousset-Hall, 1975), was inhibitory for the insect enzymes (Cohen and Casida, 1980a; Mayer *et al.*, 1980a,b). Chitin polymerization in such integument and gut crude cell-free preparations was demonstrated in a number of insect species, including *T. castaneum* (Cohen and Casida, 1980a), the striped rice borer *Chilo suppressalis* (Kitahara *et al.*, 1983), the stable fly *Stomoxys calcitrans* (Mayer *et al.*, 1980a,b), the sheep fleshfly *L. cuprina* (Turnbull and Howells, 1983), the cabbage looper *Tricholpusia ni* and the cecropia moth *Hyalophora cecropia* (Cohen and Casida, 1982). Typically, the assay is carried out in appropriate buffers using the radio-labelled UDP-GlcNAc substrate in the presence of Mg ions. Also an alternative, non-radioactive CS assay, exploiting the chitin binding property of wheat germ agglutinin (WGA), was practised (Lucero *et al.*, 2002), although apparently less sensitive as compared to the common radioactive assay (Merzendorfer, personal communication). Magnesium ions required for catalysis could be partially replaced by manganese or cobalt as divalent cations in *T. castaneum* CS preparations (Cohen and Casida, 1980a). GlcNAc is regarded as an effective allosteric element as it greatly stimulated *T. castaneum* and the root weevil *Diaprepes abbreviatus* gut (Cohen and Casida, 1980a,b; Gordon *et al.*, 1991) and *T. ni* integumental (Cohen and Casida, 1982) CS enzymes as well as enhancing the CS activity in fungal systems (Manocha and Begum, 1985; Tatsuno *et al.*, 1997). However, no such stimulation by GlcNAc was observed in homogenates of cultured epithelial cell line of the midge *Chironomus tentans* (Ludwig *et al.*, 1991) or in an integumental *H. cecropia* CS preparation (Cohen and Casida, 1982). Trypsin included in the fungal reaction mixtures stimulated CS activity and it is indicative of the zymogenic nature of the pro-enzyme that requires proteolytic activation (Cabib, 1981; Duran and Cabib, 1978; Ruiz-Herrera *et al.*, 1977; Tatsuno *et al.*, 1997). Zymogenicity of fungal CS systems was considered essential to ensure post-translational controlled activation of the enzyme after its integration into cell membranes so that premature chitin synthesis is prevented (Bartnicki-Garcia, 2006; Choi *et al.*, 1994). When trypsin was introduced in insect CS assays as a presumed replacement for a natural protease (Cohen and Casida, 1980a; Gordon *et al.*, 1991; Mayer *et al.*, 1980a,b), it became quite evident that unlike the fungal CS, the extent of the enzymatic stimulation is rather low (30–70%). The significant stimulation (up to 300%) of *M. sexta* gut CS by trypsin was clearly shown to be an indirect effect (Zimoch *et al.*, 2005). Apparently, the zymogenicity of the insect enzyme is arguable, and the slight stimulation of activity might be attributed to unmasking of active or allosteric sites in the crude enzyme preparation (Cohen, 1987a; Cohen and Casida, 1980a).

The *S. cerevisiae* Chs2 enzyme contains multiple phosphorylation sites (Martinez-Rucobo *et al.*, 2009) and in addition to hyperactivation effects by proteolysis, phosphorylation apparently plays an important role in regulating the activity and degradation of the yeast CS (Valdivia and Schekman, 2003). It was suggested that the association of calmodulin-dependent protein kinase with specific microsomal proteins might play a role in activation of *N. crassa* CS (Suresh and Subramanyam, 1997). Plausibly, putative phosphorylations, be it associated directly with the CS or with its adjacent proteins, might be relevant regulatory factors in CS insect systems as well (Fig. 5).

It appears that chitin synthesis is independent of a lipid intermediate to support and sustain GlcNAc polymerization (Cohen and Casida, 1983; Duran and Cabib, 1978). Such an intermediate, shown in a crustacean species, was involved in the formation of endogenous primer in synthesis of sugar moieties of various glycoproteins (Horst, 1983). Tunicamycin, which is an effective inhibitor of dolichyl phosphate GlcNAc (Heifetz *et al.*, 1979) and that also suppressed incorporation of chitin in epidermal tissues of the bug *T. infestans* (Quesada-Allue, 1982), had no effect on insect-derived CS activity (Cohen and Casida, 1983; Mayer *et al.*, 1981) as well as on yeast CS (Duran and Cabib, 1978) or on streptococcal hyaluronan synthase (Sugahara *et al.*, 1979). A need for a primer as template for chitin polymerization in insects was not demonstrated (Cohen and Casida, 1983), although *in vitro* crude CS systems might contain such primers. Nevertheless, a solubilised and purified yeast CS polymerized GlcNAc without inclusion of any primer (Cabib *et al.*, 1982). Alternatively, it has been suggested that the enzyme itself may serve as an initiator for starting the build-up of chitin chains.

UDP released at chitin catalysis is an effective inhibitor of insect (Cohen and Casida, 1980b, 1982; Ludwig *et al.*, 1991; Ward *et al.*, 1991) and fungal (Gooday and Trinci, 1980) CS enzymes. As its accumulation at the catalytic site might drastically reduce chitin polymerization, it was suggested that hydrolysis by UDPase might play a role in *in situ* regulation of chitin synthesis (De-Rousset-Hall and Gooday, 1975). In general, other pyrimidine di- and triphosphates (UTP, CDP and CTP), and in particular UTP, displayed considerable inhibitory effects on several insect CS enzymes (Cohen, 1987b; Ludwig *et al.*, 1991; Ward *et al.*, 1991). However, *S. calcitrans* CS was poorly inhibited (at a mM level) by UDP and UTP (Ludwig *et al.*, 1991), and the ribosyl purine di- and trinucleotides as well as cyclic AMP had no effect of *T castaneum* gut CS enzyme (Cohen and Casida, 1980b).

5.4 ASSEMBLY AND TRAFFICKING OF CATALYTIC UNITS

In the context of chitin formation, the generation of clustered CS units, their movement as well as their precise spatial and temporal localization and integration into plasma membranes have been of special research interest. Investigations using the yeast model benefited from a plethora of strains and mutants that enables thorough dissections of intricate processes related to creation of CS units and their intracellular trafficking. Conceivably, CS assembly and movement and their regulation in insect systems are similar in complexity, but still lack relevant and effective experimental tools and empirical designs. Electron microscopy studies visualized microvesicular cytoplasmic organelles, termed chitosomes, in the fungal systems (Bartnicki-Garcia *et al.*, 1978, 1979; Bracker *et al.*, 1976; Leal-Morales *et al.*, 1988). These vesicular structures (40–70 nm in diameter), which presumably stem from the endoplasmic reticulum (ER) and

Golgi complexes, are assumed to contain multi-units of the CS polymerizing enzyme, still in its zymogenic state (Ruiz-Herrera *et al.*, 1977), to be conveyed to and integrated into proper cellular locations within the plasma membranes. The existence of chitosomes *in vivo* has been a subject of controversy and regarded as artefact due to the harsh treatment required in rupturing fungal cell membranes. Moreover, while Bartnicki-Garcia stressed the validity of the exocytotic pathway of chitosomes (Bartnicki-Garcia, 2006), Sheckman and coworkers on the basis of studies with *S. cerevisiae* mutants emphasized the existence of their endocytotic origin (Chuang and Schekman, 1996). Immunological studies located two compartmental domains where CS enzyme units reside, for example in the plasma membrane and the endosomal chitosomes (Ziman *et al.*, 1996). It was suggested that the latter involves in CS recycling and serves as reservoir of CS enzymes mobilized for integration into plasma membranes following temporal and spatial regulation of chitin biosynthesis. The mechanism that regulates CS traffic via vesicular compartments is highly complex and essentially unknown, possibly requiring the mediation of Rab and other proteins in the yeast system (Ortiz and Novick, 2006).

Chitosome-like structures were observed in mites (Mothes and Seitz, 1981) and in cell-free system of *T. castaneum* (Cohen, 1982; Fig. 6). In *Mucor rouxii* and *S. cerevisiae* as well as in *T. castaneum*, coiled and extended microfibrils (50–250 nm in diameter) appeared inside and outside of the chitosomal particles (Bracker *et al.*, 1976; Cohen, 1982; Siemieniewicz *et al.*, 2007). Although chitosomal structures capable of *in vitro* chitin synthesis were detected in insects, the issue of their exact *in vivo* physiological role is still uncertain and controversial (Cohen, 1991).

According to the classical model mentioned above, CS proteins move to the cell surface *en route* of the conventional ER–Golgi complexes–chitosomes pathway (F'ig. 5). However, it has been shown for certain fungal isozymes

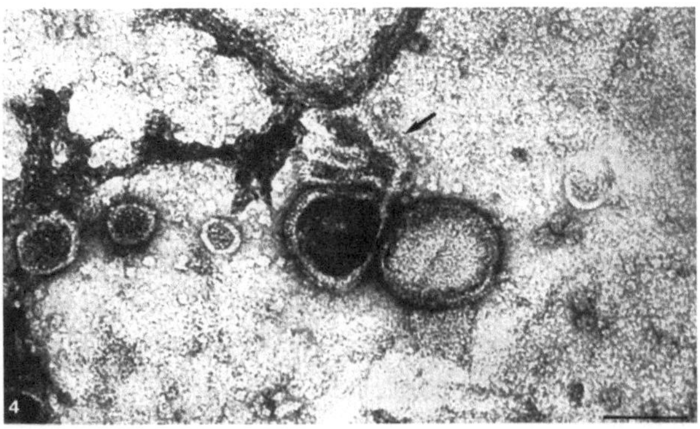

FIG. 6 Chitosome-like particles from the microsomal fraction of *T. castaneum* larvae associated with formation of chitin microfibrils (arrow).

that a non-vesicular alternative route, not yet identified, is possible (Meritxell *et al.*, 2007). Still passage of chitosomes or CS proteins (either individually or in organized clusters) from site of genesis to their proper integration into cell membranes in a precise and highly localized fashion is poorly understood. Plausibly, cytoskeletal structures play a crucial role in the organization of membrane components (Berlin, 1975). Insights from studies with plant, algal and fungal systems might provide clues relevant to such cytoskeleton-related trafficking. It has been shown that cellulose deposition in plant cells is a temporal-space process that includes integrated activities localized at the plasma membranes and areas bordering such membranes. Possibly microtubules connected with the polymerizing machinery regulates the motion of cellulose synthase complexes in cell membrane planes. There is a physical link between orientation of microtubules and the spatial organization of nascent cellulose microfibrils (Baskin, 2001; Quader, 1986; Lindeboom *et al.*, 2008). The mechanism of CS spatial localization and chitin deposition in yeast is dependent on actin cytoskeleton (Santos and Snyder, 1997). CS N-terminal myosin motor-like domains (MMDs) attached to CS suggest a direct interaction with cytoskeletal actin in fungi (Fujiwara *et al.*, 1997; Meritxell *et al.*, 2007). Genes encoding CS with an MMD fused to the CS conserved region were isolated from the filamentous fungus *Aspergillus nidulans* (Fujiwara *et al.*, 1997; Tsuizaki *et al.*, 2009). Interaction between actin and MMD is important for proper localization and function of *A. nidulans* CS (Takeshita *et al.*, 2005), yet apparently the motor activity of the fungus MMD does not mediate intracellular traffic of the CS to its target membranes (Takeshita *et al.*, 2005; Tsuizaki *et al.*, 2009). A myosin motor head domain associated with a marine bivalve mollusc CS has been recently reported (Weiss *et al.*, 2006). Such domain indicates a functional role for actin cytoskeletal elements in controlling chitin deposition and formation of mollusc shells. From experiments conducting with actin depolymerizing drugs such as cytochalasin-D and Congo red, and the herbicide amiprophos-methyl that interfere with action of microtubules, it was concluded that actin cytoskeleton serves as transport mechanism to deliver exocytotic vesicles, orientation of the polymerizing complexes and nascent microfibrils in an alga species (Quader, 1986). In summary, there is good supporting evidence that CS localization and chitin synthesis and deposition are guided and determined by their association with cytoskeletal structures.

Ultimately, a tight association of the insect CS with guiding cytoskeletal elements could determine the clustering and topological orientation of CS proteins in proper cellular localization in the plasma membrane compartments (Cohen, 1993a,b). No motor domains associated with insect CS have been so far confirmed, and unfortunately no solid data exist to substantiate such a link in insect systems. However, supporting evidence has shown that certain disruptants of microtubuli formation such as colcemid and vinblastine inhibited chitin formation in imaginal discs of the Indian meal moth *Plodia interpunctella* (Oberlander *et al.*, 1983), and colcemid inhibited moulting hormone-stimulated chitin biosynthesis in the wing pads of fifth instar locusts (Cassier and Papillon, 1991).

5.5 POST-CATALYTIC PROCESSES

5.5.1 *The translocation puzzle*

According to a consensus based on studies with fungal systems, the catalytic site of the membrane-bound CS is exposed to the cytoplasm (Braun and Calderone, 1978; Duran *et al.*, 1975; Sentandreu *et al.*, 1984), and as a result, nascent chitin polymers must be translocated from the intracellular domain to the exterior. Nevertheless, the mechanism of such translocation across plasma membranes being poorly understood, has invoked a few elucidatory hypotheses illustrated by models (Cohen, 1991; Merzendorfer, 2006). In the case of an analogous enzyme, cellulose synthase, hexagonal arrangement of six particles called 'rosette' structures has been visualized in freeze-fractured membranes by transmission microscopy (Delmer, 1987, 1999; Lindeboom, 2008). The transmembrane segments of the proteins, which are part of the cellulose biosynthesis system, form a pore through which cellulose polymers are presumably translocated. Hyaluronan, which is a common extracellular biopolymer in prokaryotes and eukaryotes that contains repeating disaccharide units of hyaluronic acid and GlcNAc, is catalysed by a large transmembrane processive glycosyl transferase enzyme, hyaluronan synthase. The topological organization of the transmembrane domains also suggests a pore-like structure (Helderman *et al.*, 2001). A membranous pore could conveniently solve the chitin translocation puzzle and was included in a recent hypothetical model that was based on the oligomeric configuration of cellulose synthase in the rosette structure (Merzendorfer, 2006). Although multiple transmembrane domains are present in insect or fungal cell membrane-integrated CS, and might in principle support a pore entity by an oligomeric pattern, no such structure has been demonstrated. However, circumstantial evidence for the apparent involvement of certain CS transmembrane segments in chitin translocation was presented (Merzendorfer, 2006). The proximity to the intracellular catalytic domain and their attachment to a conserved extracellular coil-coiled folding motif that suggests protein–protein interaction, are indicative of an unspecified mechanism for the yet unresolved riddle related to the extrusion of chitin polymers.

An alternative hypothesis dealing with the chitin translocation enigma implicates the chitosomal structures. The apparent chitosomal compartment serves as a vehicle by which clusters of packaged CS molecules are transported to the plasma membranes. However, fusion *per se* of the chitosomal vesicle with plasma membranes does not provide a solution to the chitin translocation enigma. According to one theory, chitin biosynthesis starts within the microvesicular structures (Bracker *et al.*, 1976; Cohen, 1982) that ultimately fuses with plasma membranes and exposes the already formed polymers or microfibrils to the outer surface of the cell. Problems inherent in the above hypothesis are related to the size of the microvesicle lumen, which is too small to accommodate a large chitin polymer, coupled with the availability of large UDP-GlcNAc substrate molecules at the catalytic sites. In such case, penetrability of the substrate molecules obligates a specific transport mechanism across the chitosomal membranes. The role and transmembrane topology of nucleotide sugar transporters in an intracellular structure like the Golgi apparatus was discussed by Berninsone and Hirschberg (2000).

5.5.2 *Chitin deacetylase*

Chitin deacetylase (CDA; EC 3.5.1.41) is a secreted enzyme that catalyses the hydrolysis of acetamido moieties of GlcNAc residues in chitin, thus forming the polycationic polymer of chitosan. CDA acts on nascent chitin polymers and is in tight association with CS activity (Fig. 5), or alternatively the enzyme deacetylates chitooligosaccharides produced by chitinase action. Chitin deacetylation interferes with hydrogen bonding and transforms the crystal structure of chitin (Choi *et al.*, 2000). As a result, such deacetylation process presumably modulates the physicochemical properties of chitin polymers, modifies their binding to associated macromolecules, and consequently could affect properties and architecture of chitin-based macromolecular assemblies.

Cell walls of zygomycetes fungi are characterized by containing large quantities of chitosan instead of chitin (Tan *et al.*, 1996). In other fungal species, and most likely in insects, where chitin deacetylation takes place, such modulation is relatively small and its physiological function is unclear. The first CDA, which was purified and characterized from the fungus *M. rouxii*, is a glycoprotein endo-type enzyme localized in the periplasm space (Araki and Ito, 1975).

Two binding proteins with putative CDA activity were associated with tracheal extracellular matrix and tracheal tube elongation in *Drosophila* (Lusching *et al.*, 2006; Wang *et al.*, 2006), and a chitin-binding protein was identified from the PM of *T. ni* (Guo *et al.*, 2005). So far, the only insect enzyme demonstrating chitin deacetylation activity was reported in the bertha armyworm *Mamestra configurata* recombinant PM CDA expressed in *E. coli* (Toprak *et al.*, 2008). Recently, bioinformatic search has revealed a family of multiple genes encoding putative CDA enzymes (and attached putative signal peptides) from *T. castaneum*, *D. melanogaster*, *A. gambiae* and *A. mellifera* (Dixit *et al.*, 2008). A detailed study by Arakane *et al.* (2009) listed up to nine putative CDA genes in *T. castaneum*, and found that their expression was tissue specific in epidermal cells, larval tracheae, midgut cells and imaginal appendages. Some genes were expressed throughout all the developmental periods of the insect, while others only during the feeding stages of larvae. Harnessing the RNAi knockout methodology for functionality analysis revealed that two of the CDA genes in the flour beetle interfere with larval–larval, larval–pupal and pupal–adult moults (Arakane *et al.*, 2009). It is noteworthy that CDA gene transcripts are localized apically in the cells, suggesting that polysomes involved in translating the enzyme are transported to this zone of chitin synthesis activity (Arakane *et al.*, 2009). Still research of CDA in insects is in the initial stages awaiting more studies on the enzymatic level, exact cellular localization and association with either CS or with chitinase activities as well as the extent of chitin modification and its role within the context of specific tissues, body regions and developmental processes.

5.6 CRYSTALLIZATION AND FIBRILLOGENESIS

The last step in chitin deposition involves the creation of crystalline microfibrils at the cell surface by coalescence of adjacent nascent polymer via extensive intermolecular hydrogen bonds. Disruption of these bonds by dyes such as Calcofluor white (stilbebe-type dye), Congo red (benzidine-type dye) and Primulin (thiazole dye) inhibited chitin synthesis in a number of fungal systems (Elorza *et al.*, 1983; Roncero and Duran, 1985; Roncero *et al.*, 1988; Selitrennikoff, 1985; Vermeulen and Wessels, 1986), in insects (Cohen, 1991; Zimmermann and Peters, 1987), and in a diatom algae (Herth, 1980). Albeit chitin polymerization and crystallization are tightly coupled, they are nevertheless consecutive processes as concluded from experiments using dye compounds such as Calcofluor white and Congo red (Herth, 1980; Vermeulen and Wessels, 1986). It was reported that CS activity was stimulated in the yeast (Bulawa *et al.*, 1986) and the red flour beetle (Cohen and Casida, 1990) by Calcofluor white and chitooligosaccharide-binding lectins, respectively. It was suggested that the stimulatory effect was due to disruption of polymerization followed by detachment of polymers from the catalytic sites resulting in enhanced biosynthesis of new chitin chains (Cohen and Casida, 1990).

6 Chitin hydrolysis

Chitin is degraded by digestive enzymes that break down β-1,4-glycosidic bonds. Complete chitin degradation involves a joint action of two hydrolyzing enzymes that finally generates GlcNAc monomers. An endo enzyme (chitinase) randomly hydrolyzes chitin yielding oligomers, and an exo enzyme (β-*N*-acetylglucosaminidase) exohydrolyzes dimers or terminal oligomers and polymers (Kramer and Koga, 1986). The dual action of the chitinolytic enzymes is synergistic as indicated by the increased production of monomers when acting in tandem (Fukamizo and Kramer, 1985). As chitin microfibrils are covalently or non-covalently associated with proteins (arthropods) or glucans (fungi), protease or glucanase activities are often accompany and greatly facilitate chitin hydrolysis.

In chitin-forming organisms, chitinolysis serves fundamental physiological purposes of growth and development. Chitin degradation in filamentous fungi and yeasts is crucial in maintaining normal hyphal growth, and is prerequisite for hyphal branching. Hydrolysis of chitin facilitates spore release and spore germination as well as budding and septum formation in yeast cells (Gooday, 1990a). Chitin hydrolysis assumes a central function of epidermal cells and is essential for cyclical events involved in normal shedding-off the old cuticles at time of moulting and before reshaping the exoskeletal structures, and perceptibly is under strict hormonal control in arthropods (Kramer and Koga, 1986). Integumental chitinolytic enzymes from a large number of insects including the silkmoth *Bombyx mori* (Kimura 1976), the fruit fly *Drosophila hydei* (Spindler 1976), the migratory locust *Locusta migratoria* (Zielkowsky and Spindler, 1978), the tobacco hornworm *M. sexta* (Dziadik-Turner *et al.*, 1981) and the small cabbage white *Pieris rapae* (Shi *et al.*, 2007), were purified and characterized.

Chitin degradation systems have evolved in non-chitinous organisms such as prokaryotic microflora, certain vertebrates and in vascular plants (Cohen, 1993a,b; Gooday, 1990a,b,c; Muzzarelli, 1977). Chitin is a biodegradable biopolymer readily and effectively hydrolyzed by plethora of marine and soil micro-organisms and no residues are accumulated in their respective environments. Chitin is an important nutritional source for chitinolytic micro-organisms, as well as for invertebrate and vertebrate organisms. Insectivorous plants (Amagase *et al.*, 1972), mycopathogens (Chet *et al.*, 1986; Manocha, 1987) or entomopathogens (Barreto *et al.*, 2004; Coudron *et al.*, 1984; da Silva *et al.*, 2005) use their respective chitinolytic activity to access insect prey or to penetrate arthropod cuticles and fungal cell walls.

Induction of chitinase and as well as other hydrolytic enzymes in plants is one of a coordinated and complex defence mechanism triggered in response to attack by phytopathogens and arthropod pests (Doares *et al.*, 1995; Kombrink and Somssich, 1995; Sahai and Manocha, 1993). Plants contain inducible systems that react to invasion of pathogens *inter alia* by *de novo* synthesis of pathogenesis-related (PR) proteins (Linthorst, 1991) such as glucanase, chitinase and chitosanase (Boller, 1986; Grenier and Asseli, 1990; Mauch *et al.*, 1988; Rousseau-Limouzin and Fritig, 1991; Schlumbaum *et al.*, 1986). It was demonstrated that chitosan acts as a transducing signal following interaction of a gall-producing mite and a plant (Bronner *et al.*, 1989), and externally applied polycationic chitosan or chitosan oligomers were detected inside plant cell nuclei (Hadwiger *et al.*, 1986). Their high binding affinity to DNA indicates interaction at the genomic level to activate defensive genes and to induce synthesis of PR proteins. Chitin has been regarded as a possible target site for controlling insect pests (Cohen, 1993a,b), and chitinolytic enzymes were probed as potential bioinsecticides by direct application (feeding or injection) (Fitches *et al.*, 2004; Kabir *et al.*, 2006; Rao *et al.*, 2004), and indirectly by recombinant plants (Ding *et al.*, 1998; Wang *et al.*, 2005) and baculoviruses (Gopalakrishnan *et al.*, 1995) expressing insect genes. Chitinase is widespread in *Bacillus thuringiensis*, and some strains with enhanced enzyme activity displayed higher toxicity against insect larvae (Liu *et al.*, 2002).

6.1 CHITINASES

Based on similarity in amino acid sequences, chitinases (EC 3.2.1.14) are classified as family 18 glycosyl hydrolases (Henrissat and Bairoch, 1993) catalysing the hydrolysis of β-1,4-glycosylic bonds in the chitin polymer. This widespread catalytic activity is detected in viruses, bacteria, fungi, invertebrates and vertebrates, and the enzymes are sub-divided into catalytically active and proteins that only retain chitin-binding activity and may act as lectins (Funkhouser and Aronson, 2007). Amino acid sequence alignment of bacterial, mammalian and invertebrate chitinases depicted a multi-domain architecture: (a) a catalytically active domain, (b) a C-terminal cysteine-rich chitin-binding region attached to the catalytic domain, and (c) a serine/threonine-rich segment (Kramer and Muthukrishnan, 1997). Secretion of the enzyme is apparently mediated by a typical cleavable signal peptide in the yeast *S. cerevisiae* (Kuranda and Robbins, 1991), and its existence in insect chitinases was predicted (Merzendorfer and Zinoch, 2003) and recently reported (Genta *et al.*, 2006). Like the CS enzyme, the proteolytic activation of fungal chitinases indicates their zymogenic nature (Dickinson *et al.*, 1991; Humphreys and Gooday, 1984).

The catalytic domain in all chitinases is conserved, but individual enzymes may have different combinations involving the two other segments. Apparently, these two domains are not essential for chitin hydrolysis as several insect chitinases lacking these regions are still catalytically active (Felix *et al.*, 2000; Girard and Jouanin, 1999; Wang *et al.*, 1996; Zhu *et al.*, 2001). The cysteine-rich region functions as a chitin-binding site that facilitates hydrolysis by targeting the enzyme to the stationary polymeric substrate (Venegas *et al.*, 1996). The serine/threonine region contains extensive glycosylation sites that might be involved in activation of the zymogenic enzyme or in chitinase turnover (Merzendorfer and Zinoch, 2003).

The catalytic domain of chitinases, which is based on the triose phosphate isomerase (TIM) barrel-shape structure (Lasters *et al.*, 1998; Reardon and Farber, 1995) and adapted from the 3D crystal structures of bacterial (Perrakis *et al.*, 1994; Zees *et al.*, 2009) and plant (Van Scheltinga *et al.*, 1994) chitinases is characterized by eight parallel β-strands that are surrounded by eight parallel α-helices. Topologically, this TIM-barrel catalytic construct forms a groove on the surface of chtinase proteins. A typical barrel-shaped architecture was also reported for other hydrolytic enzyme like the bee venom hyaluronidase (Markovic-Housely *et al.*, 2000) and *Clostridium* endoglucanase (Kitago *et al.*, 2007). The barrel-shaped design is most probably shared by fungal and insect chitinases. *M. sexta* chitinase as a model insect enzyme aligned with the known 3D crystal structure of *Serratia* enzyme revealed resemblance to the typical TIM $(\alpha/\beta)_8$ barrel configuration of the catalytic site (Kramer and Muthukrishnan, 1997).

The conserved active site of the enzyme resides in the fourth β strand where hydrolysis of the glycosylic bond takes place. The essential glutamic acid at its carboxyl end is the catalytic proton donor, while the *N*-acetyl residue of the substrate serves as its counterpart nucleophile. All insect chitinases contain two highly conserved amino acid sequence regions, one of which in *M. sexta* for example has a rich acidic amino acid signature sequence (FDGXDLDWEYP) located in or adjacent to the catalytic site (Lu *et al.*, 2002). Site-directed mutagenesis studies revealed that tryptophan, located between aspartate and glutamate residues in this conserved motif, plays a functional role in *M. sexta* catalytic activity (Lu *et al.*, 2002; Zhang *et al.*, 2002).

Insect chitinases with obvious physiological functions in epidermal moulting fluid and midgut tissues were studied and characterized in a number of insect species. Other chitinases with unclear functions from wasp's venom (Krishnan *et al.*, 1994) and fat body of the tsetse fly were cloned (Yan *et al.*, 2002). Basically, insect chitinases are characterized by a tripartite domain structure aforementioned as well as by extensive glycosylation sites (Shen and Jacobs-Lorena, 1997; Merzendorfer and Zinoch, 2003). Depending on the specific group, insect chitinases may have single or multiple catalytic and cysteine-rich chitin-binding domains (Shen and Jacobs-Lorena, 1999; Zhu *et al.*, 2008a). The enzymes varied in molecular weights (40–85 kDa), pH optima (4–8) and isoelectric points (pH 5–7). Although uncertain, the zymogenic nature of insect chitinases as based on apparent proteolytic activation of dipteran (*Aedes* and *Anopheles*), lepidopteran (*Manduca* and *Bombyx*) and coleopteran (*Tenebrio*) enzymes, was suggested (Bhatnagar *et al.*, 2003; Koga *et al.*, 1989, 1992; Merzendorfer and Zinoch, 2003; Royer *et al.*, 2002; Shen and Jacobs-Lorena 1997). The possible involvement of the serine/threonine-rich domain in the proteolytic activation of the zymogenic pro-enzyme was proposed (Royer *et al.*, 2002) in line with the non-zymogenic nature of the tsetse fly chitinase that is deficient of this domain (Yan *et al.*, 2002).

Chitinase genes were cloned from a number of insect species mostly lepidopterans (Choo *et al.*, 2007; Kim *et al.*, 1998; Kramer *et al.*, 1993). Full-length chitinase genes were isolated from *M. sexta* (Choi *et al.*, 1997) and *B. mori* (Abdel-Banat and Koga, 2001) and the exon organization in both genes was similar. A bioinformatic study, which took advantage on the complete genome sequences of 3 insect species (*T. castaneum*, *D. melanogaster* and *A. gambiae*) identified up to 16 different chitinase-like genes that were classified into 5 groups (Zhu *et al.*, 2008a). In a comprehensive study conducted by Zhu *et al.* (2008a) recombinant chitinases of *T. castaneum* and *D. melanogaster*, expressed in Hi-5 *T. ni* cell line expression system, were purified and characterized. Group I chitinase genes encode for proteins abundant in the moulting fluid or in the integument and is represented in each insect species by a single copy of a catalytic and chitin-binding regions spaced by a serine/threonine-rich linker domain. Similarly, groups II and III are represented by a single gene in all the above species but contain two (group II) or multiple (group III) catalytic regions. Multiple genes encode for chitinase-like proteins in groups IV and V. Group V consists of imaginal disc growth factors (IDFGs) initially reported from *Drosophila* imaginal disc cell cultures (Kawamura *et al.*, 1999). These proteins, which are structurally related to other chitinases but lack catalytic activity, play a role in proliferation and differentiation of imaginal disc cells. Chitinase-like proteins of groups I and IV from *D. melanogaster* and *T. castaneum* displayed chitinolytic activities, while group V, albeit having all the conserved catalytically essential amino acid residues, lack hydrolytic activity (Zhu *et al.*, 2008a). These group V proteins, despite lacking the chitin binding protein domain, do bind tightly to the substrate. A chitinase from the mealworm beetle *Tenebrio molitor* larval midgut was purified, characterized and sequenced following cDNA cloning (Genta *et al.*, 2006). Although it displayed high similarity to other typical family 18 chitinases, this particular gene does not encode the characteristic serine/threonine-rich linker and the chitin-binding domain. It was speculated that such a truncated gene was an adaptation to entail minimal damage to the PM due to the insect chitin-rich food source.

A comprehensive study using RNAi methodology shed light on the *in vivo* functional specialization of various *T. castaneum* chitinase-like proteins (Zhu *et al.*, 2008b). Group I and II chitinases, being associated with moulting, are involved in hydrolysis of cuticular chitin. Group III proteins apparently affect contraction of abdominal muscles and are involved in wing expansion. Proteins of group V, although lacking any catalytic activity, appear to affect adult eclosion of *T. castaneum* and are involved in the moulting process.

6.2 β-*N*-ACETYLHEXOSAMINIDASE/β-*N*-ACETYLGLUCOSAMINIDASE

β-*N*-Acetylhexosaminidases (EC 3.2.1.52) belong to family 20 glycosyl hydrolases classified according to their amino acid sequence similarities (Henrissat, 1998; Henrissat and Bairoh, 1993). The diverse enzymes catalyse the removal of 1,4-linked *N*-acetyl β-D-hexosamine residues from the non-reducing GlcNAc ends of diverse substrates like polysaccharides, oligosaccharides and their respective conjugates (glycoproteins, proteoglycans and glycolipids). The enzymes are widespread in various organisms and micro-organisms (bacteria, fungi, plants, invertebrates and vertebrates) but may assume other functions aside from chitin hydrolysis. Since the enzymes are not involved solely in chitin degradation and may cleave a broad spectrum of substrates (Aumiller *et al.*, 2006; Okada *et al.*, 2007), the terms β-*N*-acetylglucosaminidases and β-*N*-acetylhexosaminidase are considered here as interchangeable. In plants for example, the enzyme is involved in metabolism of storage glycoproteins (Strasser *et al.*, 2007) and in defence-related processes (Oikawa *et al.*, 2003); in mammals (human) lysosomal glycoproteins, proteoglycans and glycolipids are the cleaved substrates (Proia, 1988). Defects in lysosomal β-hexosaminidase in human are associated with the neurodegenerative Sandhoff and Tay-Sachs diseases (Filho and Shapiro, 2004; Lemieux *et al.*, 2006; Mark *et al.*, 2003). Available crystal structures of an increasing numbers of glycosyl hydrolase enzymes shed light on the catalytic mechanism. Three-dimensional crystal structure analyses of several microbial family 20 glycosyl hydrolases revealed a dimeric configuration with a central domain of a $(\beta/\alpha)_8$ TIM-barrel scaffold harbouring the catalytic active site (Langley *et al.*, 2008; Ling *et al.*, 2009; Mark *et al.*, 2001a; Prag *et al.*, 2000; Ramasubbu *et al.*, 2005; Tews *et al.*, 1996). A similar structural organization of human lysosomal enzymes associated with Tay-Sachs and Sandhoff disorders was also observed (Lemieux *et al.*, 2006; Mark *et al.*, 2003). The hydrolysis involves an acid–base-substrate-assisted catalysis mechanism in which the nucleophile is the carbonyl oxygen of acetamido moiety of the amino-sugar substrate and the proton donor is the active site glutamate residue (Tews *et al.*, 1996; Williams *et al.*, 2002). Hydrolysis of the enzyme-bicyclic oxazolinium ion intermediate takes place via nucleophilic attack of activated water molecules (Langley *et al.*, 2008). Aspartate, as the other critical conserved carboxylate residue, helps in stabilizing the transition state configuration and the orientation of the acetamido group (Mark *et al.*, 2001a). Although no X-ray crystal structures of insect enzymes are available, multiple sequence alignment of β-*N*-acetylhexosaminidase genes from *B. mori*, *D. melanogaster* and *S. frugiperda* shows a conserved HXGGDEVXXXCW motif at the catalytic domain with the functional conserved glutamic and aspartic acids important for catalysis (Okada *et al.*, 2007).

β-*N*-Acetylhexosaminidase genes were isolated and sequenced from a number of insect species like *B. mori* (Nagamatsu *et al.*, 1995), *M. sexta* (Zen *et al.*, 1996), the spruce budworm, *Choristoneura fumiferana*, (Zheng *et al.*, 2008), *T. castaneum* (Hogenkamp *et al.*, 2008) and from *S. frugiperda* cell line (Tomiya *et al.*, 2006). Score of expressed enzymes associated with chitin hydrolysis in *M. sexta* (Dziadik-Turner *et al.*, 1981; Koga *et al.*, 1983, 1992), *B. mori* (Koga *et al.*, 1987b, 1992; Nagamatsu *et al.*, 1995), *P. rapae* (Shi *et al.*, 2007), and the Asian corn borer *Ostrinia furnacalis* (Yang *et al.*, 2008) were purified and characterized from gut and integument tissues. Chitooligosaccharides, generated via randomly endohydrolysis of chitin polymers by chitinases in insects, are regarded as the primary target substrate for β-*N*-acetylhexosaminidase. The produced GlcNAc monomers are normally re-absorbed and may be recycled as substrates for *de novo* chitin synthesis at moulting and during metamorphosis (Nagamatsu *et al.*, 1995; Zen *et al.*, 1996). As β-*N*-acetylglucosaminidases cut down the accumulation of chitooligosaccharides that are inhibitory to chitinase activity, a synergistic effect is generated by the concurrent dual action of the enzyme systems (Kramer and Muthukrishnan, 2005). In various insect species the enzyme structure is either homodimeric (Kimura, 1976; Tomiya *et al.*, 2006; Yang *et al.*, 2008) or heterodimeric (Koga *et al.*, 1986; Kramer and Aoki, 1987; Shi *et al.*, 2007). Activity of the hydrolytic enzymes in insects was reported in integumental and gut tissues (Koga *et al.*, 1986, 1987a,b; Shi *et al.*, 2007; Yang *et al.*, 2008; Zheng *et al.*, 2008), hemolymph (Koga *et al.*, 1983), the moulting fluid

(Samuels and Reynolds, 1993; Zheng *et al.*, 2008) and in accessory glands of the Mediterranean fruit fly, *Ceratitis capitata* (Marchini *et al.*, 1989). Certain enzymes as membrane-bound proteins assume additional aforementioned functions that are related to *N*-glycan modification by removing GlcNAc residues from glycoproteins (Aumiller *et al.*, 2006; Le'onard *et al.*, 2006; Okada *et al.*, 2007; Tomiya *et al.*, 2006). Studies with the red flour beetle, *T. castaneum*, identified four genes coding for β-*N*-acetylglucosaminidases (NAGase), and their respective cDNAs were cloned and sequenced (Hogenkamp *et al.*, 2008). One enzyme (TcNAG1) is the most abundant and knocking down its corresponding transcripts, using RNAi methodology, disrupted the moulting process in all developmental stages resulting in lethal phenotypes. Although the primary role of TcNAG1 involves chitin hydrolysis, the other three genes also take part in cuticle degradation. One of the genes encoded a membrane-integrated enzyme (designated *TcFDL*), related to *D. melanogaster* FDL homolog (Le'onard *et al.*, 2006), might be involved in *N*-glycan processing or modification. Apparently, involvement of membrane-bound β-*N*-acetylglucosaminidases in glycoprotein and *N*-glycan processing is widespread in insect systems and was studied mainly in lepidopteran cell lines (Altmann *et al.*, 1995; Aumiller *et al.*, 2006; Geissler *et al.*, 2008; Tomiya *et al.*, 2006). Cell lines derived from four insect species (*S. frugiperda*, *T. ni*, *B. mori* and the forest tent caterpillar *Malacosoma disstria*) contained exoglycosidase activities that most likely are involved in degradation of baculovirus-expressed glycoproteins (Licari *et al.*, 1993). In the context of acetylhexosaminidase with apparent substrate specificity and affinity to glycoproteins in insects, it is noteworthy that *D. melanogaster* spermatozoa contain two genes encoding β-*N*-acetylhexosaminidases that are necessary for sperm–egg binding that is required for egg fertilization (Cattaneo *et al.*, 2006). An enzyme, detected in secretions of the reproductive accessory gland of the medfly, *C. capitata*, females, was functionally related to metabolism of egg surface glycoproteins (Marchini *et al.*, 1989).

7 Control and regulation of chitin metabolism

Like many catalytic events of major physiological importance, also chitin metabolism is controlled or regulated in a spatial and temporal manner. The phenomenon of moulting in arthropods is accomplished by intense synthetic and catabolic activities. The discontinuous cyclical synthesis and degradation of cuticular components including chitin entail mechanisms to regulate and coordinate such modifications. The periodical activities require the cyclical presence of endocrine factors like the moulting hormone (ecdysterone), which is the obvious major candidate involved in chitin synthesis and chitinolysis (Cohen, 1987b; Kimura, 1973; Kramer *et al.*, 1985). Increased activities of enzymes functioning in chitin biosynthesis and hydrolysis were demonstrated in various insects at the organismal, tissue and cellular levels in response to the moulting hormone. However, chitin metabolism is clearly one aspect of an interconnected complex processes that implicate a sequential cascade activation of a myriad-genes battery. Extensive detailed studies of endocrine control of CS and the hydrolytic endo and exo cleaving enzymes have been carried out. Nevertheless, also regulation at the post-translational level, well-aligned generation of CS substrate molecules as well as recruitment of cytoskeletal elements associated with trafficking of apparent CS-containing vesicles, are most likely under the umbrella of the moulting hormone multi-faceted, interconnected and coordinated effects. Despite the obvious hormonal involvement, a direct transcriptional activation of specific chitin-related genes has not been demonstrated. It general, the moulting hormone triggers concerted biochemical events specifically at the time of moulting and metamorphosis, part of which include chitin biosynthesis and deposition. *In vitro* and *in vivo* hormonal effects on chitin biosynthesis and degradation were studied using whole animal systems, isolated organs and integumental tissues, and established insect cell lines (Cohen, 1987a,b). It was shown that plasma membrane plaques in the apical area of *C. ethlius* epidermal cells, apparently engaged in chitin synthesis and deposition, are periodically degraded and formed *de novo* (Locke and Huie, 1979). This general dynamic turnover, which occurs at apolysis, appears to coincide tightly with cyclic changes of the moulting hormone levels (Bollenbacher *et al.*, 1981; Riddiford, 1991; Sehnal *et al.*, 1986). Long-term exposure of the moulting hormone was necessary to initiate and maintain chitin synthesis and deposition in various *in vitro* preparations including cockroach leg regenerates (Marks and Leopold, 1971; Sowa and Marks, 1975), lepidopteran larval integument (Kitahara *et al.*, 1983), integument fragments (Oikawa *et al.*, 1993) or imaginal discs (Oberlander *et al.*, 1978). Unlike integumental tissues, the alimentary canal is continuously engaged in synthesis of the PM chito-protein matrix throughout the active feeding stages of the insect life cycle (Peters, 1976). It was observed that the moulting hormone stimulated formation of PM in dipteran species (Becker, 1978; Peters, 1976). Although emphasis was placed on the moulting hormone, also the juvenile hormone, which modulates ecdysteroid action at the molecular level and maintains *status quo* conditions throughout insect development (Riddiford, 1994), interferes *inter alia* with chitin metabolism.

In summary, as chitin synthesis and deposition involves spatial and temporal processes of great complexity, the requirement for multi-level control and regulation is obvious, yet not always backed by solid evidence and clear mechanism (Cohen, 1993a,b). Suffice to mention transcriptional, translational and post-translational (proteolytic cleavage of zymogenic CS, CS glycosylation, phosphorylation and dephosphorylation) events. Other levels include the catalytic domain (recruitment and availability of substrate molecules, allosteric effectors, endogenous inhibitors, UDPase) as well as apparent CS compartmentation and trafficking of CS units to and integration into the plasma membrane.

7.1 CHITIN SYNTHASE—EXPRESSION PATTERNS AND HORMONAL CONTROL

Temporal CS expression, which fluctuates throughout insect development, is tissue specific and most likely depends on oscillatory levels of ecdysteroids. The beet armyworm, *S. exigua*, *CS*1 gene is expressed in integument and tracheal cells but not in fat body or in Malpighian tubuli cells (Chen *et al.*, 2007). High levels of CS transcripts were detected in the early and late larval stages and during the pupal stage. Expression of *CS*1 genes is up-regulated in the integumental epidermis of *M. sexta* where the increased amounts of transcripts parallel the period of cuticular chitin formation (Zhu *et al.*, 2002). The midgut cells, which are involved in the production of the chitinous PM, harbour the predominantly *CS*2 genes. Nevertheless, due to abundant ectodermal tracheal cells also CS1 gene transcripts were detected (Zimoch *et al.*, 2005). The CS1 transcripts in midgut tracheal cells of *M. sexta* were detectable only during moult while the gene expression was reduced during the feeding periods (Zimoch *et al.*, 2005). A mirror picture was evident in columnar midgut cells involved in PM formation, as levels of *CS*2 transcripts observed throughout the inter-moult feeding stages were reduced during the moult. The expression patterns are not clear-cut in other insect species as for example the *T. castaneum* *CS*1 gene was predominantly expressed during embryonic and pupal stages (Arakane *et al.*, 2004), the *S. frugiperda* gene was expressed in epidermal tissues during the feeding stages and in the pharate-adult and pupal stages (Bolognesi *et al.*, 2005) and *Drosophila* *CS*1 gene is considerably up-regulated at the pre-pupal stage (Gagou *et al.*, 2002). The *CS*2 gene, which is restricted to the midgut tissue, displays a more consistent expression pattern. Transcripts were not detected in the non-feeding pupal stage of *T. castaneum* (Arakane *et al.*, 2004), but are abundantly found in larvae during the feeding phase of various insect species (Bolognesi *et al.*, 2005; Hogenkamp *et al.*, 2005; Ibrahim *et al.*, 2000; Kumar *et al.*, 2008; Zimoch *et al.*, 2005). However, in many of the above species high levels of transcripts were detectable also in the non-feeding pre-pupal stage. One should always bear in mind that mere transcription activity does not always determine levels of functioning CS enzymes as turnover rates of transcripts and subsequent translational and post-translational events might also be involved and play a significant role.

Drosophila *CS*1 and *CS*2 genes, barely detected prior or during the late larval stage, were considerably up-regulated at the pre-pupal period when ecdysterone peak subsided (Gagou *et al.*, 2002). The above temporal pattern of transcriptional control suggests that both genes are ecdysterone-late ones induced by products generated by early moulting hormone genes. Transcriptional activity is induced by ecdysterone via its binding to a nuclear heterodimer composed of the ecdysterone receptor and the Ultraspiracle (USP), the *Drosophila* homolog to vertebrate retinoid X receptor (Beck *et al.*, 2009; King-Jones and Thummel, 2005; Riddiford *et al.*, 2003; Thomas *et al.*, 1993). Silencing a crustacean retinoid X receptor suggests its involvement in chitinase gene transcription (Priya *et al.*, 2009). Additional nuclear receptors, identified in *Drosophila* and *Tribolium*, may form a network with roles in controlling the ecdysterone cascade (Bonneton *et al.*, 2008; Thummel, 1995). Interestingly, apparent ecdysteroid responsive elements exist in the upstream region of both *Drosophila* genes (Merzendorfer and Zinoch, 2003). *CS*1 and *CS*2 genes are tissue and cell specific and their dissimilar temporal expression indicates different response to the moulting hormone titer. In the same midgut tissue of *Manduca* the tracheal cells expresses *CS*1 gene at the beginning of the moult, while the *CS*2 gene in columnar cells that secrete the PM matrix is transcribed during the inter-moult periods (Zimoch *et al.*, 2005). Since the circulating ecdysterone titer at the beginning of the moult is high, the concomitant low level of *CS*2 transcripts at that time suggests an induced repressed expression by the hormone.

7.2 CHITINOLYTIC ENZYMES—EXPRESSION PATTERNS AND HORMONAL CONTROL

Since degradation of chitin requires the tandem activity of chitinase and β-*N*-acetylglucosaminidase enzymes, the temporal and spatial expression of their corresponding genes is similar. Studies with lepidopteran species like *B. mori*, *Hyphantria cunea*, and *C. fumineferana* demonstrated that chitinase genes are expressed at the transition periods when degradation of cuticular chitin is required. The chitinase genes are transcribed during the transition stages of larval–larval, larval–pupal and pupal–adult stages in epidermis and gut tissues. They are up-regulated during the moulting process and not detected in the inter-moult periods (Kim *et al.*, 1998; Kramer *et al.*, 1993; Shinoda *et al.*, 2001; Zheng *et al.*, 2002, 2008). The gene for *S. frugiperda* gut chitinase, unlike its gut CS gene counterpart, is expressed during the wandering and pupal stages being involved in PM chitin degradation (Bolognesi *et al.*, 2005). This different temporal expression is coordinated to function in inter-moult and moulting periods for forming or degrading PM structures, respectively. Of the four β-*N*-acetylhexosaminidase genes expressed in the red flour beetle *T. castaneum*, only TcNAG1 plays a steady role during moulting with transcriptional peaks at larval–pupal and pupal–adult moults (Hogenkamp *et al.*, 2008). However, the highest gene expression in *M. sexta* β-*N*-acetylglucosaminidase was observed before moulting at day 6 and 7 of the fifth larval instar (Zen *et al.*, 1996), and the highest activity of *B. mori* enzyme was detected just before larval and pupal moults (Kimura, 1976). High levels of β-*N*-acetylglucosaminidase gene expression, which were observed in the spruce budworm *C. fumiferana* at apolysis of the fifth and sixth instar larvae, decreased dramatically during the inter-moult period and increased again at the larval–pupal transformation (Zheng *et al.*, 2008). An opposite pattern of *CS* transcripts was demonstrated for the expression of the chitinase gene in *S. frugiperda* gut during the larval–pupal moulting period when degradation of PM chito-protein matrix takes place (Bolognesi *et al.*, 2005). It is unknown whether the different patterns are the result of either a tandem or a separate regulatory mechanism.

Gene expression of chitinolytic enzymes, which is normally induced by ecdysteroids are suppressed by the juvenile hormone. The transcriptional stimulation or suppression may be also influenced by effects of both growth hormones on stability and turnover rates of the gene transcripts. Expression and activity of chitinolytic enzymes are enhanced by injection of ecdysteroids into isolated abdomens or ligated larvae of *B. mori* (Kimura, 1973) or *M. sexta* (Fukamizo and Kramer, 1987; Kramer *et al.*, 1993). Expression of *C. fumiferana* chitinase and β-*N*-acetylglucosaminidase genes, which were induced by the moulting hormone agonist, tebufenozide, resulted in an extra larval stage due to precocious and incomplete moult (Zheng *et al.*, 2002, 2008). Chitinolytic exochitinase and endo-splitting enzymes responded differently to changes in the moulting hormone level. Relatively low levels of the injected moulting hormone increased β-*N*-acetylglucosaminidase activity of *M. sexta*, while in contrast, high concentrations were less effective, yet enhanced chitinase activity (Fukamizo and Kramer, 1987). A similar pattern of induction was shown in the silkmoth *B. mori* when different moulting hormone levels were injected into ligated fifth instar larvae (Koga *et al.*, 1992). The integumental chitinase was immediately induced by high ecdysterone levels, whereas β-*N*-acetylglucosaminidase enzyme was gradually induced even by low hormone concentrations. This difference, coupled with the later appearance of chitinase activity, indicates a probable early induction of β-*N*-acetylglucosaminidase enzyme (Fukamizo and Kramer, 1987; Koga *et al.*, 1989; Merzendorfer and Zinoch, 2003). Induction of the above corresponding genes in *M. sexta* by the moulting hormone was suppressed by a topically applied juvenile hormone mimic (fenoxycarb) (Kramer *et al.*, 1993; Zen *et al.*, 1996). Experiments using the almond moth *Ephestia cautella* revealed that treatment of fourth instar larvae with a juvenile hormone analogue (methoprene) prevented the increased of ecdysteroid titer as well as the activities of chitinase and β-*N*-acetylglucosaminidase (Spindler-Barth *et al.*, 1986). In contrast, methoprene may have an ecdysiotropic effect, and by modulating ecdysterone titer in the mealworm *T. molitor* it indirectly induces chitinase gene expression (Royer *et al.*, 2002).

8 Inhibition of chitin metabolism

Inhibition studies of chitin synthesis and hydrolysis have academic as well as practical aspects. Using potent inhibitors have proved remarkably helpful in advancing our knowledge and understanding of intricate biochemical events associated with catalysis of chitin polymerization and hydrolysis, translocation of nascent polymers across plasma membranes and subsequent extracellular chitin fibrillogenesis. Since temporal synthesis of chitin and its degradation are tightly associated with insect growth, development and dramatic metamorphic phenomena, essential regulatory mechanisms assume a pivotal role. Inhibition studies shed light on the involvement of crucial regulatory factors that facilitate the integration of sequential and multi-lateral interacting events. As stated before, chitin is an essential structural component of insect cuticles and PM matrices as well as of cell walls of most fungi. Since interference with its synthesis or degradation should inflict detrimental lesions, chitin has become an attractive target for safe and selective pesticides (insecticides as well as fungicides) with minimal toxicity to non-target organisms such as plants and vertebrates. The search for potent inhibitors revealed diverse and unrelated groups of compounds some of which are effective as pest control agents. Obvious sites for interference include the catalytic domains of polymerization and hydrolysis, translocation of polymer chains, crystallization and fibrillogenesis, and hormonal control of chitin metabolism. Transgenic plants and baculoviruses-containing genes encoding chitin-binding lectins or chitin degrading enzymes were investigated as possible routes to control phytopathogens, zoopathogens and arthropod pests.

8.1 INHIBITION OF CHITIN POLYMERIZATION

Competitive inhibitors with structural resemblance to the substrate UDP-GlcNAc effectively block chitin polymerization and a few were commercialized as pest control agents. Some other compounds either directly or indirectly affect chitin polymerization to a certain degree. CS inhibition was studied using *in vivo* (whole organisms, isolated organs and integument tissues as well as cultured cells) or *in vitro* (cell-free preparations) systems.

8.1.1 *Pyrimidine-nucleoside peptides*

Nucleoside peptide antibiotics, which are secondary metabolites of microbial origin are powerful competitive inhibitors of fungal (Bartnicki-Garcia and Lippman, 1972; Cabib, 1991; Duran and Cabib, 1978; Hori *et al.*, 1971, 1974a,b; Müller *et al.*, 1981) and insect (Cohen, 1987b; Cohen and Casida, 1980b) chitin polymerization enzymes. These highly potent compounds have been thoroughly investigated, and their precise biochemical lesion is known. More than four decades ago, Japanese scientists isolated pyrimidine-nucleoside peptides (polyoxins A–M) from cultures of *Streptomyces cacaoi* var. *asoensis* (Isono *et al.*, 1967, 1969). Polyoxin-D can be regarded as a model inhibitor of the series and its chitin synthesis inhibition and toxicity were studied more extensively compared with the other polyoxin derivatives. Nikkomycins, which represent a series of extremely powerful CS inhibitors and are similar in chemical structure to polyoxins, were isolated later from culture broth of a related species (*Stremptmyces tendae*) (Bormann *et al.*, 1985; Dähn *et al.*, 1976). Essentially, polyoxins consist of an uridyl-ribose moiety attached to a dipeptide, and nikkomycins have in addition a pyrimidine ring attached to the dipeptidyl end (Dähn *et al.*, 1976; Kobinata *et al.*, 1980) (see structures in Fig. 7). Kinetic studies revealed that the nucleotide part of the molecule interacts with the CS catalytic site (Hori *et al.*, 1974a).

Various nucleoside peptides affect differently insect species and their administration route may determine their toxicity. This is most likely due to their accessibility at the CS active site which is related to their degree of permeability and intracellular stability. Polyoxins injected into the hemocoel were toxic to lepidopteran (Gijwijt *et al.*, 1979) and orthopteran (Vardanis, 1978) species, but it was not always effective by oral application (Cohen, 1987a). Lepidopteran species were not affected by orally applied polyoxin-D and nikkomycin (Gijwijt *et al.*, 1979; Arakawa *et al.*, 2008), yet on the other hand, polyoxin-B inhibited moulting and was toxic to several lepidopteran species including the cabbage moth *Mamestra brassicae*, the oriental armyworm *Mythimna separata*, *B. mori* and the cotton leafworm *Spodoptera litura* (Arakawa *et al.*, 2008). Dipteran species, however, were almost invariably susceptible to polyoxins and nikkomycins (Binnington, 1985; Schlüter, 1982; Tellam and Eisenmann, 2000; Turnbull and Howells, 1982). Yet, polyoxin-D and nikkomycin-Z were non-toxic to larvae of the common malaria mosquito, *A. quadrimaculatus* (Zhu *et al.*, 2006). Polyoxin-A was toxic when injected into nymphs of the grasshopper *Melanoplus sangquinipes*, and death occurred during moulting due to interference with chitin formation (Vardanis, 1976). Polyoxin-D inhibited PM formation in the blow fly *Callipohora erythrocephala* (Becker, 1980), and in several mosquito species (Shahabudin *et al.*, 1995). It clearly inhibited incorporation of GlcNAc into chitin in isolated abdominal integuments of the American cockroach, *Periplaneta americana* (Nakagawa *et al.*, 1993).

Cuticular growth was effected by the nucleoside peptide inhibitors as malformed endocuticular layers were formed following injection of polyoxin-D

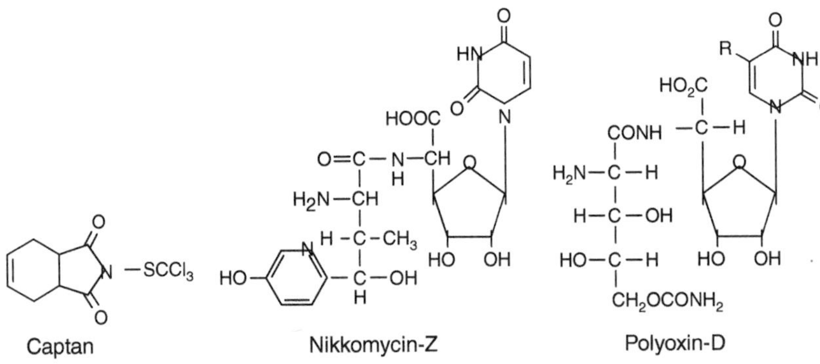

FIG. 7 Chemical structures of chitin synthase inhibitors.

into larvae of the large cabbage large, *Pieris brassicae* (Gijwijt *et al.*, 1979). A similar ultra-structural observation was reported in larvae *L. cuprina* fed with the same antibiotic (Binnington, 1985). Defective cuticles were observed in larvae of the Mexican bean beetle, *Epilachna varivestris* fed on bean leaves treated with nikkomycin (Schlüter, 1982), and the same compound inhibited moulting and disrupted the normal arrangement of pro-cuticular chitin microfibrils in the two-spotted spider mite *Tetranychus urticae* (Mothes and Seitz, 1982). Not only arthropods are affected by nucleoside peptides as nikkomycin-Z, which inhibited chitin synthesis in a mollusc species, altered the structure and function of the larval shell (Schönitzer and Weiss, 2007).

The potential of nucleoside peptides to inhibit CS by interacting directly with the catalytic site of the enzyme was studied using available fungal and insect cell-free systems (Cohen, 1987a). Unlike fungal systems, insect cell-free preparations capable of chitin polymerization were for a long time elusive allegedly due to methodological problems. The rapid loss of enzymatic activity was presumably associated with a tightly integrated plasma membrane enzyme (Cohen, 1987b; Surholt, 1974; Vardanis, 1976). Nevertheless, active CS enzymes, which were successfully extracted from a variety of insect species, were effectively inhibited by peptidyl nucleoside antibiotics. Cell-free preparations were obtained from gut tissues of various *Tribolium* species (Cohen and Casida, 1980a,b), *L. cuprina* (Turnbull and Howells, 1983), *M. brassicae* (Mitsui *et al.*, 1984) and *D. abbreviatus* (Ward *et al.*, 1991), from integumental tissues of *T. ni*, *H. cecropia* (Cohen and Casida, 1982), *C. suppressalis* (Kitahara *et al.*, 1983) and CS enzyme extracted from whole pupae of *S. calcitrans* (Mayer *et al.*, 1980a,b). Compared to polyoxins, nikkomycins are extremely potent CS inhibitors (Cohen and Casida, 1980b; Ward *et al.*, 1991). The difference in inhibition might be substantial as for example the IC_{50} values of polyoxin-D for *L. cuprina*, *T. castaneum* and *S. calcitrans* CS enzymes were 0.5 μM, 4 μM and 1 mM, respectively (Cohen, 1987a). The above discrepancies can be attributed to different crudeness of the various insect cell-free CS preparations (Cohen, 1987a). Compared with polyoxins, nikkomycins are more effective CS inhibitors, blocking chitin polymerization at low IC_{50} levels of 20 and 60 nM for *T. castaneum* and *T. ni*, correspondingly (Cohen and Casida, 1980b, 1982).

Polyoxins, being highly effective in suppressing growth of various fungal species, have been commercialized as agricultural fungicides in Japan (Hori *et al.*, 1974a; Misato, 1982). In contrast, nucleoside peptide antibiotics are practically ineffective as insect control agent. Their polar nature drastically restricts their penetration through the hydrophobic epicuticle of insects, and in addition, they might be degraded to a great extent inside the alimentary canal when ingested. Moreover, their relatively high-molecular weight prevents effective translocation across membranes of midgut cells and access to CS active sites (Cohen 1987a). Apparently, insects lack the active transport system for di- and tripeptides that facilitate movement of polyoxins and nikkomycins across fungal cell membranes (McCarthy *et al.*, 1985; Payne and Shallow,

1985). Nevertheless, it should be mentioned that a gene encoding a dipeptide transporter in *Drosophila* was reported (Roman *et al.*, 1998), but its relevance to translocation of the nucleoside peptide CS inhibitors is far from certain. The programme of developing nikkomycins as potent insecticides or miticides has been discontinued apparently due to toxicological considerations (Cohen, 2001) and reluctance of using antibiotic compounds for pest control in agriculture.

8.1.2 *Miscellaneous inhibitors*

A variety of compounds that interfere with chitin polymerization do not necessarily engage directly the CS catalytic site. Captan, a trichloromethyl sulfenyl fungicide (Fig. 7), is known to interact with sensitive sulfhydryl-containing molecules in fungal cells (Siegel, 1970). It inhibited cuticle formation in isolated cockroach leg regenerates (Marks and Sowa, 1976), in the claspers cuticle of the European corn borer (Gelman and Borkovec, 1986) and severely disrupted the synthesis of *Calliphora* PM (Becker, 1980). Captan strongly inhibited CS extracted from *T. castaneum*, *T. ni* and *H. cecropia* and was almost as effective as polyoxin-D (Cohen and Casida, 1980b, 1982). Its inhibition was alleviated by including dithiothreitol, a sulfhydryl-containing molecule, in the reaction mixture. Inhibition by related fungicides like captafol, folpet and dichlofluanid was much less effective. CS enzymes in cell-free systems prepared from the yeast *S. cerevisiae* (Cohen and Casida, 1980b), and the fungi *C. cinereus* (Brillinger, 1979) and *Sclerotiun rolfsii* (Cohen *et al.*, 1986) were insensitive to captan. This insensitivity suggests that compare to CS of insects, the fungal enzymes may have less thiol-sensitive domains associated with catalysis.

Relatively low inhibition potency of insect CS by assorted non-related compounds like buprofezin, phenylcarbamates and benzimidazoles was reported (Cohen, 1987b). Buprofezin, an insect growth regulator, which is effective in controlling hemipteran pests (Yasui *et al.*, 1985), inhibits chitin synthesis and disrupts cuticle deposition (Uchida *et al.*, 1985). Phenylcarbamate herbicides like barban is a moderate CS inhibitor (Cohen and Casida, 1980b), and a similar CS inhibition was observed with plumbagin, a natural quinine isolated from roots of the shrub *Plumbago capensis* (Kubo *et al.*, 1983). Benzimidazoles were reported to disrupt larval moulting of the silkworm *B. mori* (Kuwano *et al.*, 1982), and out of a series of structurally related compounds assayed with the *Tribolium* gut CS enzyme, structures with neryl, geranyl and citronellyl moieties were the best inhibitors (Cohen *et al.*, 1984). However, their potency was one order of magnitude less than that of polyoxin-D.

8.2 POST-CATALYSIS INHIBITION

8.2.1 *Insecticidal acylurea compounds*

An unexpected discovery of novel bioactive acylureas in the early 1970s by scientists in the Netherlands had a great impact on a renewed interest in cuticle biochemistry as well as on the introduction of commercially potent insecticides. The Dutch scientists were initially looking for potent weed control agents and combined two established herbicidal compounds, dichlobenil and diuron. The resulting acylurea compound, DU-19111, was a lead structure for the development of diflubenzuron (Dimilin) (Fig. 8), the first potent acylurea insecticide. Later, additional highly profitable commercial acylurea insect growth regulators were developed and gained widespread use, notably in integrated pest management programmes for controlling many insect pests in agriculture, forestry and public health (Cohen, 1993a,b). DU-19111 and diflubenzuron displayed poor herbicidal activity, yet they were surprisingly toxic to insect larvae by disrupting the normal formation of the cuticle via effectively blocking chitin synthesis (Mulder and Gijwijt, 1973; Verloop and Ferrell, 1977). In addition to cuticular lesions, also creation of the PM was disrupted (Becker, 1978; Clarke *et al.*, 1977; Soltani, 1984). Similarity in symptoms inflicted by polyoxin-D and acylureas was also illustrated (Gijwijt *et al.*, 1979; van Eck, 1979), and histological studies clearly showed that the model acylurea compound,

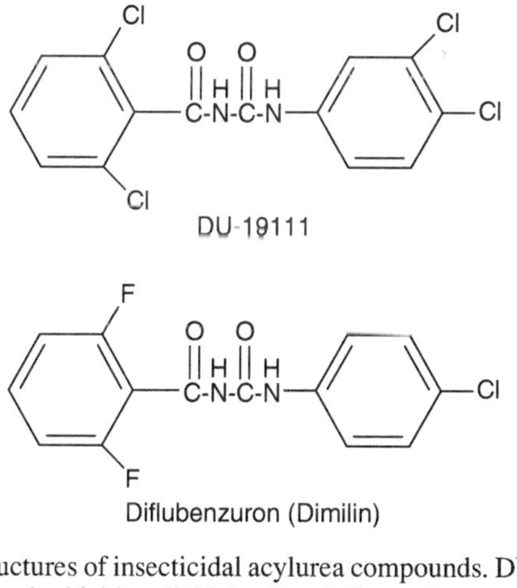

FIG. 8 Chemical structures of insecticidal acylurea compounds. DU-19111 was synthesized by combining two herbicides, dichlobenil and diuron. It serves as a lead compound for the synthesis of diflubenzuron, the first commercialized insect growth regulator that targets chitin synthesis.

diflubenzuron, induced malformed cuticles and distorted endocuticular layers (Gijwijt *et al.*, 1979; Grosscurt, 1978; Lim and Lee, 1982). Studies using whole insects or various integumental tissues confirmed the involvement of acylurea compounds in disruption of chitin synthesis by blocking, in a very short time, the incorporation of labelled glucose or amino sugars into the biopolymer (Clarke and Jewess, 1990; Deul *et al.*, 1978; Hajjar and Casida, 1979; Kitahara *et al.*, 1983; Mayer *et al.*, 1980a,b; Post and Vincent, 1973). A structure–activity relationship (SAR) study, using isolated abdominal tissues, demonstrated a positive correlation between insecticidal activities of acylureas and their corresponding chitin synthesis inhibition (Hajjar and Casida, 1979). It should be mentioned that unlike insects, acylurea compounds do not block *in vivo* chitin synthesis in fungi (Cohen, 1987a).

Studies with insect cell-free systems, in which acylurea compounds were included, illustrated no direct CS inhibition (Cohen and Casida, 1980b, 1983; Kitahara *et al.*, 1983; Mayer *et al.*, 1980a,b). The acylureas also did not affect the various sugar transformations in the biochemical pathway that generates substrate molecules (Verloop and Ferrell, 1977). Based on a large body of evidence, it was concluded that acylureas exert their direct or indirect effects at a post-catalytic stage, yet still the precise biochemical lesion inflicted is unresolved. It is no wonder that many theories have emerged to explain inhibition of chitin synthesis by such compounds. Effects on chitinase activity (Ishaaya and Casida, 1974) and interference with the endocrine system (Yu and Terriere, 1977) were regarded as secondary (Cohen and Casida, 1983; Hajjar and Casida, 1979; O'Neill *et al.*, 1977). Other presumptions for mode of action invoked generation of a remarkable potent metabolite acting at the CS target, inhibition of a protease that activate zymogenic CS, inhibition of UDPase activity that generate inhibitory UDP levels at the catalytic site and inhibition of access of the substrate UDP-GlcNAc or the allosteric effector GlcNAc to the active CS sites (Cohen, 1987a). There is no obvious solid basis for the above speculations, some of which were repudiated because an already operative enzyme is blocked and the speed by which the effects became evident following exposure to acylureas. An attractive theory brings into action elements crucial for translocating nascent chitin polymers across cell membranes. Membrane-spanning CS segments or adjacent proteins in the plasma membrane may play a key role in chitin secretion. Acylureas, as highly hydrophobic compounds, may be contained in the lipophilic plasma membrane compartment and interact there with the putative apparatus involved in chitin translocation (Cohen, 1993a,b). Recently, a pharmacological study using glibenclamide speculated that binding of diflubenzuron to a sulfonylurea receptor is the target for inhibition of chitin synthesis (Abo-Elghar *et al.*, 2004). The above suggestion, however, is in contrast to the finding that unlike diflubenzuron, glibenclamide had no toxic effect on *T. castaneum* larvae when fed orally (Merzendorfer, personal communication).

The number of theories and mere speculations to expound the exact biochemical lesion entails by insecticidal acylureas only points out our lack of fundamental knowledge regarding the detailed mechanisms involved in the interplay of the complex cellular components that act in chitin polymerization and chitin extrusion. Presumably, the above processes are also common to chitin formation in fungi, cellulose synthesis in plants and hyaluronan generation in vertebrates, and any advance in one system should greatly assist our understanding of the others.

8.2.2 *Inhibition due to hydrogen bond disruption*

Clusters of nascent chitin polymers following translocation across plasma membranes coalesce by intermolecular hydrogen bonds and form crystalline chitin polymorphs. Hydrogen bonds are also formed between chitin crystallites and glucan (in yeast and filamentous fungi), or between chitin and various cuticular and PM matrix proteins in insects. Although chitin polymerization and fibrillogenesis are coupled processes, a temporal gap between extracellular nascent chitin still at the non-crystallized state and a consecutively crystalline configuration was suggested (Herth, 1980; Vermeulen and Wessels, 1986). Polysaccharide binding dyes like Calcofluor white, Congo red and primulin, which exhibits high affinity to cellulose and chitin, are known to disrupt microfibril formation in the cellulose-forming bacterium *Acetobacter xylinum* (Benziman *et al.*, 1980; Haigler *et al.*, 1980) and in the algae *Oocystis* (Quader *et al.*, 1983) and *Poteriochromonas* (Herth, 1980). The above dyes interfere with chitin synthesis and crystallization resulting in perturbed microfibril formation in a number of yeast and filamentous fungi (Elorza *et al.*, 1983; Roncero and Duran, 1985; Roncero *et al.*, 1988; Selitrennikoff, 1984, 1985; Vermeulen and Wessels, 1986).

Unlike cuticular structures, the alimentary canal of insect is more accessible to compounds that might readily interfere with chitin synthesis and PM formation. Calcofluor white inhibited the formation of chitin microfibrils in the dipteran *C. erythrocephala* PM (Zimmermann and Peters, 1987) and disrupted the PM structure in several lepidopteran species (*T. ni*, the gypsi moth *Lymantria dispar*, the cotton bollworm *Helicoverpa zea*, the armyworm *Pesudaletia unipuncta*, the fall webworm *H. cunea*, and *S. exigua*) (Wang and Granados, 2000; Zhu *et al.*, 2007), and in the mite, *A. siro*, PM (Sobotnik *et al.*, 2008). Feeding Calcofluor white to *T. ni* and *S. exigua* larvae increased PM permeability and susceptibility to baculovirus infection apparently due to the dissociation of proteins (like the peritrophins) from chitin in the PM chito-protein matrix (Wang and Granados, 2000; Zhu *et al.*, 2007). Continuous feeding of Calcofluor white-containing diet to *S. exigua* larvae also drastically affected their development resulting in a high mortality rate (Zhu *et al.*, 2007). However, although feeding larvae of *L. cuprina* with the optical brightener affected their weight and survival it had no effect on the PM matrix structure (Tellam and Eisenmann, 2000).Utilizing chitin-binding molecules with hydrogen bond disrupting properties to affect midgut physiology by targeting PM formation, was proposed as a strategy for insect control (Wang and Granados, 2000).

8.2.3 *Inhibition by chitin-binding proteins*

Plant and invertebrate chitinases, normally have (single or multiple) chitin-binding cysteine-rich domains and several conserved aromatic amino acid charged residues that maintain protein folding (by disulfide bonds) and facilitate chitin binding and hydrolysis (Shen and Jacobs-Lorena, 1999). Cysteine-rich chitin binding multi-domains, which exist in peritrophins, associate with chitin and determine the physicochemical properties of the chito-protein matrix network of the PM (Elvin *et al.*, 1996; Hegedus *et al.*, 2009; Tellam *et al.*, 1999). Lectins are sugar-binding proteins found in plants and animals that specifically interact with certain carbohydrates moieties (Lis and Sharon, 1986; Van Damme *et al.*, 2008). Most lectins, like wheat germ agglutinin (WGA), which specifically recognize *N*-acetylglucosaminyl and *N*-acetylgalactosaminyl residues and have conserved cysteine-rich profiles and domains, are relevant in the context of post-catalysis inhibitory effects. The domain structure of plant lectins is very similar to peritrophic matrix peritrophins (Shen and Jacobs-Lorena, 1999). Due to their binding property, plant lectins may have a protective role as antifungal and insecticidal agents (Raikhel *et al.*, 1993) and were suggested to be used as selective control measures (Czapla and Lang, 1990). WGA, which binds to chitin at the tips of hyphal tubes, inhibited fungal growth of *Trichoderma viride* and *Fusarium solani* (Mirelman *et al.*, 1975); a chitin-binding lectin from pokeweed inhibited growth of germinating *N. crassa* tubes (Bramble and Gade, 1985), and a lectin derived from rhizomes of the stinging nettle, *Urtica dioica*, or purified from tobacco transgenic plants inhibited hyphal growth and spore germination in various fungi (Broekaert *et al.*, 1989; Does *et al.*, 1999). Various plant lectins, which play a role in plant defence against insect pests (Chrispeels and Raikhel, 1991; Giovanini *et al.*, 2007), are toxic or effective in delaying development of insects (Czapla and Lang, 1990; Eisemann *et al.*, 1994; Kaur *et al.*, 2006; Murdock *et al.*, 1990). A decrease in weight and change of the normal morphology of the PM were observed in *O. nubilalis* larvae fed with WGA (Harper *et al.*, 1998). The lectin, which localized heavily in the PM and microvilli, stopped insect feeding, disrupted the integrity and continuity of the chito-protein matrix layers and caused disintegration of the microvilli. WGA might act by binding to nascent chitin microfibrils or via agglutination of glycoproteins in the PM matrix (Harper *et al.*, 1998). The agglutinin from *Urtica dioica*, WGA and *Phytolacca*-derived lectins stimulated the CS enzymes in the fungi *N. crassa* (Bulawa *et al.*, 1986) and *Phycomyces blakesleeanus* (Van Parijs *et al.*, 1992) as well as in the insect *T. castaneum* (Cohen and Casida, 1990). The lectins apparently do not interact with the CS active site but rather bind to the nascent chitin polymers facilitating their detachment from the CS. Since adding monomers to growing chitin chain diminishes as the polymer becomes longer, its detachment from the CS leaves unoccupied CS active sites that polymerize the GlcNAc monomers more efficiently (Cohen and Casida, 1990).

8.3 INHIBITORS OF CHITIN HYDROLYSIS

8.3.1 *Chitinase inhibitors*

Most of the inhibitors of family 18 chitinases are natural products that include diverse structurally unrelated compounds like pseudo-trisaccharides (allosamidins), cyclic peptides (agrifin, argadin), alkaloids and methylxanthines (Andersen *et al.*, 2005). Although quite potent, the majority of the compounds (except allosamidins) display broad inhibitory spectrum, which probably has posed limits to their development as pest control agents. The best known, and the most extensively investigated inhibitor, is allosamidin (Fig. 9) isolated from mycelia of *Streptomyces* sp. or synthetically produced (Isogai *et al.*, 1989; Nishimoto *et al.*, 1991; Sakuda *et al.*, 1986, 1993; Sommers *et al.*, 1987; Takahashi *et al.*, 1992). Allosamidin is a competitive pseudo-trisaccharide inhibitor composed of two β-1,4-*N*-acetylglucosamine residues attached to

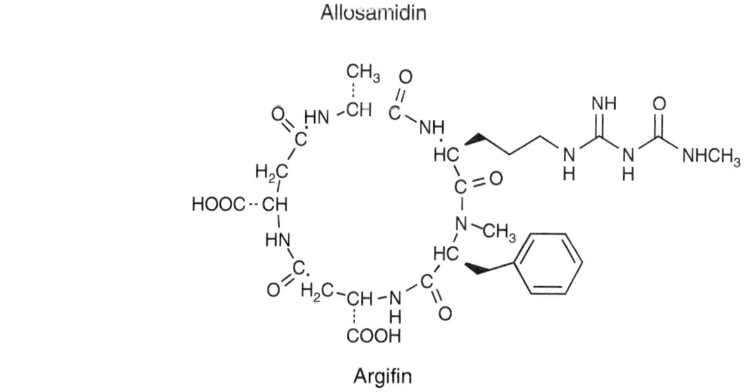

FIG. 9 Chemical structures of two microbially derived chitinase inhibitors, allosamidin and the cyclopentapeptide argifin.

the oxazoline derivative, allosamizoline (Sakuda *et al.*, 1986). Natural and synthetic allosamidins became popular inhibitors that were probed with chitinases from various sources including bacteria (Sampson and Gooday, 1998; Spindler and Spindler-Barth, 1994), malaria parasites (Shahabuddin *et al.*, 1993; Vinetz *et al.*, 2000), yeast (Nishimoto *et al.*, 1991; Sakuda *et al.*, 1990; Takahashi *et al.*, 1994), fungi (Blattner *et al.*, 1994, 1997; Bortone *et al.*, 2002; Dickinson *et al.*, 1989; Hodge *et al.*, 1996; McNab and Glover, 1991; Milewsky *et al.*, 1992; Nishimoto *et al.*, 1991), nematodes (Arnold *et al.*, 1993; Londershausen *et al.*, 1996), plants (Bokma *et al.*, 2000) and insects (Blattner *et al.*, 1997; Filho *et al.*, 2002; Isogai *et al.*, 1989; Koga *et al.*, 1987a; Londershausen *et al.*, 1996; Zhang *et al.*, 2002). In contrast to most other chitinase sources, the insect and fungal enzymes were strongly inhibited by allosamidins with IC_{50} values at the nanomolar/micromolar range. In addition to *in vitro* assays, allosamidins were examined *in vivo* for their potential toxic effects and biological activities in insects. Allosamidin caused high larval mortality in the common clothes moth *Tineola bisselliella* and in *L. cuprina* (Blattner *et al.*, 1997), and decreased nymphal survival in the aphid *Myzus persicae* (Saguez *et al.*, 2006). It prevented ecdysis and development from larva to pupa in the silkmoth (Sakuda *et al.*, 1987) and the house fly (Sommers *et al.*, 1987). However, despite being biochemically and biologically highly effective, allosamidin and a plethora of its analogues have not been developed into antifungal drugs or insect growth regulators. The inhibitors, nevertheless, became useful compounds in elucidation the mechanism of chitin hydrolysis, and in particular the understanding of substrate binding and catalysis. X-ray structures of a crystal plant (family 19) endo-chitinase (hevamine) and *Serratia* exo-chitinase complexed with allosamidin showed that the allosamizoline group binds to the centre of the enzyme active site, and such configuration apparently mimics a catalytic transition state that cannot be hydrolyzed. The role of the conserved glutamate residue as a catalytic proton donor was also confirmed (Van Aalten *et al.*, 2001; Van Scheltinga *et al.*, 1995). Also the methylxantine compound, pentoxyfylline, a moderate competitive inhibitor of chitinase, interacts at the active site of a fungal enzyme mimicking the reaction intermediate of allosamidin (Rao *et al.*, 2005a).

The cyclic pentapeptides, argifin (Fig. 9) and argadin, which were isolated from the fungal species *Gliocladium* (Nishimoto *et al.*, 2000; Omura *et al.*, 2000; Shiomi *et al.*, 2000) and *Clonostachys* (Arai *et al.*, 2000a), respectively, are powerful competitive inhibitors of family 18 chitinases from bacterial, fungal and human origin (Rao *et al.*, 2005b). The inhibition potency is similar and even higher in comparison to allosanidins as IC_{50} values of argifin and argadin against *L. cuprina* chitinase are 100 and 1 nM, respectively (Arai *et al.*, 2000b; Omura *et al.*, 2000). Crystal structure analysis of *Serratia* chitinase complexed with the cyclopentapeptides showed that the inhibitors mimic the substrate, and revealed a tight interaction of their peptide backbone with the conserved and essential side chain at the enzyme active site (Houston *et al.*, 2002a). A phage display peptide library with different peptide sizes and lengths was proposed as a screening method for discovering inhibitors that exhibit high affinity binding with *Serratia* chitinases (A and B) and inhibition constants at the nanomolar range (Petter *et al.*, 2008). X-ray crystal structure analysis of chitinase–cyclopentapeptide complexes was suggested as an approach to obtain structure-based design data for synthesizing and developing potent inhibitors (Houston *et al.*, 2002a). Unlike allosamidins, which are difficult to prepare, require many synthesis steps and as a result are expensive (Andersen *et al.*, 2005; Donohoe and Rosa, 2007), cyclopentapeptides are synthetically amenable. Based on three-dimensional crystallographic data using *Serratia* chitinase complexed with argifin, rational design of various highly potent analogues have proven feasible (Dixon *et al.*, 2009; Gouda *et al.*, 2009; Sunazuka *et al.*, 2009). The strategy combining rational molecular design based on crystallographic three-dimensional structures and solid phase synthesis of cyclic peptides should produce potent chitinase inhibitors for possible future development into viable pesticides and pharmaceuticals.

A few natural products like the cyclic dipeptide cyclo (Arg-Pro) isolated from a marine bacterial source (Izumida *et al.*, 1996a,b), styloguanidine and psammaplins obtained from marine sponges (Kato *et al.*, 1995; Tabudravu *et al.*, 2002), exhibit moderate to low inhibitory effects. Psammaplin A is a non-competitive inhibitor and a crystallographic study of its complex with *Serratia* chitinase B indicated an interaction at the active site of the enzyme (Tabudravu *et al.*, 2002). Insecticidal bioassays using the inhibitor showed mild toxic effects associated with chitinase inhibition to the diamondback moth *P. xylostella* (Tabudravu *et al.*, 2002) and on the aphid *M. persicae* (Saguez *et al.*, 2006). Several synthetic proline-containing cyclic dipeptides were reported as weak chitinase inhibitors (Houston *et al.*, 2004). The chitinase-cyclo dipeptide crystal complexes revealed that the molecular mechanism of inhibition involves an interaction of the inhibitor with amino acid residues at the active site that structurally mimics a catalytic intermediate of allosamidin (Houston *et al.*, 2002b). The design of new potent cyclic dipeptide inhibitors by modifying the side chain of the non-proline residue was suggested (Houston *et al.*, 2004).

8.3.2 β-N-*Acetylhexosaminidase inhibitors*

β-*N*-Acetylhexominidase is a target for possible design of low-molecular weight antifungal agents. A number of potent of inhibitors, some of them transition-state mimics that could serve as lead molecules, were studied (Horsch *et al.*, 1997). The compounds are competitive inhibitors, some of which are highly potent compounds with K_i values ranging from 1 to 30 nM. Like with chitinase inhibitors, available three-dimensional model of the enzyme can be used to design potent pest control agents of practical use in medicine and agriculture. In contrast to chitinases as target for inhibition, potent inhibitors of β-*N*-acetylhexosaminidases are rather limited in range and in selectivity against the insect enzyme. Certain imino-sugar compounds (Rountree *et al.*, 2009), ureido and thioureido derivatives of 2-amino 2-deoxy-D-glucose (Tournaire-Arellano *et al.*, 1998), GlcNAc-thiazoline (Knapp *et al.*, 2009), certain cyanodeoxy-glycosyl derivatives (Carmona *et al.*, 2006), the aza-sugar GalNAc-isofagomine (Mark *et al.*, 2001b), nagstatins isolated from *Streptomyces amacusaensis* (Ayogai *et al.*, 1992) and pyrostatins produced by another *Streptomyces* species (Aoyama *et al.*, 1995), were reported as inhibitors of microbial and mammalian enzymes. Compounds displaying selective inhibitory activity against fungal and insect β-*N*-acetylglucosaminidase were isolated from several strains of actinomycetes (Usuki *et al.*, 2006). One inhibitor (TNG-chitotriomycin) isolated from *Streptomyces anulatus* strongly inhibited the cotton leafworm *Prodenia litura* and certain fungal β-*N*-acetylglucosaminidases (IC_{50} at the sub-micromolar range), had no inhibitory effect on mammalian and plant enzymes (Usuki *et al.*, 2008).

9 Concluding remarks

Undoubtedly, scientific progress that includes cloning, sequencing and expression of genes associated with chitin biochemistry (CS, CDA, chitinase and β-*N*-acetylglucosaminidase), bioinformatic dissects, proteomic and functional analysis (using RNA interference tools), research combined with purification, characterization, and crystallographic studies of the pertinent enzymes considerably advanced our understanding of physiologically important mechanisms involved in chitin biosynthesis, its deposition and hydrolysis. Nevertheless, still a substantial large number of black boxes and lingering unresolved questions indicate our overall inadequate comprehension of fundamental aspects of chitin biochemistry.

The gaps in knowledge include (a) the translational and post-translational regulation and coordination of the CS and the chitinolytic enzymes in various tissues and during development; (b) the mechanism of apparent clustering of CS proteins and their precise integration into plasma membranes at apical regions of epithelial cells; (c) the exact cellular vehicle for trafficking individual or packaged CS units (chitosomes?), and function of cytoskeletal elements in spatial guidance, are far from being conclusive at least in insects; (d) the mechanism by which levels of substrate and activators are regulated at the chitin polymerization site; (e) the exact catalytic mechanism of chitin polymerization that has not yet been unequivocally resolved; (f) the catalytic machinery that determine the various polymorphs (α, β or γ) of chitin crystallites, the initiation and termination of chitin polymerization and how the final length of chitin polymers is controlled; (g) the basic enigma of chitin deposition as related to translocation of nascent polymers across cell membranes. It was suggested, but far from being experimentally confirmed, that transmembrane segments of the CS as well as neighbouring membrane proteins have a major role in the extrusion process; (h) the association and degree of association between chitin crystallites and various cuticular and PM proteins; (i) modification of chitin nascent polymers by CDA and the extent of such modification in various parts of the insect body. It is unknown how the association of proteins with the modified chitin alters the physicochemical properties of the resulting chitoprotein complex in exoskeletal and tracheal integuments; (j) the mechanism by which extracellular microfibrils are organized and oriented is unclear; (k) the biochemical mechanism that regulate and determine the different amount of chitin produced in various parts of the insect body is unidentified; (l) the exact mode of action of insecticidal acylureas. The precise cellular site and components associated with their strong inhibition of chitin synthesis is utterly unresolved. The notion of apparent interaction with cell membrane components associated with translocation of chitin polymers awaits solid experimental evidence.

Despite the impressive advent in chitin biochemistry research, the many aforementioned gaps in knowledge indicate that we have only scratched the surface in understanding many fundamental questions in this intriguing area.

In principle, the elaborate process of chitin synthesis and degradation offers selective targets for controlling arthropod pests as well as phyto- and zoopathogens. Recombinant plants and baculoviruses harbouring genes encoding chitinases as well as the enzymes *per se* were probed for controlling insect pests. Nevertheless, the catalytic step has emerged as a straightforward useful target for chemical disruption of chitin biosynthesis and hydrolysis. But in practice so far, only pyrimidine nucleoside peptides like polyoxins and the insecticidal acylurea compounds, which apparently act near the catalytic site to inhibit chitin synthesis, were developed into commercially successful pest control agents. There are a few powerful microbially derived chitinase inhibitors like allosamidins and cyclopentapeptides acting at sub-micromolar range, but their potency has not been translated into commercially viable products.

Crystallographic studies, which were largely conducted with plant and bacterial (not yet of insect origin) chitinolytic enzymes, were extremely useful for facilitating insights into the intricate catalytic machinery. In addition, using the relevant X-ray three-dimensional crystallographic information, a rational structure-based design and synthesis of new and powerful inhibitors are certainly highly attractive and hold promising potential for attaining commercial biopesticides. Unlike the soluble chitinolytic enzymes, crystallization of the membrane-bound CS is problematic. As a result it is hard to address challenging and unresolved mechanisms of catalysis and chitin translocation across plasma membranes as well as harnessing the crystallographic tools for finding potent inhibitors and explore their potential as future selective pest control agents.

Acknowledgements

I thank Dr. Hans Merzendorfer for reading the manuscript and for his useful comments and suggestions. The help of Ora Tzur from the Computer Unit in constructing certain figures is highly appreciated.

References

Abdel-Banat, B. M. A. and Koga, D. (2001). A genomic clone for a chitinase gene from the silkworm, *Bombyx mori*: structural organization identifies functional motifs. *Insect Biochem. Mol. Biol.* **31**, 497–508.

Abo-Elghar, G. E., Fujiyoshi, P. and Matsumura, F. (2004). Significance of the sulfonylurea receptor (SUR) as a target of diflubenzuron in chitin synthesis inhibition in *Drosophila melanogaster* and *Blattella germanica*. *Insect Biochem. Mol. Biol.* **34**, 743–752.

Abramczyk, D., Park, C. and Szaniszlo, P. J. (2009). Cytolocalization of the class V chitin synthase in the yeast, hyphal and sclerotic morphotypes of *Wangiella (Exophiala) dermatitis*. *Fungal Genet. Biol.* **46**, 28–41.

Altmann, F., Schwihla, H., Staudacher, E., Glössl, J. and März, L. (1995). Insect cells contain an unusual, membrane-bound β-*N*-acetylglucosaminidase probably involved in the processing of protein *N*-glycans. *J. Biol. Chem.* **270**, 17344–17349.

Amagase, S., Mori, M. and Nakayama, S. (1972). Digestive enzymes in insectivorous plants IV. Enzymatic digestion of insects by Nepenthes secretion and *Drosera peltata* extract: proteolytic and chitinolytic activities. *J. Biochem.* **72**, 765–767.

Andersen, S. O. (1979). Biochemistry of insect cuticle. *Annu. Rev. Entomol.* **24**, 24–61.

Andersen, O. A., Dixon, M. J., Eggleston, I. M. and Van Aalten, D. M. F. (2005). Natural product family 18 chitinase inhibitors. *Nat. Prod. Rep.* **22**, 563–579.

Aoyama, T., Kojima, F., Imada, C., Muraoka, Y., Naganawa, H., Okami, Y., Takeuchi, T. and Aoyagi, T. (1995). Pyrostatins A and B, new inhibitors of *N*-acetyl-β-D-glucosaminidase, produced by *Streptomyces* sp. SA-3501. *J. Enzyme Inhib.* **8**, 223–232.

Appenzeller, L., Doblin, M., Barreiro, R., Wang, H., Niu, X., Kollipara, K., Carrigan, L., Tomes, D., Chapman, M. and Dhugga, K. S. (2004). Cellulose synthesis in maize: isolation and expression analysis of the cellulose synthase (CesA) gene family. *Cellulose* **11**, 287–299.

Arai, N., Shiomi, K., Yamaguchi, Y., Masuma, R., Iwai, Y., Turberg, A., Kolbl, H. and Omura, S. (2000a). Argadin, a new chitinase inhibitor, produced by *Clonostachys* sp. FO-7314. *Chem. Pharm. Bull.* **48**, 1442–1446.

Arai, N., Shiomi, K., Iwai, Y. and Omura, S. (2000b). Argifin, a new chitinase inhibitor, produced by Gliocladium sp. 0668. II. Isolation, physico-chemical properties and structural elucidation. *J. Antibiot.* **53**, 609–614.

Arakane, Y., Hogenkamp, D. G., Zhu, Y. C., Kramer, K. J., Specht, C. A., Beeman, R. W., Kanost, M. R. and Muthukrishnan, S. (2004). Characterization of two chitin synthase genes of the red flour beetle, *Tribolium castaneum*, and alternate exon usage in one of the genes during development. *Insect Biochem. Mol. Biol.* **34**, 291–304.

Arakane, Y., Muthukrishnan, S., Kramer, K. J., Specht, C. A., Tomoyasu, Y., Lorenzen, M. D., Kanost, M. R. and Beeman, R. W. (2005). The *Tribolium castaneum* chitin synthase genes TcCHS1 and TcCHS2 are specialized for synthesis of epidermal cuticle and midgut peritrophic matrix. *Insect Mol. Biol.* **14**, 453–463.

Arakane, Y., Dixit, R., Begum, K., Park, Y., Specht, C. A., Merzendorfer, H., Kramer, K. J., Muthukrishnan, S. and Beeman, R. W. (2009). Analysis of functions of the chitin deacetylase gene family in *Tribolium castaneum*. *Insect Biochem. Mol. Biol.* **39**, 355–365.

Arakawa, T., Yukuhiro, F. and Noda, H. (2008). Insecticidal effect of a fungicide containing polyoxin B on the larvae of *Bombyx mori* (Lepidoptera: Bombycidae), and *Mamestra brassicae*, *Mythima separata*, and *Spodoptera litura* (Lepidoptera: Noctuidae). *Appl. Entomol. Zool.* **43**, 173–181.

Araki, Y. and Ito, E. (1975). A pathway of chitosan formation in *Mucor rouxii*. Enzymatic deacetylation of chitin. *Eur. J. Biochem.* **55**, 71–78.

Arnold, K., Brydon, L. J., Chappell, L. H. and Gooday, G. W. (1993). Chitinolytic activities in *Heligmosomoides polygyrus* and their role in egg hatching. *Mol. Biochem. Parasitol.* **58**, 317–324.

Arroyo-Begovich, A. and Carabez-Trejo, A. (1982). Location of chitin in cyst wall on *Entamoeba invadens* with colloidal gold tracers. *J. Parasitol.* **68**, 253–258.

Ashfaq, M., Sonoda, S. and Tsumuki, H. (2007). Developmental and tissue-specific expression of CHS1 from *Plutella xylostella* and its response to chlorfluazuron. *Pestic. Biochem. Physiol.* **89**, 20–30.

Aumiller, J. J., Hollister, J. R. and Jarvis, D. L. (2006). Molecular cloning and functional characterization of β-acetylglucosaminidase genes from Sf9 cells. *Protein Expr. Purif.* **47**, 571–590.

Ayogai, T., Suda, H., Uotani, K., Kujima, F., Aoyama, T., Horiguchi, K., Hamada, M. and Takeuchi, T. (1992). Nagstatin, a new inhibitor of *N*-acetyl-β-N-glucosaminidase produced by *Streptomyces* amakusaensis MG864-fF3. *J. Antibiot.* **45**, 1404–1408.

Bakkers, J., Kijne, J. W. and Spaink, H. P. (1999). Function of chitin oligosaccharides in plant and animal development. In: *Chitin and Chitinases* (eds Jolle's, P. and Muzzarelli, R. A. A.), pp. 71–83. Birkhauser-Verlag, Basel.

Barny, M.-A., Schoonejans, E., Economou, A., Johnston, A. W. B. and Downie, J. A. (1996). The C-terminal domain of *Rhizobium leguminosarum* chitin synthase NodC is important for function and determines the orientation of N-terminal region in the inner membrane. *Mol. Microbiol.* **19**, 443–453.

Barreto, C. C., Staats, C. C., Schrank, A. and Vainstein, M. H. (2004). Distribution of chitinases in the entomopathogen *Metarhizium anisolpliae* and effect of *N*-acetylglucosamine in protein secretion. *Curr. Microbiol.* **48**, 102–107.

Bartnicki-Garcia, S. (2006). Chitosomes: past, present and future. *FEMS Yeast Res.* **6**, 957–965.

Bartnicki-Garcia, S. and Lippman, E. (1972). Inhibition of *Mucor rouxii* by polyoxin-D: effects on chitin synthesis and morphological development. *J. Gen. Microbiol.* **71**, 301–309.

Bartnicki-Garcia, S., Bracker, C. E., Reyes, E. and Ruiz Herrera, J. (1978). Isolation of chitosomes from taxonomically diverse fungi and synthesis of chitin microfibrils *in vitro*. *Exp. Mycol.* **2**, 173–192.

Bartnicki-Garcia, S., Ruiz-Herrera, J. and Bracker, C. E. (1979). Chitosomes and chitin synthesis. In: *Fungal Wall and Hyphal Growth* (eds Burnett, J. H. and Trinci, A. P. J.), pp. 149–168. Cambridge University Press, Cambridge.

Baskin, T. I. (2001). On the alignment of cellulose microfibrils by cortical microtubules: a review and a model. *Protoplasma* **251**, 150–171.

Bassler, B. L., Yu, C., Lee, Y. C. and Roseman, S. (1991). Chitin utilization by marine bacteria. Degradation and catabolism of chitin oligosaccharides by *Vibrio furnissii*. *J. Biol. Chem.* **266**, 24276–24286.

Beck, Y., Dalaporte, C., Moras, D., Richards, G. and Billas, I. M. L. (2009). The ligand-binding domains of the three RXR-USP nuclear receptor types support distinct tissue and ligand specific hormonal responses in transgenic *Drosophila*. *Develop. Biol.* **330**, 1–11.

Becker, B. (1978). Effect of 20-hydroxy-ecdysone, juvenile hormone, Dimilin, and captan on *in vitro* synthesis of peritrophic membranes in *Calliphora erythrocephala*. *J. Insect Physiol.* **24**, 699–705.

Becker, B. (1980). Effect of polyoxin-D on *in vitro* synthesis of peritrophic membranes in *Calliphora erythrocephala*. *Insect Biochem.* **24**, 699–705.

Benson, R. L. and Friedman, S. (1970). Allosteric control of glucosamine phosphate isomerase from the adult housefly and its role in the synthesis of glucosamine 6-phosphate. *J. Biol. Chem.* **245**, 2219–2228.

Benziman, M., Haigler, G. H., Brown, R. M. Jr., White, A. R. and Cooper, K. M. (1980). Cellulose biosynthesis: polymerization and crystallization are coupled processes in *Acetobacter xylinum*. *Proc. Natl. Acad. Sci. USA* **77**, 6678–6682.

Berlin, R. D. (1975). Microtubules and the fluidity of the cell surface. *Ann. NY Acad. Sci.* **253**, 445–454.

Berninsone, P. M. and Hirschberg, C. B. (2000). Nucleotide sugar transporters of the Golgi apparatus. *Curr. Opin. Struct. Biol.* **10**, 542–547.

Bhatnagar, R. J., Arora, N., Scachidnanand, S., Shahabuddin, M., Keister, D. and Chauhan, V. S. (2003). Synthetic propeptide inhibits mosquito midgut chitinase and blocks sporogenic development of malaria parasite. *Biochem. Biophys. Res. Commun.* **304**, 783–787.

Bierstedt, A., Stankiewicz, B. A., Briggs, D. E. G. and Evershed, R. P. (1998). Quantitative and qualitative analysis of chitin in fossil arthropods using a combination of colorimetric assay and pyrolysis-gas chromatography–mass spectrometry. *Analyst* **123**, 139–145.

Binnington, K. C. (1985). Ultrastructural changes in the cuticle of the sheep blowfly, *Lucilia* induced by certain insecticides and biological inhibitors. *Tissue Cell* **17**, 131–140.

Blackwell, J., Parker, K. D. and Rudall, K. M. (1967). Chitin fibers of the diatom *Thalassiosira fluviatitis* and *Cyclotella cryptica*. *J. Mol. Biol.* **28**, 383–385.

Blattner, R., Furneaux, R. H., Kemmitt, T., Tyler, P. C., Ferrier, R. J. and Tide'n, A.-K. (1994). Syntheses of the fungicide/insecticide allosamidin and a structural isomer. *J. Chem. Soc. Perkin Trans.* **1**, 3411–3421.

Blattner, R., Gerard, P. J. and Spindler-Barth, M. (1997). Synthesis and biological activity of allosamidin and allosamidin analogs. *Pestic. Sci.* **50**, 312–318.

Bokma, E., Barends, T., Van Scheltinga, A. C. T., Dijkstra, B. W. and Beintema, J. J. (2000). Enzyme kinetics of hevamine, a chitinase from the rubber tree *Hevea brasliensis*. *FEBS Lett.* **478**, 119–122.

Bollenbacher, W. E., Smith, S. L., Goodman, W. and Gilbert, L. I. (1981). Ecdysteroid titer during larval–pupal–adult development of the tobacco hornworm, *Manduca sexta*. *Gen. Comp. Endocrinol.* **44**, 302–306.

Boller, T. (1986). Chitinase: A defense of higher plants against pathogens. In: *Chitin in Nature and Technology* (eds Muzzarelli, R., Jeuniaux, C. and Gooday, G. W.), pp. 223–230. Plenum, New York.

Bolognesi, R., Arakane, Y., Muthukrishnan, S., Kramer, K. J., Terra, W. R. and Ferreira, C. (2005). Sequences of cDNAs and expression of genes encoding chitin synthase and chitinase in the midgut of *Spodoptera frugiperda*. *Insect Biochem. Mol. Biol.* **35**, 1249–1259.

Bonneton, F., Chaumot, A. and Laudet, V. (2008). Annotation of *Tribolium* nuclear receptors reveals an increase in evolutionary rate of a network controlling the ecdysone cascade. *Insect Biochem. Mol. Biol.* **38**, 416–429.

Bormann, C., Huhn, W., Zähner, H., Rathmann, R., Hahn, H. and König, W. A. (1985). Metabolic products of microorganisms. 228. New nikkomycins produced by mutants of *Streptomyces tendae*. *J. Antibiot.* **38**, 9–16.

Bortone, k., Monzingo, A. F., Ernst, S. and Robertus, J. D. (2002). The structure of an allosamidin complex with the *Coccidiodes immitis* chitinase defines a role for a second acid residue in substrate-assisted mechanism. *J. Mol. Biol.* **320**, 293–302.

Bracker, C. E., Ruiz-Herrera, J. and Bartnicki-Garcia, S. (1976). Structure and transformation of chitin synthetase particles (chitosomes) during microfibril synthesis *in vitro*. *Proc. Natl. Acad. Sci. USA* **73**, 4570–4574.

Bramble, R. and Gade, W. (1985). Plant seed lectins disrupt growth of germinating fungal spores. *Physiol. Plant.* **64**, 402–408.

Braun, P. C. and Calderone, R. A. (1978). Chitin synthesis in *Candida albicans*: comparison of yeast and hyphal forms. *J. Bacteriol.* **133**, 1472–1477.

Brillinger, G. U. (1979). Metabolic products of microorganisms. 181. Chitin synthase from fungi, a test model for substances with insecticidal properties. *Arch. Microbiol.* **121**, 71–74.

Broekaert, W., Van Parijs, J., Leyns, F., Joos, H. and Peumans, W. J. (1989). A chitin-binding lectin from stinging nettle rhizomes with antifungal properties. *Science* **245**, 1100–1102.

Bronner, R., Westphal, E. and Dreger, F. (1989). Chitosan, a component of the compatible interaction between *Solanum dulcamara* L. and the gall mite *Eriophyes caldophthrius* Nal. *Physiol. Mol. Plant. Pathol.* **34**, 117–130.

Brydon, L. J., Gooday, G. W., Chappell, L. H. and King, T. P. (1987). Chitin in egg shells of *Onchocerca gibsoni* and *Onchocerca volvulus*. *Mol. Biochem. Parasitol.* **25**, 267–272.

Bullawa, C. E. (1993). Genetics and molecular biology of chitin synthesis in fungi. *Annu. Rev. Microbiol.* **47**, 505–534.

Bulawa, C. E., Slater, M., Cabib, E., Au-Young, J., Sburlati, A., Adair, W. L. Jr. and Robbins, P. W. (1986). The *S. cerevisiae* structural gene for chitin synthase is not required for chitin synthesis in vitro. *Cell* **46**, 213–225.

Cabib, E. (1981). Chitin structure metabolism, and regulation of biosynthesis. In: *Encyclopedia of Plant Physiology* (eds Tanner, F. A. and Loewus, F. A.), Vol. 13, pp. 395–415. Springer-Verlag, Berlin.

Cabib, E. (1987). The synthesis and degradation of chitin. In: *Advances in Enzymology and Related Areas of Molecular Biology* (ed Meister, A.), Vol. 59, pp. 59–101. John Wiley, New York.

Cabib, E. (1991). Differential inhibition of chitin synthetase 1 and 2 from *Saccharomyces cerevisiae* by polyoxin D and nikkomycins. *Antimicrob. Agents Chemother.* **35**, 170–173.

Cabib, E., Roberts, R. and Bowers, B. (1982). Synthesis of the yeast cell wall and its regulation. *Annu. Rev. Biochem.* **51**, 763–793.

Campbell, J. A., Davies, G. J., Bulone, V. and Henrissat, B. (1997). A classification of nucleotide-diphospho-sugar glycosyltransferases based on amino acid sequence similarities. *Biochem. J.* **326**, 929–939.

Candy, D. J. and Kilby, B. A. (1962). Studies on chitin synthesis in the desert locust. *J. Exp. Biol.* **39**, 129–140.

Carmona, A. T., Fialova, P., Kren, V., Ettrich, R., Martinkova, L., Moreno-Vargas, A. J., Gonzales, C. and Robina, I. (2006). Cyanodeoxy-glycosyl derivatives as substrates for enzymatic reactions. *Eur. J. Org. Chem.* **8**, 1876–1885.

Cassier, P. and Papillon, M. (1991). Ultrastructural studies of the in vitro chitin and cuticle secretions in the wing pad of *Locusta*. *Can. J. Zool.* **69**, 29–38.

Cattaneo, F., Pasini, M. E., Intra, J., Matsumoto, M., Briani, F., Hoshi, M. and Perotti, M. E. (2006). Identification and expression analysis of *Drosophila melanogaster* genes encoding β-hexosaminidase of sperm plasma membrane. *Glycobiology* **16**, 786–800.

Chang, R., Yeager, A. R. and Finney, N. S. (2003). Probing the mechanism of a fungal glycosyltransferase essential for cell wall biosynthesis. UDP-chitobiose is not a substrate for chitin synthase. *Org. Biomol. Chem.* **1**, 39–41.

Chen, X., Yang, X., Senthil, N., Kumar, S., Tang, B., Sun, X., Qiu, X., Hu, J. and Zhang, W. (2007). The class A chitin synthase of *Spodoptera exigua*: molecular cloning and expression patterns. *Insect Biochem. Mol. Biol.* **37**, 409–417.

Chet, I., Cohen, E. and Elster, I. (1986). The role of chitinase and chitin synthetase inhibitors in controlling plant pathogenic fungi. In: *Chitin in Nature and Technology* (eds Muzzarelli, R., Jeuniaux, C. and Gooday, G. W.), pp. 237–240. Plenum, New York.

Cho, Y.-W., Jang, J., Park, C. R. and Ko, S.-W. (2000). Preparation and solubility in acid and water of partially deacetylated chitins. *Biomacromolecules* **1**, 609–614.

Choquer, M., Boccara, M., Goncalves, I., Souli, M.-C. and Vidal-Cros, A. (2004). Survey of the *Botrytis cinerea* chitin synthase multigenic family through the analysis of six euascomycetes genomes. *Eur. J. Biochem.* **271**, 2153–2164.

Choi, W.-J., Santos, B., Duran, A. and Cabin, E. (1994). Are yeast chitin synthase regulated at the transcriptional or the posttranslational level? *Mol. Cell. Biol.* **14**, 7685–7694.

Choi, H. K., Choi, K. H., Kramer, K. J. and Muthukrishnan, S. (1997). Isolation and characterization of genomic clone for the gene of an insect molting enzyme, chitinase. *Insect Biochem. Mol. Biol.* **27**, 37–47.

Choo, Y. M., Lee, K. S., Kim, B. Y., Kim, D. H., Yoon, H. J., Sohn, H. D. and Jin, B. R. (2007). A gut-specific chitinase from the mulberry longicorn beetle, *Agriona germari* (Coleoptera: Cerambycidae): cDNA cloning, gene structure, expression and enzymatic activity. *Eur. J. Entomol.* **104**, 173–180.

Chrispeels, M. J. and Raikhel, N. V. (1991). Lectins, lectin genes, and their role in plant defense. *Plant Cell* **3**, 1–9.

Chuang, J. S. and Schekman, R. W. (1996). Differential trafficking and time localization of two chitin synthase proteins, Chs2p and Chs3p. *J. Cell Biol.* **135**, 597–610.

Clarke, B. S. and Jewess, P. J. (1990). The inhibition of chitin synthesis in *Spodoptera littoralis* larvae by flufenoxuron, teflubenzuron and diflubenzuron. *Pestic. Sci.* **28**, 377–388.

Clarke, L., Temple, G. H. R. and Vincent, J. F. V. (1977). The effect of chitin inhibitor—Dimilin—on the production of peritrophic membrane in the locust, *Locusta migratoria*. *J. Insect Physiol.* **23**, 241–246.

Cohen, E. (1982). In vitro chitin synthesis in an insect: formation and structure of microfibrils. *Eur. J. Cell Biol.* **16**, 284–294.

Cohen, E. (1987a). Chitin biochemistry: synthesis and inhibition. *Annu. Rev. Entomol.* **32**, 71–93.

Cohen, E. (1987b). Interference with chitin biosynthesis in insects. In: *Chitin and Benzoylphenyl Ureas* (eds Wright, J. E. and Retnakaran, A.), pp. 43–73. Junk, Dodrecht.

Cohen, E. (1991). Chitin biochemistry. In: *Physiology of the Insect Epidermis* (eds Retnakaran, A. and Binnington, K. C.), pp. 96–114. CSIRO Publications, Melbourne.

Cohen, E. (1993a). Chitin synthase: an enzyme and a selective target for pesticide action. In: *Pesticides/Environemnt: Molecular Biological Approaches* (eds Mitasui, T.,

Matsumura, F. and Yamaguchi, I.), pp. 101–113. Pesticide Science Society of Japan, Tokyo.

Cohen, E. (1993b). Chitin synthesis and degradation as targets for pesticide action. *Arch. Insect Biochem. Physiol.* **22**, 245–263.

Cohen, E. (2001). Chitin synthesis and inhibition: a revisit. *Pest Manag. Sci.* **57**, 946–950.

Cohen, E. and Casida, J. E. (1980a). Properties of *Tribolium* gut chitin synthetase. *Pestic. Biochem. Physiol.* **13**, 121–128.

Cohen, E. and Casida, J. E. (1980b). Inhibition of *Tribolium* gut chitin synthetase. *Pestic. Biochem. Physiol.* **13**, 129–136.

Cohen, E. and Casida, J. E. (1982). Properties and inhibition of insect integumental chitin synthetase. *Pestic. Biochem. Physiol.* **17**, 301–306.

Cohen, E. and Casida, J. E. (1983). Insect chitin synthetase as a biochemical probe for insecticidal compounds. In: *Pesticide Chemistry, Human Welfare and the Environment* (eds Miyamoto, J. and Kearney, P. C.), pp. 25–32. Pergamon, Oxford.

Cohen, E. and Casida, J. E. (1990). Insect and fungal chitin synthetase activity: specificity of lectins as enhancers and nucleoside peptides as inhibitors. *Pestic. Biochem. Physiol.* **37**, 249–253.

Cohen, E., Kuwano, E. and Eto, M. (1984). The use of *Tribolium* chitin synthetase assay in studying effects of benzimidazoles with a terpene moiety and related compounds. *Agric. Biol. Chem.* **48**, 1617–1620.

Cohen, E., Elster, I. and Chet, I. (1986). Properties and inhibition of *Sclerotium rolfsii* chitin synthetase. *Pestic. Sci.* **17**, 175–182.

Coudron, T. A., Kroha, M. J. and Ignoffo, C. M. (1984). Levels of chitinolytic activity during development of three entomopathogenic fungi. *Comp. Biochem. Physiol. B* **79**, 339–348.

Coutinho, P. M., Deleury, E., Davies, G. J. and Henrissat, B. (2003). An evolving hierarchial family classification for glycosyltransferases. *J. Mol. Biol.* **328**, 307–317.

Czapla, T. H. and Lang, B. A. (1990). Effect of plant lectins on the larval development of European corn borer (Lepidoptera: Pyralidae) and Southern corn rootworm (Coleoptera: Chrysomelidae). *J. Econ. Entomol.* **83**, 2480–2485.

da Silva, M. V., Santi, L., Staats, C. C., da Costa, A. M., Colodel, E. M., Driemeier, D., Vainstein, M. H. and Schrank, A. (2005). Cuticle-induced endo/exoacting chitinase CHIT 30 from *Metarhizium anisopliae* is encoded by an ortholog of chi3 gene. *Res. Microbiol.* **156**, 382–392.

Dähn, U., Hagenmaier, H., Höhne, H., König, W. A., Wolf, G. A. and Zähner, H. (1976). Stoffwechsel Produkte von Mikroorganismen. 154 Mitteilung, Nikkomycin, eine neuer Hemmstoff der Chitinsynthese bei Pilzen. *Arch. Microbiol.* **107**, 143–160.

DeAngelis, P. L. (1999). Hyaluronan synthases: fascinating glycosyltransferases from vertebrates, bacterial pathogens, and algal viruses. *Cell. Mol. Life Sci.* **56**, 670–682.

de Iannino, N. I., Pueppke, S. G. and Ugalde, R. A. (1995). Biosynthesis of the Nod factor chito-oligosaccharide in *Rhizobium fredii* is controlled by the concentration of UDP-*N*-acetyl-D-glucosamine. *Mol. Plant Microbe Interact.* **8**, 292–301.

Delmer, D. P. (1987). Cellulose biosynthesis. *Annu. Rev. Plant Physiol.* **38**, 259–290.

Delmer, D. P. (1999). Cellulose biosynthesis: exciting times for a difficult field of study. *Annu. Rev. Plant Physiol. Plant Mol. Biol.* **50**, 245–276.

De-Rousset-Hall, A. and Gooday, G. W. (1975). A kinetic study of a solubilized chitin synthetase preparation from *Coprinus cinereus*. *J. Gen. Microbiol.* **89**, 146–154.

Deul, D. J., deJong, B. J. and Kortenbach, J. A. M. (1978). Inhibition of chitin synthesis by two 1-(2, 6-disubstituted benzoyl)-3-phenylurea insecticides II. *Pestic. Biochem. Physiol.* **8**, 98–105.

Dickinson, K., Keer, V., Hitchcock, C. A. and Adams, D. J. (1989). Chitinase activity from *Candida albicans* and its inhibition by allosamidin. *J. Gen. Microbiol.* **135**, 1417–1421.

Dickinson, K., Keer, V., Hitchcock, C. A. and Adams, D. J. (1991). Microsomal chitinase activity from *Candida albicans*. *Biochim. Biophys. Acta* **1073**, 177–182.

Ding, X., Gopalakrishnan, B., Johnson, L. B., White, F. F., Wang, X., Morgan, T. D., Kramer, K. J. and Muthukrishnan, S. (1998). Insect resistance of transgenic tobacco expressing an insect chitinase gene. *Transgenic Res.* **7**, 77–84.

Dinglasan, R. R., Devenport, M., Florens, L., Johnson, J. R., McHugh, C. A., Donnelli-Doman, M., Carucci, D. J., Yates, J. R. I. I. I. and Jacobs-Lorena, M. (2009). The *Anopheles gambiae* adult midgut peritrophic matrix proteome. *Insect Biochem. Mol. Biol.* **39**, 125–134.

Dixit, R., Arakane, Y., Specht, C. A., Richard, C., Kramer, K. J., Beeman, R. W. and Muthukrishnan, S. (2008). Domain organization and phylogenetic analysis of proteins from the chitin deacetylase gene family of *Triboium castaneum* and three other species of insects. *Insect Biochem. Mol. Biol.* **38**, 440–451.

Dixon, M. J., Nathubhai, M., Andersen, O. A., van Aalten, D. M. F. and Eggleston, I. M. (2009). Solid-phase synthesis of cyclic peptides chitinase inhibitors: SAR of argifin scaffold. *Org. Biomol. Chem.* **7**, 259–268.

Doares, S. H., Syrovets, T., Weiler, E. W. and Ryan, C. A. (1995). Oligogalacturonides and chitosan activate plant defensive genes through the octadecanoid pathway. *Proc. Natl. Acad. Sci. USA* **92**, 4095–4098.

Does, M. P., Houterman, P. M., Dekker, H. L. and Cornelissen, B. J. C. (1999). Processing, targeting, and antifungal activity of stinging nettle agglutinin in transgenic tobacco. *Plant Physiol.* **120**, 421–432.

Donohoe, T. J. and Rosa, C. P. (2007). A concise and efficient synthesis of (-)-allosamizoline. *Org. Lett.* **9**, 5509–5511.

Duchesne, L. C. and Larson, D. W. (1989). Cellulose and evolution of plant life. *Bioscience* **39**, 238–241.

Duran, A. and Cabib, E. (1978). Solubilization and partial purification of yeast chitin synthetase. Confirmation of the zymogenic nature of the enzyme. *J. Biol. Chem.* **253**, 4419–4425.

Duran, A., Bowers, B. and Cabib, E. (1975). Chitin synthetase zymogen is attached to the yeast plasma membrane. *Proc. Natl. Acad. Sci. USA* **72**, 3952–3955.

Dziadik-Turner, C., Koga, D., Mai, M. S. and Kramer, K. J. (1981). Purification and characterization of two β-*N*-acetylhexosaminidases from the tobacco hornworm, *Manduca sexta* (L.) (Lepidoptera: Sphingidae). *Arch. Biochem. Biophys.* **212**, 546–560.

Eisemann, C. H., Donaldson, R. A., Pearson, R. D., Cadogan, L. C., Vuocolo, T. and Tellam, R. L. (1994). Larvicidal activity of lectins on *Lucilia cuprina*: mechanism of action. *Ent. Exp. Appl.* **72**, 1–10.

Elorza, M. V., Rico, H. and Sentandreu, R. (1983). Calcofluor white alters the assembly of chitin fibrils in *Saccharomyces cerevisiae* and *Candida albicans* cells. *J. Gen. Microbiol.* **129**, 1577–1582.

Elvin, C. M., Vuocolo, T., Pearson, R. D., East, I. J., Riding, G. A., Eisemann, C. H. and Tellam, R. L. (1996). Characterization of a major peritrophic membrane protein, peritrophin-44, from the larvae of *Lucilia cuprina*. *J. Biol. Chem.* **271**, 8925–8935.

Fan, Y., Saito, T. and Isogai, A. (2008). Preparation of chitin nanofibers from squid pen β-chitin by simple mechanical treatment under acid conditions. *Biomacromolecules* **9**, 1919–1923.

Fanelli, E., Di Vito, M., Jones, J. T. and De Giorgi, C. (2005). Analysis of chitin synthase function in a plant parasitic nematode, *Meloidogyne artiellia*, using RNAi. *Gene* **349**, 87–95.

Felix, M., Gloggler, S., Londershausen, M., Weidemann, W., Spindler, K.-D. and Spindler-Barth, M. (2000). A cDNA encoding a chitinase from the epithelial cell line of *Chironomus tentans* (Insecta, Diptera) and its functional expression. *Arch. Insect Biochem. Physiol.* **45**, 24–36.

Filho, J. A. F. and Shapiro, B. A. (2004). Tay-Sachs disease. *Arch. Neurol.* **61**, 1466–1468.

Filho, B. P. D., Lemos, F. J. A., Secundino, N. F. C., Pāscoa, V., Pereira, S. T. and Pimenta, P. F. P. (2002). Presence of chitinase and beta-*N*-acetylglucosaminidase in the *Aedes aegypti* a chitinolytic system involving peritrophic matrix formation and degradation. *Insect Biochem. Mol. Biol.* **32**, 1723–1729.

Fitches, E., Wilkinson, H., Bell, H., Bown, D. P., Gatehouse, J. A. and Edwards, J. P. (2004). Cloning, expression and functional characterization of chitinase from larvae of tomato moth (*Lacanobia oleracea*): a demonstration of the insecticidal activity of insect chitinase. *Insect Biochem. Mol. Biol.* **34**, 1037–1050.

Foster, J. M., Zhang, Y., Kumar, S. and Carlow, C. K. S. (2005). Parasitic nematodes have two distinct chitin synthases. *Mol. Biochem. Parasitol.* **142**, 126–132.

Fristrom, J. W. (1968). Hexosamine metabolism in imaginal disks of *Drosophila melanogaster*. *J. Insect Physiol.* **14**, 729–740.

Fuhrman, J. A. and Piessens, W. F. (1985). Chitin synthesis and sheath morphogenesis in *Brugia malayi* microfilariae. *Mol. Biochem. Parasitol.* **17**, 93–104.

Fujiwara, M., Horiuchi, H., Ohta, A. and Takagi, M. (1997). A novel fungal gene encoding chitin synthase with a myosin motor-like domain. *Biochem. Biophys. Res. Commun.* **236**, 75–78.

Fukamizo, T. and Kramer, K. J. (1985). Mechanism of chitin oligosaccharide hydrolysis by the binary enzyme chitinase system in insect molting fluid. *Insect Biochem.* **15**, 1–7.

Fukamizo, T. and Kramer, K. J. (1987). Effect of 20-hydroxyecdysone on chitinase and β-*N*-acetylglucosaminidase during the larval–pupal transformation of *Manduca sexta* (L.). *Insect Biochem.* **17**, 547–550.

Funkhouser, J. D. and Aronson, N. N. Jr. (2007). Chitinase family GH18: evolutionary insights from the genomic history of diverse protein family. *BMC Evol. Biol.* **7**, 96–111.

Gagou, M. E., Kapsetaki, M., Turberg, A. and Kafetzopoulos, D. (2002). Stage-specific expression of the chitin synthase DmeChSA and DmeChSB genes during the onset of *Drosophila* metamorphosis. *Insect Biochem. Mol. Biol.* **32**, 141–146.

Gardner, K. H. and Blackwell, J. (1975). Refinement of the structure of β-chitin. *Biopolymers* **14**, 1581–1595.

Geisler, C., Aumiller, J. J. and Jarvis, D. L. (2008). A *fused* lobes gene encodes the processing β-*N*-acetylglucosaminidase in Sf9 cells. *J. Biol. Chem.* **283**, 11330–11339.

Gelman, D. B. and Borkovec, A. B. (1986). The pharate adult clasper as a tool for measuring chitin synthesis and for identifying new chitin synthesis inhibitors. *Comp. Biochem. Physiol.* **85C**, 193–197.

Genta, F. A., Blanes, L., Cristofoletti, P. T., do Lago, C. L., Terra, W. R. and Ferreira, C. (2006). Purification, characterization and molecular cloning of the major chitinase from *Tenebrio molitor* larval midgut. *Insect Biochem. Mol. Biol.* **36**, 789–800.

Gijwijt, M. J., Deul, D. H. and deJong, B. J. (1979). Inhibition of chitin synthesis by benzoyl-phenylurea insecticides. III. Similarity in action in *Pieris brassicae* (L.) with polyoxin-D. *Pestic. Biochem. Physiol.* **12**, 87–94.

Giovanini, M. P., Saltzmann, K. D., Puthoff, D. P., Gonzalo, M., Ohm, H. W. and Williams, C. E. (2007). A novel wheat gene encoding a putative chitin-binding lectin is associated with resistance against Hessian fly. *Mol. Plant Pathol.* **8**, 69–82.

Girard, C. and Jouanin, L. (1999). Molecular cloning of a gut-specific chitinase cDNA from the beetle *Phaedon cochleariae. Insect Biochem. Mol. Biol.* **29**, 549–556.

Glaser, L. and Brown, D. H. (1957). The synthesis of chitin in cell free extracts of *Neurospora crassa. J. Biol. Chem.* **228**, 729–742.

Gooday, G. W. (1979). Chitin synthesis and differentiation in *Coprinus cinereus.* In: *Fungal Wall and Hyphal Growth* (eds Burnett, J. H. and Trinci, A. P. J.), pp. 203–223. Cambridge University Press, Cambridge.

Gooday, G. W. (1990a). Physiology of microbial degradation of chitin and chitosan. *Biodegradation* **1**, 177–190.

Gooday, G. W. (1990b). Inhibition of chitin metabolism. In: *Biochemistry of Cell Walls and Membranes in Fungi* (eds Kuhn, P. J., Trinci, A. P. S., Goosey, M. W. and Copping, L. G.), pp. 61–79. Springer-Verlag, Berlin.

Gooday, G. W. (1990c). The ecology of chitin degradation. In: *Advances in Microbial Ecology* (ed Marshall, K. C.), Vol. 11, pp. 387–430. Plenum Press, New York.

Gooday, G. W. and de Rousset- Hall, A. (1975). Properties of chitin synthetase from *Cuprinus cinereus. J. Gen. Microbiol.* **89**, 137–145.

Gooday, G. W. and Trinci, A. P. J. (1980). Wall structure and biosynthesis in fungi. Eukaryotic Microbial Cell. Soc. Gen. Microbiol. Symp., Cambridge Vol. 30. Cambridge University Press, Cambridge, pp. 207–252.

Gopalakrishnan, B., Muthukrishnan, S. and Kramer, K. J. (1995). Baculovirus-mediated expression of a *Manduca sexta* chitinase gene: properties of the recombinant protein. *Insect Biochem. Mol. Biol.* **25**, 255–265.

Gouda, H., Sunazuka, T., Iguchi, K., Sugawara, A., Hirose, T. Noguchi, Y., *et al.* (2009). Computer-aided rational molecular design of argifin-derivatives with increased inhibitory activity against chitinase-B from *Serratia marcescens. Bioorg. Med. Chem. Lett.* **19**, 2630–2633.

Grenier, J. and Asselin, A. (1990). Some pathogenesis-related proteins are chitosanases with lytic activity against fungal spores. *Mol. Plant Microbe Interac.* **3**, 401–407.

Grosscurt, A. C. (1978). Diflubenzuron: Some aspects of its ovicidal and larvicidal mode of action and an evaluation of its practical possibilities. *Pestic. Sci.* **9**, 373–386.

Guo, W., Li, G., Pang, Y. and Wang, P. (2005). A novel chitin-binding protein identified from the peritrophic membrane of the cabbage looper, *Trichoplusia ni. Insect Biochem. Mol. Biol.* **35**, 1224–1234.

Gwinn, J. F. and Stevenson, J. R. (1973). Role of acetylglucosamine in chitin synthesis in crayfish. II. Enzymes in the epidermis for incorporation of acetylglucosamine into UDP-acetylglucosamine. *Comp. Biochem. Physiol.* **45B**, 777–785.

Hadwiger, L. A., Kendra, D. F., Fristensky, B. W. and Wagoner, W. (1986). Chitosan both activates genes in plants and inhibit RNA synthesis in fungi. In: *Chitin in Nature and Technology* (eds Muzzarelli, R., Jeuniaux, C. and Gooday, G. W.), pp. 209–214. Plenum, New York.

Haigler, C. H., Brown, R. M. Jr. and Benziman, M. (1980). Calcofluor white ST alters the in vivo assembly of cellulose microfibrils. *Science* **210**, 903–906.

Hajjar, N. P. and Casida, J. E. (1979). Structure–activity relationships of benzoylphenyl ureas as toxicants and chitin synthesis inhibitors in *Oncopeltus fasciatus. Pestic. Biochem. Physiol.* **11**, 33–45.

Harper, M. S., Hopkins, T. L. and Czapla, T. H. (1998). Effect of wheat germ agglutinin on formation and structure of the peritrophic membrane in European corn borer (*Ostrinia nubilalis*) larvae. *Tissue Cell* **30**, 166–176.

Harris, M. T., Lai, K., Arnold, K., Martinez, H. F., Specht, C. A. and Fuhrman, J. A. (2000). Chitin synthase in the filarial parasite, *Brugia malayi*. *Mol. Biochem. Parasitol.* **111**, 351–362.

Hegedus, D., Erlandson, M., Gillot, C. and Toprak, U. (2009). New insights into peritrophic matrix synthesis, architecture, and function. *Annu. Rev. Entomol.* **54**, 285–302.

Heifetz, A., Keenan, R. W. and Elbein, A. D. (1979). Mechanism of action of tunicamycin on UDP-GlcNAc: dolichyl phosphate GlcNAc-1-phospahte transferase. *Biochemistry* **18**, 2186–2192.

Heldermon, C., DeAngelis, P. L. and Weigel, P. H. (2001). Topological organization of the hyaluronan synthase from *Streptococcus pyogenes*. *J. Biol. Chem.* **276**, 2037–2046.

Henrissat, B. (1998). Glycosidase families. *Biochem. Soc. Trans.* **26**, 153–156.

Henrissat, B. and Bairoch, A. (1993). New families in classification of glycosyl hydrolases based on amino acid similarities. *Biochem. J.* **293**, 781–788.

Herth, W. and Zugenmaier, P. (1979). The site of β-chitin fibril formation in centric diatoms. II. The chitin-forming cytoplasnic structures. *J. Ultrast. Res.* **68**, 16–27.

Herth, W. (1980). Calcofluor white and congo red inhibit chitin microfibril assembly of *Poterioochromonas*: evidence for a gap between polymerization and microfibril formation. *J. Cell Biol.* **87**, 442–450.

Herth, W. and Zugenmaier, P. (1977). Ultrastructure of the chitin fibrils of the centric diatom *Cyclotella cryptica*. *J. Ultrastr. Res.* **61**, 230–239.

Herth, W. and Barthlott, W. (1979). The state of β-chitin fibril formation in centric diatoms. I. Pores and fibril formation. *J. Ultrastr. Res.* **68**, 6–15.

Hodge, A., Gooday, G. W. and Alexander, I. J. (1996). Inhibition of chitinolytic activities from tree species and asorted fungi. *Phytochemistry* **41**, 77–84.

Hogenkamp, D. G., Arakane, Y., Zimoch, L., Merzendorfer, H., Kramer, K. J., Beeman, R. W., Kanost, M. R., Specht, C. A. and Muthukrishnan, S. (2005). Chitin synthase genes in *Manduca sexta*: characterization of a gut-specific transcript and differential tissue expression of alternately spliced mRNAs during development. *Insect Biochem. Mol. Biol.* **35**, 529–540.

Hogenkamp, D. G., Arakane, Y., Kramer, K. J., Muthukrishnan, S. and Beeman, R. W. (2008). Characterization and expression of the β-*N*-acetylhexosaminidase gene family of *Tribolium castaneum*. *Insect Biochem. Mol. Biol.* **38**, 478–489.

Hopkins, T. L. and Harper, M. S. (2001). Lepidopteran peritrophic membranes and effects of dietary wheat germ agglutinin on their formation and structure. *Arch. Insect Biochem. Physiol.* **47**, 100–109.

Hori, M., Kakiki, K., Suzuki, S. and Misato, T. (1971). Studies on the mode of action of polyoxins. Part III. Relation of polyoxin structure to chitin synthetase inhibition. *Agric. Biol. Chem.* **35**, 1280–1291.

Hori, M., Eguchi, J., Kakiki, K. and Misato, T. (1974a). Studies on the mode of action of polyoxins. VI. Effect of polyoxin B on chitin synthesis in polyoxin-sensitive and resistant strains of *Alternaria kikuchiana*. *J. Antibiot.* **27**, 260–266.

Hori, M., Kakiki, K. and Misato, T. (1974b). Interaction between polyoxin and active center of chitin synthetase. *Agric. Biol. Chem.* **38**, 699–705.

Horsch, M., Mayer, C., Sennhauser, U. and Rast, D. M. (1997). β-*N*-acetylhexosaminidase: A target for the design of antifungal agents. *Pharmacol. Ther.* **76**, 187–218.

Horst, M. N. (1983). The biosynthesis of crustacean chitin. Isolation and characterization of a polyprenol-linked intermediates from brine shrimp microsomes. *Arch. Biochem. Biophys.* **223**, 254–263.

Houston, D. R., Shiomi, K., Arai, N., Omura, S., Peter, M. G., Turberg, A., Synstad, B., Eijsink, V. G. H. and van Aalten, D. M. F. (2002a). High-resolution structures of a chitinase complexed with natural product cyclopentapeptide inhibitors: Mimicry of carbohydrate substrate. *Proc. Natl. Acad. Sci. USA* **99**, 9127–9132.

Houston, D. R., Eggleston, I., Synstad, B., Eilsink, V. G. H. and van Aalten, D. M. F. (2002b). The cyclo dipeptide CI-4 [cyclo-(L-Arg-D-Pro)] inhibits family 18 chitinases by structural mimicry of a reaction intermediate. *Biochem. J.* **368**, 23–27.
Houston, D. R., Synstad, B., Eijsink, V. G. H., Stark, M. J. R., Eggleston, I. M. and van Aalten, D. M. F. (2004). Structure-based exploration of cyclic dipeptide chitinase inhibitors. *J. Med. Chem.* **47**, 5713–5720.
Humphreys, A. M. and Gooday, G. W. (1984). Properties of chitinase activities from *Mucor mucedo*: Evidence for a membrane-bound zymogenic form. *J. Gen. Microbiol.* **130**, 1359–1366.
Ibrahim, G. H., Smartt, C. T., Kiley, L. M. and Christensen, B. M. (2000). Cloning and characterization of a chitin synthase cDNA from the mosquito *Aedes aegypti*. *Insect Biochem. Mol. Biol.* **30**, 1213–1222.
Iconomidou, V. A., Willis, J. H. and Hamodrakas, S. J. (1999). Is β-pleated sheet the molecular conformation which dictates formation of helicoidal cuticle? *Insect Biochem. Mol. Biol.* **29**, 285–292.
Imai, T., Watanabe, T., Yui, T. and Sugiyama, J. (2003). The directionality of chitin biosynthesis: a revisit. *Biochem. J.* **374**, 755–760.
Ishaaya, I. and Casida, J. E. (1974). Dietary TH 6040 alters composition and enzyme activity of housefly larval cuticles. *Pestic. Biochem. Physiol.* **4**, 484–490.
Isogai, A., Sato, M., Sakuda, S., Nakayama, J. and Suzuki, A. (1989). Structure of demethylallosamidin as an insect chitinase inhibitor. *Agr. Biol chem.* **53**, 2825–2826.
Isono, K., Nagatsu, J., Koninata, K., Kasai, K. and Suzuki, S. (1967). Studies on polyoxins, antifungal antibiotics. Part V. Isolation and characterization of polyoxins C, D, E, F, G, H, and I. *Agric. Biol. Chem.* **31**, 190–199.
Isono, K., Asahi, K. and Suzuki, S. (1969). Studies on polyoxins, antifungal antibiotics. XIII. The structures of polyoxins. *J. Am. Chem. Soc.* **91**, 7490–7505.
Izumida, H., Imamura, N. and Sano, H. (1996a). A novel chitinase inhibitor from a marine bacterium, *Pseudomonas* sp. *J. Antibiot.* **49**, 76–80.
Izumida, H., Nishijima, M., Takadera, T., Nomoto, A. M. and Sano, H. (1996b). Chitinase inhibitors cyclo (Arg-Pro) against cell separation of *Saccharomyces cerevisiae* and the morphological change in *Candida albicans*. *J. Antibiot.* **49**, 829–831.
Jang, M.-K., Kong, B.-G., Jeong, Y.-I., Lee, C. H. and Nah, J.-W. (2004). Physicochemical characterization of α-chitin, β-chitin, and γ-chitin separated from natural resources. *J. Polym. Sci.* **42**, 3423–3432.
Jaworsky, E., Wang, L. and Margo, G. (1963). Synthesis of chitin in cell-free extracts of *Prodenia eridania*. *Nature* **198**, 790.
Kabir, K. E., Sugimoto, H., Tado, H., Endo, K., Yamanaka, A., Tanaka, S. and Koga, D. (2006). Effect of *Bombyx mori* chitinase against Japanese pine sawyer (*Monochamus alternatus*) adults as a biopesticide. *Biosci. Biotechnol. Biochem.* **70**, 219–229.
Kameda, T., Miyazawa, M., Ono, H. and Yoshida, M. (2005). Hydrogen bonding structure and stability of α-chitin studied by 13C solid-state NMR. *Macromol. Biosci.* **5**, 103–106.
Kamst, E. and Spaink, P. H. (1999). Functional domains in the chitin oligosaccharide synthase NodC and related β-polysaccharide synthases. *Trends Glycosci. Glycotechnol.* **11**, 187–199.
Kato, T., Shiruzi, Y., Izumida, H., Yokoyama, A. and Endo, M. (1995). Styloguanidines, new chitinase inhibitors from the marine sponge *Stylotella aurantium*. *Tetrahedron Lett.* **36**, 2133–2136.
Kato, N., Mueller, C. R., Fuchs, J. F., Wessely, V., Lan, Q. and Christensen, B. M. (2006). Regulatory mechanisms of chitin biosynthesis and role of chitin in peritrophic matrix formation in the midgut of adult *Aedes aegypti*. *Insect Biochem. Mol. Biol.* **36**, 1–9.

Kaur, M., Singh, K., Rup, P. J., Saxena, A. K., Khan, R. H., Ashraf, M. T., Kamboj, S. S. and Singh, J. (2006). A tuber lectin from *Arisaema Helleborifolium* Schott with anti-insect activity against melon fruit fly, *Bactrocera cucurbitae* (Coquillett) and anti-cancer effect on human cancer. *Arch. Biochem. Biophys.* **445**, 156–165.
Kawamura, K., Shibata, T., Saget, O., Peel, D. and Bryant, P. J. (1999). A new family of growth factors produced by the fat body and active on *Drosophila* imaginal disc cells. *Development* **126**, 211–219.
Kawasaki, T., Tanaka, M., Fujie, M., Usami, S., Sakai, K. and Yamada, T. (2002). Chitin synthesis in chlorovirus CVK-2-infected *Chlorella* cells. *Virology* **301**, 123–131.
Kim, M. G., Shin, S. W., Bae, K.-S., Kim, S. C. and Park, H.-Y. (1998). Molecular cloning of chitinase cDNAs from the silkworm, *Bombyx mori* and the fall webworm, *Hyphantria cunea*. *Insect Biochem. Mol. Biol.* **28**, 163–171.
Kimura, S. (1973). The control of chitin deposition by ecdysterone in larvae of *Bombyx mori*. *J. Insect Physiol.* **19**, 2177–2188.
Kimura, S. (1976). The chitinase system in the cuticle of the silkworm *Bombyx mori*. *Insect Biochem.* **6**, 479–482.
King-Jones, K. and Thummel, C. S. (2005). Nuclear receptors—A perspective from *Drosophila*. *Nat. Rev. Genet.* **6**, 311–323.
Kitago, Y., Karita, S., Watanabe, N., Kamiya, M., Aizawa, T., Sakka, K. and Tanaka, I. (2007). Crystal structure of Cel44A, a glycoside hydrolase family 44 endoglucanase from *Clostridium thermocellum*. *J. Biol. Chem.* **282**, 35703–35711.
Kitahara, K., Nakagawa, Y., Nishioka, T. and Fujita, T. (1983). Cultured integument of *Chilo suppressalis* as a bioassay system of insect growth regulators. *Agric. Biol. Chem.* **47**, 1583–1589.
Knapp, S., Fash, D., Abdo, M., Emge, T. J. and Rablen, P. R. (2009). GlcNAc-thiazoline conformations. *Bioorg. Med. Chem.* **17**, 1831–1836.
Kobinata, H., Uramoto, M., Nishii, M., Kusakabe, H., Nakamura, G. and Isono, K. (1980). Neopolyoxin A, B, and C, new chitin synthetase inhibitors. *Agric. Biol. Chem.* **44**, 1709–1711.
Koga, D., Mai, M. S. and Kramer, K. J. (1983). Comarative biochemistry of insect exo-β-N-acetyl-D-glucosaminadase from the tobacco hornworm, *Manduca sexta* L. *Comp. Biochem. Physiol. B.* **74**, 515–520.
Koga, D., Nakashima, M., Matsukura, T., Kimura, S. and Ide, A. (1986). Purification and properties of β-*N*-acetyl-D-glucosaminidase from alimentary canal of the silkworm, *Bombyx mori*. *Agric. Biol. Chem.* **50**, 2357–2368.
Koga, D., Isogai, A., Sakuda, S., Matsumoto, S., Suzuki, A., Kimura, S. and Ide, A. (1987a). Specific inhibition of *Bombyx mori* chitinase by allosamidin. *Agric. Biol. Chem.* **51**, 471–476.
Koga, D., Shimazaki, C., Yamamoto, K., Inoue, K., Kimura, S. and Ide, A. (1987b). β-*N*-acetylglucosaminidase from integument of the silkworm, *Bombyx mori*: comparative biochemistry with the pupal alimentary canal enzyme. *Agric. Biol. Chem.* **51**, 1679–1681.
Koga, D., Fujimoto, H., Funakoshi, T., Utsumi, T. and Ide, A. (1989). Appearance of chitinolytic enzymes in integument of *Bombyx mori* during the larval–pupal transformation. Evidence for zymogenic forms. *Insect Biochem.* **19**, 123–128.
Koga, D., Funakoshi, T., Mizuki, K., Ide, A., Kramer, K. J., Zen, K.-C., Choi, H. and Muthukrishnan, S. (1992). Immunoblot analysis of chitinolytic enzymes in integument and molting fluid of the silkworm, *Bombyx mori*, and the tobacco hornworm, *Manduca sexta*. *Issect Biochem. Mol. Biol.* **22**, 305–311.
Kombrink, E. and Somssich, I. E. (1995). Defense responses of plants to pathogens. *Adv. Bot. Res.* **21**, 1–34.

Kramer, K. J. and Aoki, H. (1987). Chitinolytic enzymes from pupae of the red flour beetle, *Tribolium castaneum*. *Comp. Biochem. Physiol.* **86B**, 613–621.

Kramer, K. J. and Koga, D. (1986). Insect chitin. Physical state, synthesis, degradation and metabolic regulation. *Insect Biochem.* **16**, 851–877.

Kramer, K. J. and Muthukrishnan, S. (1997). Insect chitinases: molecular biology and potential use a biopesticides. *Insect Biochem. Mol. Biol.* **27**, 887–900.

Kramer, K. J. and Muthukrishnan, S. (2005). Comprehensive molecular insect science. In: *Biochemistry and Molecular Biology* (eds Gilbert, L. I., Iatrou, K. and Gill, S.), Vol. 4, pp. 111–144. Elsevier, Oxford.

Kramer, K. J., Dziadik-Turner, C. and Koga, D. (1985). Chitin metabolism in insects. In: *Comprehensive Insect Physiology, Biochemistry and Pharmacology* (eds Kerkut, G. A. and Gilbert, L. I.), Vol. 3, pp. 75–115. Pergamon, Oxford.

Kramer, K. J., Corpuz, L., Choi, H. K. and Muthukrishnan, S. (1993). Sequence of cDNA and expression of the gene encoding epidermal and gut chitinases of *Manduca sexta*. *Insect Biochem. Mol. Biol.* **23**, 691–701.

Krishnan, A., Nair, P. N. and Jones, D. (1994). Isolation, cloning, and characterization of new chitinase stored in active form in chitin-lined venom reservoir. *J. Biol. Chem.* **269**, 20971–20976.

Kubo, I., Uchida, M. and Klocke, J. A. (1983). An insect ecdysis inhibitor from the African medicinal plant, *Plumbago capensis* (Plumbaginaceae); a naturally occurring chitin synthetase inhibitor. *Agric. Biol. Chem.* **47**, 911–913.

Kumar, M. N. V. R. (2000). A review of chitin and chitosan applications. *React. Funct. Polym.* **46**, 1–27.

Kumar, N. S., Tang, B., Chen, X., Tian, H. and Zhang, W. (2008). Molecular cloning, expression pattern and comparative analysis of chitin synthase gene B in *Spodoptera exigua*. *Comp. Biochem. Physiol. B* **149**, 447–453.

Kuranda, M. J. and Robbins, P. W. (1991). Chitinase is required for cell separation during growth of *Saccharomyces cerevisiae*. *J. Biol. Chem.* **266**, 19758–19767.

Kurita, K. (2001). Controlled functionalization of the polysaccharide chitin. *Prog. Polym. Sci.* **26**, 1921–1971.

Kuwano, E., Sato, N. and Eto, M. (1982). Insecticidal benzimidazoles with a terpenoid moiety. *Agric. Biol. Chem.* **46**, 1715–1716.

Langley, D. B., Harty, D. W. S., Jacques, N. A., Hunter, N., Guss, J. M. and Collyer, C. A. (2008). Structure of *N*-acetyl-β-D-glucosaminidase (GcnA) from the endocarditis pathogen *Streptococcus gordonii* and its complex with the mechanism-based inhibitor NAG-thiazoline. *J. Mol. Biol.* **377**, 104–116.

Lanusa, D. M., Carbajal, J. A. and Borras, R. (1996). Identification of surface coat carbohydrates in *Blastocystis hominis* by lectin probes. *Int. J. Parasitol.* **26**, 527–532.

Lasters, I., Wodak, S. J., Alard, P. and van Cutsem, E. (1988). Structural principles of parallel beta-barrels in proteins. *Proc. Natl. Acad. Sci. USA* **85**, 3338–3342.

Leal-Morales, C. A., Bracker, C. E. and Bartnicki-Garcia, S. (1988). Localization of chitin synthetase in cell-free homogenates of *Saccharomyces cerevisiae*: chitosomes and plasma membrane. *Proc. Natl. Acad. Sci. USA* **85**, 8516–8520.

Lehane, M. J. (1997). Peritrophic matrix structure and function. *Annu. Rev. Entomol.* **42**, 525–550.

Lemieux, M. J., Mark, B. L., Cherney, M. M., Withers, S. G., Mahuran, D. J. and James, N. G. (2006). Crystallographic structure of human β-hexosaminidase A: Interpretation of Tay-Sachs mutations and loss of GM_2 galglioside hydrolysis. *J. Mol. Biol.* **359**, 913–929.

Lenardon, M. D., Whitton, R. K., Munro, C. A., Marshal, D. G. and Neil, A. R. (2007). Individual chitin synthase enzymes synthesize microfibrils of differing structure

at specific locations in the *Candida albicans* cell wall. *Mol. Microbiol.* **66**, 1164–1173.
Le'onard, R., Rendic', D., Rabouille, C., Wilson, I. B. H., Pte'at, T. and Altmann, F. (2006). The *Drosophila* fused lobes gene encodes an *N*-acetylglucosaminidase involved in N-glycan processing. *J. Biol. Chem.* **281**, 4867–4875.
Lesage, G. and Bussey, H. (2006). Cell wall assembly in *Saccharomyces cerevisiae. Microbiol. Mol. Biol. Rev.* **70**, 317–343.
Licari, P. J., Jarvis, D. L. and Bailey, J. E. (1993). Insect cell hosts for baculovirus expression vectors contain endogenous exoglycosidase activity. *Biotechnol. Prog.* **9**, 146–152.
Lim, S.-J. and Lee, S.-S. (1982). The toxicity of diflubenzuron on *Oxya japonica* (Willemse) and its effect on moulting. *Pestic. Sci.* **13**, 537–544.
Lindeboom, J., Mulder, M. B., Vos, J. W., Ketelaar, T. and Emons, A. M. C. (2008). Cellulose microfibril deposition: coordinated activity at the plant cell membrane. *J. Microsc.* **231**, 192–200.
Ling, Z., Suits, M. D. L., Bingham, R. J., Bruce, N. C., Davies, G. J., Fairbanks, A. J., Moir, J. W. B. and Taylor, E. J. (2009). The X-ray crystal structure of an Arthrobacter protophormiae endo-β-*N*-acetylglucosaminidase reveals a (β/α)8 catalytic domain, two ancillary domains and active site residues key for transglycosylation activity. *J. Mol. Biol.* **389**, 1–9.
Linthorst, H. J. M. (1991). Pathogenesis-related proteins in plants. *Crit. Rev. Plant Sci.* **10**, 123–150.
Lis, H. and Sharon, N. (1986). Lectins as molecules and as tools. *Annu. Rev. Biochem.* **55**, 35–67.
Liu, M., Cai, Q. X., Liu, H. Z., Zhang, B. H., Yan, J. P. and Yuan, Z. M. (2002). Chitinolytic activities in *Bacillus thuringiensis* and their synergistic effects on larvicidal activity. *J. Appl. Microbiol.* **93**, 374–379.
Locke, N. (1991). Insect epidermal cell. In: *Physiology of the Insect Epidermis* (eds Binnington, K. and Retnakaran, A.), pp. 1–22. CSIRO, Melbourne.
Locke, M. and Huie, P. (1979). Apolysis and the turnover of plasma membrane plaques during cuticle formation in an insect. *Tissue Cell* **11**, 277–291.
Londershausen, M., Turberg, A., Bieseler, B., Lennartz, M. and Peter, M. G. (1996). Characterization and inhibitor studies of chitinases from the parasitic blowfly (*Lucilia cuprina*), a tick (*Boophilus microplus*), an intestinal nematode (*Haemonchus contortus*) and a bean (*Phaseolus vulgaris*). *Pestic. Sci.* **48**, 305–314.
Lu, Y., Zen, K.-C., Muthukrishnan, S. and Kramer, K. J. (2002). Site-directed mutagenesis and functional analysis of active acidic amino acid residues D142, D144 and E146 in *Manduca sexta* (tobacco hornworm) chitinase. *Insect Biochem. Mol. Biol.* **32**, 1369–1382.
Lucero, H. A., Kuranda, M. J. and Bulik, D. A. (2002). A nonradioactive high throughput assay for chitin synthase activity. *Analyt. Biochem.* **305**, 97–105.
Ludwig, M., Spindler-Barth, M. and Spindler, K. D. (1991). Properties of chitin synthase in homogenates from *Chironomus* cells. *Arch. Insect Biochem. Physiol.* **18**, 251–263.
Lusching, S., Bätz, T., Armbruster, K. and Krasnow, M. A. (2006). *serpentine* and *vermiform* encode matrix proteins with chitin binding and deacetylation domains that limit tracheal tube length in *Drosophila. Curr. Biol.* **16**, 186–194.
Manocha, M. S. (1987). Cellular and molecular aspects of fungal host-mycoparasite interactions. *Z. Pflanzenkr. Pflanzenschutz.* **94**, 432–444.
Manocha, M. S. and Begum, A. (1985). Properties of chitin synthase from mucoraceous hosts of a mycoparasite. *Can. J. Microbiol.* **31**, 6–12.

Marchini, D., Bernini, L. F. and Dallai, R. (1989). β-*N*-acetylhexosaminidase in the secretion of the female reproductive accessory glands of *Ceratitis capitata* (Diptera). *Insect Biochem.* **19**, 549–555.

Mark, B. L., Vocadlo, D. J., Knapp, S., Triggs-Raine, B. L., Withers, S. G. and James, M. N. G. (2001a). Crystallographic evidence for substrate-assisted catalysis in a bacterial β-hexosaminidase. *J. Biol. Chem.* **276**, 10330–10337.

Mark, B. L., Vocadlo, D. J., Zhao, D., Knapp, S., Withers, S. G. and James, M. N. G. (2001b). Biochemical and structural assessment of the 1-N-azasugar GalNAc-isofagomine as a potent family 20 β-*N*-acetylhexosaminidase inhibitor. *J. Biol. Chem.* **276**, 42131–42137.

Mark, B. L., Mahuran, D. J., Cherney, M. M., Zhao, D., Knapp, S. and James, M. N. G. (2003). Crystal structure of human β-hexosaminidase B: Understanding the molecular basis of Sandhoff and Tay-Sachs disease. *J. Mol. Biol.* **327**, 1093–1109.

Markovic′-Housely, Z., Miglierini, G., Soldatova, L., Rizkallah, P. J., Müller, U. and Schirmer, T. (2000). Crystal structure of hyaluronidase, a major allergen of bee venom. *Structure* **8**, 1025–1035.

Marks, E. P. and Leopold, R. A. (1971). Deposition of cuticular substances *in vitro* by leg regenerates from the cockroach, *Leucophaea maderae* (D.). *Biol. Bull.* **140**, 73–83.

Marks, E. P. and Sowa, B. A. (1976). Cuticle formation in vitro. In: *The Insect Integument* (ed Hepburn, H. R.), pp. 339–357. Elsevier, Amsterdam.

Martinez-Rucobo, F. W., Eckhardt-Strelau, L. and Terwisscha van Scheltinga, A. C. (2009). Yeast chitin synthase 2 activity is modulated by proteolysis and phosphorylation. *Biochem. J.* **417**, 547–554.

Mauch, F., Hadwiger, L. A. and Roller, T. (1988). Antifungal hydrolases in pea tissue. I. Purification and characterization of two chitinases and two β-1, 3-glucanases differently regulated during development and in response to fungal infection. *Plant Physiol.* **87**, 325–533.

Maue, L., Meissner, D. and Merzendorfer, H. (2009). Purification of an active, oligomeric chitin synthase complex from the midgut of the tobacco hornworm. *Insect Biochem. Mol. Biol.* **39**, 654–659.

Mayer, R. T., Chen, A. C. and DeLoach, J. R. (1980a). Characterization of a chitin synthase from the stable fly *Stomoxys calcitrans* (L.). *Insect Biochem.* **10**, 549–556.

Mayer, R. T., Meola, S. M., Coppage, D. L. and DeLoach, J. R. (1980b). Utilization of imaginal tissues from pupae of the stable fly for the study of chitin synthesis and screening chitin synthesis inhibitors. *J. Econ. Entomol.* **73**, 76–80.

Mayer, R. T., Chen, A. C. and DeLoach, J. R. (1981). Chitin synthesis inhibiting insect growth regulators do not inhibit chitin synthase. *Experientia* **37**, 337–338.

McCarthy, P. J., Troke, P. F. and Gull, K. (1985). Mechanism of action of nikkomycin and the peptide transport system of *Candida albicans*. *J. Gen. Microbiol.* **131**, 775–780.

McNab, R. and Glover, L. A. (1991). Inhibition of *Neurospora crassa* cytosolic chitinase by allosamidin. *FEMS Microbiol. Lett.* **82**, 79–82.

Merzendorfer, H. (2006). Insect chitin synthases: a review. *Comp. Biochem. Physiol. B* **176**, 1–15.

Merzendorfer, H. and Zinoch, L. (2003). Chitin metabolism in insects: structure, function and regulation of chitin synthases and chitinases. *J. Exp. Biol.* **206**, 4393–4412.

Meritxell, R., Bartnicki-García, S., González-Prieto, J. M., Sánchez-León, E., Verdín-Ramos, J. A., Beltrán-Aguilar, A. and Freitag, M. (2007). Spitzenkörper localization and intracellular traffic of green fluorescent protein-labeled CHS-3 and CHS-6 chitin synthases in living hyphae of *Neurospora crassa*. *Eukaryot. Cell* **6**, 1853–1864.

Milewsky, S., O'Donnell, R. W. and Gooday, G. W. (1992). Chemical modification studies of the active centre of *Candida albicans* chitinase and its inhibition by allosamidin. *J. Gen. Microbiol.* **138**, 2545–2550.
Miller, R. F. (1991). Chitin paleoecology. *Biochem. Syst. Ecol.* **19**, 401–411.
Minke, R. and Blackwell, J. (1978). The structure of α-chitin. *J. Mol. Biol.* **120**, 167–433.
Mirelman, D., Galun, E., Sharon, N. and Lotan, R. (1975). Inhibition of fungal growth by wheat germ agglutinin. *Nature* **256**, 414–416.
Misato, T. (1982). Present status and future aspects of agricultural antibiotics. *J. Pestic. Sci.* **7**, 301–305.
Mitsui, T., Nobusawa, C. and Fukami, J.-I. (1984). Mode of inhibition of chitin synthesis by diflubenzuron in the cabbage armyworm, *Mamestra brassicae* L. *J. Pestic. Sci.* **9**, 19–26.
Moreira, M. F., dos Santos, A. S. Marottaa, H. R., *et al.* (2007). A chitin-like component in *Aedes aegypti* eggshells, eggs and ovaries. *Insect Biochem. Mol. Biol.* **37**, 1249–1261.
Mothes, U. and Seitz, K.-A. (1982). A possible pathway of chitin synthesis as revealed by electron microscope in *Tetranychus urticae* (Acari, Tetranychidae). *Cell Tissue Res.* **214**, 443–448.
Mothes, U. and Seitz, K. A. (1982). Action of the microbial metabolite and chitin synthesis inhibitor nikkomycin on the mite *Tetranychus urticae*; an electron microscopy study. *Pestic. Sci.* **13**, 426–441.
Mulder, R. and Gijwijt, M. J. (1973). The laboratory evaluation of two promising new insecticides which interfere with cuticle deposition. *Pestic. Sci.* **4**, 737–745.
Müller, H., Furter, R., Zähner, H. and Rast, D. M. (1981). Metabolic products of microorganisms. 203. Inhibition of chitosomal chitin synthetase and growth of *Mucor rouxii* by nikkomycin Z, nikkomycin X, and polyoxin A: a comparison. *Arch. Microbiol.* **130**, 195–197.
Murdock, L. L., Huesing, J. E., Nielsen, S. S., Pratt, R. C. and Shade, R. E. (1990). Biological effects of plant lectins on the cowpea weevil. *Phytochemistry* **29**, 85–89.
Muzzarelli, R. A. A. (1977). Chitin. Pergamon, Oxford, 309pp.
Nagahama, H., Higuchi, T., Jayakumar, R., Furuike, T. and Tamura, H. (2008). XRD studies of β-chitin from squid pen with calcium solvent. *Int. J. Biol. Macromol.* **42**, 309–313.
Nagahashi, S., Sudoh, M., Ono, N., Sawada, R. Yamaguchi, E., *et al.* (1995). Characterization of chitin synthase 2 of *Saccharomyces cerevisiae*. Implication of two highly conserved domains as possible catalytic sites. *J. Biol. Chem.* **270**, 13961–13967.
Nagamatsu, Y., Yanagisawa, I., Kimoto, M., Okamoto, E. and Koga, D. (1995). Purification of a chitoologosaccharidolytic β-*N*-acetylglucosaminidase from *Bombyx mori* larvae during metamorphosis and the nucleotide sequence of its cDNA. *Biosci. Biotechnol. Biochem.* **59**, 219–225.
Nakagawa, Y., Matsumura, F. and Hashino, Y. (1993). Effect of diflubenzuron on incorporation of [^{3}H]-*N*-acetylglucosamine ([^{3}H]NAGA) into chitin in the intact integument from the newly molted American cockroach *Periplaneta americana*. *Comp. Biochem. Physiol.* **106C**, 711–715.
Neville, A. C. (1967). Chitin orientation in cuticle and its control. *Adv. Insect Physiol.* **4**, 269–286.
Neville, A. C. (1975). Biology of Arthropod Cuticle. Springer-Verlag, New York, p. 448.
Neville, A. C. (1986). The physics of helicoids. Multidirectional plywood structures in biological systems. *Phys. Bull.* **37**, 74–76.
Neville, A. C. and Luke, B. M. (1969). A two-system model for chitin-protein complexes in insect cuticles. *Tissue Cell* **1**, 689–707.

Nishimoto, Y., Sakuda, S., Takayama, S. and Yamada, Y. (1991). Isolation and characterization of new allosamidins. *J. Antibiot.* **44**, 716–722.

Oberlander, H., Van Ferkovich, S. M., Van Essen, F. and Leach, C. E. (1978). Chitin biosynthesis in imaginal discs cultured in vitro. *Wilhelm Roux's Arch. Dev. Biol.* **185**, 95–98.

Oberlander, H., Lynn, D. E. and Leach, C. E. (1983). Inhibition of cuticle production in imaginal disks of *Plodia interpunctella* (cultured in vitro): Effects of colcemid and vinblastine. *J. Insect Physiol.* **29**, 47–53.

Oikawa, N., Nakagawa, Y., Soya, Y., Nishimura, E., Kurihara, N., Ueno, T. and Fujita, T. (1993). Enhancement of *N*-acetylglucosamine incorporation into the cultured intergument of Chilo suppressalis by molting hormone and dibenzoylhydrazine insecticides. *Pestic. Biochem. Physiol.* **47**, 165–170.

Oikawa, A., Itoh, E., Ishihara, A. and Iwamura, H. (2003). Purification and characterization of β-*N*-acetylhexosaminidase from maize seedlings. *J. Plant Physiol.* **160**, 991–999.

Okada, T., Ishiyama, S., Sezutsu, H., Usami, A., Tamura, T., Mita, K., Fujiyama, K. and Seki, T. (2007). Molecular cloning and expression of two novel β-*N*-acetylglucosaminidases from silkworm *Bombyx mori*. *Biosci. Biotechnol. Biochem.* **71**, 1626–1635.

Omura, S., Arai, N., Yamaguchi, Y., Masuma, R., Iwai, Y., Namikoshi, M., Turberg, A., Kolbl, H. and Shiomi, K. (2000). Argifin, a new chitinase inhibitor, produced by *Gliocladium* sp. FTD-0668. I. Taxonomy, fermentation, and biological activities. *J. Antibiot.* **53**, 603–608.

O'Neill, M. P., Holman, G. M. and Wright, J. E. (1977). β-Ecdysone levels in pharate pupae of the stable fly, *Stomoxys calcitrans* and interaction with the chitin inhibitor diflubenzuron. *J. Insect Physiol.* **23**, 1243–1244.

Ortiz, D. and Novick, P. J. (2006). Ypt32p regulates the translocation of Chs3p from the internal pool to the plasma membrane. *Eur. J. Cell Biol.* **85**, 107–116.

Pahlic, M. and Stevenson, J. R. (1978). Glucosamine-6-phosphate synthesis in crayfish epidermis. *Comp. Biochem. Physiol.* **60B**, 281–285.

Payne, J. W. and Shallow, D. A. (1985). Studies on drug targeting in the pathogenic fungus *Candida albicans*: peptide transport mutants resistant to polyoxins, nikkomycins and bacilysin. *FEMS Microbiol. Lett.* **28**, 55–60.

Pear, J. R., Kawagoe, Y., Schreckengost, W. E., Delmer, D. P. and Stalker, D. M. (1996). Higher plants contain homologs of the bacterial celA genes encoding the catalytic subunit of cellulose synthase. *Proc. Natl. Acad. Sci. USA* **93**, 12637–12642.

Perrakis, A., Tews, I., Dauter, Z., Oppenheim, A. B., Chet, I., Wilson, K. S. and Vorgias, C. E. (1994). Crystal structure of a bacterial chitinase at 2.3Å resolution. *Structure* **2**, 1169–1180.

Peters, W. (1976). Investigation on the peritrophic membranes of Diptera. In: *The Insect Integument* (ed Hepburn, H. R.), pp. 515–543. Elsevier, Amsterdam.

Petter, C., Scholz, C., Wessner, H., Hansen, G., Henklein, P., Watanabe, T. and Höhne, W. (2008). Phage display screening for peptidic chitinase inhibitors. *J. Mol. Recognit.* **21**, 401–409.

Post, L. C. and Vincent, W. R. (1973). A new insecticide inhibits chitin. *Naturwissenschaften* **60**, 431–432.

Poulicek, M. and Jeuniaux, C. (1991). Chitin biodegradation in marine environments: An experimental approach. *Biochem. Syst. Ecol.* **19**, 385–394.

Poulicek, M., Voss-Foucart, M. F. and Jeuniaux, C. (1986). Chitinoproteic complexes and mineralization in mollusk skeletal structures. In: *Chitin in Nature and Technology* (eds Muzzarelli, R. A. A., Jeuniaux, C. and Gooday, G. W.), pp. 7–12. Plenum, New York.

Prag, G., Papanikolau, Y., Tavlas, G., Vorgias, C. E., Petratos, K. and Oppenheim, A. B. (2000). Structure of chitobiase mutants complexed with the substrate di-*N*-acetyl-D-glucosamine: the catalytic role of the conserved acidic pair, aspartate 539 and glutamate 540. *J. Mol. Biol.* **300**, 611–617.

Priya, T. A. J., Li, F., Zhang, J., Wang, B., Zhao, C. and Xiang, J. (2009). Molecular characterization and effect of RNA interference of retinoid X receptor (RXR) on E75 and chitinase expression in Chinese shrimp *Fenneropenaeus chinensis*. *Comp. Biochem. Physiol. B* **153**, 121–129.

Proia, R. L. (1988). Gene encoding the human β-hexosaminidase β chain: extensive homology of intron placement in the α- and β-chain genes. *Proc. Natl. Acad. Sci. USA* **85**, 1883–1887.

Quader, H. (1986). Cellulose microfibrils orientation in *Oocystis solitaria*: proof that microtubules control the alignment of the terminal complexes. *J. Cell Sci.* **83**, 223–234.

Quader, H., Robinson, D. G. and Van Kempen, R. (1983). Cell wall development in *Oocystis solitaria* in the presence of polysaccharide binding dyes. *Planta* **157**, 317–323.

Quesada-Allue, L. A. (1982). The inhibition of insect chitin synthesis by tunicamycin. *Biochem. Biophys. Res. Commun.* **105**, 312–319.

Raikhel, N. V., Lee, H,.-I. and Broekaert, W. F. (1993). Structure and function of chitin-binding proteins. *Annu. Rev. Plant Physiol. Plant Mol. Biol.* **44**, 591–615.

Ramasubbu, N., Thomas, L. M., Ragunath, C. and Kaplan, J. B. (2005). Structure analysis of dispersin B, a biofilm-releasing glycoside hydrolase from the periodontopathogen *Actinobacillus actinomycetemcomitans*. *J. Mol. Biol.* **349**, 475–486.

Rao, R., Fiandra, L., Giordana, B., de Eguileor, M., Congiu, T., Burlini, N., Arciello, S., Corrado, G. and Pennacchio, F. (2004). AcMNPV ChiA protein disrupts the peritrophic membrane and alters midgut physiology of *Bombyx mori* larvae. *Insect Biochem. Mol. Biol.* **34**, 1205–1213.

Rao, F. V., Andersen, O. A., Vora, K. A., DeMartino, J. A. and van Aalten, D. M. F. (2005a). Methylxanthine drugs are chitinase inhibitors: investigation of inhibition and binding modes. *Chem. Biol.* **12**, 973–980.

Rao, F. V., Houston, D. R., Boot, R. G., Aerts, J. M. F. G., Hodkinson, M., Adams, D. J., Shiomi, K., Omura, S. and van Aalten, D. M. F. (2005b). Specificity and affinity of natural product cyclopentapeptide inhibitors against *A. fumigatus*, human, and bacterial chitinases. *Chem. Biol.* **12**, 65–76.

Reardon, D. and Farber, G. K. (1995). The structure and evolution of α/β barrel proteins. *FASEB J.* **9**, 497–503.

Revol, J.-F. and Marchessault, R. H. (1993). *In vitro* chiral nematic ordering of chitin crystallites. *Int. J. Biol. Macromol.* **15**, 329–335.

Riddiford, L. M. (1991). Hormonal control of sequential gene expression in insect epidermis. In: *Physiology of the Insect Epidermis* (eds Binnington, K. and Retnakaran, A.), pp. 46–54. CSIRO, Melbourne.

Riddiford, L. M. (1994). Cellular and molecular actions of juvenile hormone I. General considerations and premetamorphic actions. *Adv. Insect Physiol.* **24**, 213–274.

Riddiford, L. M., Hiruma, K., Zhou, X. and Nelson, C. A. (2003). Insights into the molecular basis of the hormonal control of molting and metamorphosis from *Manduca sexta* and *Drosophila melanogaster*. *Insect Biochem. Mol. Biol.* **33**, 1327–1338.

Roman, G., Meller, V., Wu, K. A. and Davis, R. L. (1998). The opt1 gene of *Drosophila melanogaster* encodes a proton-dependent dipeptide transporter. *Am. J. Physiol. Cell Physiol.* **275**, c857–c869.

Roncero, C. and Duran, A. (1985). Effect of Calcofluor white and Congo red on fungal cell wall morphogenesis. In vivo activation of chitin polymerization. *J. Bacteriol.* **163**, 1180–1185.

Roncero, C., Valdivieso, M. H., Ribas, J. C. and Duran, A. (1988). Isolation and characterization of *Saccharomyces cerevisiae* mutants resistant to Calcofluor white. *J. Bacteriol.* **170**, 1950–1954.

Rountree, J. S. S., Butters, T. D., Wormald, M. R., Boomkamp, S. D., Dwek, R. A., Asano, N., Ikeda, K., Evinson, E. L., Nash, R. J. and Fleet, G. W. J. (2009). Design, synthesis, and biological evaluation of enantiomeric β-*N*-acetylhexosaminidase inhibitors LABNAc and DABNAc as potential agents against Tay-Sachs and Sandhoff disease. *Chem. Med. Chem.* **4**, 378–392.

Rousseau-Limouzin, M. and Fritig, B. (1991). Induction of chitinases, 1, 3-β-glucanases and other pathogenesis-related proteins in sugar beet leaves upon infection with *Cercospora beticula. Plant Physiol. Biochem.* **29**, 105–117.

Royer, V., Fraichard, S. and Bouhin, H. (2002). A novel putative insect chitinase with multiple catalytic domains: hormonal regulation during metamorphosis. *Biochem. J.* **366**, 921–928.

Rudall, K. M. and Kenchington, W. (1973). The chitin system. *Biol. Rev.* **49**, 597–636.

Ruiz-Herrera, J., Lopez-Romero, E. and Bartnicki-Garcia, S. (1977). Properties of chitin synthetase in isolated chitosomes from yeast cells of *Mucor rouxii. J. Biol. Chem.* **252**, 3338–3343.

Saguez, J., Dubois, F., Vincent, C., Laberche, J.-C., Sangwan-Norreel, B. S. and Giordanengo, P. (2006). Differential aphicidal effects of chitinase inhibitors on the polyphagous homopteran *Myzus persicae* (Sulzer). *Pest Manag. Sci.* **62**, 1150–1154.

Sahai, A. S. and Manocha, M. S. (1993). Chitinases of fungi and plants: their involvement in morphogenesis and host–parasite interaction. *FEMS Microbiol. Rev.* **11**, 317–338.

Sakuda, S., Isogai, A., Matsumoto, S. and Suzuki, A. (1986). The structure of allosamidin, a novel insect chitinase inhibitor, produced by *Streptomyces* sp. *Tetrahedron Lett.* **27**, 2475–2478.

Sakuda, S., Isogai, A., Matsumoto, S. and Suzuki, A. (1987). Search for microbial insect growth regulator. II. Allosamidin, a novel insect chitinase inhibitor. *J. Antibiot.* **40**, 296–300.

Sakuda, S., Nishimoto, Y., Ohi, M., Watanabe, M., Takayama, S., Isogai, A. and Yamada, Y. (1990). Effect of demethylallosamidin, a potent yeast chitinase inhibitor, on the cell division of yeast. *Agric. Biol. Chem.* **54**, 1333–1335.

Sakuda, S., Isogai, A., Suzuki, A. and Yamada, Y. (1993). Chemistry and biochemistry of the chitinase inhibitors, allosamidins. *Actinomycetologica* **7**, 50–57.

Sampson, M. N. and Gooday, G. W. (1998). Involvement of chitinases of *Bacillus thuringiensis* during pathogenesis in insects. *Microbiology* **144**, 2189–2194.

Samuels, R. I. and Reynolds, S. E. (1993). Molting fluid enzymes of the tobacco hornworm, *Manduca sexta*: Timing of proteolytic and chitinolytic activity in relation to pre-ecdysial development. *Arch. Insect Biochem. Physiol.* **24**, 33–44.

Santos, B. and Snyder, M. (1997). Targeting of chitin synthase 3 to polarized sites in yeast requires Chs5p and Myo2p. *J. Cell Biol.* **136**, 95–110.

Saxena, I. M., Brown, R. M. Jr., Fevre, M., Geremia, R. A. and Henrissat, B. (1995). Multidomain architecture of β-glycosyltransferases: Implications for mechanism of action. *J. Bacteriol.* **117**, 1419–1424.

Schlüter, U. (1982). Ultrastructural evidence for inhibition of chitin synthesis by nikkomycin. *Wilhelm Roux Arch. Dev. Biol.* **191**, 205–207.

Schlumbaum, A., Mauch, F., Vögelli, H. and Boller, T. (1986). Plant chitinases are potent inhibitors of fungal growth. *Nature* **324**, 365–367.

Schonitzer, V. and Weiss, I. M. (2007). The structure of mollusc larval shells formed in the presence of the chitin synthase inhibitor nikkomycin Z. *BMC Struct. Biol.* **7**, 71.

Sehnal, F., Delbecque, J.-P., Maroy, P. and Mala, J. (1986). Ecdysteroid titres during larval life and metamorphosis of *Galleria mellonella*. *Insect Biochem.* **16**, 157–162.

Selitrennikoff, C. P. (1984). Calcofluor white inhibits *Neurospora* chitin synthetase activity. *Exp. Mycol.* **8**, 269–272.

Selitrennikoff, C. P. (1985). Chitin synthetase activity of *Neurospora crassa*: Effects of Primulin and Congo red. *Exp. Mycol.* **9**, 179–182.

Sentandreu, R., Martinez-Ramon, A. and Ruiz-Herrera, J. (1984). Localization of chitin synthase in *Mucor rouxii* by an autoradiographic method. *J. Gen. Microbiol.* **130**, 1193–1199.

Shahabuddin, M., Toyoshima, T., Aikawa, M. and Kaslow, D. C. (1993). Transmission-blocking activity of a chitinase inhibitor and activation of malarial parasite chitinase by mosquito protease. *Proc. Natl. Acad. Sci. USA* **90**, 4266–4270.

Shahabuddin, M., Kaidoh, T., Aikawa, M. and Kaslow, D. C. (1995). *Plasmodium gallinaceum*: Mosquito peritrophic matrix and the parasite-vector compatibility. *Exp. Parasitol.* **81**, 386–393.

Shen, Z. and Jacobs-Lorena, M. (1997). Characterization of a novel gut-specific chitinase gene from the human malaria vector *Anopheles gambiae*. *J. Biol. Chem.* **272**, 28895–28900.

Shen, Z. and Jacobs-Lorena, M. (1999). Evolution of chitin-binding proteins in invertebrates. *J. Mol. Evol.* **48**, 341–347.

Shi, Y., Jiang, Z., Han, P., Zheng, G.-X., Song, K.-K. and Qing, -Xi (2007). Purification and some properties of β-*N*-acetyl-D-glucosaminidase from the cabbage butterfly (*Pieris rapae*). *Biochimie* **89**, 347–354.

Shibuya, N. and Minami, E. (2001). Oligosaccharide signaling for defense responses in plant. *Physiol. Mol. Plant Pathol.* **59**, 223–233.

Shinoda, T., Kobayashi, T., Matsui, M. and Chinzei, Y. (2001). Cloning and functional expression of chitinase cDNA from the common cutworm *Spodoptera litura*, using recombinant baculivirus lacking the virus-encoded chitinase gene. *Insect Biochem. Mol. Biol.* **31**, 521–532.

Shiomi, K., Arai, N., Iwai, Y., Turberg, A., Kolbl, H. and Omura, S. (2000). Structure of argifin, a new chitinase inhibitor produced by *Gliocladium* sp. *Tetrahedron Lett.* **41**, 2141–2143.

Siegel, M. R. (1970). Reaction of certain trichloromethyl sulfenyl fungicides with low molecular weight thiols. *In vitro* studies with glutathione. *J. Agric. Food Chem.* **18**, 819–822.

Siemieniewicz, K. W., Kajla, M. K. and Schrempf, H. (2007). Elucidating the biosynthesis of chitin filaments and their configuration with specific proteins and electron microscopy. *Micromol. Biosci.* **7**, 40–47.

Silvert, D. J. and Fristrom, J. W. (1980). Biochemistry of imaginal disks: Retrospect and prospect. *Insect Biochem.* **10**, 341–355.

Sobotnik, J., Alberti, G., Weyda, F. and Hubert, J. (2008). Ultrastructure of the digestive track in *Acarus siro* (Acari: Acaridae). *J. Morph.* **269**, 54–71.

Soltani, N. (1984). Effects of ingested diflubenzuron on the longevity and the peritrophic membrane of adult mealworms (*Tenebrio molitor* L.). *Pestic. Sci.* **15**, 221–225.

Somers, P. J. B., Yao, R. C., Doolin, L. E., McGowan, M. J., Fukuda, D. S. and Mynderse, J. S. (1987). Method for the detection and quantitation of chitinase inhibitors in fermentation broths; isolation and insect life cycle effect of A82516. *J. Antibiot.* **40**, 1751–1756.

Sowa, B. A. and Marks, E. P. (1975). An in vitro system for the quantitative measurement of chitin synthesis in the cockroach; Inhibition by TH 4060 and polyoxin D. *Insect Biochem.* **5**, 855–859.

Spindler, K. D. (1976). Initial characterization of chitinase and chitobiase from the integument of *Drosophila hydei*. *Insect Biochem.* **6**, 663–667.

Spindler, K. D. and Spindler-Barth, M. (1994). Inhibition of chitinolytic enzymes from *Streptomyces griseus* (bacteria), *Artemia salina* (crustracea), and a cell line from *Chironomus tentans* (insecta) by allosamidin and isoallosamidin. *Pestic. Sci.* **40**, 113–120.

Spindler-Barth, M., Shaaya, E. and Spindler, K. D. (1986). The level of chitinolytic enzymes and ecdysteroids during larval–pupal development in *Ephestia cautella* and their modifications by a juvenile hormone analogue. *Insect Biochem.* **16**, 187–190.

Stankiewicz, B. A., Briggs, D. E. G., Evershed, R. P., Flannery, M. B. and Wuttke, M. (1997). Preservation of chitin in 25-million-year-old fossils. *Science* **276**, 1541–1543.

Stevenson, J. R. and Hettick, B. P. (1980). Metabolism of chitin precursors by crayfish tissues during chitin synthesis. *J. Exp. Zool.* **214**, 37–48.

Strasser, R., Bondili, J. S., Schoberer, J., Svoboda, B., Liebminger, E., Glössl, J., Altmann, F., Steinkellner, H. and Mach, L. (2007). Enzymatic properties and subcellular localization of Arabidopsis β-*N*-acetylhexosaminidase. *Plant Physiol.* **145**, 5–16.

Sugahara, K., Schwartz, N. B. and Dorfman, A. (1979). Biosynthesis of hyaluronic acid by *Streptococcus*. *J. Biol. Chem.* **254**, 6252–6261.

Sunazuka, T., Sugawara, A., Iguchi, K., Hirose, T., Nagai, K., Noguchi, Y. Saito, Y., *et al.* (2009). Argifin; efficient solid phase total synthesis and evaluation of analogues of acyclic peptide. *Bioorg. Med. Chem.* **17**, 2751–2758.

Suresh, K. and Subramanyam, C. (1997). A putative role for calmodulin in activation of *Neurospora crassa* chitin synthase. *FEMS Microbiol. Lett.* **150**, 95–100.

Surholt, B. (1974). Studies *in vivo* and *in vitro* on chitin synthesis during the larval–adult moulting cycle of the migratory locust, *Locusta migratoria* L. *J. Comp. Physiol.* **102**, 135–147.

Tabudravu, J. N., Eijsink, V. G. H., Gooday, G. W., Jaspars, M., Komander, D., Legg, M., Synstad, B. and van Aalten, D. M. F. (2002). Psammaplin A, a chitinase inhibitor isolated from the Fijian marine sponge *Aplysinella rhax*. *Bioorg. Med. Chem.* **10**, 1123–1128.

Takahashi, S., Terayama, H. and Kuzuhara, H. (1992). Total synthesis of (-)-allosamidin, an insect chitinase inhibitor, employing chitin as a key starting material. *Tetrahedron Lett.* **33**, 7565–7568.

Takahashi, S., Terayama, H., Kuzuhara, H., Sakuda, S. and Yamada, Y. (1994). Preparation of demethylallosamidin isomer having a N,N′-diacetylchitobiosyl moiety and its potent inhibition against yeast chitinase. *Biosci. Biotechnol. Biochem.* **58**, 2301–2302.

Takeo, S., Fujise, M., Akiyama, T., Habuchi, H. Itano, N., *et al.* (2004). *In vivo* hyaluronan synthesis upon expression of the mammalian hyaluronan synthase gene in *Drosophila*. *J. Biol. Chem.* **279**, 18920–18925.

Takeshita, N., Ohta, A. and Horiuchi, H. (2005). CsmA, a class V chitin synthase with myosin motor-like domain, is localized through direct interaction with the actin cytoskeleton in *Aspergillus nidulans*. *Mol. Biol. Cell* **16**, 1961–1970.

Tan, S. C., Tan, T. K., Wong, S. M. and Khor, E. (1996). The chitosan yield of zygomycetes at their optimum harvesting time. *Carbohyd. Polym.* **30**, 239–242.

Tanaka, K., Katsura, N., Saku, T. and Kasuga, S. (1988). Composite texture of chitin and keratin in an animal organ, *Lingula seta*. *Polymer J.* **20**, 119–123.

Tatsuno, K., Yamada-Okabe, H., Takagi, M., Arisawa, M. and Sudoh, M. (1997). Properties of yeast-expressed *Aspergillus nidulans* chitin synthase B which is essential for hyphal growth. *FEMS Microbiol. Lett.* **149**, 279–284.

Tellam, R. L. and Eisemann, C. (2000). Chitin is only a minor component of the peritrophic matrix from larvae of *Lucilia cuprina*. *Insect Biochem. Mol. Biol.* **30**, 1189–1201.

Tellam, R. L., Wijffels, G. and Willadsen, P. (1999). Peritrophic matrix proteins. *Insect Biochem. Mol. Biol.* **29**, 87–101.
Tellam, R. L., Vuocolo, T., Johnson, S. E., Jarmey, J. and Pearson, R. D. (2000). Insect chitin synthase. cDNA sequence, gene organization and expression. *Eur. J. Biochem.* **267**, 6025–6042.
Tews, I., Perrakis, A., Oppenheim, A., Dauter, Z., Wilson, K. S. and Vorgias, C. E. (1996). Bacterial chitobiase structure provides insights into catalytic mechanism and the basis of Tay-Sachs disease. *Nat. Struct. Biol.* **3**, 638–648.
Tharanathan, R. N. and Kittur, F. S. (2003). Chitin—The undisputed biomolecule of great potential. *Crit. Rev. Food Sci. Nutr.* **43**, 61–87.
Thomas, H. E., Stunnenberg, H. G. and Stewart, A. F. (1993). Heterodimerization of the *Drosophila* ecdysone receptor with retinoid X receptor and ultraspiracle. *Nature* **362**, 471–475.
Thummel, C. S. (1995). From embryogenesis to metamorphosis: The regulation and function of *Drosophila* receptor superfamily members. *Cell* **83**, 871–877.
Tomiya, N., Narang, S., Park, J., Abdul-Rahman, B., Choi, O., Singh, S., Hiratake, J., Sakata, K., Betenbaugh, M. J., Palter, K. B. and Lee, Y. C. (2006). Purification, characterization, and cloning of a *Spodoptera frugiperda* Sf9 β-N-acetylhexosaminidase that hydrolyzes terminal *N*-acetylglucosamine on the N-glycan core. *J. Biol. Chem.* **281**, 19545–19560.
Toprak, U., Baldwin, D., Erlandson, M., Gillott, C., Hou, X., Coutu, C. and Hegedus, D. D. (2008). A chitin deacetylase and putative insect intestinal lipases are components of the *Mamestra configurata* (Lepidoptera: Noctuidae) peritrophic matrix. *Insect Mol. Biol.* **17**, 573–585.
Tournaire-Arellano, C., Younes-El Hage, S., Vales, P., Caujolle, R., Sanon, A., Bories, C. and Loiseau, P. (1998). Synthesis and biological evaluation of ureido and thioureido derivatives of 2-amino-2-deoxy-D-glucose and related aminoalcohols as *N*-acetyl-β-D-hexosaminidase inhibitors. *Carbohyd. Res.* **314**, 47–63.
Tsuizaki, M., Takeshita, N., Ohta, A. and Horiuchi, H. (2009). Myosin motor-like doman in the class VI chitin synthase CsmB is essential to its function in *Aspergillus nidulans*. *Biosci. Biotechnol. Biochem.* **73**, 1163–1167.
Turnbull, I. F. and Howells, A. J. (1982). Effects of several larvicidal compounds on chitin biosynthesis by isolated larval integuments of the sheep blowfly *Lucilia cuprina*. *Aust. J. Biol. Sci.* **35**, 491–503.
Turnbull, I. F. and Howells, A. J. (1983). Integumental chitin synthase activity in cell-free extracts of larvae of the Australian sheep blowfly, *Lucilia cuprina*, and two other species of Diptera. *Aust. J. Biol. Sci.* **36**, 251–262.
Uchida, M., Asai, T. and Sugimoto, T. (1985). Inhibition of cuticle deposition and chitin biosynthesis by a new growth regulator, buprofezin, in *Nilaparvata lugens* Stal. *Agric. Biol. Chem.* **49**, 1233–1234.
Usuki, H., Nitoda, T., Okuda, T. and Kanzaki, H. (2006). Screening and partial characterization of inhibitors of insect β-*N*-acetylglucosaminidase. *J. Pestic. Sci.* **31**, 41–46.
Usuki, H., Nitoda, T., Ichikawa, M., Yamaji, N., Iwashita, T., Komura, H. and Kanzaki, H. (2008). TMG-chitotriomycin, an enzyme inhibitor specific for insect and fungal β-*N*-acetylglucosaminidase, produced by actinomycete *Streptomyces anulatus* NBRC 13369. *J. Am. Chem. Soc.* **130**, 4146–4152.
Valdivia, R. H. and Schekman, R. (2003). The yeasts Rho1p and Pkc1p regulate the transport of chitin synthase III (Chs3p) from internal stores to the plasma membrane. *Proc. Natl. Acad. Sci. USA* **100**, 10287–10292.
Van Aalten, D. M. F., Komander, D., Synstad, B., Gaseidnes, S. S., Peter, M. G. and Eijsink, V. G. H. (2001). Structural insights into the catalytic mechanism of family 18 exo-chitinase. *Proc. Natl. Acad. Sci. USA* **98**, 8979–8984.

Van Damme, E. J. M., Lannoo, N. and Peumans, W. J. (2008). Plant lectins. *Adv. Bot. Res.* **48**, 107–209.

Van Dellen, K. L., Bulik, D. A., Specht, C. A., Robbins, P. W. and Samuelson, J. C. (2006). Heterologous expression of an *Entamoeba histolytica* chitin synthase in *Saccharomyces cerevisiae*. *Eukaryot. Cell* **5**, 203–206.

Van Eck, W. H. (1979). Mode of action of two benzoylphenyl ureas as inhibitors of chitin synthesis in insects. *Insect Biochem.* **9**, 295–300.

Van Scheltinga, A. C. T., Kalk, K. H., Beintema, J. J. and Dijkstra, B. W. (1994). Crystal structures of hevamine, a plant defence protein with chitinase and lysozyme activity, and its complex with an inhibitor. *Structure* **2**, 1181–1189.

Van Scheltinga, A. C. T., Armand, S., Kalk, K. H., Isogai, A., Henrissat, B. and Dijkstra, B. W. (1995). Stereochemistry of chitin hydrolysis by a plant chitinase/lysozyme and X-ray structure of a complex with allosamidin evidence for substrate assisted catalysis. *Biochemistry* **34**, 15619–15623.

Van Parijs, J., Joosen, H. M., Peumans, W. J., Geuns, J. M. and Van Lacre, A. J. (1992). Effect of the *Urtica dioica* agglutinin on germination and cell wall formation of *Phycomyces blakesleeanus* Burgeff. *Arch. Microbiol.* **158**, 19–25.

Vardanis, A. (1976). An *in vitro* assay system for chitin synthesis in insect tissue. *Life Sci.* **19**, 1949–1956.

Vardanis, A. (1978). Polyoxin fungicides: Demonstration of insecticidal activity due to inhibition of chitin synthesis. *Experientia* **34**, 228–229.

Venegas, A., Goldstein, J. C., Beauregard, K., Oles, A., Abdulhayoglu, N. and Fuhrman, J. A. (1996). Expression of recombinant microfilarial chitinase and analysis of domain function. *Mol. Biochem. Parasitol.* **78**, 149–159.

Verloop, A. and Ferrell, C. D. (1977). Benzoylphenyl ureas—A new group of larvicides interfering with chitin deposition. In: *Pesticide Chemistry in the 20th Century* (ed Plimmer, J. R.), pp. 237–270. Am. Chem. Soc., Washington, DC.

Vermeulen, C. A. and Wessels, J. G. H. (1986). Chitin biosynthesis by fungal membrane preparation. Evidence for transient non-crystalline state of chitin. *Eur. J. Biochem.* **158**, 411–415.

Veronico, P., Gray, L. J., Jones, J. T., Bazzicalupo, P., Arbucci, S., Cortese, M. R., Di Vito, M. and De Giorgi, C. (2001). Nematode chitin synthases: gene structure, expression and function in *Caenorhabditis elegans* and the plant parasitic nematode *Meloidogyne artiellia*. *Mol. Genet. Genomics* **266**, 28–34.

Valenzuela, J. G., Specht, C. A., Aravind, A., Langer, R. C., Ribeiro, J. M. C. and Kaslow, D. C. (2000). Chitinase of the avian malaria parasite *Plasmodium gallinaceum*, a class enzymes necessary for parasite invasion of the mosquito midgut. *J. Biol. Chem.* **275**, 10331–10341.

Wagner, G. P. (1994). Evaluation of multifunctionality of the chitin system. *EXS* **69**, 559–577.

Wagner, G. P., Lo, J., Laine, R. and Almeder, M. (1993). Chitin in the epidermal cuticle of vertebrate (*Paralipophrys trigloides*, Blenniidae, Teleostei). *Cell. Mol. Life Sci.* **49**, 317–319.

Wang, P. and Granados, R. R. (2000). Calcofluor disrupts the midgut defense system in insects. *Insect Biochem. Mol. Biol.* **30**, 135–143.

Wang, X., Ding, X., Gopalakrishnan, B., Morgan, T. D., Johnson, L., White, F. F., Muthukrishnan, S. and Kramer, K. J. (1996). Characterization of a 46 kDa insect chitinase from transgenic tobacco. *Insect Biochem. Mol. Biol.* **26**, 1055–1064.

Wang, J., Chen, Z., Du, J., Sun, Y. and Liang, A. (2005). Novel insect resistance in *Brassica napus* developed by transformation of chitinase and scorpion toxin genes. *Plant Cell Rep.* **24**, 549–555.

Wang, S., Jayaram, S. A., Hemphälä, J., Senti, K.-A., Tsarouhas, V., Jin, H. and Samakovlis, C. (2006). Septate-junction-dependent luminal deposition of chitin deacetylase restricts tube elongation in the *Drosophila* trachae. *Curr. Biol.* **16**, 180–185.

Ward, H. D., Alroy, J., Lev, B. I., Keusch, G. T. and Pereira, M. E. A. (1985). Identification of chitin as a structural component of *Giardia* cysts. *Infect. Immun.* **49**, 629–634.

Ward, G. B., Mayer, R. T., Feldlaufer, M. F. and Svoboda, J. A. (1991). Gut chitin synthase and sterols from the larvae of *Diaprepes abbreviatus* (Coleoptera: Curculionidae). *Arch. Insect Biochem. Physiol.* **18**, 105–117.

Wei, J. and DeAngelis, P. L. (2003). Analysis of two active sites of the hyaluronan synthase and the chondroitin synthase of *Pasteurella multocida*. *Glycobiology* **13**, 661–671.

Weiss, I. M., Schönitzer, V., Eichner, N. and Sumper, M. (2006). The chitin synthase involved in marine bivalve mollusk shell formation contains a myosin domain. *FEBS Lett.* **580**, 1846–1852.

Williams, S. J., Mark, B. R., Vocadlo, D. J., James, M. N. G. and Withers, S. G. (2002). Aspartate 313 in the *Streptomyces plictatus* hexosaminidase plays a critical role in substrate-assisted catalysis by orienting the 2-acetamido group and stabilizing the transition state. *J. Biol. Chem.* **277**, 40055–40065.

Yabe, T., Yamada-Okabe, T., Nakajima, T., Sudoh, M., Arisawa, M. and Yamada-Okabe, M. (1998). Mutational analyses of chitin synthase 2 of *Saccharomyces cerevisiae*. Identification of additional amino acid residues involved in its catalytic activity. *Eur. J. Biochem.* **258**, 941–947.

Yan, J., Cheng, Q., Narashimhan, S., Li, C.-B. and Aksoy, S. (2002). Cloning and functional expression of a fat body-specific chitinase cDNA from the tsetse fly, *Glossina moristans moristans*. *Insect Biochem. Mol. Biol.* **32**, 979–989.

Yang, Q., Liu, T., Liu, F., Qu, M. and Qian, X. (2008). A novel β-*N*-acetyl-D-hexosaminidase from the insect *Ostrinia furnacalis* (Guene'e). *FEBS J.* **275**, 5690–5702.

Yarden, O. and Yanofsky, C. (1991). Chitin synthase 1 plays a major role in cell wall biogenesis in *Neurospora crassa*. *Genes Dev.* **5**, 2420–2430.

Yasui, M., Fukada, M. and Maekawa, S. (1985). Effects of buprofezin on different developmental stages of the greenhouse whitefly, *Trialeurodes vaporariorum* (Westwood) (Homoptera: Aleyrodidae). *Appl. Entomol. Zool.* **20**, 340–347.

Yeager, A. R. and Finney, N. S. (2004a). The first direct evaluation of the two-active site mechanism for chitin synthase. *J. Org. Chem.* **69**, 613–618.

Yeager, A. R. and Finney, N. S. (2004b). Second-generation dimeric inhibitors of chitin synthase. *Bioorg. Med. Chem.* **12**, 6451–6460.

Yu, S. J. and Terriere, L. C. (1977). Ecdysone metabolism by soluble enzymes from three species of Diptera and its inhibition by the insect growth regulator TH-6040. *Pestic. Biochem. Physiol.* **7**, 48–55.

Zees, A. C., Pyrpassopoulos, S. and Vorgias, C. E. (2009). Insights into the role of the (α + β) insertion in the TIM-barrel catalytic domain, regarding the stability and the enzymatic activity of chitinase A from *Serratia marcescens*. *Biochim. Biophys. Acta* **1794**, 23–31.

Zen, K.-C., Choi, H. K., Krishnamachary, N., Muthukrishnan, S. and Kramer, K. J. (1996). Cloning, expression, and hormonal regulation of an insect β-*N*-acetylglucosaminidase gene. *Insect Biochem. Mol. Biol.* **26**, 435–444.

Zhang, J. and Zhu, K. Y. (2006). Characterization of a chitin synthase cDNA and its increased mRNA level associated with decreased chitin synthesis in *Anopheles quadrimaculatus* exposed to diflubenzuron. *Insect Biochem. Mol. Biol.* **36**, 712–725.

Zhang, H., Huang, X., Fukamizo, T., Muthukrishnan, S. and Kramer, K. J. (2002). Site-directed mutagenesis and functional analysis of an active tryptophan of insect chitinase. *Insect Biochem. Mol. Biol.* **32**, 1477–1488.

Zhang, Y., Foster, J. M., Nelson, L. S., Ma, D. and Carlow, C. K. S. (2005). The Chitin synthase genes chs-1 and chs-2 are essential for *C. elegans* development and responsible for chitin deposition in the eggshell and pharynx, respectively. *Develop. Biol.* **285**, 330–339.

Zheng, Y., Zheng, S., Cheng, X., Ladd, T., Lingohr, E. J., Krell, P. J., Arif, B. M., Retnakaran, A. and Feng, Q. (2002). A molt-associated chitinase cDNA from the spruce budworm, *Choristoneura fumiferana. Insect Biochem. Mol. Biol.* **32**, 1813–1823.

Zheng, Y.-P., Krell, P. J., Doucet, D., Arif, B. M. and Feng, Q.-L. (2008). Cloning, expression, and localization of a molt-related β-*N*-acetylglucosaminidase in the spruce budworm, *Choristoneura fumiferana. Arch. Insect Biochem. Physiol.* **68**, 49–59.

Zhu, X., Zhang, H., Fukamizo, T., Muthukrishnan, S. and Kramer, K. J. (2001). Properties of *Manduca sexta* and its C-terminal deletions. *Insect Biochem. Mol. Biol.* **31**, 1221–1230.

Zho, Y.-C., Specht, C. A., Dittmer, N. T., Muthukrishnan, S., Kanost, M. R. and Kramer, K. J. (2002). Sequence of a cDNA and expression of the gene encoding a putative epidermal chitin synthase of *Manduca sexta. Insect Biochem. Mol. Biol.* **32**, 1497–1506.

Zhu, K. Y., Heise, S., Zhang, J., Anderson, T. D. and Starkey, S. R. (2007). Comparative studies on effects of three chitin synthesis inhibitors on common malaria mosquito (Diptera: Culicidae). *J. Med. Entomol.* **44,** 1047–1053.

Zhu, R., Liu, K., Peng, J., Yang, H. and Hong, H. (2007). Optical brightener M2R destroys the peritrophic membrane of *Spodoptera exigua* (Lepidoptera: Noctuidae) larvae. *Pest Manag. Sci.* **63**, 296–300.

Zhu, Q., Arakane, Y., Beeman, R. W., Kramer, K. J. and Muthukrishnan, S. (2008a). Characterization of recombinant chitinase-like proteins of *Drosophila melanogaster* and *Tribolium castaneum. Insect Biochem. Mol. Biol.* **38**, 467–477.

Zhu, Q., Arakane, Y., Beeman, R. W., Kramer, K. J. and Muthukrishnan, S. (2008b). Functional specialization among insect chitinase family genes revealed by RNA interference. *Proc. Natl. Acac. Sci. USA* **105**, 6650–6655.

Zielkowsky, R. and Spindler, K. D. (1978). Chitinase and chitobiase from the integument of *Locusta migratoria*: Characterization and titer during the fifth larval instar. *Insect Biochem.* **8**, 67–71.

Ziman, M., Chuang, J. S. and Schekman, R. W. (1996). Chs1p and Chs3p, two proteins involved in chitin synthesis, populate a compartment of the *Saccharomyces cerevisiae* endocytic pathway. *Mol. Biol. Cell* **7**, 1909–1919.

Zimmermann, D. and Peters, W. (1987). Fine structure and permeability of peritrophic membranes of *Calliphora erythrocephala* (Meigen) (Insecta: Diptera) after inhibition of chitin and protein synthesis. *Comp. Biochem. Physiol.* **86B**, 353–360.

Zimoch, L. and Merzendorfer, H. (2002). Immunolocalization of chitin synthase in tobacco hornworm. *Cell Tiss. Res.* **308**, 287–297.

Zimoch, L., Hogenkamp, D. G., Kramer, K. J., Muthukrishnan, S. and Merzendorfer, H. (2005). Regulation of chitin synthesis in the larval midgut of *Manduca sexta. Insect Biochem. Mol. Biol.* **35**, 515–527.

Diverse Strategies of Protein Sclerotization in Marine Invertebrates: Structure–Property Relationships in Natural Biomaterials

Daniel J. Rubin,* Ali Miserez,† and J. Herbert Waite‡

*School of Engineering and Applied Sciences, Harvard University, Cambridge, Massachusetts, USA
†School of Materials Science and Engineering, Nanyang Technological University, Singapore, Singapore
‡Molecular, Cell and Developmental Biology Department, University of California, Santa Barbara, California, USA

ADVANCES IN INSECT PHYSIOLOGY VOL. 38
ISBN 978-0-12-381389-3
DOI: 10.1016/S0065-2806(10)38003-9

1 Introduction

Sclerotization refers generally to biological processes that cause tissues to become hard and stiff (Greek σκληρτης meaning hardness, stiffness or toughness). Vertebrate hard tissues are usually endowed with stiffness by mineralization but there are also non-mineralized pathways. In most insects, for example, the integument (cuticle), egg cases (chorion) and silks are sclerotized as organic composites during their maturation into functional structures. Sclerotization of arthropod integument is often assumed to represent a highpoint in the evolution of load-bearing tissues due to the rate of maturation, low density, tunability and robustness of mechanical properties, resistance to moisture and microbial attack, among others (Hepburn, 1976; Vincent and Wegst, 2004). With the exception of the mineralized crustacean carapaces, most arthropod sclerotization exhibits a peculiarly aromatic bias. Perhaps *biases* would be more accurate in so far as the involvement of two chemically distinct precursors has been discerned (Fig. 1): (A) the monophenolic bias and (B) the diphenolic or catecholic bias. The first involves oxidative protein

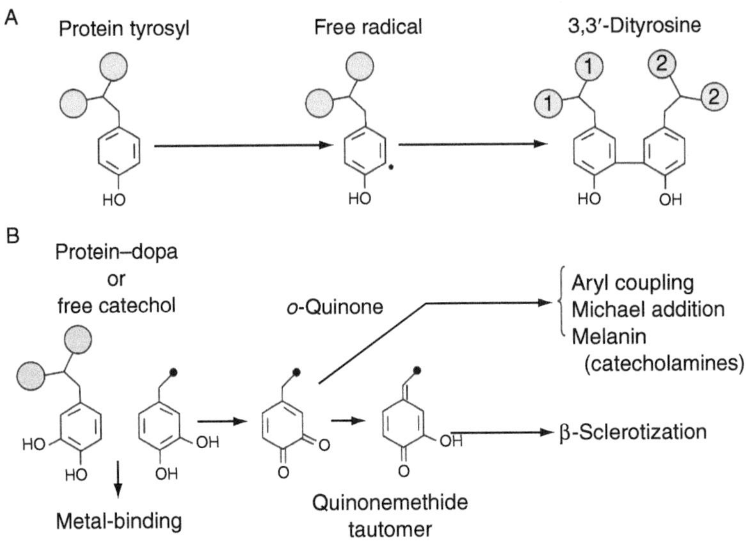

FIG. 1 Roadmap of sclerotization based on aromatic cross-linking. (A) Monophenolic pathway in which tyrosyl residues undergo a one-electron oxidation catalyzed by peroxidase to form di- and tri-tyrosyl cross-links. The ball-shaped substituents represent the protein backbone. (B) Catecholic pathways that can go in many different directions that include metal binding before oxidation, ring-ring coupling, nucleophilic attack (Michael attack by lysine, histidine, glucosamine, cysteine), melanin formation and ring-sidechain coupling (β-sclerotization), all after a two-electron oxidation. The large ball substituents denote protein backbone, whereas the smaller ones are small functional groups, for example, *N*-acetylethyl, carboxyl, β-alanyl.

tyrosyl coupling and leads to the formation of dityrosine cross-links as in the resilin component of exoskeleton (Andersen, 2003). It is not controversial. In the second, catecholic precursors are low molecular weight compounds or tethered to protein as 3,4-dihydroxyphenyl-L-alanine (Dopa) and can be directed along many different pathways including melanization, cross-link formation, polymerization and metal ion binding (Andersen, 2005; Waite, 1990). Given the many potential pathways and wide-ranging effects of this type, it is not surprising that some aspects of both chemistry and function remain warmly contested. Recent work is only beginning to reveal how truly tangled this biochemical web is (Arakane *et al.*, 2009).

Few would contest that catechol-mediated sclerotization has achieved a high-point in the exquisitely adapted mechanical properties of insect exoskeletons. But insects have no monopoly of sclerotization, nor of how ingeniously it is used. The phylogenetic distribution of aromatic sclerotization outside arthropods is very extensive (Waite, 1990). Apparently both the tyrosine and catechol-mediated pathways extend back to before the metazoa. Dityrosines occur in the yeast cell wall (Briza *et al.*, 1990) and both dityrosine and the catecholic amino acid Dopa are associated with protein sclerotization in the oocyst wall of *Eimeria* (parasitic Protista) (Belli *et al.*, 2003). There is convincing evidence for catechol-mediated sclerotization in the egg capsules of turbellarians (Phylum Platyhelminthes) (Huggins and Waite, 1993) and trematodes (Waite and Rice-Ficht, 1992) and even in oviparous sharks (Phylum Chordata) (Rusaouen, 1976). Dopa-containing proteins from sclerotized structures other than egg capsules have been detected in the perisarc and sheath of hydroids, in gorgonins (Phylum Cnidaria) (Waite, unpublished data) (Knight, 1970) and in tube cement of a polychaete worm (Phylum Annelida) (Zhao *et al.*, 2005).

Simple citation does little justice to the breathtaking diversity and beauty of these non-arthropod structures. Therefore, a more visual homage is offered for some of the examples such as the perisarc (exoskeleton) of the hydroid *Aglaophenia latirostris*, the eggcase of the commensal turbellarian *Bdelloura candida*, the tube dwelling of the sandcastle worm *Phragmatopoma californica* and the egg capsule of the swell shark *Cephaloscyllium ventriosum* (Fig. 2A–D). *Bdelloura*, in particular (Fig. 2B, *left*), is ideal for visualizing the stages of sclerotization. A flatworm has been specifically stained for Dopa in a whole body mount during vitellogenesis when albumen and Dopa-containing eggshell precursor proteins are accumulated in anticipation of fertilization (Huggins and Waite, 1993). Following fertilization, the eggshell is assembled and the vitelline network is completely depleted of precursors (Fig. 2B, *middle*). The encapsulated eggs are finally attached to the book gills of horseshoe crab *Limulus polyphemus*, whereupon they eventually hatch and the capsules slowly decompose (Fig. 2B, *right*).

Our aim here is to present recent progress in the aromatic sclerotization of non-arthropod exoskeletal materials using a few model organisms that have allowed significant insights to be made. It is our contention that arthropod and non-arthropod sclerotization strategies are not mutually exclusive and that

FIG. 2 Phylogenetic diversity of sclerotization. (A) *Aglaophenia*, a common intertidal hydroid (Cnidaria). The feather-like sprigs are colonies of polyps ensheathed by a sclerotized perisarc which is a composite of chitin, polyphenols and protein. (B) *Bdelloura candida* is a commensal marine flatworm (Turbellaria, Platyhelminthes). In anticipation of egg-laying, worms accumulate eggshell precursors in the vitellaria (shell glands), which are here stained with Arnow's reagent to highlight Dopa-containing proteins (i). Dopa proteins are relocated to the nascent egg-capsule (ii), and the eggs are deposited with stalks onto the book gills of *Limulus polyphemus*. (C) Sandcastle worm *Phragmatopoma californica* (Polychaeta, Annelida) uses Dopa-rich proteins for glueing together the sandgrains that comprise its tube-like home. (D) The egg capsule of a swell shark *Cephaloscyllium ventriosum* (Chondrichthyes, Chordata) contains collagens in a plywood-like arrangement cross-linked by catechols. Scale bar is 1 cm. (See Color Plate 1 at the back of the book.)

research efforts on the two could cross-fertilize each other more than in the past. In addition, the sclerotization of both arthropod and non-arthropod exoskeletons offers important 'bio-inspired' insights for the next generation of technological load-bearing materials in a world suddenly alert to issues of efficiency and waste management (Burillo *et al.*, 2002; Neville, 1993).

We limit the focus of this presentation to four case studies: mussel byssus, the jaws of two polychaetes and squid beak. Several significant themes will be presented in the following order: (1) the stiffness of a material is influenced by the abundance and organization of proteins with stiff domains, (2) high catechol densities do not necessarily reflect a high degree of sclerotization, (3) protein coupled catecholate and imidazole ligands can induce sclerotization by metal binding, and (4) the covalent association of proteins and catechols can be incrementally tuned to differentially dehydrate chitin. Given that our model organisms are marine invertebrates and continually bathed by a saline medium, their sclerotization pathways probably evolved along different lines than those in terrestrial insects. The known properties of seawater, particularly its high dielectric constant and short Debye screening length (< 1 nm), would tend to favour water-independent interactions such as covalent and coordination bonding over electrostatic and van der Waals forces during the macromolecular condensation associated with sclerotization in seawater (Israelachvili, 1985). Hydrophobic interactions would also be more potent in the saline medium (Creighton, 1989; Israelachvili, 1985).

2 The mussel byssus

Sclerotization of load-bearing structures occurs extensively in the phylum Mollusca. Below, we highlight two sclerotized molluscan structures—one from the common mussel, *Mytilus edulis* (Class Bivalvia) and one from the Humboldt squid, *Dosidicus gigas* (Class Cephalopoda) (Section 5). Mussels belonging to two bivalve superfamilies, Mytilidae and Dreissenidae, both practice sophisticated sclerotization strategies based on Dopa-containing proteins. Dopa-proteins are present in the byssus, periostracum, shell matrix and possibly the hinge ligament (Waite, 1990) and arise by a poorly understood post-translational modification of targeted tyrosine residues. Of these only the byssus has been studied in any biochemical detail and exhibits wide ranging mechanical properties.

The byssus serves as a holdfast to keep the mussel securely tethered to a hard substratum as well as to dampen wave impact. It is first expressed in pediveligers soon after the free swimming veliger settles from the water column (Yonge, 1962). Byssus resembles a bundle of threads, which are made in succession, one at a time, throughout the lifetime of the mussel (Fig. 3A). The threads—typically 3–5 cm long and 50–200 μm in diameter depending on the species—are distally tipped by an adhesive plaque and converge proximally into a bundle by a rachis-like structure called a stem (Fig. 3B). The stem inserts into the soft living tissue at the base of the foot. The foot is responsible for producing each thread, which it does by a fascinating 5-min reaction injection moulding process using a groove on the ventral side of the foot as a die for moulding new threads (Waite, 1992). The threads produced have uniform diameters but the lengths can vary as needed from 1 to 8 cm in *Mytilus californianus*, for example.

2.1 THREAD CORE THEMES: DOMAIN STIFFNESS, HISTIDINE–METAL COMPLEXATION AND DOPA–HISTIDINE CROSS-LINKS

Byssal threads from *Mytilus* are known to have three functional domains each with a distinct protein composition: the distal and proximal thread core, and the cuticle. The thread core is perhaps the premier example of a structure in which

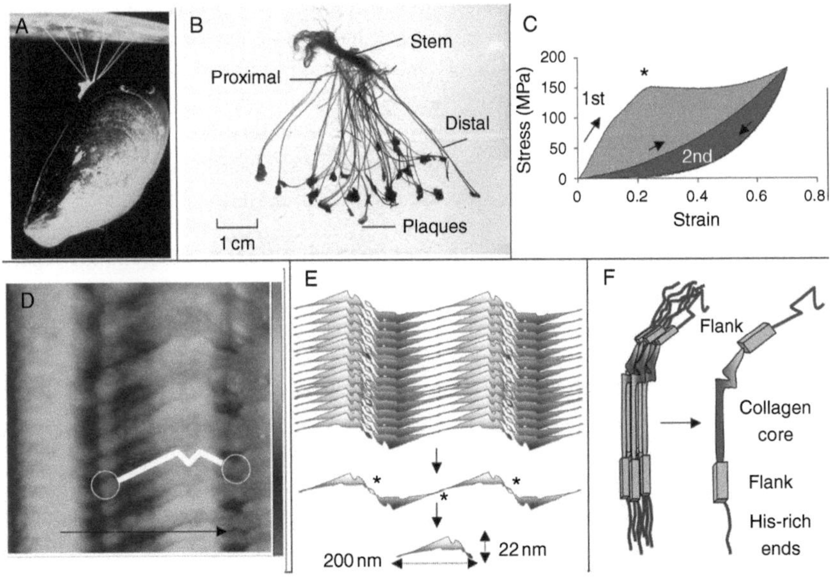

FIG. 3 Sclerotization of mussel byssus. (A) *Mytilus edulis*, the common, mussel hanging from a glass substratum by its byssus. (B) Byssus removed from a mussel and showing the different portions known as stem, proximal and distal thread and adhesive plaque. (C) Cyclic stress–strain in natural seawater of a distal byssal thread portion (strain rate = 0.2 min^{-1}) at 18 °C. If the first cycle exceeds the yield point (*), the modulus of the second cycle, which immediately follows the first, is significantly reduced. Substantial recovery of stiffness occurs after an hour or more of rest. Modulus is initial stress/strain. Engineering strain = [final length − initial length]/initial length. (D) Atomic force micrograph in tapping mode of assembled preCOLs just beneath the cuticle in a byssal thread of *M. galloprovincialis*. Arrow indicates direction of thread axis. (E) Deconstruction of preCOL assembly showing smectic register and C_2 symmetry with inversion (*), and a bent preCOL bundle. (F) A preCOL bundle of seven is further deconstructed to a single preCOL. (See Color Plate 2 at the back of the book.)

biochemical gradients in protein distribution are known to have mechanical consequences. The most obvious mechanical consequence is a 10–100-fold gradient in Young's modulus (from 500 to 50 MPa in *M. edulis*) decreasing along the distal to proximal axis (Waite *et al.*, 2004). This modulus or stiffness is characteristic of stress–strain cycles of pristine threads only. If a distal thread segment is extended beyond the yield point, but below the breaking stress, the stiffness of the next cycle is reduced by as much as 80% (Fig. 3C). Curiously, threads can achieve 100% recovery of initial stiffness during a 24-h resting period, but recovery of 75–80% in about 1 h is not uncommon in some species such as *M. californianus* (Carrington and Gosline, 2004).

The thread core is a fibre-filled composite in which liquid crystalline precursors rapidly self-assemble into very long microfibres separated by a sparse matrix (Hassenkam *et al.*, 2004; Sagert and Waite, 2009). Precursors or preCOLs (from *pre*pepsinized *col*lagens) are increasingly compared to liquid crystals because of their initial fluidity and their mesogenic shape—a stiff bent centre flanked by extended and flexible sticky ends (Hassenkam *et al.*, 2004) (Fig. 3D). Ultrastructural examinations of both the foot tissue and thread cores by transmission electron and atomic force microscopy reveal smectic organization, that is, a laterally aligned register of mesogens (mesogen = single molecule in a liquid crystal) in both with C_2 symmetry and inversion centres (Fig. 3D and E). More particularly, each discernable mesogen represents a six- or seven-member bundle of preCOLs with all tilts pointing in the same direction (Hassenkam *et al.*, 2004). These can be deconstructed to a single preCOL with a kinked triple helical collagen and additional flanking substituents (Fig. 3F). The flanking regions on either side of the collagen consist of two distinct parts: an inner variable structural domain such as β-pleated silk, polyglycine strands, or elastin, and outer N- and C-terminal histidine-rich domains endowed with several residues of Dopa (Harrington and Waite, 2007).

There are three distinct preCOLs in the thread core: preCOL-D, preCOL-P and preCOL-NG. The first two are distributed in a roughly complementary manner along the thread, whereas the third is uniformly distributed (Fig. 4A). A schematic diagram of preCOL-D (Fig. 4B) better illustrates the domain structure and highlights the fact that the primary molecular feature distinguishing the three preCOL types is the flanking sequence. Not surprisingly, flanking domains influence the local stiffness of anisotropic scaffolds as they come in three grades: compliant (2 MPa elastin domain of preCOL-P), intermediate (150 MPa Gly-rich domain of preCOL-NG) and very stiff (10 GPa silk domain of preCOL-D) (Fig. 4C). As preCOL-NG (NG = non-gradient) is uniformly distributed along the thread axis, it is largely the relative distribution of preCOL-D and -P that defines local stiffness, but how? The mechanical gradient was previously modelled by the density and the parallel or serial arrangement of flanking domains (Waite *et al.*, 2004): by these measures, the distal thread end would be stiffest due to the high density of silk domains in preCOL-D, whereas the compliance at the proximal end is imposed by preCOL-P with its elastin domains.

PreCOL organization alone reveals little to nothing about how preCOLs are cohesively held together. Numerous studies have provided molecular details about the preCOL self-assembly process: upon secretion from storage in regulated secretory granules at pH 5–6, preCOLs are released into the ventral groove as a smectic mesophase and there moulded into threads (Harrington and Waite, 2007, 2008a,b). Whereas the charges on the histidine ($pK_a = 6.5$) clusters at each end of the termini of the preCOLs provide repulsive interactions between the stacks of lateral arrays during storage at pH 5–6, at the pH of seawater ~8.2, histidine in secreted preCOLs is effectively neutralized, and repulsion is lost. In fact, in the presence of metal ions, the clustered imidazoles further provide a strong interaction by shared metal ion binding. Elevated levels ($\leq 1\%$ dry weight) of Zn^{2+} and Cu^{2+} in the thread core show little tendency to bind the protonated imidazoles of histidines at pH 5, however, at pH 8 they are strongly complexed.

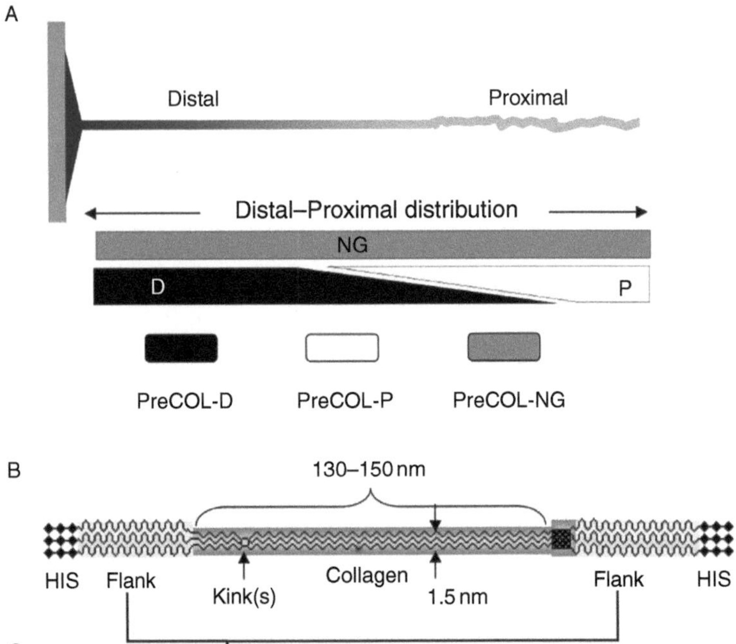

C

PreCol	Flank prototype	Sequence prototype	Stiffness MPa
D	Dragline silk	$\{[A_{10}]_7(GGX)_3\}_{11}$	10,000
P	Elastin	$(XGXPG)_{10}$	2
NG	Poly-[GGX]	$(GGGX)_{70}$	150
	Core prototype		
All	Collagen	$(Gly\text{-}X\text{-}Y)_{170}$	1500

FIG. 4 PreCOL distribution and preCOL domain structure. (A) Gradient distribution of preCOL-D, -P and -NG along the length of a byssal thread. Note the marked bias of D and P towards the distal and proximal ends, respectively. NG is uniformly distributed. (B) Linear domain structure of a preCOL in which His refers the histidine-(and Dopa)-rich domains. The kink denotes the location of [Gly-X-Y] sequence aberrations that cause the preCOLs to bend. The flanking domains can contain either silk-, elastin-, or polyGly-like domains. (C) The sequence motifs and mechanical properties of each flanking domain is summarized and contrasted to the central collagen domain.

Anyone who has constructed a fusion protein with His-tag for purposes of a one-step purification on a metal-immobilized affinity column discovers this. As preCOLs meet end-to-end via the imidazole–metal complexes, preCOLs form long trains *in situ* and *in vitro* (Harrington and Waite, 2007).

The metal-binding structures between preCOLs have yet to be directly characterized; however, as in another biocomposite detailed later (Section 4), the structure of a metal-mediated histidine complex characterized in the jaw of *Nereis* was shown by X-ray absorption spectroscopy to be roughly tetrahedral with a Zn^{2+} nucleus, and with three imidazoles and a chloride as the fourth ligand per nucleus (Lichtenegger *et al.*, 2003). In byssus, the metal complexes are below the detection limit of X-ray absorption near edge spectroscopy, but the effects of metal removal on thread stiffness are immediately apparent: EDTA treatment reduced stiffness of distal threads by about half (Vaccaro and Waite, 2001) and titrating distal threads incrementally from pH 8 (seawater) to $\sim$3 exhibits the largest drops in stiffness at pH near the pK_a of histidine—recalling that the imidazolium form cannot bind metal (Harrington and Waite, 2007).

An interpretation of end-to-end preCOL cross-linking that is consistent with the initial stiffness, stress-softening and post-yield recovery of distal threads is outlined in Fig. 5. Parallel fibres from two types of assembled preCOLs

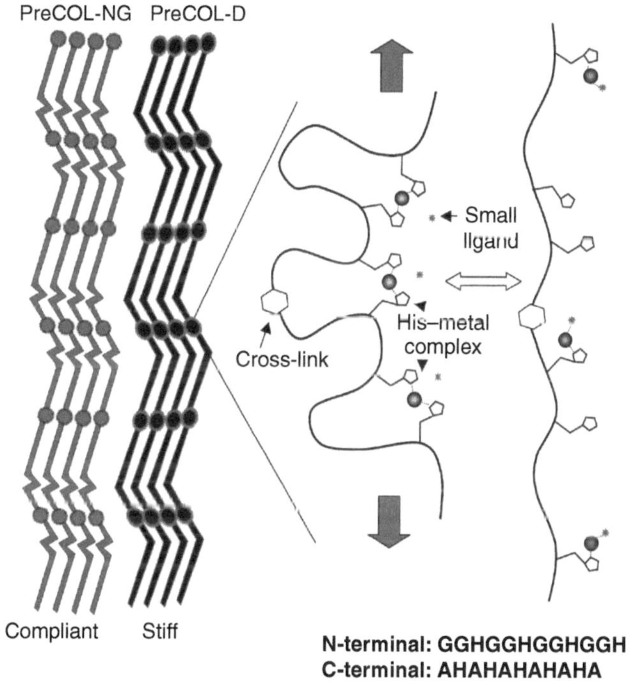

FIG. 5 Schematic interplay between different preCOL fibres and their histidine-rich couplings. At the distal end of the byssus, fibrils of preCOL-D and preCOL-NG run parallel to one another. preCOL-NG fibres are more compliant than preCOL-D fibres because their gly-rich flanking domains are much less stiff than the silk flanks of preCOL-D. Under loading, preCOL-D fibres bear most of the weight until their His–metal connections snap, whereas in preCOL-NG fibres the NG flanks extend elastically. Providing that the 'cross-link' remains intact, the His–metal connections should find each other again during the time dependent recovery. Typical His-rich sequences in preCOL-NG are shown below to underscore that 'His-rich' is not intended to imply sequence identity.

(D and NG) are subjected to loading and resist extension to 20% strain, at which yield occurs (Fig. 3C). We attribute yield to breakage of multiply liganded Zn or Cu centres in the histidine-rich domains in stiff preCOL-D as the flanking domains of preCOL-NG are compliant enough to play along. It is appropriate to note that the ligand–metal interaction between the imidazole of histidine and the metal ion is not a covalent bond; rather it is a coordinate bond in which the electron-rich ligand deposits non-bonding electrons into empty orbitals (usually 3d orbitals) of the metal. As each imidazole is pulled away from the metal, a smaller temporary ligand such as chloride or water takes its place. Apparently, the histidine–metal complexes dissipate energy so well that the central collagen domains are completely insulated from deformation during extension (Harrington *et al.*, 2009). Little additional force is initially required as these hidden loops are stretched out. The force, however, increases as the chains become straight and final breakage is resisted by a covalent cross-link, possibly the Dopa–histidine adduct described below (Fig. 5). Recovery of initial stiffness is more of a mystery but presumably involves elastic recoil in preCOL-NG elements (Harrington and Waite, 2007).

The N- and C-terminal histidine-rich domains of preCOLs contain the only Dopa residues in the proteins—usually 2–3 per chain terminus in preCOL-D (Qin *et al.*, 1997). Apparently these residues are oxidized to Dopaquinones after which they form covalent cross-links between the preCOLs. Locating cross-links is difficult because thread cores cannot be dissolved by any denaturants; however, following thread hydrolysis in hot 6 M HCl, significant levels of histidyl-Dopa adducts (MH^+ 351 Da) can be captured from the hydrolysates and detected by tandem mass spectrometry (Harrington, 2008). The curious strategy of a two-tiered cross-linking chemistry in which molecules are first brought together by highly favourable histidine–Cu/Zn metal–ligand binding equilibria, and followed by covalent bond formation is quite ingenious. Under tension, both would strongly resist deformation, but the histidine–metal complexes would resist first and yield reversibly (Schmitt *et al.*, 2002), thereby adding measurably to material toughness. Deformation and rupture of the Dopa–histidine based cross-links would occur much later during deformation and no recovery would be possible.

The reaction between Dopa and histidine at the preCOL ends inspired an elegant *in vitro* experiment in which Dopa and histidine were attached to opposite ends of two synthetic complementary DNA oligomers (Liu *et al.*, 2006). In the single-stranded DNA, no cross-links were formed upon oxidation; in contrast in the hybridized antiparallel dimer in which the Dopa- and histidine-modified 3′ and 5′ DNA ends are next-door neighbours, entropic barriers are easily overcome and high yield cross-linking was observed. Given that the collagens of preCOLs hybridize to trimers and that Dopa and histidine residues are concentrated at the ends, it is quite plausible to propose that Dopa and histidine are well placed for intra- and intermolecular cross-linking.

Histidine is emerging as an unusually versatile player: (1) its pK_a provides the pH-dependent trigger to turn off repulsion and (2) to bind metals, thereby bringing together different preCOLs. Finally, (3) once the chains have been assembled, at least some of the histidines don one last 'hat' to couple covalently with nearby Dopaquinone residues. There could be yet another role. Model compounds synthesized from histidine-rich peptides complexed to Cu^{2+} exhibit significant oxidase-like activity (Brown and Kodadek, 2001). In other words, metal-binding by the histidine-rich domains could be adapted not only to reversibly couple chains but to be the catalyst of their own permanent coupling with Dopa.

2.2 THE BYSSAL COATING: DOPA–METAL COMPLEXATION

Covering all exterior parts of the mytilid byssus is a cuticle roughly 3–5 μm thick. Cuticle structure in all species of *Mytilus* consists of a granule-filled composite material. In *M. galloprovincialis*, the electron dense granules have an average diameter of 0.8 μm and a volume fraction of 0.5 (Holten-Andersen *et al.*, 2007; Vitellaro-Zuccarello, 1981) (Fig. 6A and B). We assume the function of this cuticle is to protect the thread core from abrasion as well as from microbial degradation. The former is deduced from the fact that the

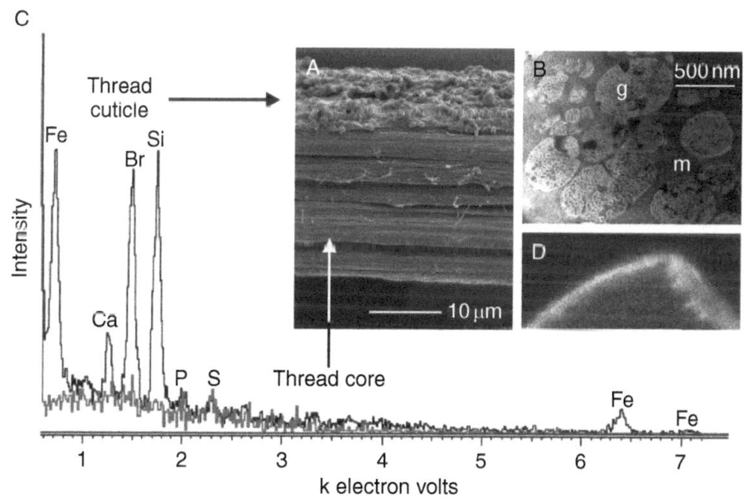

FIG. 6 Byssal cuticle architecture and chemistry. (A) Scanning electron micrograph highlighting the granular outer cuticle, and upon stripping off the cuticle the fibrous core is exposed. (B) Transmission electron micrograph of cuticle thin section stained with osmium tetroxide; G, granule; m, matrix. (C) Energy dispersive X-ray spectrum of both the cuticle and core of (D) Raman microprobe analysis of a transverse section of the cuticle and core at 460–600 nm. Scale is the same as in (A). (See Color Plate 3 at the back of the book.)

hardness and stiffness of hydrated cuticle is 4–5 times greater than the core as measured by nanoindentation (Holten-Andersen *et al.*, 2007); the latter from the observation that exposed collagen sequences (from preCOLs) are induction factors for the expression and secretion of collagenases by marine *Vibrio* species (Merkel *et al.*, 1975). The concept of a hard, stiff coating on a compliant core is intriguing in part because it seems ludicrous—a bit like expecting a hard waxy outer coat to provide protection for rubber tubing: the waxy coating would shatter upon extension. Cuticle shattering, however, does not occur even when threads are strained to 70% in *M. galloprovincialis* and 100% in *M. californianus* (Holten-Andersen *et al.*, 2009a). Catastrophic cuticle failure, observed as macrocracking, is averted by two features: granule deformation and formation of an extensive network of uniformly distributed microcracks at interfaces between the matrix and the granules (Holten-Andersen *et al.*, 2007). Granule deformation, which is measured as a change in aspect ratio, only happens up to 30% extension and is immediately reversible. Microcrack formation, which can be followed by inspecting the exterior of whole threads by environmental SEM, also appears to be reversible, at least prior to macrocrack formation. Reversibility has been proposed to result from the breaking and reformation of sacrificial bonds in the matrix (Harrington *et al.*, 2010).

The chemical analysis of byssal cuticle is incomplete but metal ions such as Ca^{2+} and Fe^{3+} and proteins are known to be consistently prominent (Holten-Andersen *et al.*, 2009b) (Fig. 6B). Mussel footprint protein-1 (mfp-1) is the major cuticle protein and has between 10 and 20 mol% Dopa. Mfp-1 is moderately large ($\sim$110 kDa in *M. edulis*) and highly repetitive with over 70 tandem repeats of the decapeptide sequence [A-K-P-Y°-S-O*-O-T-Y*-K] in which O, O*, Y° and Y* denote 4-hydroxyproline, 3,4-dihydroxyproline, usually tyrosine but occasionally Dopa, and always Dopa, respectively (Filpula *et al.*, 1990; Taylor *et al.*, 1994a,b) (Fig. 7A). Despite the highly conserved repeat motif, the protein is an extended random coil in solution (Deacon *et al.*, 1998; Haemers *et al.*, 2005).

Addition of Fe^{3+} to mfp-1 in solution at pH 7 results in the formation of highly stable coloured complexes: red at high Dopa to Fe^{3+} ratios and purple at ratios of roughly 1:1 (Taylor *et al.*, 1996) (Fig. 7B). The colours are indicative of charge transfers as reported in tris-catecholato–Fe complexes and confirmed by resonance Raman shifts between 500 and 700 cm^{-1} (Taylor *et al.*, 1996). Although only moderately interesting as coordination polymers in solution, these results take on much greater significance if this chemistry is pertinent to iron binding *in situ*. Raman microprobe analysis of biological samples with 1-μm spatial resolution makes it possible to directly test for Dopa–iron complexes in the cuticle. The complexes do indeed occur in the cuticle as predicted (Fig. 6D), and, when scanned at a Raman shift frequency of 460 cm^{-1} (corresponding to catechol-to-metal charge transfer), the density of Dopa–iron complexes in the granules appears to be about twice as high as in the matrix (Harrington *et al.*, 2010).

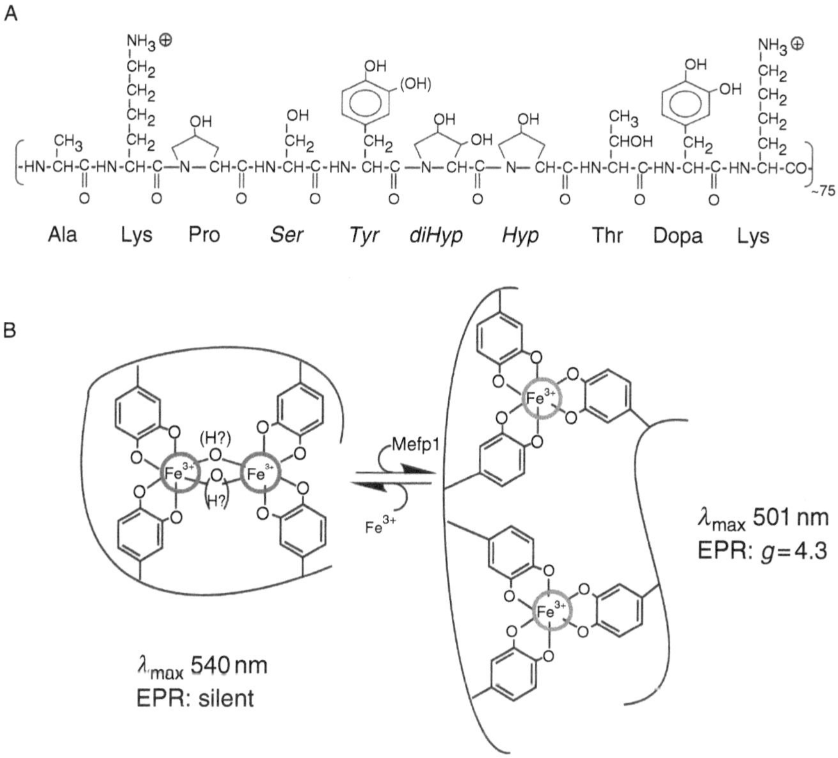

FIG. 7 Byssal cuticle protein. (A) Mfp-1, the protein constituent of cuticle, has been isolated and characterized from the mussel foot. Following a short non-repetitive N-terminal sequence, the proteins from *M. edulis* and *M. galloprovincialis* consist of 70 or more tandem repeats of a decapeptide sequence as shown. Tyrosine in the ninth position is always hydroxylated to Dopa, whereas Tyr-5 is converted less than 50% of the time. (B) At pH 7, mfp-1 is complexed to Fe^{3+} via the catecholate groups of Dopa in two modes: the red EPR active tris-catecholato-iron at high Dopa to iron ratios and the purple oxo-bridged bis-catecholato-iron species at lower ratios (Taylor *et al.*, 1996). The lines connecting the catechols represent the backbone of the same or different proteins. (See Color Plate 4 at the back of the book.)

A reasonable current interpretation is that the cuticle is cross-linked by bis- and tris-complexes of Dopa ligands with Fe^{3+} (Fig. 7B); the mfp-1 interactions with iron are nearly as strong as the avidin–biotin complex and as reversible. Single molecule analysis by atomic force microscopy suggests that such complexes are stronger than any non-covalent interaction and are fully reversible (Lee *et al.*, 2006). Since both the catechol and iron are redox active at physiological conditions, some redox exchange between dopa and the iron nucleus must eventually be unavoidable (Sever *et al.*, 2004), and lead to covalent cross-links.

2.3 THE BYSSAL PLAQUE: CATECHOL PROTECTION AND ADHESION

Much of the research on mussel byssus has been driven by the apparent prowess and versatility of adhesion in mussels. Mussel attachment plaques provide a theatre for studying adhesion and considerable effort has gone into characterizing the seven known plaque proteins in *Mytilus* species (Waite *et. al.*, 2005; Lin *et al.*, 2007). All plaque proteins contain Dopa, which ranges from as little as 0.5 mol% in preCOLs to nearly 30 mol% in mfp-5. The distribution of Dopa-rich proteins in the plaque appears to be prejudiced towards the surface of the substratum. Amino acid composition analysis of serial sections of mature plaques reveals that Dopa levels increase exponentially from about 1 to 2 mol% in the distal thread to ~ 15 mol% near the interface (Fig. 8). Only two proteins approach or exceed this magnitude of Dopa—mfp-3 (20 mol%) and mfp-5 (28 mol%). Independent localization of mfp-3, mfp-5 and later of mfp-6 at the plaque–substratum interface was achieved by *in situ* matrix-assisted laser desorption ionization (MALDI) typically associated with mass spectrometry. Accordingly, matrix is applied to a plaque footprint on mica or glass and ionized at 337 nm with a pulsed ultraviolet laser (Zhao and Waite, 2006a; Zhao *et al.*, 2006) (Fig. 9A). Desorbed ions are accelerated by physical extraction and detected by a time of flight mass analyser. Mfp-3 and mfp-5 were detected in footprints by MALDI at masses of 5–7 and 9–10 kDa, respectively; the Dopa-rich sequence of mfp-5 is shown in Fig. 9B. Presumably the reason that Dopa levels near the plaque interface only reach 15 mol% is that other Dopa-poor proteins, for example, mfp-6 are also present.

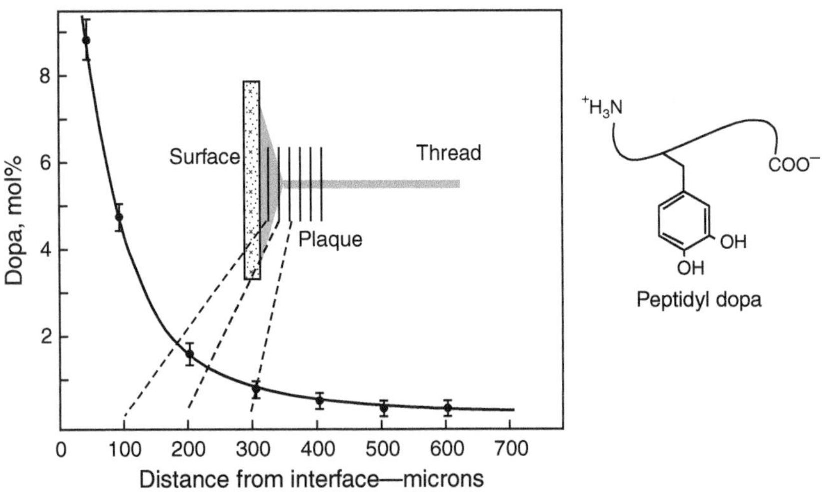

FIG. 8 Distribution of Dopa in distal byssus of *M. galloprovincialis* upon approaching the plaque–substratum interface. Dopa was measured following acid hydrolysis (6 M HCl 110 °C *in vacuo* at 24 h) of pooled micron thick sections in incremental steps of ~ 100.

The presence and persistence of unoxidized Dopa in the footprint proteins begs the question, how is it protected? Given the tendency of catechols to undergo facile oxidation to quinones (*o*-quinones and quinonemethides) followed by coupling reactions to form cross-links or large melanin-like polymers, one wonders why this is not happening to Dopa in plaques. Indeed, the presence of catecholoxidase activity in the byssus has been known for some time (Waite, 1985). Although a small amount of Dopa (~0.5 mol%) in the plaque is transformed to 5,5′-diDopa (McDowell *et al.*, 1999) and *S*-cysteinyl-Dopa cross-links (Zhao and Waite, 2005, 2006b), the bulk appears to be earmarked for interfacial interactions and metal ion complexes as described in the section on cuticle. Dopa-based interfacial interactions on metal oxides, for example, titanium oxide and iron oxide as studied by atomic force microscopy are particularly revealing. Using cantilever tips with a single tethered Dopa, Lee *et al.* (2006) measured a force to break of 0.9 nN on TiO_2. Similar reversibility was absorbed on mica using the surface forces apparatus (Lin *et al.*, 2007). This adhesion is about half the force required to break a covalent bond but reforms spontaneously and is reversible through hundreds of cycles. Oxidation of Dopa to Dopaquinone reduced the adhesive force by 90% (Lee *et al.*, 2006). If good

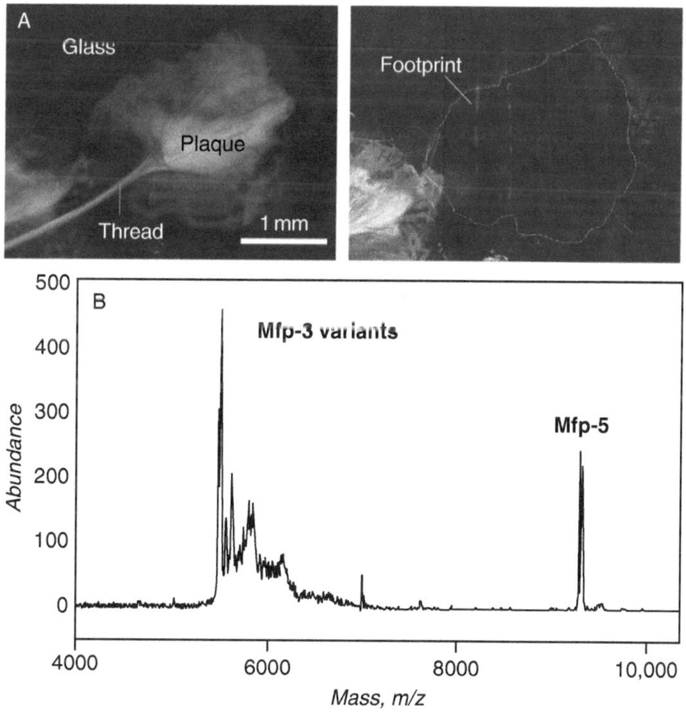

FIG. 9 (continued)

C

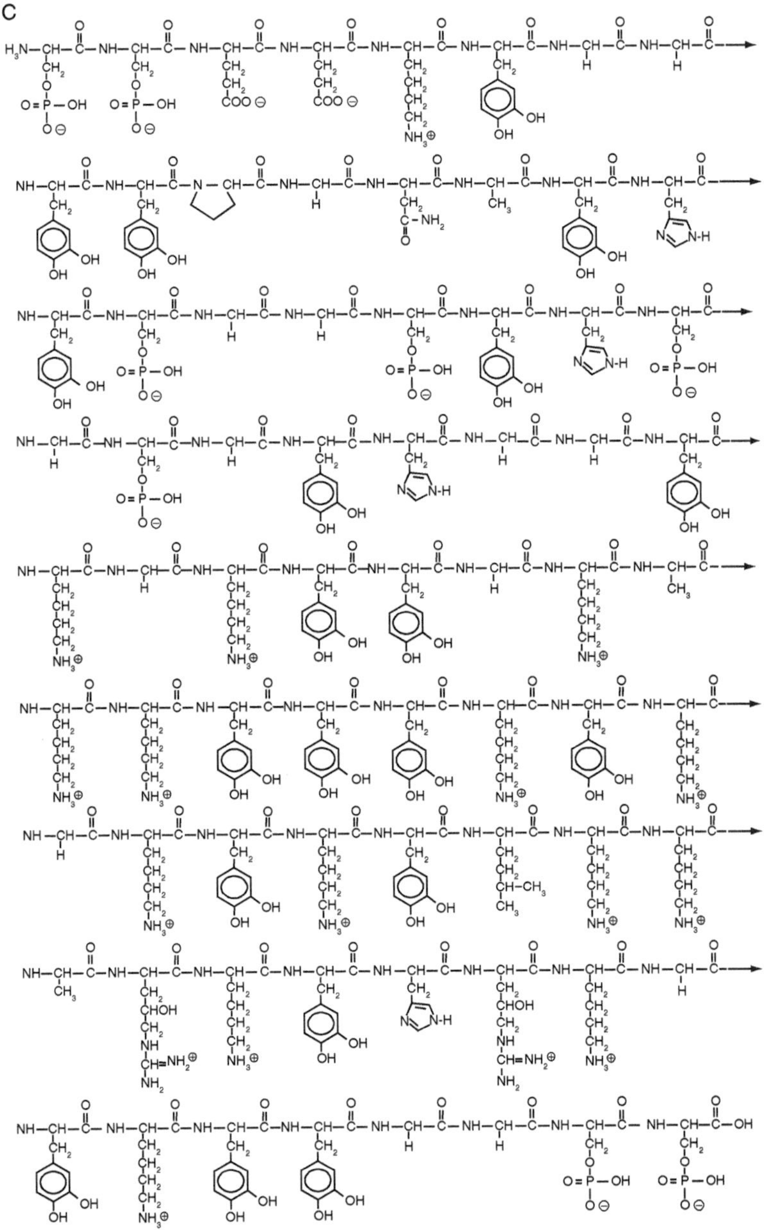

FIG. 9 Interfacial chemistry of distal plaques. (A) Plaque footprints were prepared from plaques deposited on clean glass or mica surfaces by scraping away with a clean single edged razor blade; matrix was applied to the footprint and dried before performing MALDI TOF mass spectrometry. (B) Most reproducible protein ions in the plaque footprints are mfp-3 and mfp-5. (C) Chemical sequence of mfp-5 showing the abundant Dopa and lysine residues (Waite and Qin, 2001). (See Color Plate 5 at the back of the book.)

adhesion is to be achieved, Dopaquinones are undesirable at the interface with a mineral, and the mussel probably takes measures to prevent oxidation from happening. These measures are worth knowing more about.

In addition to forming bidentate chelate complexes with surface metal oxides via Dopa, plaque proteins also coordinate metal ions. Fe^{3+} was detected in plaques using electron paramagnetic resonance (Sever *et al.*, 2004), although given the poor spatial resolution of EPR, whether this iron is associated with the cuticle coating or is present within the plaque needs to be re-explored by Raman microprobe analysis.

3 *Glycera* jaws

Although mostly soft and sinuous, some polychaete worms are equipped with sharp jaws necessary for feeding, grasping and burrowing. In the carnivorous *Glycera dibranchiata* ('blood worm') four such jaws are found at the extremity of the worm's proboscis, which is rapidly everted during attack (Fig. 10A). Following prey penetration, *Glycera* jaws are particularly well adapted for injecting venom—they are hollow like syringe needles (Fig. 10B). Given the need to penetrate both carapaces and benthic sediments, the jaws have adapted impressive physical properties, most notably hardness and wear resistance. Gibbs and Bryan (1980) pioneered research on polychaete jaws, and their work represents the first report of Cu and Zn loading in a biological tissue. Further investigation of polychaete jaws with current mechanical, structural and chemical tools has revealed a complex hierarchical architecture and provided novel design paradigms for hard-sclerotized tissues with little or no mineral content.

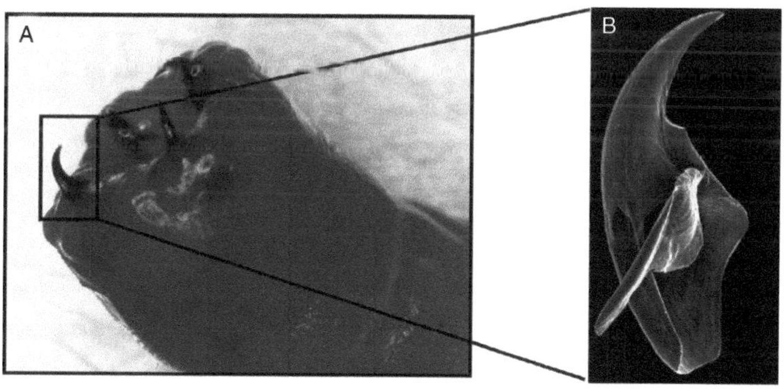

FIG. 10 *Glycera dibranchiata*. (A) Everted proboscis of the polychaete *Glycera* illustrating the presence of four sharp jaws. (B) SEM micrograph of a jaw displaying the hollow interior venom canal at lower left (Moses *et al.*, 2007). (See Color Plate 6 at the back of the book.)

3.1 JAW COMPONENTS: MELANIN, HISTIDINE-RICH PROTEINS AND COPPER

Biochemical knowledge of *Glycera* jaws has been gained using a variety of analytical and spectroscopic techniques. These assays were conducted on jaws hydrolyzed under strong acidic and basic conditions, or degraded in concentrated alkaline hydrogen peroxide (Moses *et al.*, 2006). Based on complementary elemental imaging and synchrotron X-ray diffraction studies (Lichtenegger *et al.*, 2002), a full picture of the jaw chemistry has emerged. *Glycera* jaw is a biocomposite primarily composed of three components: proteins ($\sim$50 wt%), eumelanin pigment ($\sim$40 wt%) and copper (10%) in both mineralized ($Cu_2(OH)_3Cl$, atacamite) and unmineralized forms. The overall amino acid composition is strongly biased towards glycine (Gly) and histidine (His), with relative residue amounts of 50 and $\sim$25 mol%, respectively. Copper distribution is highly inhomogeneous and enriched towards the edges and tip. Histidine concentration also increases towards the tip, while glycine shows the opposite trend. Further subtleties of Cu and atacamite spatial distributions and implication on jaw biomechanics are presented in Section 3.3. Measurements of the AA composition at regular intervals during long hydrolysis treatment indicate that His is chemically less accessible than Gly, suggesting that His residues may be chemically bound to other jaw components.

The highly cross-linked nature of *Glycera* jaws makes its constituent proteins inherently insoluble to strong extracting buffers, and this has so far precluded progress in the characterization of their protein sequences. Sequence data on the jaw proteins of a closely related species, *Nereis*, has on the other hand been obtained and is presented in Section 4.1.

3.2 JAW MICROARCHITECTURE: SCLEROTIZATION AND HIERARCHICAL ORDERING

The most striking feature of *Glycera* jaw microstructure is the sub-surface localization of atacamite fibres. These fibres have a diameter in the range 30–50 nm and are $\sim$500 nm long. The fibres run parallel to the surface of the jaw (Lichtenegger *et al.*, 2002; Pontin *et al.*, 2007) (Fig. 11A). Additional observations using backscattered electron imaging (BEI) and electron dispersive spectroscopy (EDS) mapping have refined this finding (Fig. 11B). From the edge of the jaw to a depth of about 30 μm, the jaw is comprised of four distinct microstructural layers: (i) In the outer layer, 3–5 μm thick, Cu ions are present in non-mineralized form. (ii) Within the second domain, roughly 5–15 μm thick, one finds the mineralized atacamite fibres; (iii) The third domain is chemically similar to the outer layer, but is significantly thicker (10–20 μm); (iv) At a depth of 30–40 μm one reaches the 'bulk' of the jaw, lacking significant Cu deposits. These compositional differences, correlated with layers $\sim$10 μm, impress upon the researcher the necessity of examining structures at appropriate resolutions when considering mechanisms of hardening. The physical properties of these layers will be discussed in the following section.

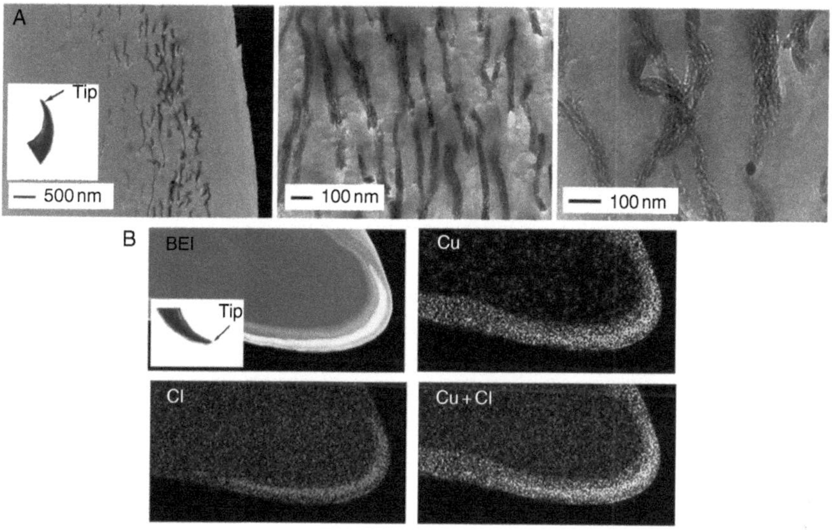

FIG. 11 Microstructural characteristics of *Glycera* jaws. (A) TEM images of longitudinal sections at various magnifications indicate the presence of elongated fibres, determined by X-ray diffraction to be Cu-based atacamite mineral (longitudinal section of tip). (B) Backscattered electron imaging (BEI) micrograph and EDS mapping of Cu and Cl, showing that the atacamite-rich region is located a few microns from the surface, while the outermost layer is composed of unmineralized Cu (Pontin *et al.*, 2007). (See Color Plate 7 at the back of the book.)

A second essential characteristic of *Glycera* jaw is that rather than being a finely dispersed pigment, *Glycera* melanin provides a cohesive structural scaffold network with which the other jaw components interact. The primary substituents of eumelanin, 5,6 dihydroxyindole and 5,6-dihydroxyindole-2-carboxylic acid are depicted in Fig. 12B. When jaws are subjected to extended acid hydrolysis all components except melanin are removed. Unexpectedly, the overall shape of the jaw is conserved (Moses *et al.*, 2006) (Fig. 12A). Jaw melanin is deposited in the form of 200 nm thick sheets, oriented perpendicularly to the jaw axis, and running contiguously throughout the jaw structure (Fig. 12C). Other components are localized between the melanin sheets in the form of protein fibres or atacamite mineral fibres (near the edge). Together, the presence of the melanin scaffold, the high histidine content and the presence of copper ions form a tightly constructed network in which Cu can mediate strong coordination with imidazole ligands on the one side and the dihydroxyindoles of the melanin scaffold on the other (Froncisz *et al.*, 1980; Liu and Simon, 2005). Both of these interactions are presumed to be present in *Glycera* jaws, providing chemical and structural stability, the details of which will be discussed in the following section. A third

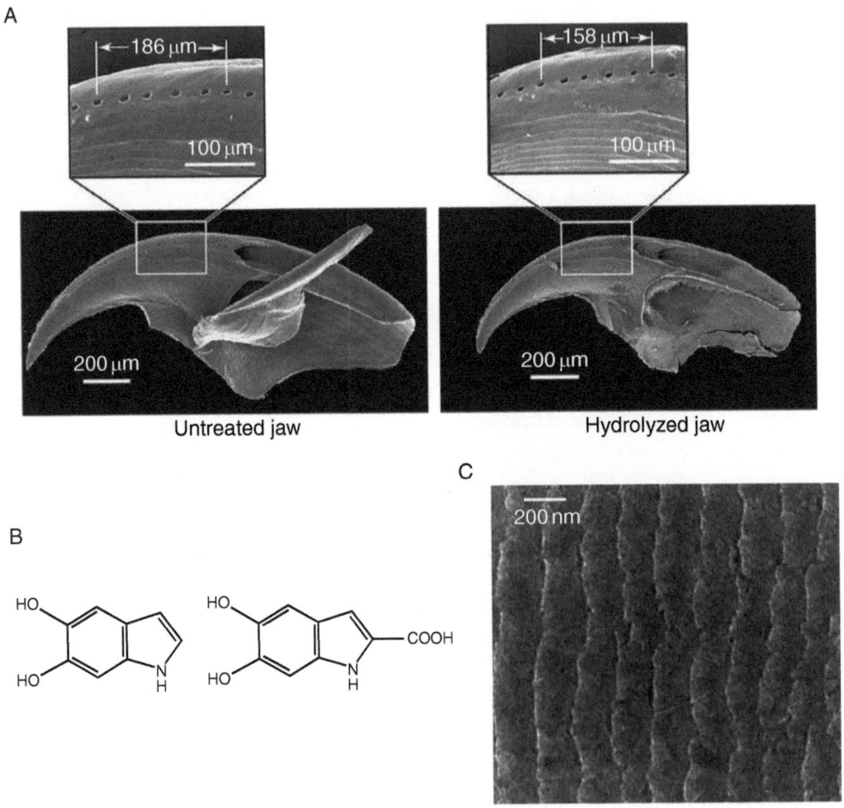

FIG. 12 Melanin scaffold of *Glycera* jaw. (A) SEM images of native and hydrolyzed jaws, both images taken at the same magnification. Shrinkage induced by hydrolysis is measured by the contraction of distance between the venom pores. The jaw structure (including venom pores) is completely retained in absence of protein. (B) Chemical structure of eumelanin primary subunits: 5,6-dihydroxyindole (left) and 5,6-dihydroxyindole-2-carboxylic acid coupled to one another through the 2-, 3-, 4- and 7-ring positions. *Cis*-diols present in eumelanin also chelate metal in a similar fashion to catechols. (C) TEM micrograph of melanin sheets running across the jaws. Melanin sheets are contiguous. Reproduced from Moses *et al.* (2006, 2007).

possible chemical interaction that is expected to contribute to network strength is the covalent coupling of quinone entities of the melanin network to histidine residues, similar to the stabilization of insect cuticles by histidine–catechol adducts (Kerwin *et al.*, 1999; Kramer *et al.*, 2001; Schaefer *et al.*, 1987).

Aspects of the biogenesis of *Glycera* jaws continue to be mysterious. For example one might hypothesize that melanin, because it is contiguous throughout the jaw, providing shape and rigidity, self-assembles first. In this scenario, the histidine-rich proteins and copper ions might be laid upon the framework later as a means of refining the chemical and physical properties of the jaw. But what then directs the melanin self-assembly? Alternatively, Cu^{2+} ions can direct assembly of melanin into well-defined scaffold (Liu and Simon, 2005; Stepien *et al.*, 1989). Therefore, a second hypothesis regarding jaw formation is that Cu, when associated with histidine-rich protein scaffold, triggers the deposition of melanin and the assembly of the jaw as a complete composite (Moses *et al.*, 2006). *Glycera* jaw stands alone as a clear example of a tissue hardened by melanin formation.

3.3 STRUCTURE-TO-PROPERTY RELATIONSHIPS IN *GLYCERA* JAW

A combination of electron microscopy and nano-scale surface probe mechanical testing has revealed that the jaws exhibit interrelated chemical and mechanical gradients at the micro-scale (Lichtenegger *et al.*, 2002; Moses *et al.*, 2006, 2008; Pontin *et al.*, 2007). A summary of these data is presented in Fig. 13. By probing the four domains described in the previous section by nanoindentation (Fig. 13A), and complementing these data by nano-scratch (Fig. 13B and D), and nano-wear testing (Fig. 13C), it has been established that the highest wear performance is associated with regions enriched with *non-mineralized* Cu ions directly on the jaw surface. Remarkably, the level of hardness and modulus attained in this domain surpass those achieved on highly cross-linked synthetic polymers (Broomell *et al.*, 2007). The atacamite-rich domain, on the other hand, exhibits lower wear and scratch resistance, suggesting that they might serve a different function. One role hypothesized is that the atacamite fibres provide increased bending stiffness. The differential composition and properties of the outer, secondary and tertiary layers underscore the importance of hierarchical ordering within biocomposites. While each layer may be considered a composite of melanin, protein and metal, the jaw as a whole can be considered a stratified composite encompassing a wear-resistant outer shell, a reinforced, rod-like interior and a second wear-resistant inner shell. These intricacies in construction are undoubtedly necessary in achieving the robust whole-jaw properties that are required by *Glycera* for their niche-specific survival activities.

As *Glycera* is a marine worm, jaw properties were also compared under hydrating conditions to test functionality in the native state. When hydrated, the layer-to-layer relative properties are amplified. While each domain shows significantly reduced hardness, friction and wear resistance, the decrease is much more drastic in the atacamite-rich region (Moses *et al.*, 2008) and in the region free of Cu. The scratch resistance, for instance (expressed in terms of scratch depth at a given normal force), only decreases by 20% in the hydrated state in the unmineralized Cu regions, while the reduction is 100% for the atacamite-rich regions (Fig. 13D). These results clearly show the critical role of unmineralized Cu in proper functionality of the jaw under hydrated conditions. By forming strong molecular interactions with His residues as well as with the melanin scaffold, it likely prevents water ingress, possibly by occupying sites otherwise accessible to water molecules, and in turn minimizes excessive plasticization.

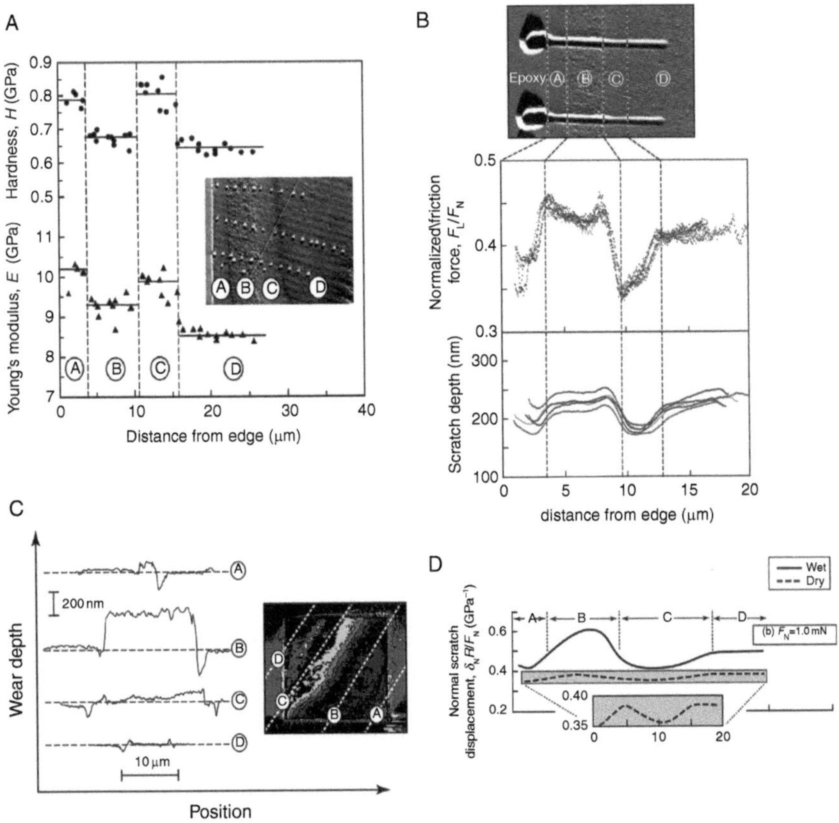

FIG. 13 Surface-probe nanomechanical investigations of *Glycera* jaws. (A) Nanoindentation profiles of the four microstructural domains present from surface to bulk, conveying that the hardest and stiffest regions are the first and third layer (those that contain unmineralized copper). (B) Nano-scratch results support the hypothesis layers one and three are most wear resistant. Across the four domains, higher scratch resistance is associated with lower normalized scratch force F_L/F_N. Layer two, which contains atacamite fibres, is the least scratch resistant. (C) Nano-wear boxes encompassing the four domains. Wear depth profiles were obtained across lines shown on the right-hand side, wear depth is larger in atacamite-rich regions. (D) Considering normalized scratch forces on hydrated specimens as compared to dry specimens, the 'hard–soft–hard–soft' theme remains, but differences in properties are magnified. The change is manifest most strikingly in the atacamite-rich second layer. Reproduced from Pontin *et al.* (2007) and Moses *et al.* (2008). (See Color Plate 8 at the back of the book.)

Lastly, the melanin scaffold also plays an essential role in the structural stability of the jaw as was shown by performing nano-mechanical tests on hydrolyzed jaws (i.e. jaws in which all components, save melanin, were depleted) (Moses *et al.*, 2006). The hardness and modulus is about half that of intact jaws, which remains quite remarkable considering that proteins and metal ions, thought to be responsible for tying the structure together, have been completely removed from the jaw. In the hydrated state, there is an additional reduction of approximately 20% of the hardness and modulus. Although eumelanin chemical structure is complex and not fully understood (Wakamatsu *et al.*, 2003), it is essentially a complex and stacked macromolecular network of hexamerically coupled *o*-dihydroxyindoles (Borges *et al.*, 2001; Pezzella *et al.*, 2007), hence providing a relevant comparison with insect cuticles in which sclerotization partially occurs via covalent bonding of low-molecular phenolic compounds (Arakane *et al.*, 2009; Sugumaran, 1991). In some respects, the disposition of melanin in *Glycera* jaw is reminiscent of molluscan shell mineralization, for example. Here a sparse protein matrix provides a scaffold onto which mineral formation can nucleate and grow. Removal of all protein with bleach leads to shell embrittlement but does not change shell shape. Perhaps, the combination of a histidine-rich protein and Cu ions provides an effective scaffold for melanin nucleation and growth. This strategy could provide useful paradigms for polymer chemists seeking to synthesize novel biocomposites (Meredith and Sarna, 2006).

4 *Nereis* jaws

The burrowing polychaete, *Nereis virens*, is native to the North Atlantic region. While *Nereis* and *Glycera* are close relatives and intertidal neighbours, *Nereis* and *Glycera* jaws exhibit very significant differences. To begin, *Nereis* has two jaws, rather than four, and they are at least twice as large as those of *Glycera* (Fig. 14A and B). Structurally, the jaw is not built in 'stacks' but is fibrillar (Fig. 14C). Furthermore, *Nereis* jaws are replete with zinc, a stark contrast to the copper-rich jaws of *Glycera*. A final difference is the lack of melanin, a crucial component in both the structure and function of *Glycera* jaws. Instead, *Nereis* jaws contain upwards of 90% protein by dry weight. Despite the difference in number, size and composition, the two jaw types possess similar physical properties—a striking example of nature's ability to create tools of similar function from disparate materials by iteration over evolutionary time-scales.

4.1 JAW COMPONENTS: HISTIDINE- AND TYROSINE-RICH PROTEIN, HALOGENATION AND ZINC

N. virens jaws are comprised of 90% proteins (dry weight), 8% halogens (Cl, Br, I) and 2% Zn (Birkedal *et al.*, 2006; Broomell *et al.*, 2007; Lichtenegger *et al.*, 2005). The amino acid composition of the proteins is reminiscent of those in *Glycera* jaw, with a high Gly (35 mol%) and His (22 mol%) content as measured after acid hydrolysis. Critical components are distributed along the jaw with significant concentration gradients, as illustrated in Fig. 15A and B. Histidine concentration is graded from base (5 mol%) to tip

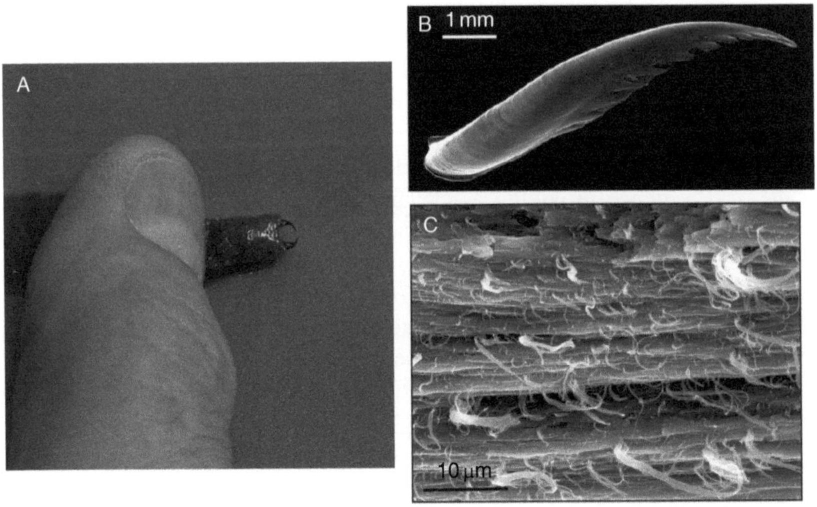

FIG. 14 *Nereies virens*: Head, jaw and microstructure. (A) *Nereis*' proboscis with two large everted jaws. (B) SEM image of a full jaw depicting curved outer edge and serrated inner surface. Reproduced from Broomell *et al.* (2008). (C) SEM micrograph of a fracture surface illustrating the thin-fibrillar structure of the jaw core near the tip.

(30 mol%) (Lichtenegger *et al.*, 2003; Waite *et al.*, 2004), whereas alanine exhibits an inverse gradient, thereby indicating the presence of at least two proteins with distinct spatial distributions along the jaw. Zinc, which is present as the Zn^{2+} ion, is concentrated towards the tip (10 wt% vs. 0.2 wt% at the base) and is strongly correlated with a chlorine gradient (Fig. 15B). X-ray absorption near edge spectroscopic analysis suggests that the increasing and correlating gradients of His, Zn and Cl are related to the presence of Zn-insulin-like coordination bonds ($ZnHis_3$ Cl), illustrated in Fig. 16A and B, where Zn ions act as a roughly tetrathedral coordination centre for sclerotization by interactions with three imidazole rings and a Cl ion.

As mentioned earlier, protein extraction from mature jaws remains challenging given the intractability of highly sclerotized tissue. Despite this inherent difficulty, critical progress has been achieved in elucidating *Nereis* jaw protein sequence by a combination of traditional biochemical and molecular approaches. The first jaw proteins were extracted from jaw pulp tissue and isolated. One of the proteins, Nvjp-1 (*N. virens* jaw protein 1), contains three variants with molecular weight in the range 35–38 kDa, its amino acid composition closely agrees with that of tip hydrolysates (36% Gly/26% His) hence implicating it as a major protein of the jaw tip. Limited internal sequence was obtained from peptides produced by Nvjp-1 proteolysis. The cDNA deduced full sequence of Nvjp-1 reveals that although histidine-rich, it does not contain any well-defined repetitive motifs (Broomell *et al.*, 2008).

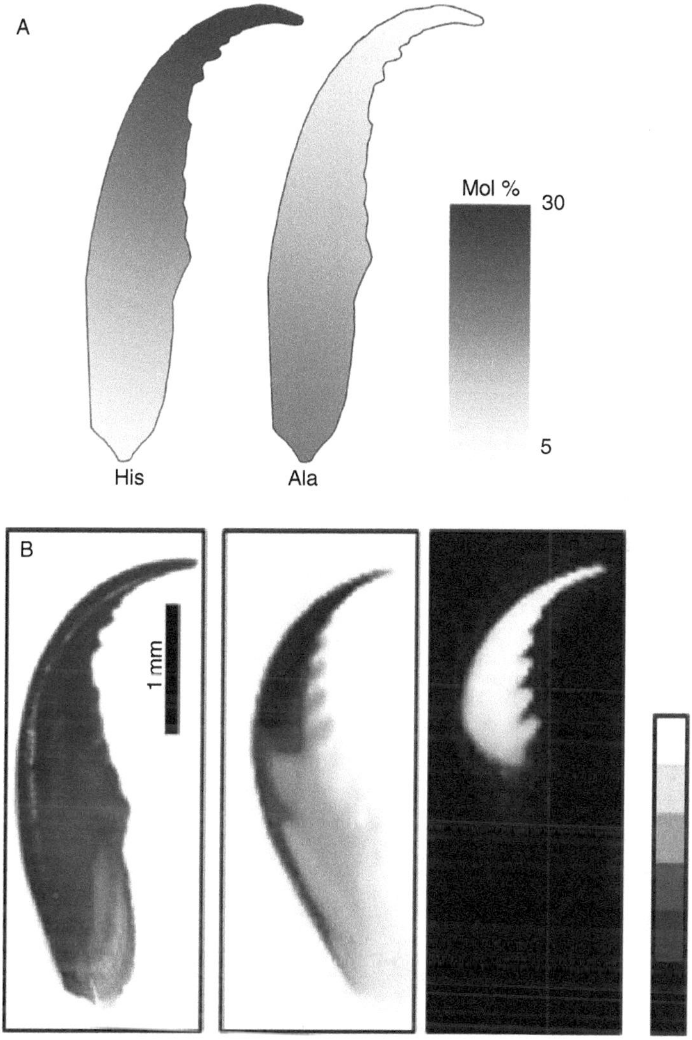

FIG. 15 Molecular gradients in *Nereis* jaws. (A) Distribution of histidine and alanine contents in biopsied samples taken from jaws at regular distal-to-proximal intervals (in relative mol% of total amino acid content post-hydrolysis). Histidine, a preferred ligand for complexes with Zn^{2+} ions is enriched towards the tip whereas alanine (the role of which is currently unknown), shows the opposite trend. (B) Zn gradient revealed by X-ray fluorescence microscopy (bright regions indicate higher Zn concentration). Zn increases towards the tip in parallel to the increase of histidine. Reproduced from Lichtenegger *et al.* (2003) and Broomell *et al.* (2007). (See Color Plate 9 at the back of the book.)

Halogens also display graded concentration with significant enrichment towards the edge of the jaw. The chemical binding environment of the halogens has been elucidated by high-resolution X-ray electron spectroscopy (XPS)

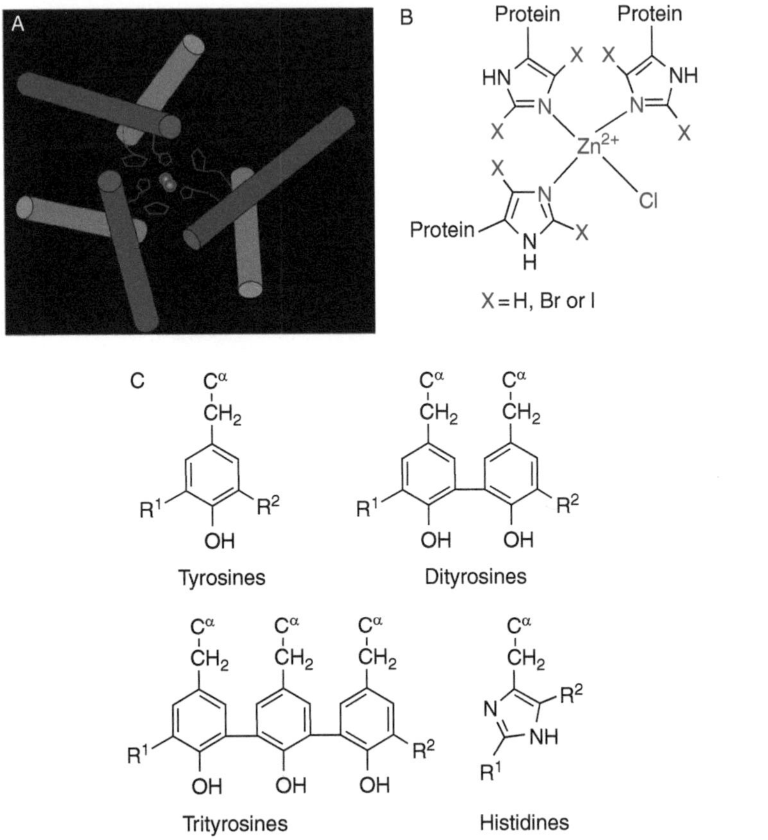

FIG. 16 Coordination complexes and halogenated amino acids in *Nereis* jaws. (A, B) Schematic of insulin-like coordination between Zn, imidazole rings and Cl. This is the best fit from the X-ray database for minerals and metalloproteins to X-ray transmission data for *Nereis* jaws. It may represent the main mechanism/complex by which *Nereis* jaw garners its robust physical properties. (C) Halogenated amino acid and cross-linking chemistry between halogenated Tyr residues. R_1 and R_2 represent iodine and/or bromine. While tyrosine cross-links have been observed, the specific role of post-translational halogenation is still unknown. Reproduced from Waite *et al.* (2004) and Birkedal *et al.* (2006). (See Color Plate 10 at the back of the book.)

investigations (Khan *et al.*, 2006) and by detailed analysis of TOF-ESI-MS of jaw hydrolysate products separated by successive rounds of chromatography (Birkedal *et al.*, 2006). Together, these data show that halogens occur as post-translational modifications, forming various combinations of previously unidentified halogenated His and Tyr amino acids (Fig. 16C). One role of these modified amino acids, supported by tandem mass spectrometric analysis, is in di- and tri-tyrosine cross-links. While the variations in hardness and modulus do not appear correlated to the presence of halogens, these cross-linked regions do encase the jaw (Birkedal *et al.*, 2006), suggesting that they may play a role as a barrier to either chemo-enzymatic degradation or mechanical wear. Alternatively, given the presence of halogenated amino acid residues near the non-fibrillar surface coating of the jaw, their physico-chemical role may be to increase hydrophobicity of the tissue to prevent water penetration.

4.2 JAW MICROARCHITECTURE: SCLEROTIZED PROTEIN FILAMENTS

Structural studies of *Nereis* jaws have been assessed using electron microscopy and a variety of X-ray diffraction techniques under a high-intensity synchrotron X-ray beam. The major structural characteristic of the tissue is that proteins form non-crystalline fibres that are preferentially oriented along the long axis of the jaw (Fig. 14C). According to small-angle X-ray scattering (SAXS) data, the fibre diameter is estimated to be 100–150 nm (Lichtenegger *et al.*, 2003). The outer, halogen-rich coating layer, on the other hand, is not fibrillar. The assembly of proteins into fibres has been studied using recombinantly expressed *Nvjp-1* under various physico-chemical conditions; these experiments show that self-assembly is strongly triggered by variations in metal ion concentration, pH and ionic strength (Broomell *et al.*, 2008). Notably, the solubility of *Nvjp-1* strongly decreases upon Zn^{2+} addition, and the formation of high-aspect ratio fibres only occurs in the presence of Zn ions. Fibrillogenesis is also strongly favoured at high pH, a condition that also induces increasing α-helix and β-sheet domains in the secondary structure. The formation of fibrillar structure in conjunction with secondary structural changes to β-sheet conformation suggests that *Nvjp-1* shares similarities with amyloid proteins.

4.3 STRUCTURE-TO-PROPERTY RELATIONSHIPS IN *NEREIS* JAW

Analogous to *Glycera* jaws, *Nereis* jaws display graded mechanical properties: the jaws are stiffer and harder towards the edge and the tip of the jaw, a gradient that is directly correlated to chemical gradients of Zn and His. The effect of Zn ions on the structural properties was further explored by chelating Zn from exposed surfaces using EDTA and measuring the modulus and hardness before and after Zn chelation by nanomechanical testing (Fig. 17A and B) (Broomell *et al.*, 2006). Hardness and modulus values in regions that are depleted of their native Zn content decrease by more than three-fold. However, Zn can be reintegrated into the structure by incubating the jaw surface with a solution of $ZnCl_2$ at slightly basic pH, thereby recovering almost 90% of the mechanical properties. Since Zn is mostly present as the metal ion coordinated by His-ligands in the sclerotized tissue, this critical experiment establishes the biological relevance of the known ability of His–Zn complexes to form robust and *reversible* interactions *in vitro* (Schmitt *et al.*, 2002), and highlights the largely unexplored potential of structural organic/metal composites that are strengthened by metal/imidazole complexes (Srivastava *et al.*, 2008).

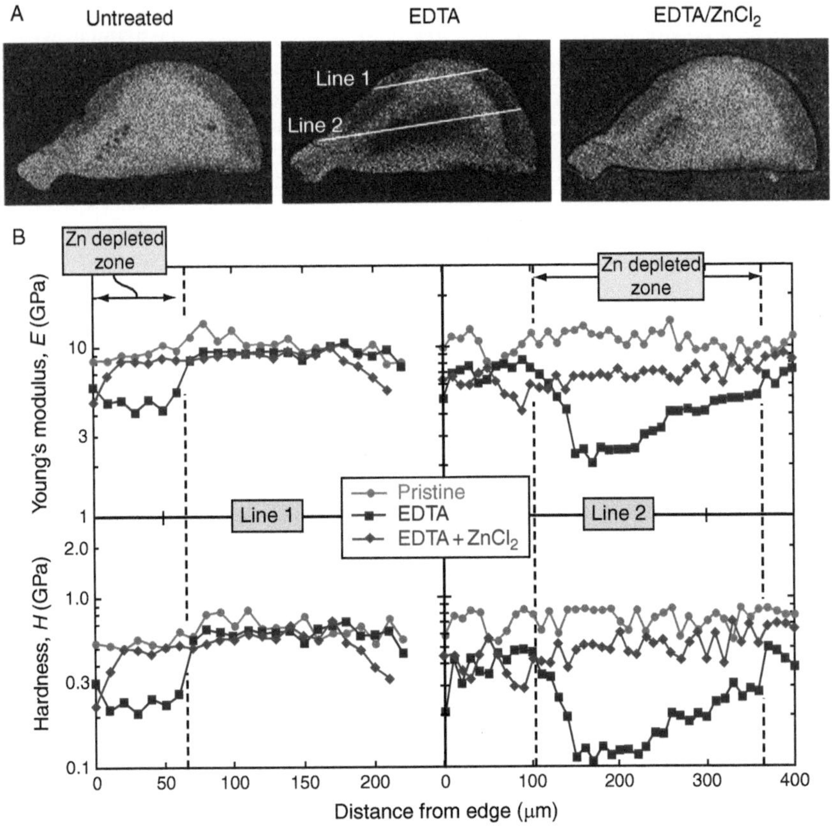

FIG. 17 Mechanical properties of *Nereis* jaws by nanoindentation. (A) Hardness and modulus profiles were assessed on (1) pristine jaw, (2) jaw with chelated Zn (by EDTA treatment), and (3) jaw reconstituted with Zn using $ZnCl_2$ solution. Lines 1 and 2 depict nanoindentation trajectories. It is evident that Zn ions may be removed from and restored into the jaw. (B) Mechanical consequences of Zn depletion and reintroduction demonstrate that the removal and reapplication of Zn^{2+} can diminish and restore the hardness and stiffness of a sclerotized tissue. Reproduced from Broomell *et al.* (2008). (See Color Plate 11 at the back of the book.)

Interestingly, not only Zn has the ability to restore the chelated macromolecular network, but other transition metals (Cu and Mn) can be surrogates for Zn with partial or complete efficacy (Fig. 18) (Broomell *et al.*, 2008). Cu, in particular, allows for nearly full hardness and modulus recovery under dry conditions, although its efficiency under hydrated conditions is not as high as for Zn. Mn can also reform coordination bonds, but its efficiency is limited by its lower concentration after the reconstitution treatment. A critical condition for hardening, however, appears to be that the His concentration be sufficient for the transition metals to successfully integrate into the depleted regions of the jaw.

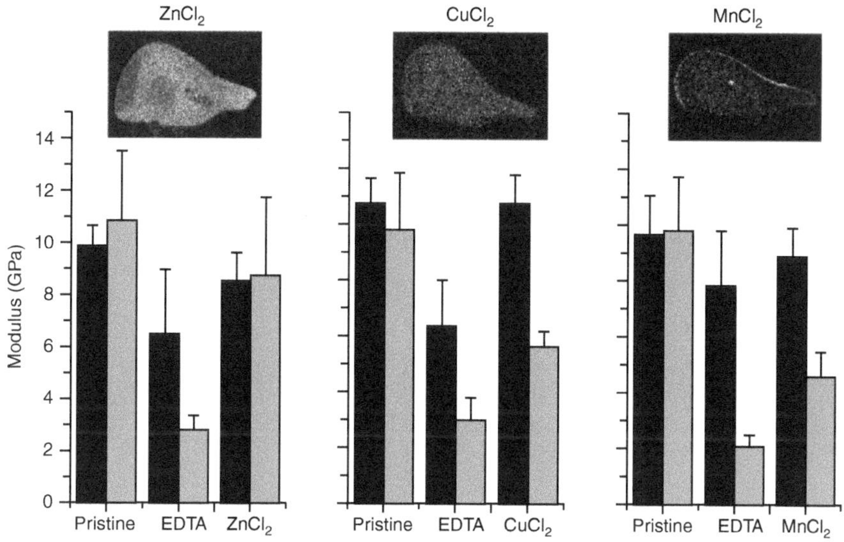

FIG. 18 Introduction of native and foreign metals in to *Nereis* jaw post-Zn chelation. Jaws were incubated in $ZnCl_2$, $CuCl_2$ and $MnCl_2$. Surrogate transition metals showed a similar recovery in material properties although Mn^{2+} has less affinity to infiltrate the structure to form complexes, providing an additional 'lever' to tune novel materials in which imidazole–metal chelation complexes are utilized. Each graph represents data that was compiled from at least two jaws tested at each stage of the treatment regimen. Black and grey bars represent results for tests in air and tests in Milli-Q water, respectively. The error bars represent standard deviations. Reproduced from Broomell *et al.* (2008). (See Color Plate 12 at the back of the book.)

The effect of hydration is also highly dependent on the metal content (Broomell *et al.*, 2008). At high metal concentration, statistical analysis showed that the mechanical properties are essentially retained in the hydrated state, while both the hardness and modulus are reduced three-fold under hydrated conditions when metal has been chelated. This emphasizes another critical feature of metal–protein sclerotization, namely that non-mineralized metal incorporation can be used by marine organisms to manipulate both stiffness and hardness by means of coordinate bonding as well as dehydration. Reminiscent of mussel byssal threads (Section 2.1) is the two-stage cross-linking process involving first metal-complexation by histidine-rich protein fibres, followed by the formation of covalent di-tyrosine cross-links.

5 *Dosidicus* beak

D. gigas (Humboldt Squid) are marine predators encountered along the Eastern Pacific from Chile to Baja California—with increasing incursions up to the Northern Californian coast in recent years. Their aggressive feeding behaviour

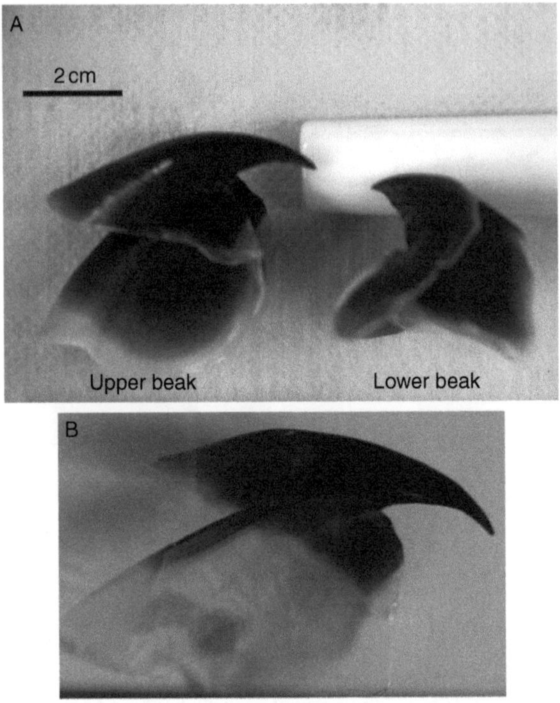

FIG. 19 *Dosidicus gigas* beaks dissected from the buccal mass. (A) Upper and lower beak clearly depicts a gradient in colour from translucent/white to opaque/black along the proximal to distal axis. (B) Upper beak after a sagittal cut along the beak with a diamond saw, illustrating the V-shape morphology with external and internal regions meeting at the sharp tip. (See Color Plate 13 at the back of the book.)

is partially enabled by the presence of a hard, tough beak situated at the extremity of the digestive system (Fig. 19A). The beak is a robust tool that has been reported to be able to crush crab shells. Unlike the systems previously described, the beak of *D. gigas* is devoid of metals, minerals and halogens. It is completely organic, owing its physical properties to chitin, protein, phenolic compounds and water, exclusively. *D. gigas* beaks were initially chosen as a model system due to their large size, on the order of a 3–7 cm (Clarke, 1986), which enables a greater variety of characterization tools to study their biochemistry and their biomechanical properties.

5.1 COMPONENTS: PROTEIN, CHITIN AND CROSS-LINKED PIGMENT

Compositional analysis and X-ray diffraction investigations have revealed that *D. gigas* beak is a biocomposite of proteins and α-chitin (Miserez *et al.*, 2007). α-Chitin, a fundamental component of insect cuticle largely absent in our discussion thus far, is composed of *N*-acetylglucosamine monomers arranged with β-1,4 linkages resulting in strong, highly linear, heavily hydrogen bonded fibres. Moving from wing to rostrum, there are various degrees of hydration along which the water content is inversely correlated with the chitin and protein content (Miserez *et al.*, 2008). Minerals and metal ions, on the other hand, are absent. The distribution of the components is spatially graded: the base of the beak intimately embedded within the buccal tissue is composed of highly hydrated chitin (water content 70 wt%) and a few percent protein, while the composition of the hard, tanned rostrum is dominated by densely cross-linked proteins. When hydrated, the rostrum is made of ~72 wt% cross-linked proteins (with cross-links accounting for at least 15% of the weight), ~13% chitin and ~15% water (Fig. 20A) (Miserez *et al.*, 2008).

The overall amino acid composition of the proteins in the highly tanned region is rich in Gly (27 mol%), Ala (15 mol%) and His (11 mol%). The relatively large His content is again noteworthy and has critical implications for the cross-linking chemistry, as detailed below. Not surprisingly given the highly sclerotized nature of *D. gigas* beaks, its constituent proteins are insoluble to most aggressive extraction cocktails. However, some proteins from the mature beak have been successfully extracted in denaturing buffers such as 5% acetic acid/6 M urea. SDS-PAGE reveals a family of proteins with molecular weights in the range 50–15 kDa, two of which were selected for further study, Dgbp-3 and Dgbp-5 (for *D. gigas* beak protein) (Fig. 21). Amino acid analysis of these soluble proteins indicates little consistency with that from the hydrolyzed beaks (Fig. 20B). In particular, the Gly content is clearly lower than in the beak (15 mol% at most vs. 30 mol%), whereas that of His is also significantly smaller (4–5 mol% vs. 10 mol%), suggesting that these proteins do not directly participate in protein cross-linking. In order to detect possible chitin interactions with these proteins, lectin-binding assays can be used (Goodarzi and Turner, 1996). Lectins are particularly useful in binding glycosylated proteins with high specificity. By labelling lectin with a biotin probe, specific glycosylation is readily monitored on blotted membranes by subsequent exposition with horseradish peroxidase. In the case of chitin, biotinylated *Datura stramonium agglutinin* (DSA) can be employed thanks to its specificity with *N*-acetylglucosamine. Both dgbp-3 and -5, on the other hand, are shown by a lectin-binding assay to be linked to *N*-acetylglucosamine (Fig. 21), suggesting an important role as a chitin anchor in the biocomposite. The potential roles of DGBP-3 and DGBP-5 as chitin anchors are preliminary and have to be confirmed, as has the amino acid sequence of the proteins. This is an on-going investigation and full sequence of DGBPs has not been elucidated yet.

As stated above, the fully tanned portion of the beak under hydrated conditions consists of at least 15% by weight of pigmented, insoluble cross-links. This can be colorimetrically visualized by subjecting beaks to the catechol-specific Arnow Stain (Fig. 22A) (Arnow, 1937). The detection of abundant catechols is strikingly reminiscent of insect cuticles and was also observed in other squid species

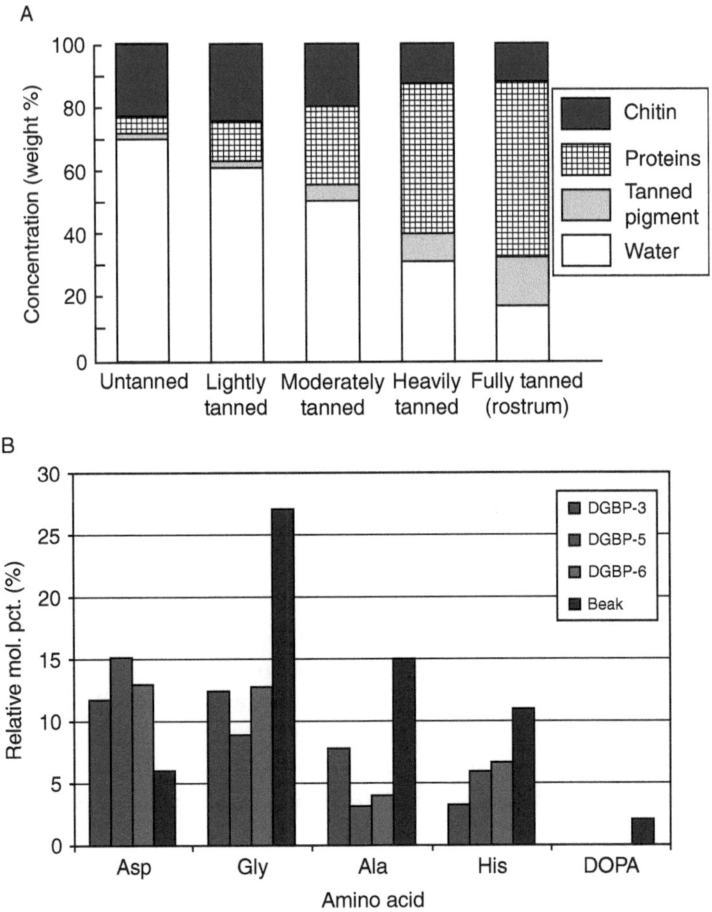

FIG. 20 Compositional gradient and amino acid bias. (A) Compositional gradients in *D. gigas* beaks. Whereas the untanned region of the beak is roughly % water, with little protein, the rostrum is only 15% water with approximately 72% protein and cross-linked pigment. In contrast, chitin concentration remains fairly constant. Reproduced from Miserez *et al.* (2008). (B) Relative amino composition of full beak and extractable proteins. There is no apparent similarity between the mature beak, rich in Gly, His and Ala and Dgbp-3, -5 and -6, which are more homogenous in composition, with an enrichment of Asp.

(unpublished data). In order to explore the details of cross-link chemistry, portions of the beak were subjected to acid hydrolysis (6 M HCl at 110 °C), and each portion produced both soluble and insoluble fractions. For the fraction solubilized by hydrolysis, phenylboronate affinity chromatography was used to selectively capture a family of catecholic compounds from the hydrolysate. These adducts were subsequently separated by RP-HPLC and analysed by tandem mass

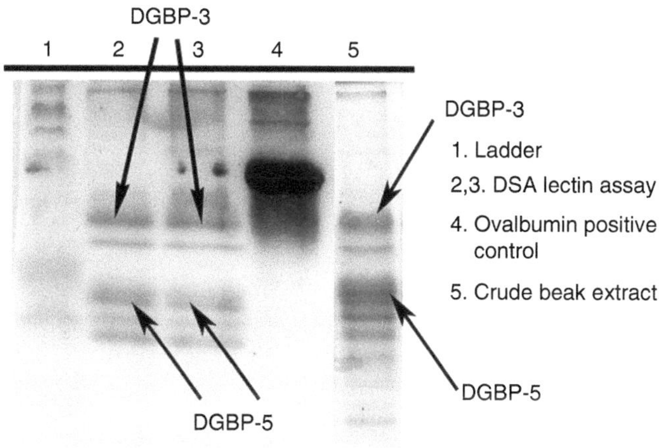

FIG. 21 Lectin-binding proteins in *Dosidicus* beak. A DSA (*Datura stramonium agglutinin*) lectin blot, known to bind glucosamine, was used to assay beak proteins for possible chitin-binding sites. A biotylated DSA conjugate was used, which was subsequently exposed to horseradish peroxidase before final detection with a chloronapthol reagent. Many beak proteins including Dgbp-3 and Dgbp-5 are decorated with glucosamine. (See Color Plate 14 at the back of the book.)

spectroscopy and ^{1}H NMR analyses. Fragmentation patterns of the separated products show that most are based on the coupling of 4-methyl catechol (4MC) and histidine, along with a smaller subset of Dopa and histidine adducts (Miserez *et al.*, 2008). The family of catechol–histidine adducts is depicted in Fig. 22B. The smallest of these cross-links corresponds to His–4MC adduct, with the proper bond assignment between the catecholic (2 position) and imidazole (ε^2 position) rings obtained by ^{1}H NMR (Miserez *et al.*, 2008). The family of catechol–histidine adducts is depicted in Fig. 22B. The smallest of these cross-links corresponds to His–4MC adduct, with the proper bond assignment between the catecholic (2 position) and imidazole (ε^2 position) rings obtained by ^{1}H NMR (Miserez *et al.*, unpublished data). The largest cross-link is associated with the linear coupling of His–4MC–His–4MC residues, with the bond location assumed to follow that deduced from the His–4MC adduct.

In a parallel study, the insoluble fraction post-hydrolysis was analysed. By subjecting this insoluble residue directly to MALDI time-of-flight mass spectrometry without the usual matrix, the insoluble fraction was fragmented into a number of ions smaller than 200 Da. The most abundant of these corresponded to the mass of His and Dopa ions, suggesting that these two residues are coupled in the formation of a dense cross-linked network. Notably, these molecular compounds were not detected in the untanned regions of the beak, thus implicating their critical role in the tanning and hardening process. There is a fundamental interest in identifying the catecholic precursors extracted from beak hydrolysates. Again using Arnow's Stain on thin frozen sections sliced perpendicular to the beak axis,

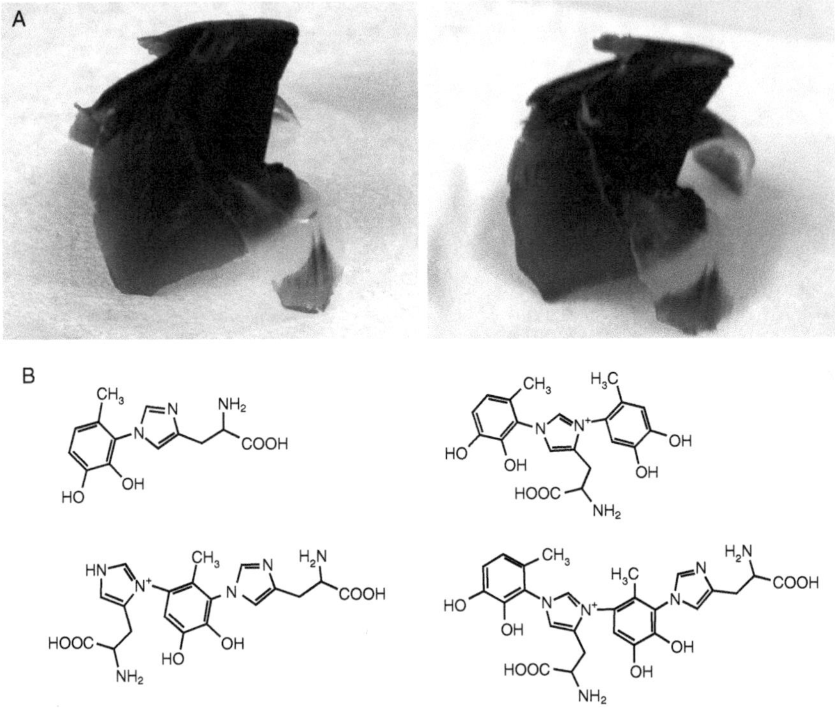

FIG. 22 Catecholic cross-linking in *D. gigas* beaks. (A) Arnow's Stain of a beak indicates localization of catecholic compounds in tanned regions, which appear dark red after staining. Although beak pigmentation obscures the red colour of Arnow, it is clear that the concentration of catecholic compounds is enriched in a graded fashion from wing to tip. (B) Chemical structure of hydrolysates isolated by affinity chromatography and RP-HPLC, as deduced from ESI-MS/MS and ^{1}H NMR. Soluble cross-links are composed of di-, tri- and tetramer adducts of His and 4MC. Assuming that the His is originally present in the backbone of a major beak protein, the trimer and tetramer structures are implicated as cross-link structures. (See Color Plate 15 at the back of the book.)

colouration suggests that the catecholic agents are stockpiled in specific glands of the buccal mass (Fig. 23A and B). Current work shows that these compounds can be extracted from the glands and subsequently isolated by gel-filtration chromatography (Fig. 23C), and their identification is underway. Once completed, this will shed light on whether the catecholic precursors are low-molecular-weight compounds (as in insect cuticles), or are bound as Dopa to a peptide backbone thus resembling other sclerotized tissues of marine organisms such as mussel thread, discussed in Section 2.1. Based on the detection of Dopa and Dopa–histidine adducts in hydrolyzed squid beak, we originally speculated that 4MC was derived from Dopa by bond cleavage between the alpha and beta carbons. However, this behaviour has not been observed in other Dopa peptides, thus raising the peculiar possibility that squids bolster the sclerotization of Dopa proteins with secreted 4MC.

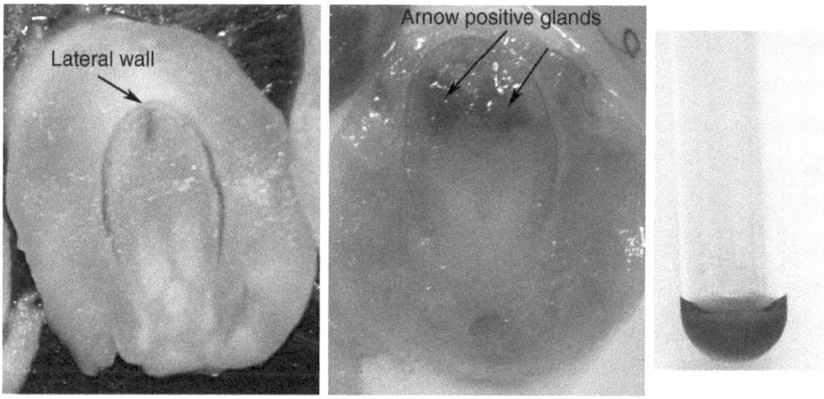

FIG. 23 The buccal mass and its catechol-rich glands. (A) Cross-section of the buccal mass, including location of the proximal region of the lateral wall (inverted U-shape). (B) Glands situated in close proximity with the lateral wall are Arnow positive suggesting that they stockpile catecholic molecules (small molecules or protein-bound catechols, i.e. DOPA). (C) Extracted compounds from the glands after late elution from gel-filtration chromatography are Arnow positive suggesting that catechols are either 'free' or are bound to small proteins/peptides. (See Color Plate 16 at the back of the book.)

5.2 BEAK MICROARCHITECTURE: AN ORGANIC LAMELLAR COMPOSITE

Comprehensive descriptions of the morphology and the biomechanics of the whole beak/buccal mass assembly of cephalopods have been presented by Kear (1994) and Uyeno and Kier (2005). At the macroscopic scale, the beak is made of two sub-structures, the upper and the lower portions (Fig. 19A), which slide by each other during feeding action but are not directly connected (Uyeno and Kier, 2005). The upper and lower portions differ in shape (Clarke, 1986) but share a similar overall morphology; they roughly exhibit a hollow V-shape structure (shown in Fig. 19B for the upper beak) with a sharp, dark and hard rostrum (tip) at the distal end, and a proximal region consisting of transparent wings and lateral walls that are embedded within a muscular ball of connective tissue called the buccal mass. The buccal assembly is comprised of four major muscles that ensure the complex movement, stability and biting power of the beak without relying on any direct contact between the lower and upper parts. Of significant importance for beak growth is the presence of a single layer of elongated cells at the beak/buccal mass interface, the beccublasts, which play a dual role of secreting the beak precursors and connecting the mandibular muscles to the beak (Dilly and Nixon, 1976).

Electron micrographs of fracture surfaces reveal a layered microstructure, with individual lamellae approximately 2–5 μm thick, traversing the wall and oriented along the beak axis (Fig. 24) (Miserez *et al.*, 2007). The lamellae, however, do not terminate at the external surface; instead, an outer surface layer

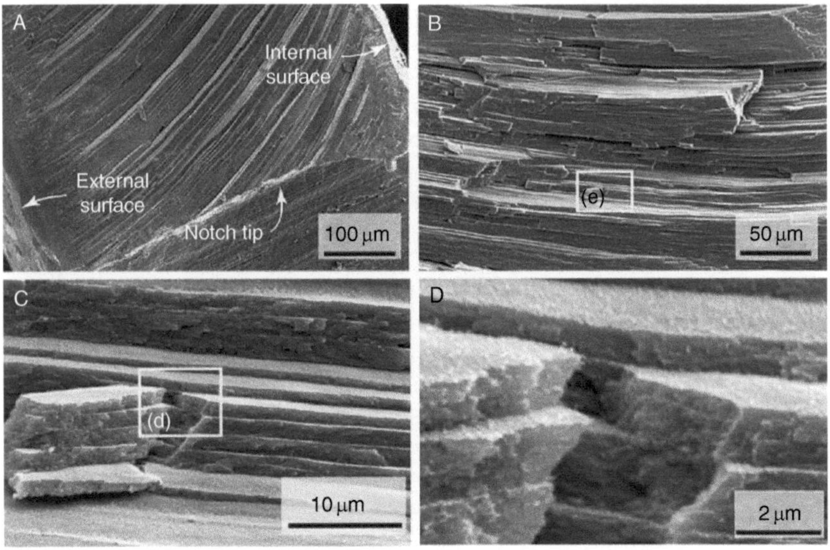

FIG. 24 Lamellar microstructure in *D. gigas* beaks revealed by SEM micrographs on fracture surfaces. The tip of the beak is oriented above the plane of the paper. Reproduced from Miserez *et al.* (2007).

oriented perpendicular to the lamellae is present and possibly provides an extra barrier against water ingression.

Soaking the beaks in alkaline peroxide (1 N NaOH and 5% hydrogen peroxide at 70 °C) solubilizes the proteins, cross-links and pigments, leaving an intact chitin network (Fig. 25A). While the overall structure shrinks and is softened (when hydrated) after such treatment, the general shape remains. What is left is a chitinous, entangled fibrillar network that provides the structural scaffold of the beak. Scanning electron micrographs of this chitinous scaffold at various magnifications are presented in Fig. 25B. Leaflets of chitin are made of smaller elongated fibres with a diameter of 25–35 nm, a range that agrees with values inferred from Guinier's approximation of SAXS data on native beaks (Miserez *et al.*, 2007). The latter experiments also indicate that chitin fibres are preferentially oriented in the rostrum, but feature a more isotropic distribution in the proximal regions. The structural arrangement and physical interactions at chitin/protein interfaces are currently unknown.

5.3 STRUCTURE-TO-PROPERTY RELATIONSHIPS IN *DOSIDICUS* BEAKS

The *D. gigas* beak shares a critical design strategy with polychaete worm jaws: all exhibit distinct mechanical gradients from the proximal base where they are attached to the buccal muscular tissue to the rostrum or tip. Given the much

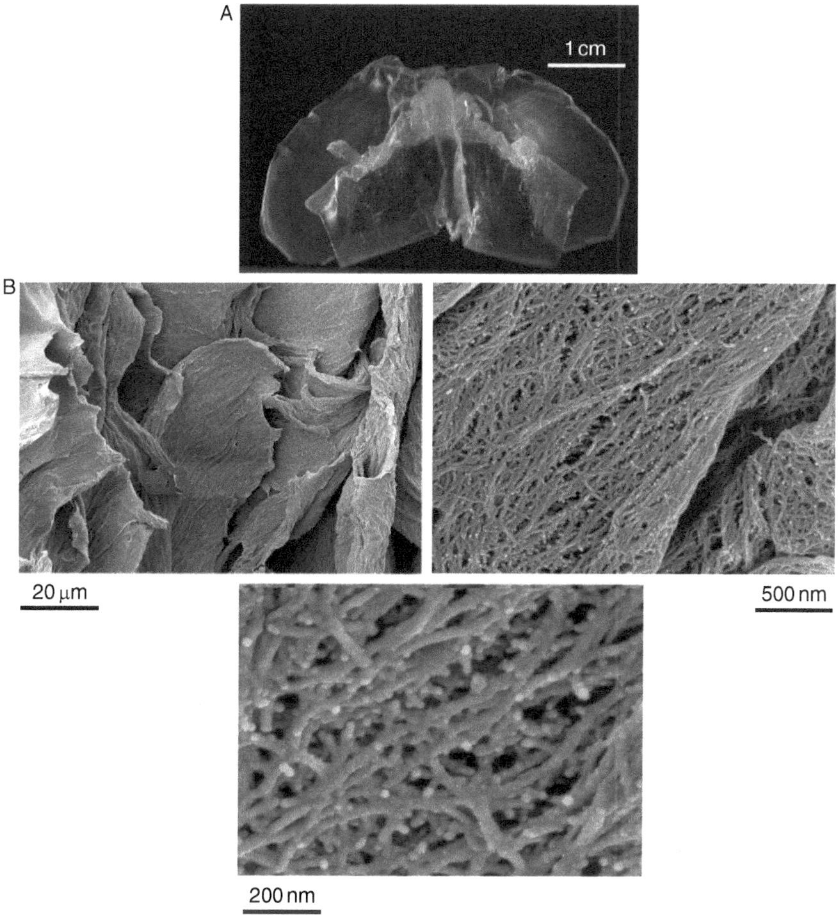

FIG. 25 *Dosidicus* beak after alkaline peroxide treatment. (A) Frontal view of a full lower beak showing that the residual shape is retained after pigment and protein removal. (B) SEM micrographs at increasing magnification of the degraded beak the after fixation at various magnifications, illustrating the residual scaffold made of leaflets containing intertangled chitin nanofibres. (See Color Plate 17 at the back of the book.)

larger size of the squid and of its beak, these gradients are large enough to be measured by standard uniaxial tensile testing as well as by nanoindentation. Strictly speaking, elastic modulus values inferred from tensile experiments and automated indentations are not fully equivalent (Cheng and Cheng, 2004; Gouldstone *et al.*, 2007). However, measuring these properties on samples for which both techniques could be employed indicates that the values are comparable, hence allowing one to use data gathered from the two techniques for comparative purposes. The elastic (Young) modulus of the soft, untanned portion is around 50–60 MPa as measured by unaxial tensile testing under hydrated conditions (Miserez *et al.*, 2008). The stiffness sharply increases going towards the tip, to reach values in the range 4–6 GPa for heavily tanned regions, here measured by nanoindentation with a fluid-cell tip (Fig. 26) (modulus vs. protein + cross-link content). Hence, there is a variation in stiffness that spans 2 orders of magnitude from tip to the base in the beak, a design which is critical in minimizing contact damage between highly mechanically mismatched tissues (Benjamin *et al.*, 2006; Suresh, 2001; Waite *et al.*, 2004). When the samples are fully desiccated, the gradient is attenuated. Here the modulus of the tip reaches 8–10 GPa, but the untanned and peroxidized sections have a modulus as high as 4–5 GPa.

The large size of the beaks has allowed the measure of resistance to crack propagation of the structure, a critical property for assessing structural integrity but lacking in the insect literature due to the much smaller size of the cuticles. The fracture toughness is equivalent to that of the toughest synthetic polymers, which is quite remarkable considering that the beak is nearly twice as hard and that the resistance to crack propagation is usually inversely correlated to the hardness in many structural materials. The lamellar microstructure observed in the beaks is a recurrent design in many structural tissues, including both mineralized and non-mineralized structures (Kamat *et al.*, 2000; Miserez *et al.*, 2008; Neville, 1993), and is well established to toughen the tissue against

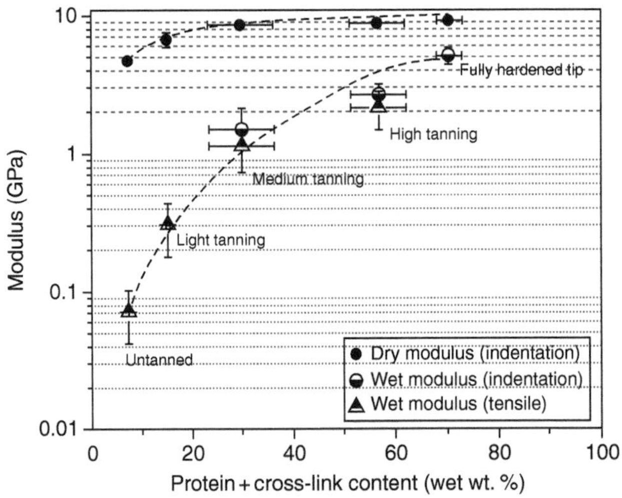

FIG. 26 Stiffness gradients in *D. gigas* beaks, expressed as modulus versus (protein and cross-link content). Note the 2 orders of magnitude change in modulus from base (untanned material) to tip (fully tanned material) in hydrated samples. In desiccated conditions, the gradient is largely attenuated. Although both dehydration and covalent cross-linking add to stiffness, from base to tip the difference in modulus is only a factor of 2. This assay provides evidence for the importance of the dehydration on the structure and properties of the protein as well as the chitin.

catastrophic crack propagation, in a fashion quite similar to what is known in layered structural composites (Hutchinson and Suo, 1992).

It is of fundamental interest to discuss stiffness and hardness values of the beaks in the light of the physico-mechanical characteristics of chitin. Given the increasing usage of chitin and chitosan in tissue engineering applications, it is rather surprising to find little reliable data on its intrinsic mechanical characteristics. Using X-ray diffraction experiments under (and using a cellulose standard to ascertain their values), Nishino *et al.* (1999) have deduced a *dry* elastic modulus of 41 and 65 GPa for α-chitin and chitosan, respectively. Literature values for the hydrated chitosan scaffold are more scattered, perhaps reflecting the fact that various processing methods are employed to prepare chitosan films (Notin *et al.*, 2006). The discrepancy could result from materials with large differences in porosity content, or the fact that hydration level is not accurately controlled during the experiments. Both factors will strongly influence the mechanical response of the scaffold. Regardless of these uncertainties, reported values of modulus for wet chitosan are in the range 10–15 MPa in the hydrated state, and for porous networks (Madihally and Matthew, 1999; Wu *et al.*, 2005a, 2006) seem consistent. As is the case for many biological tissues, the elastic modulus varies over a few orders of magnitude between fully dry and wet conditions. This has important implications for the observed mechanical gradients of *Dosidicus* beaks. Since measured modulus values are bound between the values reported for dry and wet chitosan/chitin networks, it implies that the structural properties are tuned by fine control of hydration in the tissue, and possibly by a reduction of pore volume fraction during hardening. Whether water is expelled during cross-linking itself or by the impregnation of hydrophobic proteins through the porous network is critical but remains open for discussion. Experiments to tackle this issue could involve X-ray scattering methods and nanomechanical instrumentation on individual fibres. It is anticipated that such clarification could provide critical lessons for biomimetic processing of chitin/protein biocomposites with graded properties.

6 Selective comparison of sclerotization strategies

As the purpose of this review is to shed light on the diversity of sclerotization strategies by comparison across invertebrate species, let us first revisit the impetus for fundamental research in natural biomaterials. The chemistry and micro- and macrostructure of these materials have been tuned for particular functions over millions of years and provide excellent case studies for functional material design. Beyond functionality, the materials presented here proffer valuable principles usually lacking in engineered consumer materials—programmed obsolescence and degradation. Using biological structures as model systems, carefully constructed next-generation materials may be designed for ‘cradle-to-cradle’ use; a term coined by McDonough and

Braungart Design Chemistry (MBDC) Consultants. While the cradle-to-cradle philosophy is broad, providing an entirely new conception of industry, a worthy first step is the creation of materials that are non-toxic and degradable by natural processes including environmental wear and microbial degradation. In doing so, necessary nutrients are not depleted, but are instead returned to the ecosystem over reasonable timescales (McDonough and Braungart, 2002). This is in stark contrast to current polymer-based materials that may be energy intensive to manufacture, toxic and resistant to degradation (Burillo *et al.*, 2002), adding unnecessary pressure to the survival struggle of plants and animals and further threatening biodiversity.

Whether these biomaterials are eggshells, holdfasts, ligaments, mandibles, jaws or beaks, they are all adapted to cope with deformation by external forces. Essentially, each material is a potentially novel design for tackling the same problem—not unlike separating teams of scientists and engineers, giving them a goal, and analysing the results. Inherent in the similarities and differences will be fundamentally distinct, ground level strategies for reaching the set goal. It is our belief that there is insightful research in both arthropod and marine invertebrate sclerotization, and that systematic comparison (although in no way entirely comprehensive) has value for research as well as bio-inspired modelling and engineering. An in-depth biochemical comparison of the many, varied cuticular structures across arthropod species is beyond the scope of this chapter and can be found elsewhere (Andersen, 2010; Vincent and Wegst, 2004); however, for the purpose of continuity, a description of the general anatomy of insect cuticle, and three specialized arthropod structures—*Aedes aegypti* chorion, *Atta sexdens* mandible and *Schistocerca gregaria* wing ligaments—are presented for the benefit of comparison to aforementioned marine structures.

6.1 HARDENING AND PIGMENTATION: AN INTRODUCTION TO INSECT CUTICLE CHEMISTRY AND MICROSTRUCTURE

Insect cuticle is a layered, fibrous composite of chitin, water, protein, catechol, lipid and occasionally metal and mineral, secreted by a single layer of epidermal cells (Vincent and Wegst, 2004). It is remarkable in that it naturally serves many purposes including repelling water, resisting wear, and reversibly adhering to substrates while remaining light and flexible for flight (Vincent and Wegst, 2004). The impressive range in material properties—Young's modulus of 1 MPa to 20 GPa—is largely based on the relative amounts of each cuticle component. Its foundation is a semi-crystalline array of α-chitin strands that arrange to form nanofibres (on the order of 3 nm in diameter and 0.3 μm long) (Atkins, 1985). In part due to its hydrophilic nature, in the presence of water it is liquid crystalline, a property which likely aids in cuticle self-assembly but does not provide the robust properties desired in mature cuticle (Murray and Neville, 1998). It is then strengthened by infusion with proteins and catechol-based small molecules that can form intermolecular cross-links. The relative contributions of different hardening mechanisms have been discussed for some time.

Nearly 70 years ago, Pryor (1940) proposed a mechanism of 'tanning' to describe the colouration and hardening of biological tissues. His work on the ootheca of *Blatta orientalis* described a chemically irreversible process of covalent linkages based on the coupling of oxidized phenolic compounds to the amino acid side chains of proteins, rendering the material more robust and resistant to chemical and enzymatic degradation. Passing further back in time, the initial discovery of the necessity of oxygen in the pigmentation of *Tibicen septemdecim* (cicada) cuticle was established as early as 1911 by Gortner through a series of oxygen-deprivation experiments including submersion in water and displacement of oxygen by carbon dioxide (Gortner, 1911). We know now that both pigmentation and hardening are intimately related, yet, today, nearly 100 years later, both the mechanisms and roles of oxidative coupling in insect cuticle are not fully understood.

However, disentangling the components of fully sclerotized tissues as a means of understanding their assembly is particularly challenging and often disheartening. For many years, two hypotheses, both of which are relevant and substantiated, have been contested. The first, mentioned above, is the quinone tanning hypothesis (Pryor, 1940). In this scenario, *N*-acetylcatecholamines in the cuticle are oxidized to orthoquinones or *para*-quinone methides by oxidases (laccases or tyrosinases) (Suderman *et al.*, 2006). The electrophilic nature of these molecules causes them to be spontaneously attacked by neighbouring nucleophiles, most notably the ε^2-nitrogen of histidine, via a Michael addition to the β-carbon of a *para*-quinone methide (β-sclerotization) or a ring carbon (often position 5) of an *ortho*-quinone (Xu *et al.*, 1997). Indeed, histidine–catechol adducts have been purified and characterized by solid-state NMR and mass spectrometric studies by multiple authors and are implicated in stabilization of the cuticle (Kerwin *et al.*, 1999; Kramer *et al.*, 2001; Schaefer *et al.*, 1987). In post 'catecholation', it is supposed that a second redox cycle results in the coupling of the *N*-acetylcatecholamine-linked peptide with other *N*-acetylcatecholamines and/or other proteins (Suderman *et al.*, 2006). At this point, the covalent link between adjacent protein chains would be complete, the motion of the backbone would be fixed, and the material as a whole would be more rigid. Although phenolic compounds linked to two other cuticle components, have not been isolated, the abundance of histidine–catechol adducts renders their presence as cross-links in native cuticle highly probable.

The second hypothesis, set forth by Vincent and Hillerton (1979) argues that the increase in robustness is not due to chemical cross-linking itself, but the dehydrating effect, and resultant hydrogen bonding, produced by infiltrating the protein–chitin composite with catecholamines. Dehydration could work by a number of mechanisms including the occupation of hydration sites, the creation of a hydrophobic, phenolic network in which cuticular proteins are trapped, or increasing the hydrophobicity of the proteins themselves (Vincent and

Ablett, 1987; Vincent and Hillerton, 1979). As with cross-linking, there is also much evidence for the contribution of dehydration in the hardening of insect cuticle. Studies as early as 1940 by Fraenkel and Rudall show the innate loss of water and consequential strengthening of the cuticle, while more recent studies by Vincent and Andersen show that forced dehydration increases strength, and disruption of secondary structure by submersion in formic acid returns the cuticle to a rubbery state (Andersen, 1964; Fraenkel and Rudall, 1940; Vincent, 1980). These discoveries have lead researchers to implicate the role of protein secondary structure in sclerotization through hydrogen-bond stabilized β-sheet interactions (Hackman and Goldberg, 1979; Iconomidou *et al.*, 2005). The strength of protein–protein and chitin–protein interaction is based on the density of hydrogen bonds that are formed in the absence of water. Indeed, hydrogen bonding is a major contributor in the properties of synthetic polymers as well, including the familiar silk-like Nylon-66 which hydrogen bonds in a similar fashion to β-sheets, carbonyl oxygen to amide hydrogen. A speculative analysis of chitin–protein interfacial energy by Vincent and Wegst suggests that a lower bound might be approximated by assuming that the shear strength of a single H-bond is of the order of 30 pN, and calculating that the area around each bond is about 10^{-18} m^2, results in a shear strength of ~30 MPa (Vincent and Wegst, 2004).

More specifically, a theme identified among a wide range of cuticle proteins is the presence of the so-called R&R consensus sequence (Hamodrakas *et al.*, 2002; Rebers and Willis, 2001). This consensus, which is predicted to form β-pleated sheets, shows strong evidences to be involved in protein/chitin interactions. Based on the geometric spacing of the amide nitrogens on α-chitin (1.032 nm) and carbonyl oxygens of an anti-parallel β-sheet protein backbone (0.69 nm), it is purported that these components hydrogen bond at 2.064 nm intervals (every two repeats in α-chitin and three repeats in protein β-sheet). This serves to reduce the water content of the cuticle, increasing Young's modulus (Vincent and Wegst, 2004). Furthermore, His residues appear to be ideally located in the R&R consensus to form cross-linking reactions (Iconomidou *et al.*, 2005). Given the high density of His–catechol cross-link interactions in *Dosidicus* beaks, pursuing whether the R&R consensus also exists in *Dosidicus* (and cephalopods in general) would be of significant interest. Lastly, a proof of concept by Nitta *et al.* (2004) shows that the simple addition of a polyphenolic substrate (epigallocatechin gallate) to a solution of xyloglucan caused gelation through the controlled dehydration of the polysaccharides, an *in vitro* suggestion that polyphenols are able to stabilize polysaccharides by water displacement. For a comprehensive review of known insect cuticular proteins and their conserved sequences, see the work of Andersen *et al.* (1996).

As untangling the sclerotization puzzle through biochemical analysis has proven difficult, some researchers have chosen to study sclerotized systems through genetic analysis. Work on *Drosophila melanogaster* and *Tribolium castaneum* presented by Arakane et al. (2005, 2009) and Wright (1987) show that, because sclerotin precursors are created through many enzymatic steps from tyrosine, observing the effects of enzyme inactivation can provide insight to the importance of a given sclerotin. For example, Arakane et al. recently used double stranded RNA to downregulate the function of aspartate 1-decarboxylase (ADC) and 3,4-dihidroxyphenylalanine decarboxylase (DDC) to assay the effect on cuticle physical properties of knocking out NBAD (*n*-β-alanyldopamine), a major catechol used in cuticle cross-linking. While inhibiting DDC was lethal, inhibition of ADC resulted in excessive pigmentation and a weakly cross-linked cuticle as evidenced by an increased viscous behaviour in sinusoidally stressed elytra, further supporting the importance of covalent cross-linking in bulk cuticle properties, as well as the discrepancy between polyphenolic content and sclerotization (Arakane *et al.*, 2009).

The question is not whether the diphenols are oxidized and polymerize within the cuticle to cause pigmentation and hardening, but whether the result, hardening, is based on chemical cross-linking, or the dehydration of the other components, namely chitin. As both are undoubtedly taking place, deconvoluting the role of each is not a simple task. Suffice it to say that, in arthropod cuticle, both are relevant, while the effects of hydration and protein secondary structure appear dominant. In either case, the finished product is undeniably robust, lightweight and diverse with a density of approximately 1.3 g/cm^3 and stiffness ranging from approximately 1 MPa for resilin, 1 kPa 50 MPa in soft cuticles and 1–20 GPa in hard cuticles (Vincent and Wegst, 2004).

6.2 CASE STUDIES

Aedes aegypti (mosquito) chorion and *Atta sexdens* (leaf-cutter ant) mandible, and *Schistocerca* (locust) wing ligaments were selected for brief review to highlight aspects of each sclerotized structure. Mosquito chorion, a member of the monophenolic sclerotization family, is entirely proteinaceous and is stabilized and sclerotized by di- and tri-tyrosine cross-links created by chorion peroxidase (CPO) as well as phenoloxidase (Li and Li, 2006). Leaf-cutter ant mandibles, on the other hand, are composed primarily of chitin and protein, but are fortified with zinc (Zn) and manganese (Mn), which is incorporated post-ecdysis in mature ants (Schofield *et al.*, 2003). Lastly, locust wing ligaments highlight the importance of protein secondary structure (or lack thereof), in elastomeric tissues. While these strategies are not exclusive to the given organism, di- and tri-tyrosine cross-links were described above in *Nereis* jaws (Birkedal *et al.*, 2006), zinc incorporation has been noted in a number of species, maximized in the scorpion stingers at 25% of dry weight (Schofield, 2001), and elastomeric connective tissues are found in mammalian vasculature as well as in scallop ligaments (Alexander, 1966; Hoeve and Flory, 1958). The three examples are significant adaptations to the general arthropod sclerotization scheme outlined in Section 6.1.

6.2.1 Aedes aegypti *chorion*

The chorion of mosquito eggs is an entirely proteinaceous structure, produced by follicle cells (Raikhel and Lea, 1991). The shell itself is composed of two layers, the endochorion and exochorion, which are distinctly different in topology (smooth and electron dense and fibrillar/lamellar, by electron microscopy, respectively) (Monnerat *et al.*, 1999; Soumar and Ndiaye, 2005; Valle *et al.*, 1999), suggesting different protein composition. In recent work by Li and Li (2006), three proteins were characterized from the endochorion layer, 15a1, 15a2 and 15a3.

Post-oviposition, the chorion of mosquitoes resembles a round white mass that darkens over time (Chapman, 1998). At this stage, it is susceptible to dehydration as well as denaturants, suggesting that there is little-to-no covalent cross-linking present (Li and Li, 2006). Between the three proteins, there is one highly conserved region, [K-C-G-A-N-L-L-V-G-C-A-P-S-V-A-H-V-P-C-V-P], containing three cysteine residues. It is purported that these cysteine residues are responsible for the initial pre-tanning structural integrity of the chorion (albeit weak). The pigment and strength that develops over time is due to the action of oxidases—chorion peroxidase (CPO) and phenoloxidase—responsible for the creation di- and tri-tyrosine cross-links and melanization (Han, *et al.*, 2000; Li *et. al.*, 1996). If one recalls the discussion of mussel byssal thread core in which metal binding and covalent cross-linking are implicated as a fast and slow mechanism for sclerotization, one sees a similar adaptation here. While the disulfide bonds can be formed very quickly upon switching to an oxidizing environment, the covalent tyrosine cross-links, which bear the end result of rendering the chorion robust and resistant to desiccation, are 'bought' the time to form. To our knowledge, no physical properties have been determined for mosquito chorion proteins. One hypothesis is that, like resilin (discussed in Section 6.2.3), which is also stabilized by tyrosine cross-links, the structures will have partial elastomeric behaviour. However, for fear of stretching and damaging the fragile embryo, it is likely to be significantly more stiff, which, in combination with a recoverable yield, will generate a large hysteresis, rendering the chorion a shock-absorbing shell adapted to take numerous low-mid-energy impacts over time, similar to the egg-capsules of the marine whelk, *Busycon* (Rapoport and Shadwick, 2002). Stiffness will ultimately depend not on cross-link type, but rather cross-link density.

6.2.2 Atta sexdens *mandible*

Leaf-cutter mandibles, conversely, are not solely proteinaceous. They are assumed to contain the normal components of insect cuticle—protein, chitin and *N*-acetylcatecholamines—and have, in addition, a significant fraction of Zn (Schofield, 2001). As one might expect, this system, partially reminiscent of *Nereis* jaw, is adapted for chewing. Rather than act as a shock-absorber with a recoverable yield, leaf-cutter mandibles are hard and, most importantly wear resistant. While direct measurements of wear-resistance have not been made, it is known that the hardness of *Atta* mandible is directly correlated to the presence of Zn (of which Zn^{2+} is the most stable form). Indeed, hardness, measured by nanoindentation, increases nearly three-fold during zinc accumulation in the adult animal (Schofield *et al.*, 2002). Furthermore, there is (perhaps unexpectedly) a temporal separation between colouration and hardening. While the mandibles are fully grown, they are not fully hardened for 3–4 weeks (Schofield *et al.*, 2002). It is worth noting that this is not a unique adaptation, Zn has also been detected in a wide variety of arthropods, spanning many orders, with an apparent correlation to herbivory (Hillerton and Vincent, 1982). How the Zn is transported and utilized upon arrival remain important unanswered questions.

The correlation of Zn, Cl, Ca and Mn was also explored in *Atta*. It appears that Zn and Cl were strongly correlated, while Mn had an inverse spatial distribution to the others (Schofield *et al.*, 2003). In light of the results by Broomell *et al.* (2008) in which *Nereis* jaws were depleted and reinfused with Zn, Cu and Mn, it is perplexing as to why Mn and Zn would be oppositely correlated. Furthermore, with such a high content of Zn, it is unlikely that all Zn is complexed by proteins and Cl, as in *Nereis*, yet it is also unlikely to be present except as nanoparticles as X-ray and electron diffractometry have detected no crystalline Zn-rich regions (Schofield *et al.*, 2002). While it is clear that Zn concentration is a necessary adaptation to harden *Atta* mandibles for natural function, many questions remain to be answered regarding the mechanism by which this additional hardening is achieved.

6.2.3 Schistocerca gregaria *wing-hinge ligaments*

Our final case study centres on a rubbery protein present throughout the cuticles of many arthropods (Andersen and Weis-Fogh, 1964). First characterized by Weis-Fogh in the early 1960s, resilin is a nearly perfect elastomer that is capable of stretching to over 100% strain and is evolutionarily optimized so as to not dissipate energy in the rapid wing oscillations ($\sim 20\ s^{-1}$ in *Schistocerca*) necessary for flight (Weis-Fogh, 1960). As the wings are forced to an initial extreme position, the majority of the energy is recovered as the wing transitions to the opposite position in a rubber band-like behaviour that leaves no lasting deformation (Weis-Fogh, 1960). Note that wing-hinge ligaments also contain lamellar chitin at a dry weight of $\sim 2\%$. Incubation of rubbery cuticle with proteases or in hydrolytic conditions completely destroys the mechanical properties of the tissue, providing evidence that the protein is the necessary component for function (Gosline 1978; Weis-Fogh, 1960).

Rubbers consist of three-dimensional networks of molecular chains that are kinked, but not tangled, and held together by scattered intermolecular cross--links (Treolar, 1958). As it is strained, entropy decreases as order is imposed on the system, irreversibly breaking the material if cross-links are compromised.

Resilin itself is rich in glycine (35–40%), proline (7–10%), lacking in cysteine, methionine and tryptophan and forms a series of β-turns, creating an extended β-spiral structure stabilized by di- and tri-tyrosine cross-links (Andersen, 1964; Ardell and Andersen, 2001; Bailey and Weis-Fogh, 1961). A pro-resilin gene sequence, characterized by Ardell and Andersen (2001), suggests that resilin is a large, secreted protein with many repeats of a Gly-, Pro-, Ser- and Tyr-based sequence (Ardell and Andersen, 2001). While the combination of gly and pro maintains the extended conformation, the roughly equidistant placement of tyrosines, every 14–16 residues in repeat domains, create an equal cross-link density throughout the material. As resilin is stretched, the loosely coiled secondary structure is elongated and the material becomes birefringent (Weis-Fogh, 1960). Just as in rubber, deformation is resisted by entropic forces not favourable secondary-structure interactions.

More recently, Andersen described the differential cross-link density of resilin taken from several body parts of *S. gregaria*. While hinge ligaments are nearly entirely resilin, trace amounts of di- and tri-tyrosine cross-links, indicative of resilin, were found throughout cuticles but are likely not intimately related to overall function (Andersen, 2004). Between the extremes, one finds many resilin-like areas that are considered 'transitional' between elastic and hard cuticles (Andersen and Weis-Fogh, 1964). This suggests that either the proteins present in cuticle are modified forms of resilin, or the cuticle is a graded blend of resilin and non-resilin proteins, the ratio of which switches according to cuticle structure (Andersen, 2004). As wing-hinge ligaments are nearly entirely proteinaceous and contain di- and tri-tyrosine linked proteins, they are similar to *Aedes* chorion. However, the elastic properties of resilin, which stem from the general lack of intermolecular interactions—excepting covalent cross-links—cause it to have little-to-no hysteresis. While it is hypothesized herein that the chorion is visco-elastic designed to take many impacts over time, dissipating the energy without harming the fragile embryo within, resilin is designed to store and release energy elastically, minimizing any excess exertion by the locust.

6.3 A COMPARISON OF ARTHROPOD, MOLLUSCAN AND ANNELID SCLEROTIZATION

To compare and contrast the abovementioned information on both arthropod and marine invertebrate sclerotization, several organisms, including but not limited to those previously described, are listed in Table 1. Each organism is listed alongside the selected material of study, organic precursors, inorganic precursors, cross-links and enzymes necessary for maturation. The insect species chosen for comparison are leaf-cutter ant (*Atta sexdens*) mandibles, mosquito (*Aedes aegypti*) egg capsules and locust wing-hinge ligaments (*S. gregaria*); *Atta* mandibles to expand the comparison in biting functionalities between evolutionary dissimilar organisms, *Aedes* chorion to highlight pure protein sclerotization stabilized by monophenolic cross-linking, and *Schistocerca* hinge ligaments to convey necessary protein features for biopolymer elasticity.

TABLE 1

A comparison of arthropod, molluscan and polychaete biocomposites with respect to organic components, inorganic components, stabilization themes and necessary enzymes

Organism	Structure	Organic components	Inorganic components	Stabilization themes	Necessary enzymes
Mytilus edulis	Byssal thread	Dopa-containing protein	Metal-ions (Fe^{3+})	Dopa-based covalent cross-linking Metal chelation Domains stiffness	Byssal catecholoxidase
Glycera dibranchiata	Jaw	Eumelanin Gly + His-rich proteins	Metal-mineralized (atacamite) Metal-ions (Cu^{2+})	Mineralization Metal chelation Melanin-directed covalent cross-linking	Peroxidase
Nereis virens	Jaw	Gly + His-rich protein	Metal-ions (Zn^{2+}) Halogens—iodine, bromine, chlorine	Imidazole–Zn chelate complexes Dityrosine and halo-dityrosine cross-links	Peroxidase Haloperoxidase
Dosidicus gigas	Beak	Chitin Gly + His-rich Dopa-containing protein 4-Methyl catechol (4MC)	None	Dopa–His cross-links 4MC-based covalent cross-linking	Phenoloxidase
Aedes aegypti	Eggshell	Protein	None	Di-/tri-tyrosine cross-links Disulfides	Chorion peroxidase Phenoloxidase
Atta sexdens	Mandible	Protein Chitin catecholamines	Metal mineralization—zinc Metal-ions (Zn^{2+}) Halogens—chlorine	Unknown metal interactions Assumed covalent cross-links	Phenoloxidase
Schistocerca gregaria	Wing-hinge ligament	Protein Chitin	None	Di-/tri-tyrosine cross-links	Peroxidase

In comparing the organisms, structures, components and enzymes, one sees that stabilization can occur through several distinct mechanisms: (1) covalent cross-linking by mono- and diphenols, (2) incorporation of metal ions in chelate complexes, (3) incorporation of mineralized metals, and (4) domain stiffness of proteins. Omitted is the effect of dehydration which is inherent in all condensed systems that have been discussed: Covalent cross-linking and metal chelation may directly block hydration sites; mineralization necessarily excludes most water from the mineralized portion of the tissue; domain stiffness, referring to silk-like domains in mussel preCOLs as well as cuticular proteins, are both stabilized by hydrogen-bonded β-sheet structures. Of course, in each case the removal of water is coupled with the generation of a new molecular interaction, resuming the aforementioned discussion of stabilization versus dehydration. Moreover, from the small subset of examples presented herein, one sees that sclerotization can occur with and without inorganic components—transition metals, minerals and halogens—with and without polysaccharides—α-chitin, or with and without protein secondary structure, while in all cases, the microstructure of the material is honed for its eventual function. Distilling the topic to reveal a single sclerotization strategy instead exposes evolution's many ingenious designs. What remains consistent is the manifold use of and versatility of aromatic chemistry.

Covalent cross-linking is present in all materials mentioned above, though the precise mechanism and structure of cross-links varies. These interactions are made possible by oxidizing enzymes, which catalyze the dehydrogenation of diphenols to quinones, a necessary step in the covalent cross-links described throughout this review (Locke, 1969). Types of phenoloxidases include tyrosinases (EC 1.14.18.1)—able to catalyze the oxidation of both mono- and diphenolic substrates via two-electron transfer, laccases (EC 1.10.3.2) and catecholoxidases (EC 1.10.3.1)—adapted to catalyze diphenols exclusively, and peroxidases (EC 1.11.1.7)—able to catalyze the catechol to quinone transformation via one-electron transfer (Hopkins and Kramer, 1992). While there is overlap in function, these enzymes may be spatially and temporally segregated, thereby reserving each to a 'main' function: tyrosinases in wound healing, laccases in tanning, and peroxidases in tyrosine cross-linking (Hopkins and Kramer, 1992). For the purposes of this review, the generalized term 'phenoloxidase' is used in Table 1 in cases where no further information is available and the presence of certain cross-links in the mature structure presupposes the presence of an oxidase. *Note*: haloperoxidases share single electron oxidation with peroxidases, but are responsible for the addition of halogens (in the case of *Nereis*, addition of halogens to tyrosine and histidine).

The phenolic biases, mono- and di-, are exemplified by a comparison of *Aedes* chorion, *Nereis* jaw and *Schistocerca* hinge ligaments to *Dosidicus* beak and *Mytilus* thread. Furthermore, considering chorion, hinge ligaments and jaw, it is evident that tyrosine cross-links can be present in both 'rubbery' and hard systems. Similarly, a comparison of beak and thread suggests that diphenolic sclerotization can be used in stretchy (thread) and hard (beak) structures. Again, the cross-link density is the true determinant of the mechanical properties.

A second bias in the use of phenolic molecules is whether they are attached to protein (as in Dopa-containing proteins) or free to diffuse. Currently, it appears that freely diffusible catechols (*N*-acetylcatechols) are present in terrestrial arthropod hard cuticles, but not necessarily in elastomeric arthropod structures or in the comparatively simpler marine invertebrate species. Indeed, in *Mytilus* and *Nereis* the cross-link precursor (tyrosine and/or DOPA) is bound to protein. A possible explanation for the use of peptide-bound catechols in marine species is that free catechols would be too readily lost by diffusion into the water medium. Surely, in the case of *Mytilus* where the byssal thread is injection-moulded, free catechol would diffuse from the coating while sclerotization occurred. Perhaps unfortunately, the use of peptide-bound catechols resolves the structure to a fixed ratio of catechol–protein/polysaccharide, whereas in arthropods, *N*-acetylcatechols can be consistently pumped into the cuticle, dehydrating and repairing the structure (Lai-Fook, 1966). Note that *Mytillus* uses gradients of Dopa in byssal proteins when necessary, so it is not correct to equate Dopa–proteins with a lack of versatility (Sun and Waite, 2005). However, it is our belief that this ability is a major evolutionary impetus for the shift to diffusible catechols in terrestrial arthropods. However, abundance of 4MC and lack of extractable Dopa-containing proteins in *Dosidicus* beak, the hardest wholly organic material discussed herein, leads one to believe that the squid *is* using diffusible catechols (Miserez *et al.*, unpublished observation). This would provide the first example of diffusible catechol used to sclerotized a marine invertebrate tissue and would confer another parallel between cuticle and beak.

Metal–ligand complexation has been characterized in byssal threads and *Nereis* and *Glycera* jaws and is implicated in the mandibles of *Atta*. Broomell *et al.* (2008) showed for the first time the correlation in hardness of an organic structure due solely to depletion and incorporation of metal ions whereas Srivastava *et al.* (2008) demonstrated the same phenomenon in metal infused imidazole-rich hydrogels. From a biomimetic perspective, this is a particularly intriguing finding in that it can be incorporated comparatively simply. Poly-vinyl imidazole can be readily synthesized and incorporated to a porous structure of choice, after which metal ions can be infused. While these three structures are adapted similarly for biting and chewing, metal chelation is also present in byssal threads utilizing the recoverable nature of chelation complexes. This is a second case in which a sclerotization strategy is employed for two purposes, increasing hardness and increasing toughness. Moreover, a last lever for next-generation material synthesis is the choice of metal incorporated, as Broomell *et al.* showed the variable potential of copper, zinc and manganese to re-establish the hardness of *Nereis* jaw.

Mineralization, not discussed here at length, is a common strategy for creating hard biomaterials (Brooker *et al.*, 2002; Weiner and Zaslansky, 2004). In aquatic organisms, consider the carapaces of crustaceans, which are composed of layered assemblies of α-chitin and protein that initiate the formation of both crystalline and amorphous $CaCO_3$ (calcite) (Raabe *et al.*, 2005; Roer and Dillaman, 1984). In terrestrial organisms, consider that human enamel is 95% mineralized (hydroxyapatite) and is the most wear-resistant biomaterial known to date (Lichtenegger *et al.*, 2002). Although these systems are well studied, the body of knowledge on non-mineralized hard tissues used for feeding is less developed, but has grown significantly in the last decade. Of course, 'non-mineralized' tissues are not necessarily wholly devoid of mineral, there is a gradient of relative mineralization (Broomell *et al.*, 2007). In the case of *Glycera* jaw, both copper ions (Cu^{2+}) and mineralized copper in the form of atacamite are used to fortify the jaw representing up to 10% of the dry weight. Intriguingly, the most mineralized portion of the jaw is not located on the outside as is the case with enamel, but as a set of fortifying rods that run lengthwise under the surface of the jaw. These reinforcements are implicated in providing bending strength to resist the pressure exerted by feeding and burrowing (Lichtenegger *et al.*, 2002). *Atta* mandible contains a similar quantity of copper, 16% dry weight, but has not been detected as mineral (Schofield, 2001).

Lastly, creating proteins with differing domain stiffness, or no domain stiffness at all in the case of resilin, can be utilized as a materials engineering strategy. Perhaps, silks are the archetypal example of β-sheet stacking and its resultant enhancement of toughness, but the discussion here revolved more closely around the preCOL D, preCOL NG and preCOL P of the *Mytilus* thread core and the associated gradient of mechanical properties (Waite *et al.*, 2004). As is readily attempted in block co-polymer studies, creating fibres of differential stiffness based on thermodynamic stability of secondary structure properties is both feasible and environmentally benevolent as it would create easily degradable fibres. In this arena, the roadblock appears to be in physical processing of the protein fibres (Weiguo *et al.*, 2009). Drawing rates and buffer conditions are particularly important in the creation of robust fibres. This stands as added evidence that the structure of matrix proteins in insect cuticle strongly influences the properties of the mature material. If instead one is interested in creating a bioelastomer, one can look to resilin and purposefully engineer a biomaterial with little-to-no secondary structure and instead control stiffness and extensibility strictly through cross-link density.

7 Conclusions

This review has aimed to increase awareness of sclerotization strategies in non-mineralized tissues through a comparison of the biochemical components and microstructures of marine invertebrate and arthropod hard tissues. Highlighted above are marine invertebrate load-bearing structures that are wholly organic (*D. gigas* beak), metal-chelating (*M. edulis* threads and *N. virens* jaws) and partially mineralized (*G. dibranchiata*). From the study of these organisms, several significant insights regarding the effects of biochemical composition and structure on material properties, namely hardness and stiffness, were made and, when relevant, applied to known data on sclerotized arthropod structures.

(1) As found in the mussel byssal thread, the stiffness of a biomaterial can be significantly varied by the organization and density of rigid and flexible domains in the fibrous core. The gradation and orientation of preCOLs is directly responsible for the half stiff, half stretchy behaviour of the byssal thread. Conversely, it is the absence of defined secondary structure that is responsible for the elastomeric behaviour of arthropod resilin. (2) High catechol concentrations do not necessarily result in increased stiffness. In the byssal plaque of *M. edulis*, the catechol concentration increases as the mussel byssal plaque approaches the substratum, but the plaque and thread are not rendered any harder or stiffer. Instead, the catecholic gradient is part of an adhesive strategy. By an unknown mechanism, the DOPA residues present on mfp-3 and mfp-5 are protected from oxidation and thus are preserved for surface interaction at the attachment site. (3) Protein-based ligands, such as catecholate and imidazole, are used as cross-link surrogates to induce sclerotization by multiply binding metals. In polychaete jaws and mussel thread coating and probably as well in insect mandible and scorpion stinger (Schofield, 2001), metal ion–ligand complexes could be used in the whole or in discrete layers of the structure to increase local hardness and/or wear resistance. Histidine (imidazole-based pK_a = ~6.5) metal-binding sequences are surprisingly widespread which is likely related to the effect of physiological pH on the charge of the side-chain. (4) The covalent association of proteins and catechols can be incrementally tuned to differentially dehydrate chitin resulting in a manifold increase in hardness and stiffness. From work on *D. gigas* beak, it is evident that the mixture of proteins and catechols infused into chitin leaflets and cross-linked, enables the adjustment of stiffness over a range of 2 orders of magnitude. As this system most closely mimics insect cuticle, it is our hope that further progress in cross-link characterization and synchrotron X-ray studies will shed new light on the quinone-tanning/dehydration sclerotization debate. (5) Lastly, each of these systems displays the use of gradients. These may be in protein domains, ligand and metal densities, or desolvation strategies, and are all carefully manipulated to create stiffness gradients as a means of mitigating differences in modulus between mechanically mismatched tissues or surfaces. Lastly, these studies also reiterate the importance of studying material properties at an appropriate size scale. In the case of *Glycera* jaws, the wear-resistant outer layer is as small as 3 μm in some areas. Bulk testing would not have been able to provide information on the properties of different layers.

Although there is sometimes correlation between catechol content and stiffness, it is simplistic to assume a causal relationship. The true value of the insights learned from these model organisms is in the diversity of mechanisms for tissue hardening. Indeed, the mechanisms and tissues are not equal—each is tuned to optimize different functions. Each structure utilizes a set of components to finely tune the macroscopic properties of that material. For creating hard tissues, Nature's toolbox includes the incorporation of covalent cross-linking, addition of metal and ligand pairs, manipulation of protein domain stiffness, controlled desolvation and the possibility of many strategies yet to be uncovered. With such diversity, it would be edifying to better understand why evolution is driven towards one sclerotization strategy rather than another. One hypothesis is that the reliance on metal ions is resource-dependent. Mussels and polychaetes have ample access to mineral particulates from which to obtain their metal ions (Davies and Simkiss, 1996). Squids in contrast are mesobathic hunters where metal ions and mineral particulates are scarcer.

Acknowledgements

We thank the editors, Dr. Jerome Casas and Dr. Stephen J. Simpson for the invitation to contribute to this series. Grants from NIH and NSF provided critical funding for this research. The writing of this overview was divided into three manageable details with Rubin responsible for Sections 6 and 7 and the collation of the document, Miserez responsible for Sections 3–5 and Waite writing the introduction and Section 2.

References

Alexander, R. McN (1966). Rubber-like properties of the inner hinge-ligament of pectinidae. *J. Exp. Biol.* **44**, 119–130.

Andersen, S. O. (1964). The crosslinks in resilin identified as dityrosine and trityrosine. *Biochim. Biophys. Acta* **93**, 213–215.

Andersen, S. O. (2003). Structure and function of resilin. In: *Elastomeric Proteins* (eds Shewry, P. R., Tatham, A. S. and Bailey, A. J.), pp. 1–23. Cambridge University Press, Cambridge, UK.

Andersen, S. O. (2004). Regional differences in degree of resilin cross-linking in the desert locust, *Schistocera gregaria. Insect Biochem. Mol. Bio.* **34**, 459–466.

Andersen, S. O. (2005). Cuticular sclerotization and tanning. In: *Comprehensive Molecular Insect Science* (eds Gilbert, L. I., Iatrou, K. and Gill, S.)Vol. 4, pp. 145–170. Elsevier, Oxford, UK.

Andersen, S. O. (2010). Insect cuticular sclerotization: a review. *Insect. Biochem. Mol. Biol.* **40,** 166–178.

Andersen, S. O. and Weis-Fogh, T. (1964). Resilin, a rubber like protein in arthropod cuticle. *Adv. Insect Physiol.* **2**, 1–65.

Andersen, S. O., Peter, M. G. and Roepstorff, P. (1996). Cuticular sclerotization in insects. *Comp. Biochem. Physiol. B* **113**, 689–705.

Arakane, Y., Muthukrishnan, S., Beeman, R. W., Kanost, M. R. and Kramer, K. J. (2005). Laccase 2 is the phenoloxidase gene required for beetle cuticle tanning. *Proc. Natl. Acad. Sci. USA* **102**, 11337–11342.

Arakane, Y., Lomakin, K., Beeman, R. W., Muthukrishnan, S., Gerhke, S. H., Kanost, M. R. and Kramer, K. J. (2009). Molecular and functional analyses of amino acid decarboxylases involved in cuticle tanning in *Tribolium castaneum. J. Biol. Chem.* **284**, 16584–16594.

Ardell, D. H. and Andersen, S. O. (2001). Tentative identification of a resilin gene in *Drosophila melanogaster*. *Insect Biochem. Mol. Biol.* **31**, 965–970.
Arnow, L. (1937). The formation of dopa by the exposure of tyrosine solutions to ultraviolet radiation. *J. Biol. Chem.* **118**, 531–537.
Atkins, E. D. T. (1985). Conformations of polysaccharides and complex carbohydrates. *Proc. Int. Symp. Biom. Struct. Int.* **8**, 375–387. (Supplement to the *Journal of Bioscience*).
Bailey, K. and Weis-Fogh, T. (1961). Amino acid composition of a new rubber-like protein, resilin. *Biochim. Biophys. Acta* **48**, 452–459.
Belli, S. I., Wallach, M. G., Luxford, C., Davies, M. J. and Smith, N. C. (2003). Roles of tyrosine-rich precursor glycoproteins and dityrosine- and 3,4-dihydroxyphenylalanine-mediated protein cross-linking in development of the oocyst wall in the coccidian parasite *Eimeria maxima*. *Eukaryot. Cell* **2**, 456–464.
Benjamin, M., Tuomi, H., Ralphs, J. R., Bydder, G., Best, T. M. and Milz, S. (2006). Where tendons and ligaments meet bone: attachment sites ('entheses') in relation to exercise and/or mechanical load. *J. Anat.* **208**, 471.
Birkedal, H., Khan, R. K., Slack, N., Broomell, C., Lichtenegger, H. C., Zok, F. W., Stucky, G. D. and Waite, J. H. (2006). Halogenated veneers: protein cross-linking and halogenation in the jaws of *Nereis*, a marine polychaete worm. *ChemBioChem* **7**, 1392–1399.
Borges, C. R., Roberts, J. C., Wilkins, D. G. and Rollins, D. E. (2001). Relationship of melanin degradation products to actual melanin content: application to human hair. *Anal. Biochem.* **290**, 116–125.
Briza, P., Ellinger, A., Winkler, G. and Breitenbach, M. (1990). Characterization of a DL-dityrosine-containing macromolecule from yeast ascospore walls. *J. Biol. Chem.* **265**, 15118–15123.
Brooker, L. R., Lee, A. P., Macey, D. J. and Webb, J. (2002). Molluscan and other marine teeth. In: *Encyclopedia of Materials Science and Technology* (ed Calvert, P.), pp. 5186–5189. Elsevier, Amsterdam.
Broomell, C., Mattoni, M., Zok, F. W. and Waite, J. H. (2006). Critical role of zinc in hardening of *Nereis* jaws. *J. Exp. Biol.* **209**, 3219–3225.
Broomell, C. C., Khan, R. K., Moses, D. N., Miserez, A., Pontin, M. G., Stucky, G. D., Zok, F. W. and Waite, J. H. (2007). Mineral minimization in nature's alternative teeth. *J. R. Soc. Interface* **4**, 19–31.
Broomell, C. C., Zok, F. W. and Waite, J. H. (2008). Role of transition metals in sclerotization of biological tissue. *Acta Biomat.* **4**, 2045–2051.
Brown, K. C. and Kodadek, T. (2001). Protein crosslinking mediated by metal ion complexes. *Metal Ions Biol. Syst.* **38**, 352–384.
Burillo, G., Clough, R. L., Czvikovszky, T., Guvend, O., Le Moel, A., Liu, W., Singh, A., Yang, J. and Zaharescu, T. (2002). Polymer recycling: potential application of radiation technology. *Radiation Physics and Chemistry* **64**, 41–51.
Carrington, E. and Gosline, J. M. (2004). Mechanical design of mussel byssus: load cycle and strain rate dependence. *Am. Malacol. Bull.* **18**, 135–142.
Chapman, R. F. (1998). *The Insects: Structure and Function*. 4th ed. Cambridge University Press, Cambridge, UK, pp. 295–342.
Cheng, Y. T. and Cheng, C. M. (2004). Scaling, Dimensional Analysis and Indentation Measurements. *Mater. Sci. Eng.* **44**, 91–149.
Clarke, M. R. (1986). A Handbook for the Identification of Cephalopod Beaks. Clarendon Press, Oxford, UK.
Creighton, T. E. (1989). Proteins—Structures and Molecular Properties. Freeman and Company, New York.

Davies, N. A. and Simkiss, K. (1996). The uptake of zinc from artificial sediments by *Mytilus edulis*. *J. Mar. Biol. Assoc., UK* **76**, 1073–1079.
Deacon, M. P., Davis, S. S., Waite, J. H. and Harding, S. E. (1998). Structure and mucoadhesion properties of mussel glue protein in dilute solution. *Biochemistry* **37**, 14108–14112.
Dilly, P. N. and Nixon, M. (1976). The cells that secret the beaks in octopods and squids (Mollusca, Cephalopoda). *Cell Tissue Res.* **167**, 229–241.
Filpula, D. R., Lee, S. M., Link, R. P., Strausberg, S. and Strausberg, R. L. (1990). Structural and functional repetition in a marine mussel adhesive protein. *Biotech. Prog.* **6**, 171–177.
Fraenkel, G. and Rudall, K. M. (1940). A study of the physical and chemical properties of the insect cuticle. *Proc. R. Soc. Lond. Ser. B* **129**, 1–35.
Froncisz, W., Sarna, T. and Hyde, J. S. (1980). Cu^{2+} probe of metal–ion binding sites in melanin using electron paramagnetic resonance spectroscopy. I. Synthetic melanins. *Arch. Biochem. Biophys.* **202**, 289–303.
Gibbs, P. E. and Bryan, G. W. (1980). Metals in nereid polychaetes—The contribution of metals in the jaws to the total-body burden. *J. Mar. Biol. Assoc., UK* **60**, 205.
Goodarzi, M. T. and Turner, G. A. (1996). A lectin-binding assay for the rapid characterization of the glycosylation of purified glycoproteins. In: *The Protein Protocols Handbook* (ed Walker, J. M.), pp. 619–625. Humana Press.
Gortner, R. (1911). Studies on melanin II: the pigmentation of the adult periodical cicada (*Tibicen septendecim* L.). *J. Biol. Chem.* **10**, 89–95.
Gosline, J. M. (1978). Hydrophobic interaction and a model for the elasticity of elastin. *Biopolymers* **17**, 677–695.
Gouldstone, A., Chollacoop, N., Dao, M., Li, Ju., Minor, A. M. and Shen, Y. L. (2007). Indentation across size scales and disciplines: recent developments in experimentation and modeling. *Acta Mater.* **55**, 4015–4039.
Hackman, R. H. and Goldberg, M. (1979). Some conformational studies of *Calliphora vicina* larval cuticular protein. *Insect Biochem.* **9**, 557–561.
Haemers, S., van der Leeden, M. C. and Frens, G. (2005). Coil dimensions of the mussel adhesive protein mefp-1. *Biomaterials* **26**, 1231–1236.
Hamodrakas, S. J., Willis, J. H. and Iconomidou, V. A. (2002). A structural model of the chitin-binding domain of cuticle proteins. *Insect Biochem. Mol. Biol.* **32**, 1577–1583.
Han, Q., Li, G. and Li, J. (2000). Purification and characterization of chorion peroxidase from *Aedes aegypti* eggs. *Arch. Biochem. Biophys.* **378**, 107–115.
Harrington, M. J. (2008). Structure–function relationships in the proteinaceous fibers of the mussel byssus. Thesis, University of California Santa Barbara.
Harrington, M. J. and Waite, J. H. (2007). Holdfast heroics: comparing *Mytilus californianus*. *J. Exp. Biol.* **210**, 4307–4318.
Harrington, M. J. and Waite, J. H. (2008a). How nature modulates a fiber's mechanical properties: mechanically distinct fibers drawn from natural mesogenic block copolymer variants. *Adv. Mater.* **20**, 1–5.
Harrington, M. J. and Waite, J. H. (2008b). pH-Dependent locking of giant mesogens in fibers drawn from mussel byssal collagens. *Biomacromolecules* **9**, 1480–1486.
Harrington, M. J., Gupta, H. S., Fratzl, P. and Waite, J. H. (2009). Collagen insulated from tensile damage by domains that unfold reversibly: in situ X-ray investigation of mechanical yield and damage repair in the mussel byssus. *J. Struct. Biol.* **167**, 47–54.
Harrington, M. J., Masic, A., Holten-Andersen, N., Waite, J. H. and Fratzl, P. (2010). Ironclad fibers: a metal-based biological strategy for hard flexible coatings. *Science* **328**, 216–220.

Hassenkam, T., Gutsmann, T., Hansma, P., Sagert, J. and Waite, J. H. (2004). Giant bent core mesogens in the thread forming process of marine mussels. *Biomacromolecules* **5**, 1351–1355.

Hepburn, H. R. (ed) (1976). In: *The Insect Integument*, Elsevier Scientific Publishing Company, Amsterdam.

Hillerton, J. F. and Vincent, J. F. V. (1982). The specific location of zinc in insect mandibles. *J. Exp. Biol.* **101**, 333–336.

Hoeve, C. A. J. and Flory, P. J. (1958). The elastic properties of elsastin. *J. Am. Chem. Soc.* **80**, 6253–6256.

Holten-Andersen, N., Fantner, G. E., Hohlbauch, S., Waite, J. H. and Zok, F. W. (2007). Protective coatings on extensible biofibers. *Nat. Mater.* **6**, 669–672.

Holten-Andersen, N., Zhao, H. and Waite, J. H. (2009a). Stiff coatings on compliant biofibers: the cuticle of *Mytilus californianus* byssal threads. *Biochemistry* **48**, 2752–2759.

Holten-Andersen, N., Toprak, M., Stucky, G. D., Zok, F. W. and Waite, J. H. (2009b). Metals and the integrity of a biological coating: the cuticle of mussel byssus. *Langmuir* **25**, 3323–3326.

Hopkins, T. L. and Kramer, K. J. (1992). Insect cuticle sclerotization. *Ann. Rev. Entomol.* **37**, 273–302.

Huggins, L. G. and Waite, J. H. (1993). Eggshell formation in *Bdelloura candida*, an ectoparasitic turbellarian of the horseshoe crab *Limulus polyphemus*. *J. Exp. Zool.* **265**, 549–557.

Hutchinson, J. W. and Suo, Z. (1992). Mixed-mode cracking in layered materials. *Adv. Appl. Mech.* **29**, 63–191.

Iconomidou, V. A., Willis, J. H. and Hamodrakas, S. J. (2005). Unique features of the structural model of 'hard'cuticle proteins: implications for chitin–protein interactions and cross-linking in cuticle. *Insect Biochem. Mol. Biol.* **35**, 553–560.

Israelachvili, J. N. (1985). Intermolecular and Surface Forces. Academic Press, London.

Kamat, S., Su, X., Ballirini, R. and Heuer, A. H. (2000). Structural basis for the fracture toughness of the shell of the conch *Strombus gigas*. *Nature* **405**, 1036–1040.

Kear, A. J. (1994). Morphology and function of the mandibular muscles in some coleoid cephalopods. *J. Mar. Biol. Assoc., UK* **74**, 801–822.

Kerwin, J. L., Turecek, F., Xu, R., Kramer, K. J., Hopkins, T. L., Gaitlin, C. L. and Yates, J. R. (1999). Mass spectrometric analysis of catechol–histidine adducts from insect cuticle. *Anal. Biochem.* **268**, 229–237.

Khan, R. K., Stoimenov, P. K., Mates, T. E., Waite, J. H. and Stucky, G. D. (2006). Exploring gradients of halogens and zinc in the surface and subsurface of *Nereis* jaws. *Langmuir* **22**, 8465–8471.

Knight, D. P. (1970). Sclerotization of the calyptoblastic hydroid *Laomedea flexuosa*. 1. Identification and localization of dopamine in the hydroid. *Tissue Cell* **2**, 467–477.

Kramer, K. J., Kanost, M. R., Hopkins, T. L., Jiang, H., Zhu, Y., Xu, R., Kerwin, J. L. and Turecek, F. (2001). Oxidative conjugation of catechols with proteins in insect skeletal systems. *Tetrahedron* **57**, 385–392.

Lai-Fook, J. (1966). The repair of wounds in the integument of insects. *J. Insect Physiol.* **12**, 195–226.

Lee, H., Scherer, N. F. and Messersmith, P. B. (2006). Single-molecule mechanics of mussel adhesion. *Proc. Natl. Acad. Sci. USA* **103**, 12999–13003.

Li, J. S. and Li, J. (2006). Major chorion proteins and their crosslinking during chorion hardening in *Aedes aegypti* mosquitoes. *Insect Biochem. Mol. Biol.* **36**, 954–964.

Li, J., Hodgeman, B. A. and Christensen, B. M. (1996). Involvement of peroxidase in chorion hardening in *Aedes aegypti*. *Insect Biochem. Mol. Biol.* **26**, 309–317.

Lichtenegger, H. C., Schoberl, T., Barti, M. H., Waite, J. H. and Stucky, G. D. (2002). High abrasion resistance with sparse mineralization: copper biomineral in worm jaws. *Science* **298**, 389–392.

Lichtenegger, H. C., Schöberl, T., Ruokolainen, J. T., Cross, J. O., Heald, S. M., Birkedal, H., Waite, J. H. and Stucky, G. D. (2003). Zinc and mechanical prowess in the jaws of *Nereis*, a marine worm. *Proc. Natl. Acad. Sci. USA* **100**, 9144–9149.

Lichtenegger, H. C., Cross, J. O., Heald, S. M., Waite, J. H. and Stucky, G. D. (2005). Distribution and role of trace metals in *Glycera* worm jaws studied with synchrotron microbeam techniques. *Chem. Mater.* **17**, 2927–2931.

Lin, Q., Gourdon, D., Sun, C. J., Holten-Andersen, N., Anderson, T. H., Waite, J. H. and Israelachvili, J. N. (2007). Adhesion mechanisms of the mussel foot proteins mfp-1 and mfp-3. *Proc. Natl. Acad. Sci. USA* **104**, 3782–3786.

Liu, Y. and Simon, J. D. (2005). Metal–ion interactions and the structural organization of *Sepia* eumelanin. *Pigment Cell Res.* **18**, 42–48.

Liu, B., Burdine, L. and Kodadek, T. (2006). Chemistry of periodate-mediated cross-linking of 3,4-dihydroxylphenylalanine-containing molecules to proteins. *J. Am. Chem. Soc.* **128**, 15228–15235.

Locke, M. (1969). The localization of a peroxidase associated with hard cuticle formation in an insect, *Calpodes ethlius*, Stoll, Lepidoptera, Hesperiidae. *Tissue Cell* **1**, 555–574.

Madihally, S. and Matthew, H. (1999). Porous chitosan scaffolds for tissue engineering. *Biomaterials* **20**, 1133–1142.

McDonough, W. and Braungart, M. (2002). Cradle to Cradle: Remaking the Way We Make Things. Northpoint Press, New York.

McDowell, L. M., Burzio, L. A., Waite, J. H. and Schaefer, J. (1999). REDOR Detection of cross-links formed in mussel byssus under high flow stress. *J. Biol. Chem.* **274**, 20293–20295.

Meredith, P. and Sarna, T. (2006). The physical and chemical properties of eumelanin. *Pigment Cell Res.* **19**, 572–594.

Merkel, J. R., Dreisbach, J. H. and Ziegler, H. B. (1975). Collagenolytic activity of some marine bacteria. *Appl. Microbiol.* **29**, 145–151.

Miserez, A., Li, Y., Waite, J. H. and Zok, F. W. (2007). Jumbo squid beaks: inspiration for design of robust organic composites. *Acta Biomater.* **3**, 139–149.

Miserez, A., Schneberk, T., Sun, J. C., Zok, F. W. and Waite, J. H. (2008). The transition from stiff to compliant materials in squid beaks. *Science* **319**, 1816–1819.

Monnerat, A. T., Soares, M. J., Lima, J. B., Rosa-Freitas, M. G. and Valle, D. (1999). *Anopheles albitarsis* eggs: ultrastructural analysis of chorion layers after permeabilization. *J. Insect Physiol.* **45**, 915–922.

Moses, D. N., Mattoni, M., Slack, N. L., Waite, J. H. and Zok, F. (2006). Role of melanin in mechanical properties of *Glycera* jaws. *Acta Biomater.* **2**, 521–530.

Moses, D. N., Pontin, M. G., Waite, J. H. and Zok, F. W. (2008). Effects of hydration on mechanical properties of a highly sclerotized tissue. *Biophys. J.* **94**, 3266–3272.

Murray, S. B. and Neville, A. C. (1998). The role of pH, temperature and nucleation in the formation of cholesteric liquid crystal spherulites from chitin and chitosan. *Int. J. Biol. Macromol.* **22**, 137–144.

Neville, A. C. (1993). Biology of Fibrous Composites. Cambridge University Press, Cambridge, UK.

Nishino, T., Matsui, R. and Nakame, K. (1999). Elastic modulus of the crystalline region of α-chitin and chitosan. *J. Polym. Sci. B* **37**, 1191–1196.

Nitta, Y., Fang, Y., Takemasa, M. and Nishinari, K. (2004). Gelation of xyloglucan by addition of epigallocatechin gallate as studied by rheology and differential scanning calorimetry. *Biomacromolecules* **5**, 1206–1213.

Notin, L., Viton, C., David, L., Alcouffe, P., Rochas, C. and Domard, A. (2006). Morphology and mechanical properties of chitosan fibers obtained by gel-spinning: influence of the dry-jet-stretching step and ageing. *Acta Biomater.* **2**, 387–402.

Pezzella, A., Panzella, L., Natangelo, A., Arzillo, M., Napolitano, A. and d'Ischia, M. (2007). 5,6-Dihydroxyindole tetramers with ''anomalous'' interunit bonding patterns by oxidative coupling of 5,5′,6,6′-Tetrahydroxy-2,7′-biindolyl: emerging complexities on the way toward an improved model of eumelanin buildup. *J. Org. Chem.* **72**, 9225–9230.

Pontin, M. G., Moses, D. N., Waite, J. H. and Zok, F. W. (2007). A nonmineralized approach to abrasion-resistant biomaterials. *Proc. Natl. Acad. Sci. USA* **104**, 13559–13564.

Pryor, M. G. M. (1940). On the hardening of the ootheca of *Blatta orientalis*. *Proc. R. Soc. Lond. Ser. B* **128**, 378–393.

Qin, X. X., Coyne, K. J. and Waite, J. H. (1997). Tough tendons: mussel byssus has collagen with silk-like domains. *J. Biol. Chem.* **272**, 32623–32627.

Raabe, D., Sachs, C. and Romano, P. (2005). The crustacean exoskeleton as an example of a structurally and mechanically graded biological nanocomposite material. *Acta Mater.* **53**, 4281–4292.

Raikhel, A. S. and Lea, A. O. (1991). Control of follicular epithelium development and vitelline envelope formation in the mosquito; Role of juvenile hormone and 20-hydroxyecdysone. *Tissue Cell* **23**, 577–591.

Rapoport, H. S. and Shadwick, R. E. (2002). Mechanical characterization of an unusual elastic biomaterial from the egg capsules of marine snails (*Busycon* spp.). *Biomacromolecules* **3**, 42–50.

Rebers, J. E. and Willis, J. H. (2001). A conserved domain in arthropod cuticular proteins binds chitin. *Insect Biochem. Mol. Biol.* **31**, 1083–1093.

Roer, R. and Dillaman, R. (1984). The structure and calcification of the crustacean cuticle. *Am. Zool.* **24**, 893–909.

Rusaouen, M. (1976). The dogfish shell gland, a histochemical study. *J. Exp. Mar. Biol. Ecol.* **23**, 267–283.

Sagert, J. and Waite, J. H. (2009). Hyperunstable matrix proteins in the byssus of *Mytilus galloprovincialis*. *J. Exp. Biol.* **212**, 2224–2236.

Schaefer, J., Kramer, K. J., Garbow, J. R., Jacob, G. S., Stejskal, E. O., Hopkins, T. L. and Speirs, R. D. (1987). Aromatic cross-links in insect cuticle: detection by solid-state 13C and 15N NMR. *Science* **235**, 1200–1204.

Schmitt, L., Ludwig, M., Gaub, H. E. and Tampe, R. (2002). A metal-chelating microscopy tip as a new toolbox for single-molecule experiments by atomic force microscopy. *Biophys. J.* **78**, 3275–3285.

Schofield, R. M. S. (2001). Metals in cuticular structures. In: *Scorpion Biology and Research* (eds Brownell, P. and Polis, G.), pp. 234–256. Oxford University Press, Oxford, UK.

Schofield, R. M. S., Nesson, M. H. and Richardson, K. A. (2002). Tooth hardness increases with zinc-content in mandibles of young adult leaf-cutter ants. *Naturwissenschaften* **89**, 579–583.

Schofield, R. M. S., Nesson, M. H., Richardson, K. A. and Wyeth, P. (2003). Zinc is incorporated into cuticular ''tools'' after ecdysis: the time course of the zinc distribution in ''tools'' and whole bodies of an ant and a scorpion. *J. Insect Physiol.* **49**, 31–44.

Sever, M. J., Weisser, J. T., Monahan, J., Srinivasan, S. and Wilker, J. J. (2004). Metal mediated cross-linking in the generation of a marine-mussel adhesive. *Angew. Chem.* **43**, 448–450.

Soumar, M. L. and Ndiaye, M. (2005). Ultrastructural studies of mosquito ovogenesis. *Tissue Cell* **37**, 117–124.

Srivastava, A., Holten-Andersen, N., Stucky, G. and Waite, J. H. (2008). Ragworm jaw-inspired metal ion cross-linking for improved mechanical properties of polymer blends. *Biomacromolecules* **9**, 2873–2880.

Stepien, K. B., Dworzanski, J. P., Bilinska, B., Porebska-Budny, M., Holleck, A. M. and Wilczok, T. (1989). Catecholamine melanins: structural changes induced by copper ions. *Biochim. Biophys. Acta* **997**, 49–54.

Suderman, R. J., Dittmer, N. T., Kanost, M. R. and Kramer, K. J. (2006). Model reactions for insect cuticle sclerotization: cross-linking of recombinant cuticular proteins upon their laccase-catalyzed oxidative conjugation with catechols. *Insect Biochem. Mol. Biol.* **36**, 353–365.

Sugumaran, M. (1991). Molecular Mechanisms of Sclerotization. In: *Physiology of Insect Epidermis* (eds Binnington, K. and Retnakaran, K.), CSIRO Publications, Melbourne, Australia.

Sun, C. J. and Waite, J. H. (2005). Mapping chemical gradients within and along a fibrous structural tissue: mussel byssal threads. *J. Biol. Chem.* **280**, 39332–39336.

Suresh, S. (2001). Graded materials for resistance to contact deformation and damage. *Science* **292**, 2447–2451.

Taylor, S. W., Waite, J. H., Ross, M. M., Shabanowitz, J. and Hunt, D. F. (1994a). *Trans*-2,3-*cis*-3,4-dihydroxyproline in the tandemly repeated consensus decapeptides of an adhesive protein from *Mytilus edulis*. *J. Am. Chem. Soc.* **116**, 10803–10804.

Taylor, S. W., Luther, G. W. and Waite, J. H. (1994b). Polarographic and spectrophotometric investigation of Fe(III) complexation to DOPA-containing peptides and proteins from *Mytilus edulis*. *Inorg. Chem. Chem.* **33**, 5819–5824.

Taylor, S. W., Chase, D. B., Emptage, M. H., Nelson, M. J. and Waite, J. H. (1996). Ferric ion complexes of a Dopa-containing adhesive protein from *Mytilus edulis*. *Inorg. Chem.* **35**, 7572–7577.

Treolar, L. R. G. (1958). The Physics of Rubber Elasticity. 2nd ed. The Clarendon Press, Oxford, UK.

Uyeno, T. A. and Kier, W. M. (2005). Functional morphology of the cephalopod buccal mass: a novel joint type. *J. Morph.* **264**, 211–222.

Vaccaro, E. and Waite, J. H. (2001). Yield and post yield behavior of mussel byssal thread: a self-healing biomolecular material. *Biomacromolecules* **2**, 906–911.

Valle, D., Monnerat, A. T., Soares, M. J., Rosa-Greitas, M. G., Pelajo-Machado, M., Vale, B. S., Lenzi, H. L., Galler, R. and Lima, J. B. (1999). Mosquito embryos and eggs: polarity and terminology of chorionic layers. *J. Insect Physiol.* **45**, 701–708.

Vincent, J. F. V. (1980). Insect cuticle—a paradigm for natural composites. In: *The Mechanical Properties of Biological Materials* (eds Vincent, J. F. V. and Currey, J. D.), Cambridge University Press, Cambridge, UK.

Vincent, J. F. V. and Ablett, S. (1987). Hydration and tanning in insect cuticle. *J. Insect Physiol.* **33**, 973–979.

Vincent, J. F. V. and Hillerton, J. E. (1979). The tanning of insect cuticle—a critical review and a revised mechanism. *J. Insect Physiol.* **25**, 653–658.

Vincent, J. F. V. and Wegst, U. (2004). Design and mechanical properties of insect cuticle. *Arthropod Struct. Dev.* **33**, 187–199.

Vitellaro-Zuccarello, L. (1981). Ultrastructural and cytochemical study on the enzyme gland of the foot of a mollusc. *Tissue Cell* **13**, 701–713.

Waite, J. H. (1985). Catecholoxidase in the byssus of the common mussel. *J. Mar. Biol. Assoc., UK* **65**, 359–371.

Waite, J. H. (1990). The phylogeny and chemical diversity of quinone-tanned glues and varnishes. *Comp. Biochem. Physiol.* **97**, 19–29.

Waite, J. H. (1992). The formation of mussel byssus: anatomy of a natural manufacturing process. *Results Probl. Cell Differ.* **19**, 27–54.

Waite, J. H. and Rice-Ficht, A. C. (1992). Eggshell precursor proteins of *Fasciola hepatica*, II. Microheterogeneity in vitelline protein B. *Mol. Biochem. Parasitol.* **54**, 143–151.

Waite, J. H. and Qin, X.-X. (2001). Polyphenolic phosphoprotein from the adhesive pads of the common mussel. *Biochemistry* **40**, 2887–2893.

Waite, J. H., Lichtenegger, H. C., Stucky, G. D. and Hansma, P. (2004). Exploring the molecular and mechanical gradients in structural bioscaffolds. *Biochemistry* **43**, 7653–7662.

Waite, J. H., Holten-Andersen, N., Jewhurst, S. A. and Sun, C. J. (2005). Mussel adhesion: finding the tricks worth mimicking. *J. Adhes.* **81**, 297–317.

Wakamatsu, K., Fujikawa, K., Zucca, F. A., Zecca, L. and Ito, S. (2003). The structure of neuromelanin as studied by chemical degradative methods. *J. Neurochem.* **86**, 1015–1023.

Weiguo, Q., Teng, W., Cappello, J. and Wu, X. (2009). Wet-spinning of recombinant silk-elastin-like protein polymer fibers with high tensile strength and high deformability. *Biomacromolecules* **10**, 602–608.

Weiner, S. and Zaslansky, P. (2004). Teeth: structure/mechanical design strategy. In: *Encyclopedia of Materials Science and Technology* (eds Buschow, K. H., Flemings, M. C., Kramer, E. J., Cahn, R. W., Ilschner, B. and Mahajan, S.), pp. 1–5. Elsevier, Amsterdam.

Weis-Fogh, T. (1960). A rubber-like protein in insect cuticle. *J. Exp. Biol.* 889–907.

Wright, T. R. F. (1987). The genetics of biogenic amine metabolism, sclerotization and melanization in *Drosophila melanogaster*. *Adv. Genet.* **24**, 127–222.

Wu, L. Q., Ghodssi, R., Elabd, Y. A. and Payne, G. F. (2005a). Biomimetic pattern transfer. *Adv. Funct. Mater.* **15**, 189–195.

Wu, L. Q., McDermott, M. K., Zhu, C., Ghodssi, R. and Payne, G. F. (2006). Mimicking biological phenol reaction cascades to confer mechanical function. *Adv. Funct. Mater.* **16**, 1967–1974.

Xu, R., Huang, X., Hopkins, T. L. and Kramer, K. J. (1997). Catecholamine and histidyl protein cross-linked structures in sclerotized insect cuticle. *Insect Biochem. Mol. Biol.* **27**, 101–108.

Yonge, C. M. (1962). The primitive significance of the byssus in the Bivalvia and its effects in evolution. *J. Mar. Biol. Assoc., UK* **42**, 113–125.

Zhao, H. and Waite, J. H. (2005). Coating Proteins: structure and cross-linking in Fp-1 from the green shell mussel *Perna canaliculus*. *Biochemistry* **44**, 15915–15923.

Zhao, H. and Waite, J. H. (2006a). Linking adhesive and structural proteins in the attachment plaque of *Mytilus californianus*. *J. Biol. Chem.* **281**, 26150–29158.

Zhao, H. and Waite, J. H. (2006b). Proteins in load bearing junctions: histidine-rich copper binding protein of mussel byssus. *Biochemistry* **45**, 26150–29158.

Zhao, H., Sun, C. J., Stewart, R. J. and Waite, J. H. (2005). Cement proteins of the tube building polychaete *Phragmatopoma californica*. *J. Biol. Chem.* **280**, 42938–42944.

Zhao, H., Robertson, N. B., Jewhurst, S. A. and Waite, J. H. (2006). Probing the adhesive footprints of *Mytilus californianus* byssus. *J. Biol. Chem.* **281**, 11090–11096.

Insect Cuticular Surface Modifications: Scales and Other Structural Formations

Helen Ghiradella

Department of Biology, The University at Albany, Albany, New York, USA

Abbreviation

2D pc	two-dimensional photonic crystal
3D pcs	three-dimensional photonic crystals

ADVANCES IN INSECT PHYSIOLOGY VOL. 38
ISBN 978-0-12-381389-3
DOI: 10.1016/S0065-2806(10)38006-4

1 Introduction

The biology of an arthropod is defined by and depends on the cuticle, the non-living exoskeleton that is secreted by the single layer of epidermal cells that constitutes the living part of the body wall. Indeed, the cuticle may be thought of as a sort of ''body stocking'' which lines all topologically external surfaces of the animal: on a global scale, it projects out to cover all protuberances—wings, limbs, mouthparts, and others—and it pushes in to line the tracheal system, gut, gland ducts, and the like. On a cellular level, a single epidermal cell can push out an extension to make a scale, bristle, hair, papilla, or any of an array of other cuticular structures, or it can draw in to form a *chordotonal organ*, which may serve as a mechanical or acoustical receptor or participate, perhaps indirectly, in infrared reception (Schmitz *et al.*, 2001). Or the cell can draw in to form a *tracheole*, one of the capillaries of the tracheal system.

To accomplish all this, cuticle is and must be an adaptable and versatile building material. Andersen (2009a) reviews briefly cuticle as a material, and its roles in the forms and functions of the exoskeleton (Andersen, 2009b, see also Chapter 3). It is light, tough, and strong and can be made rigid, flexible, elastic, rubbery, solid or porous, isotropic or anisotropic, as the needs require. In bulk form, it provides protection and support and can also be the source of various forms of coloration. But it can also be sculpted into fine surface structures that serve a variety of functions. The multiplicity of these is bewildering, but fortunately for our understanding, most of them seem to be variations on a few common themes. Understanding a subset of these will allow us to understand the general principles of such structures, what is known of their development and, perhaps more important, indicate questions, perspectives, and directions for future research. For this reason, rather than attempt an encyclopedic overview of all their forms, the author has chosen to concentrate on those she knows best, the bristles and scales that adorn so many arthropod integuments. We will also consider the cuticular patterns of a couple of relatively new entries, a Collembolan and a member of the Class Pauropoda.

To start, we remind the reader that arthropod ''cuticle'' is an entire class of materials that can all be tailored as described earlier with respect to their material properties. A main focus here will be the development of this cuticle into the precise and complex patterns that characterize integumental outgrowths. In connection with this development, one might ask how much information is needed to specify a scale, especially one that is specialized, for example, for producing a structural colour. We suspect that it is less than one might imagine, at least in the genome. Where else could the information be harboured and controlled? Emergent properties of the chemical and/or physical processes that are set in motion as development proceeds must contribute in large part to the final pattern formation. In particular, it is becoming clear that two major players in this cellular development are (1) the cytoskeleton, in particular actin filaments and bundles, and (2) the smooth endoplasmic reticulum (SER). We also believe that relatively simple chemical and physical processes (e.g. elastic buckling) participate in the moulding process. And at all times there must be a conversation between the cell, its local environment, and the systems that must interpret and express in cuticle the resulting consensus.

Where possible, the author is going to rely on review articles in order to save on what would otherwise be an astronomical number of references and detailed points. Two excellent and extensive references in the general field of insect cuticle are Neville (1975, 1993); two others that focus on structural colours but include much basic information on the underlying cuticular structures are Kinoshita and Yoshioka (2005) and Kinoshita (2008). Ghiradella (1998) presents an overview of scales and bristles (although some of the interpretations are old) and, in a review of integumental outgrowths of arthropods specialized for walking on water, Bush *et al.* (2008) present special analyses of the pertinent structures; these authors also review engineering concepts concerning design for cuticular interaction with fluids and fluid interfaces. (There are also numerous reviews in the burgeoning field of insect colour production—see Chapter 5.)

2 General classes of cuticular outgrowths

Figure 1 presents an overview of just a few of the many possible structures that ornament cuticle; we start with a short review of each.

2.1 HAIRS

Hairs (also termed *microchaetes*, *microsetae*, and *microtrichs)* (Fig. 1A) are fine, tapered projections that are relatively short (on the order of a few microns long) and slender (usually less than a micron in diameter). They lack sockets and are rarely ornamented. Their development is similarly simple: the cell extrudes a process, sometimes a cilium, filled with microtubules and/or microfilaments and secretes the cuticle around this template (Mitchell and Peterson, 1989; Mitchell *et al.*, 1983, 1990; Whitten, 1973). According to Whitten (1973), the hairs are usually associated with kinetosomes, which appear in all insect cells except those of the tracheal system. Zhang *et al.* (2009) remark that fine filopodia may also result from actin modelling; these may form the templates for some of the very fine hairs that some arthropods produce (*vide infra*).

Thorpe and Crisp (1949) describe the structure and function of lawns of recurved hairs that characterize the surfaces of certain aquatic insects (see Mill, 1998b, for a more general review of gills and other specializations of aquatic insects). These serve as *plastrons*, or anti-wetting devices. Because they are unwettable, they maintain permanent air films that act as physical gills, enabling the animals to stay submerged in oxygenated streams, often for their entire lives. Other possible functions of hairs include protection from abrasion, control of

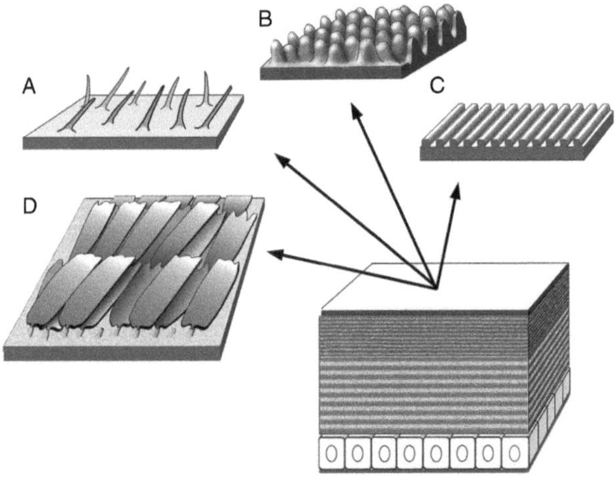

FIG. 1 A patch of unspecialized cuticle, showing the thin outer layer, the epicuticle, and the inner layers, which latter may vary greatly from place to place. A single layer of epidermal cells (resting, as always, on a sheet of basement membrane) is responsible for producing, shaping, and when necessary resorbing the cuticle. (A–D) show some of the possible external modifications of any arthropod exoskeleton. (A) There may be a covering of hairs, or *microchaetes*, simple extensions lacking sockets or other significant sculpting. (B) The surface may be produced into a series of papillae that reduce glare and increase the efficiency of light absorption. (C) The surface may be sculpted into any of a number of patterns including, for example, fine ridges that serve as diffraction gratings. (D) The surface may be ornamented by *macrochaetes*, scales or bristles that have sockets and are often ornately patterned. Modified from Ghiradella and Butler (2009).

boundary layer properties on wings and other body surfaces, and filtering of food (Palmer, 1998, presents a review of taxa and structures specialized in this direction).

2.2 PAPILLAE AND DIFFRACTION GRATINGS

Papillae (sometimes called ''nipples'') (Fig. 1B) are more or less conical protrusions secreted around microvilli that the cell raises on its surface. On insect eyes these effectively provide a gradual change in refractive index from that of air (by definition 1) to that of cuticle (varied, but generally in the 1.5 range). With no sharp interface between the two media, light passes from one to the other without reflection or refraction, that is, without loss. On the insect eye this lack of reflection translates as less glare to signal ''eye'' to a potential predator and also allows for better harvesting of the light, thereby increasing sensitivity (useful to any nocturnal animal). On wings, these arrays render ''clearwing'' moths less visible to potential predators because their transparent wings are barely visible against the background (see Yoshida, 2005, for review of anti-reflective mechanisms in insects; more recently, Stavenga *et al.* (2010)

suggest that such papillae on the scale-free dorsal regions of *Graphium* spp. wings may facilitate the conduction of light through the blue-green wing membrane to the ventral scales, which then reflect the light back, thereby enhancing the wing colour (and adding polarization effects when viewed obliquely)).

Microvilli raised by the epidermal lining of the midgut apparently serve as templates for the porous cuticular *peritrophic membrane* that lines the region (Neville, 1975). As we shall see, in some butterfly scales microvilli may be involved in the formation of the regularly porous arrays that we call two-dimensional photonic crystals (2D pcs).

Insect mastery of structural colour production makes it almost inevitable that in some cases cuticle be thrown into fine rulings that serve as diffraction gratings (Fig. 1C).

Papillae and diffraction gratings are only two of the myriad arrays of possible surface scuptings and we present them only as examples. Figures 36 and 37 present another example, the hexagonally patterned surface of the integument of members of the class Collembola (*vide infra*).

2.3 BRISTLES AND SCALES

Bristles and scales (macrochaetes, macrosetae, macrotrichs) (Fig. 1D) are basically the same structure and are characterized by being larger (on the order of tens of microns in length and width) and more ornate than hairs. At their bases they taper to stalks, or *petioles*, which are inserted into sockets. If they are innervated, bristles (and occasionally scales) can serve as sensory receptors in addition to any other functions they may have. Bristles are generally cylindrical in form while scales are flattened plates, but a better distinction devolves on whether there is ornamentation wrapped around the whole structure (bristle) or present on just one side or surface (scale). Bristles are common on all arthropods, whereas scales are restricted to several insect orders and occasional representatives of other arthropod groups. The Lepidoptera are of course distinctive in their rich tapestry of scaly investiture which, among other things, is in large part responsible for their brilliant and patterned colours.

Macrochaete functions abound: the following list was partially abstracted from Ghiradella (1998) (but also see Winterton, 2009). They serve as mechanoreceptors of various sorts (see Keil, 1997, 1998 and French and Torkkeli, 2009, for reviews of bristle mechanoreceptor structure and function). They produce and disperse pheromones and scents (Boppré and Vane-Wright, 1989; Egelhaaf *et al.*, 1992; Grant and Brady, 1973; Huxley and Barnard, 1988; Sellier, 1971a,b; Vane-Wright, 1972; Wasserthal and Wasserthal, 1977). They carry colours that serve in courtship, warning, camouflage, and in species and gender recognition (see Chapter 5). They protect against spider webs and sundew plants and other challenges (Eisner and Shepherd, 1966; Eisner *et al.*, 1964; Hölldobler and Wilson, 1986), and they absorb sound, thereby reducing for flying moths the chances of predation by bats (Nachtigall, 1974). They aid in flight by streamlining larger fliers and by acting as extensions that increase wing area (Huxley and Barnard, 1988); in some cases they also affect boundary layers, thereby increasing lift and thrust (Nachtigall, 1974). They serve as insulation against heat loss (Casey and Hegel, 1981; Heinrich, 1987; Kingsolver and Moffat, 1982) or, conversely, they control body temperature by absorbing or reflecting infrared solar radiation (Kingsolver, 1988; Miaoulis and Heilman, 1998). In some cases, they serve as filters to trap food (Palmer, 1998).

To this admittedly incomplete list, Bush *et al.* (2008) have added some additional perspectives, among them the function of channelling water droplets off to provide rain-proofing or channelling mist to make the moisture available to the insect. They report that on macrochaetes, nano-grooves parallel to the direction of flow can apparently reduce drag by easing slip, while grooves orthogonal to the flow direction provide more grip and thereby more thrust. Indeed their review suggests a whole series of fluid mechanics questions the authors would like addressed with respect to macrochaete structure and function.

3 Wings and scales

Figure 2 presents a fragment of butterfly wing, together with its covering of scales. As is typical in most (but not all) cases there seem to be at least two layers, larger *cover* scales and smaller *ground* scales, the latter only partly visible under the former. In fact, the scale bases are arranged in single rows, alternating CGCGCG; double layer appearance is because the smaller scales are tucked under their larger neighbours.

Scales can occur anywhere on an insect body, but most studies have focused on those of the wings, perhaps because of their tendency to carry the brightest colours on the animal. As Ghiradella (1998) remarks, "A butterfly wing expresses several patterns at the same time: (1) the general shape of the wing and the pattern of its venation; (2) the layout of the colour patterning, characteristic of each species and yet varying with sex, morph, wing (front or hind), and wing surface (dorsal, or ventral); (3) the layout of the scales and their sockets on the wing; and (4) the ultrastructure of the individual scales." This latter category includes scales with *structural colours,* caused by the interaction of light with the fine architecture, rather than with any chemical pigments of the scales.

Nijhout (1991) presents an excellent discussion of the control and development of patterning at the level of the wing. Here, we will confine our discussion to macrochaetes and particularly to scales, which have been extensively studied from the viewpoint of their patterning. Ghiradella (1998) presents an overview of the general processes by which an individual epidermal cell differentiates into a macrochaete; she then goes on to discuss the ultrastructure of a "standard" (unspecialized) scale and of some of the variations thereof. Here, we will start first with an overview of scale ultrastructure and then discuss the process

FIG. 2 *Urania fulgens*, part of a wing, showing the typical "cover/ground" scale arrangement, in which larger (cover) scales overlap and partially cover smaller (ground) ones. These scales have squared apices, but rounded or lobed ones are even more common. The arrow points to a socket near the (broken) edge of the wing. Scale bar, 50 μm.

by which the epidermal cell forms this ultrastructure. We will try to keep all this in the larger context of what we now know about pattern formation capabilities of epidermal cells.

4 Introduction to scale structure

Figure 3 presents a diagrammatic view of a fragment of a more or less unspecialized scale (compare Figs. 5 and 6), together with some of the more common variants on this form. The scale starts development as a cylindrical bristle but then flattens so that the final form of its cuticle is typically that of a flat *envelope* with what we will call an "upper" surface (visible to the outer world) and a "lower" surface (facing the wing). The lower surface is usually featureless, but the upper is typically characterized by a series of longitudinal folds (*ridges*) connected at intervals by orthogonal struts (*crossribs*). The "rooftrees" of the ridges are usually ornamented by slender overlapping folds (*lamellae*), and running orthogonally to these (and down the ridge sides) are series of finer folds (*microribs*), some of which continue across the crossribs to the adjacent

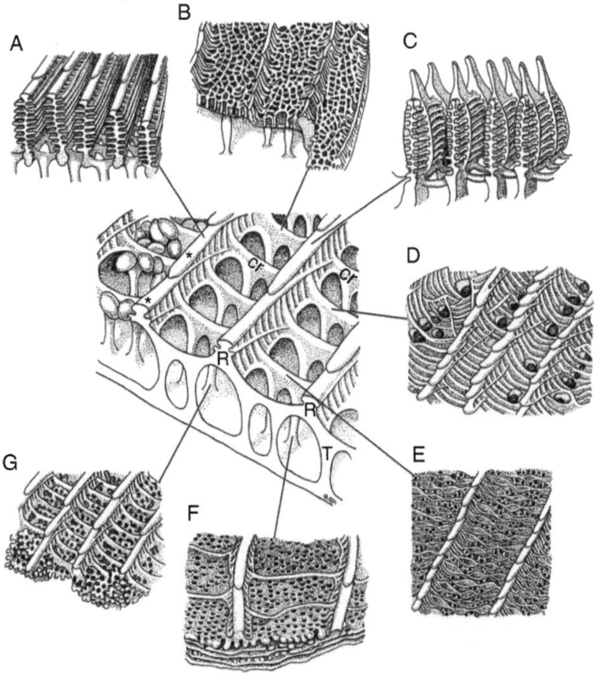

FIG. 3 A new look at an old figure: possible structural variations on a generic scale. The centre presents a cut-away view of a fragment of a more or less unspecialized wing scale. The scale is basically a flattened sac or envelope, one side of which is usually towards the wing while the other is exposed to the outer world. (For convenience we shall designate these as the ''upper and lower surfaces'', although their exact positions may vary from wing to wing.) The upper surface is usually thrown into longitudinal folds or *ridges* (R), joined at intervals by *crossribs* (cr), the two sets framing a series of *windows* opening into the scale interior. The lower surface is usually flat and featureless. The ridges typically have overlapping ''rooftrees'', or *lamellae* (*); from these, fine folds, or *microribs* run down the sides, more or less orthogonal to the lamellae, and sometimes across the crossribs. Upper and lower surfaces are typically joined by internal pillars, or *trabeculae* (T), and in some pierid butterfly scales there may be pigment granules or *beads* (here shown in the lefthand segment). Any part of this scale may be modified to produce any of a range of specializations (not an exhaustive list). (A) The ridges can become taller than usual and their lamellae stack up to produce a series of thin films that act as multilayer interference mirrors. (B) The window regions may display a series of fine alveoli that form in essence a 2D photonic crystal with associated optical properties. (C) The ridge structure can ''rock back'' so that the lamellae are now vertical and it is the microribs that are acting as the reflective multilayers. (D) The microribs can extend across from ridge to ridge and essentially close the windows. (E) The region between the ridges may be filled with a ''plates and pores'' structure that we take to be essentially a variant of the 2D pc structure shown in (B). (F) The scale interior, or *lumen*, can be filled with stacks of laminae that form a second type of multilayer interference mirror. (G) The scale interior can be filled with a lattice that is effectively a form of 3D photonic crystal, again with associated optical properties. Reprinted with permission from Ghiradella (1998).

ridges. Ridges and crossribs typically frame series of *windows* that allow views into the scale lumen, which latter may contain pillars (*trabeculae*) that connect upper and lower surfaces and, in some cases, beads that hang from the superstructure and that contain pigments whose deployment may absorb and control reflection in wavelengths ranging from ultraviolet to blue and orange (Giraldo and Stavenga, 2008; Giraldo *et al.*, 2008).

This, then, is a ''typical'' scale. Any of these structures may be modified in any of several ways to produce variants to suit the functions mentioned in Section 2.3. Since most studies have focused on those that produce structural colours, we will consider more closely a few of these. Here again, these are meant as representative examples, not an exhaustive list (see Chapter 5, for more details).

Figure 3A: The ridges can be taller so that their lamellae form a stack of thin films that collectively form a multilayer interference mirror or ''multilayer reflector'' (Figs. 7 and 14); this mechanism is responsible for such structural colours as the brilliant blues of the genus *Morpho* and others, for the ultraviolet reflection in certain male Pierids, and for the greens in some Papilionids.

Figure 3B: The windows can present a series of ''alveoli'' that form a 2D pc (Figs. 11–13) that may channel or direct incoming light to, for example, reflective structures within the interior of the scale. Vukusic and Hooper (2005) describe a particularly sophisticated system in which the pc directs incoming skylight to a Bragg reflector at the bottom of the scale lumen. The system is fluorescent and apparently reemits the incoming light at wavelengths better suited to the eyes of conspecific observers.

Figure 3C: The reflective system on the ridges can ''rock back'' so that it is the microribs that are serving as the multilayer reflectors; the lamellae may or may not still be present. Other variations in lamellar angle also exist: in the Papilionid, *Trogonoptera brookianus*, the lamellae do not slant, but run longitudinally, parallel to the scale surface, while in some Brassolids, they slant steeply enough so that they and the microribs (still roughly orthogonal to each other) are both presumably reflective, but in different directions: to our knowledge, this system has not been studied optically.

Figure 3D: The microribs can run from ridge to ridge, essentially obliterating the windows (Fig. 8); such scales typically have a ''satiny'' sheen.

Figure 3E: The window structure can be modified into a ''pores and plates'' configuration (Figs. 14 and 15) that the author believes is a variant of the structure in Fig. 1B, although not so regular as to qualify as a 2D pc.

Figure 3F: The scale lumen may be filled with laminae[1] that act as thin-film mirrors (Figs. 16 and 17).

[1] Early in her career, the author referred to such internal thin films as ''laminae'' to distinguish them from the ridge lamellae, which are also thin films, but which she did not consider the same structure. At one point, a rather authoritarian reviewer insisted that both classes of structure be called lamellae, specifically ''ridge lamellae'' and scale ''body lamellae''. It is amply clear now that they are *not* the same structure, and I am reverting to my previous linguistic habit of distinguishing between lamellae (ridges) and laminae (scale body).

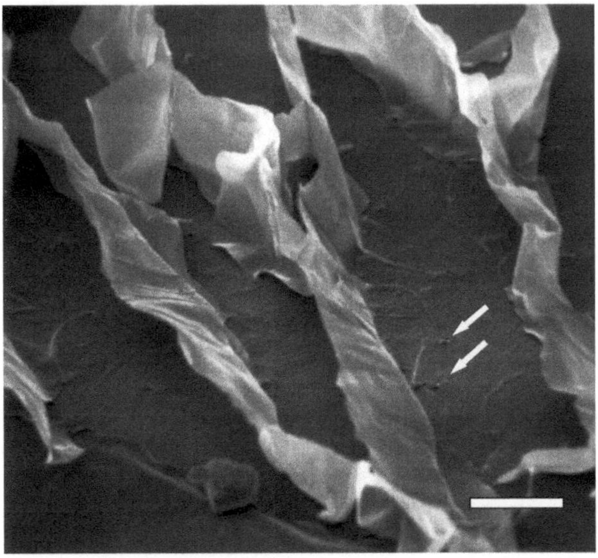

FIG. 4 *Elbella polyzona*, wing fragment with several ''maché'' scales. These appear to lack all inner layers so that they are thin and crinkly and mat irregularly together to scatter light and produce a frosted effect. Arrows point to their comparably flimsy sockets. Scale bar, 50 μm.

Figure 3G: The scale lumen may contain iridescent three-dimensional photonic crystal (3D pcs) reflectors (Figs. 19–22; Fig. 23 shows the same in a bristle).

In some cases, the scale may dispense entirely with its internal structures (Fig. 4) and become essentially a thin envelope that can crumple against its neighbours to form a sort of ''scale maché'' that produces a frosted effect on the wing.

5 Basic scale patterning

Having now an idea of basic scale structure and of a few of its possible modifications, let us look at actual examples of some of the forms described in Fig. 3. Figure 5 presents a surface view of part of an essentially unspecialized scale. The ridges are topped with slanting lamellae (*), which pattern seems to be the more common case. Whatever the slant of the lamellae, the microribs usually run orthogonally down the ridges from them and some of these run across the crossribs to mount the neighbouring ridges. In the present scale, the windows are a bit more irregular than usual (compare Fig. 18), but they still show the basic rectangular shaping.

Figure 6 presents part of a rather spare scale that has shed a fair amount of its superstructure. Ridges, lamellae, microribs, and crossribs are present, but the windows are particularly large and provide a view into the interior of the scale,

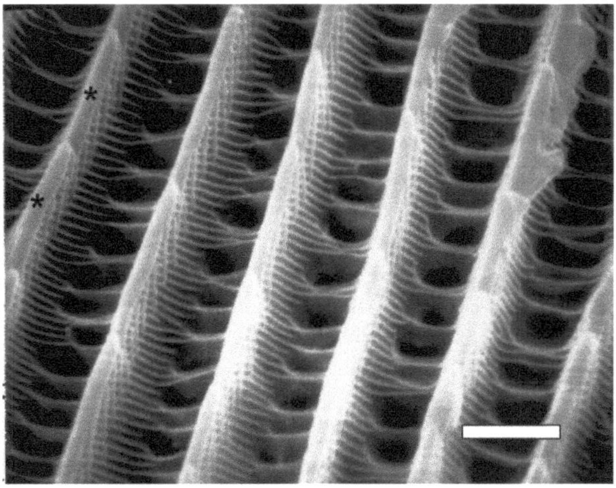

FIG. 5 *Prepona phaedra,* part of a more or less unspecialized ground scale. The ridges are topped by slanting lamellae (*) and the microribs run down, more or less orthogonally, some continuing across along the crossribs to the next ridge. Scale bar, 1 μm.

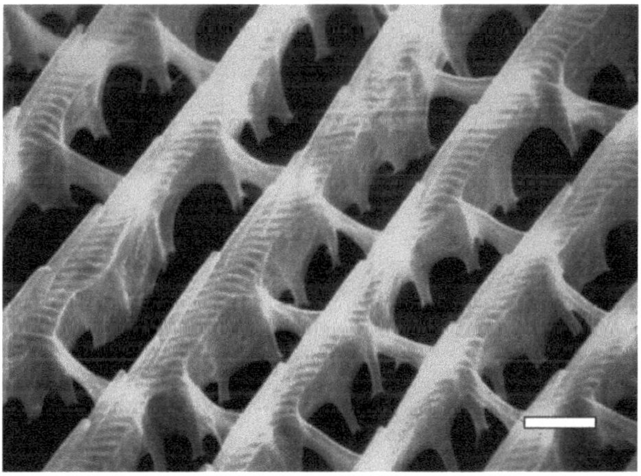

FIG. 6 *Limenitis astyanax*, part of cover scale. In this rather ''stripped down'' scale, the windows are particularly open, providing a clear view of the scale interior with its trabeculae, some of which appear webbed. Scale bar, 1 μm.

where the trabeculae can be seen joining the upper and lower surfaces. This economy of structure is carried to extremes in the so-called ''glass scales'' (e.g., on *Morpho* wings), where the scale consists only of the bottom surface and the ridges, the latter supported by flimsy trabeculae (e.g., see Ghiradella, 2005, Fig. 11).

5.1 RIDGE/MICRORIB VARIANTS

Figure 7 shows a scale whose ridges carry stacked thin-film lamellae but whose crossrib and microrib structures are more or less standard, although the windows disappear near the scale base where the ridges are closer together. The loss of windows where scales taper, either at their bases or at their apices, is common and may reflect a ''failure'' of part of the pattern formation system where the geometry changes (getting narrower in this case).

Figure 8 presents a scale such as is figured in Fig. 3D. The top figure shows a typical socket and scale base, but a closer look (bottom) shows that the microribs have completely taken over the spaces between the ridges. In such scales, the envelope is typically thicker than usual and the scale is flattened, so that the lumen is reduced to fine spaces under the ridges.

The scale in Fig. 9, as that in Fig. 8, has the microribs filling in the windows but all the elements are displaying a particularly florid patterning. We can only guess at why and how the scale has developed this presentation.

Figure 10 presents another rococo variation on the basic structure, again involving the window region, this time in a papilionid butterfly (as a group,

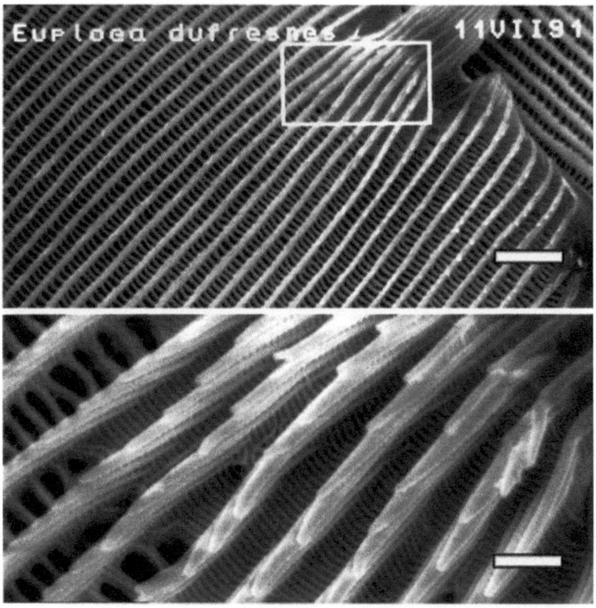

FIG. 7 *Euploea desfresnes*, scale base and part of petiole. The main body of the scale has the standard ridge/crossrib relationship, but a closer look reveals that near the base, where the ridges are closer together, the microribs run from ridge to ridge so that there are no windows. The ridges show the stacked lamellae that characterize the thin-film reflector structures. Scale bars, 5 μm (top), 2 μm (bottom). Reprinted from Ghiradella (2005).

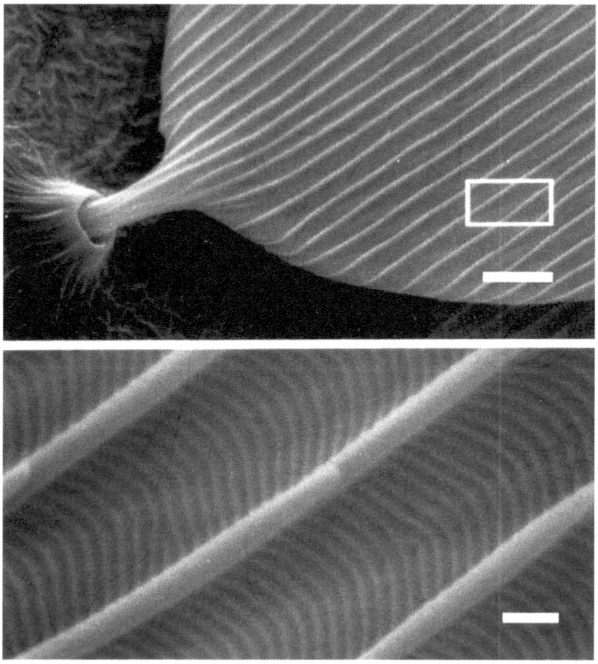

FIG. 8 *Lampoptera curiosa*, scale socket and base. As the enlargement shows (bottom), the microribs run from ridge to ridge, precluding any windows. Such scales typically also have thickened envelopes that are flattened, essentially obliterating the scale lumen. The wing membrane itself is textured, a common feature in many insects. Scale bars, 5 μm (top), 1 μm (bottom).

Papilionids seem to depart from the standard crossrib pattern: see Fig. 11). The questions ''why'' and ''how'' expressed with regard to Fig. 9 also apply here. Note the fine ''finishing'' of the pore edges, more or less typical of insect cuticular systems. (Once again, the pattern gets muted near the scale petiole (upper figure, bottom right).)

So far, we have viewed scale structures whose formation involves the outer envelope: ridges, crossribs, microribs, and some variations thereof. Any developmental pattern formation programs must account for simple ridges and those with the stacked lamellae, for lamellae that may be tilted or not or may be replaced by stacked microribs, for standard windows and those that are reduced or netted, and so forth. We ask the reader to note these questions, to which we will return in Section 8.

5.2 TWO-DIMENSIONAL PHOTONIC CRYSTALS

Figure 11 shows fragments of two scales from a Papilionid, *Papilio zalmoxis*. The ground scale (right) shows the papilionid equivalent of windows: the crossribs form a netted, rather than reticular, pattern between the ridges.

FIG. 9 *Caligo ilioneus*. In this scale also, the microribs have displaced the windows, but all structures have a somewhat rococo aspect. Scale bar, 1 μm.

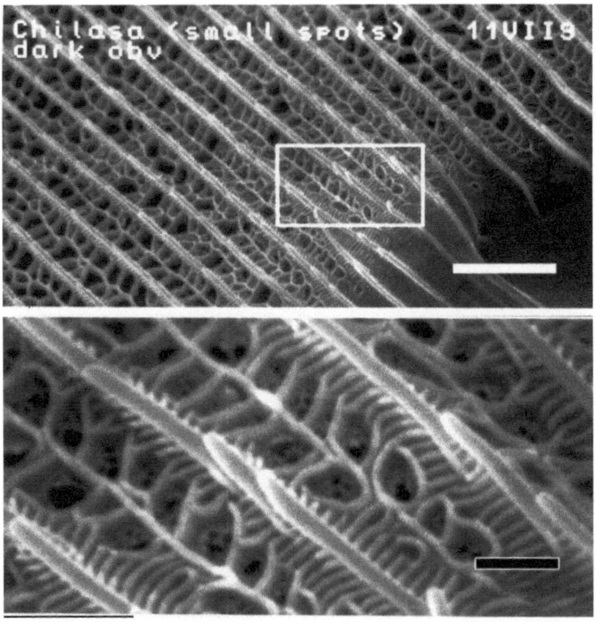

FIG. 10 *Chilasa* sp. Yet another variety of detailing, this time in a papilionid scale (detail enlarged at bottom). This may be a variant on the typical netted papilionid window (see Fig. 11). Scale bars, 5 μm (top), 1 μm (bottom).

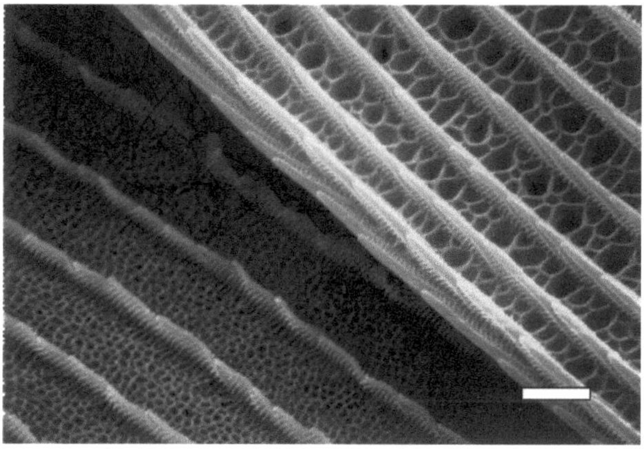

FIG. 11 *Papilio zalmoxis*, two scales. On the right is a ground scale, which shows the reticulated window network that for some reason seems characteristic of the Papilionidae. The cover scale (left) displays the alveoli diagrammed in Fig. 3 B; these form a 2D photonic crystal that may be part of a particularly sophisticated system (see Vukusic and Hooper, 2005). Scale bar, 2 μm.

In the cover scale (left), the pattern consists of the aforementioned alveoli that form a 2D pc. Figure 12 presents an internal and inverted view of such a scale, the pc is a slab that fills the top half of the scale lumen. The lower surface here is slightly thickened; it consists of the three-layer Bragg reflector mentioned earlier for this scale (Vukusic and Hooper, 2005). The trabeculae (whole at left, fractured elsewhere) join the upper and lower scale surfaces.

Figure 13 shows a frontal transmission electron microscope (TEM) section of a 2D pc slab, one similar to those in Figs. 11 and 12. Woven fibrils, presumably of chitin, determine the pattern. As mentioned earlier, this structure is similar in form and scale to the peritrophic membrane, the porous cuticular mesh secreted over a microvillar template by the epidermal cells lining the midgut (see Binnington *et al.*, 1998; Neville, 1975, for illustration and discussion). We presume that the 2D pc is likely to be formed in the same way: the epidermal cell would erect an array of microvilli and weave the cuticular microfibrils around and between them. In this case, the cuticle secretion presumably continues longer to produce a thick 2D pc slab, rather than the thin, but (we believe) related structure we are about to consider in the androconial scale (Figs. 14 and 15).

Figure 14 presents fragments of two scales from a UV reflective pierid male. The fractured UV reflective scale (top) shows clearly the lamellae of the thin-film mirror on the ridges as well as the interior scale trabeculae and the fractured lower surface. The other scale is an *androconial* scale, of the type present in the black wing margins of the same male. It shows the "pores and plates" structure that the author believes to be a variant of the 2D pc. A whole mount (Fig. 15) presents the supporting fibril structure, which also shows similarity to the

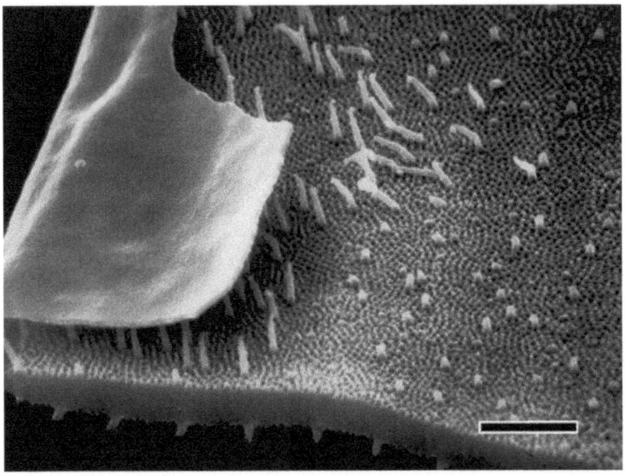

FIG. 12 *Papilio zalmoxis*, fractured 2D photonic crystal scale, inverted. Seen from below the crystal appears to be a single slab that fills the whole scale. Fractured trabeculae project from the crystal underside, joining the lower surface (left) where the scale is still intact. At bottom, tips of broken ridges project from the exterior of the upper surface. Scale bar, 4 μm.

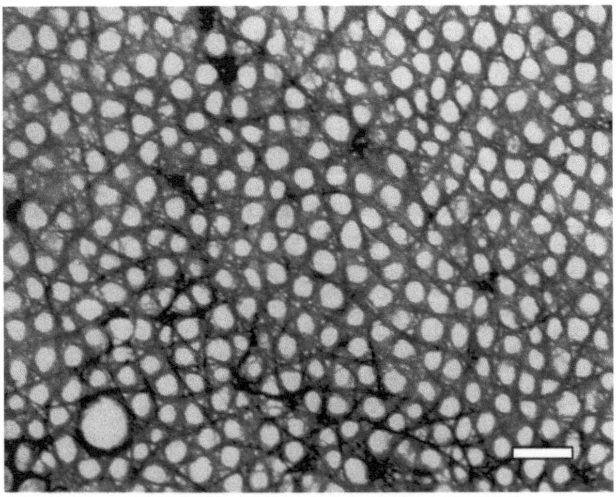

FIG. 13 *Papilio zalmoxis*, 2D photonic crystal, frontal section, TEM. Cuticular fibrils, presumably chitin, are woven into a pattern reminiscent of that in insect peritrophic membrane, which latter is patterned around an array of microvilli erected by a midgut epidermal cell for the purpose (Neville, 1975). Scale bar, 0.5 μm.

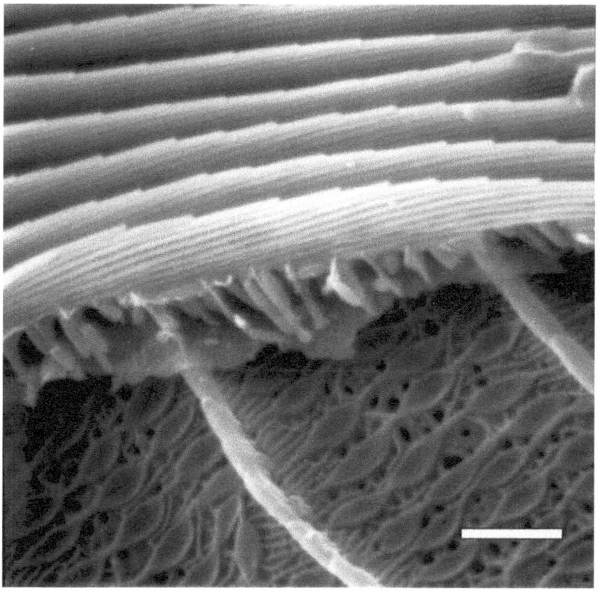

FIG. 14 *Colias eurytheme* male, parts of two scales. The fractured top scale shows ridges with thin-film lamellae (these reflecting in the ultraviolet), as well as the trabeculae in the scale lumen and part of the bottom surface. At bottom is a so-called *androconial* scale, found in the black wing borders of the males; this shows the "plates and pores" structure diagrammed in Fig. 3 E. Scale bar, 2 μm. Reprinted with permission from Ghiradella (1998).

underpinnings of the 2D pc in the *P. zalmoxis* scale (Fig. 13). Here, however, the fibrils are periodically splayed out, which apparently provides enough support to stabilize the plate regions through the exigencies of development and eclosion.

Although much of the 2D pc structure is situated within the scale, we believe that its presumed patterning on cell surface microvilli qualifies it as a "surface" structure. Let us continue on into the scale lumen and turn our attention to two more internal specializations, the laminae and the 3D pcs.

5.3 INTERNAL STRUCTURES

Figures 16 and 17 present views of internal stacks of multilayer laminae. In *Urania riphaeus* (Fig. 16), the ridges have slightly stacked lamellae and the crossribs are sketchy. But the entire lumen of the scale is filled with stacks of thin-film laminae that constitute an internal multilayer reflector. In the *Papilio* scale (Fig. 17), the upper view shows that while there is a faint hint of ridge and crossrib structure, the entire system is modified into a series of coarsely cup-shaped concavities. This type of structure is known in at least one case to

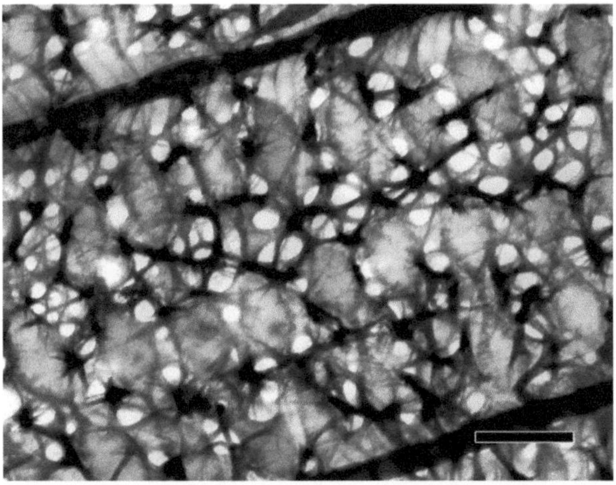

FIG. 15 *Colias eurytheme*, whole mount of part of an androconial scale, TEM, showing two ridges and the surface between them. As in the 2D photonic crystal shown in Figs. 11–13, the structure seems supported by woven fibrils. Where these splay out, the nascent cuticle is supported and remains as the lenticular plates. Where not, it falls through, leaving the visible pores. Scale bar, 0.5 μm. Reprinted with permission from Ghiradella (1998).

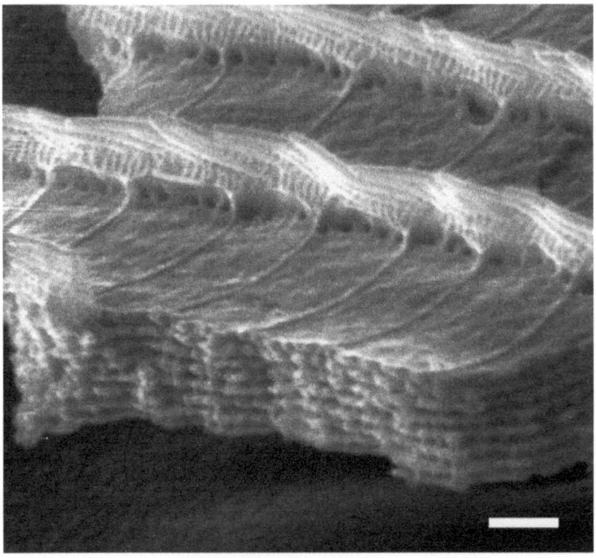

FIG. 16 *Urania ripheus*, fractured reflective scale. The ridges are more or less unspecialized and the crossribs are minimized, but the interior is filled with a stack of laminae that serve as a thin-film mirror. Scale bar, 1 μm.

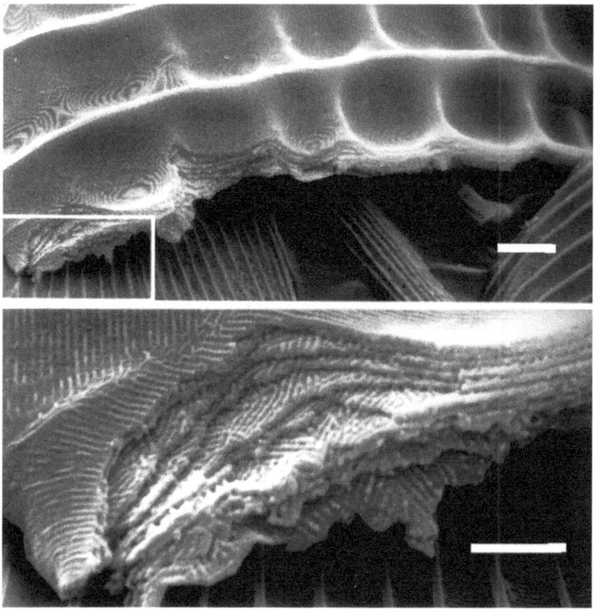

FIG. 17 *Papilio* sp., fractured iridescent scale, showing the reflective internal laminae that provide the iridescent colour. The whole scale shape is a substantial modification of the standard ridge/crossrib structure, common occurrence among the Papilionidae. Scale bars, 5 μm (top), 2 μm (bottom). Reprinted with permission from Ghiradella (1998).

produce an iridescent green colour that results from combining blue and yellow that reflect, respectively, from the more vertical and more horizontal parts of the concavities (Vukusic *et al.*, 2000). The reflective elements, multilayer laminae with what appear to be spacers, show clearly in the lower view.

Lycaenid butterflies often have reflective scales that appear to have perforated laminae. Figure 18 shows a cover and a ground scale from *Lycaena rubita*. The ground scale (left) is more or less standard, but the cover scale clearly has an internal structure that superficially resembles a 2D pc but may be, in fact, closer in form to the internal laminae shown in Figs. 16 and 17. Wilts *et al.* (2009) show that a higher degree of perforation is correlated with a decrease in the width of the reflection spectra, which control, together with tailoring of other aspects of scale structure and material properties, provides the animal with a wide range of spectral tuning options. Similar scales are described (among others) in Kertész *et al.* (2006) and in Prum *et al.* (2006); these authors make the point that unlike the 2D pc slabs described in the *P. zalmoxis* reflective scales, these structures invade the ridges and are at least partly contained within them. In fact, in at least some cases they seem intermediate between 2D and 3D pcs, which latter are also common in many scales (and bristles) (Figs. 19–23).

Figure 19 shows part of two scales from the Lycaenid *Thecla herodotus*. In this and other similar Lycaenids, the scales are neither cover nor ground but in

FIG. 18 *Lycaena rubita* cover (c) and ground (g) scales. Both scales have somewhat standard ridges, but the cover scale has reduced crossribs; in its lumen are layers of what look like perforated laminae. Such scales present structures that appear intermediate between multilayers and 3D photonic crystals. Scale bar, 1 μm.

FIG. 19 *Thecla herodotus,* two iridescent scales. In this and in related species, the distinction between cover and ground scales is blurred: all are overlapped by those in the previous rows so that only their distal portions show (this overlap may be the reason for the scales' length). The reflective 3D photonic crystallites (see Fig. 20) only occupy the visible distal parts of the scales. (Note the puckering of the mounting medium around the scales: this is an example of *elastic buckling*, to which we refer during the discussion of development.) Scale bar, 20 μm. Reprinted with permission from Ghiradella (1998).

fact overlap in such a way that the distal regions are visible. Here, these distal regions clearly show internal bodies that are in fact 3D pcs (compare Fig. 20). (In passing, let us call attention to the sinusoidal folds in the mounting medium in Fig. 19—we will need to refer to them later.)

Figure 20 presents a closer look at this type of crystallite system. The crystallites have different orientations so that the direction of reflection varies from one to the next, presenting in the intact wing a spangled appearance that mimics (at least to human eyes) that of sun-dappled leaves. In an early report on this sort of structure, Morris (1975) interpreted the crystal structure as simple cubic (SC), but on the basis of the existence of both hexagonal and square packing in sectioned profiles. Ghiradella and Radigan (1976) proposed instead a face centred cubic (fcc) structure. Michielsen and Stavenga (2008) demonstrated that in at least some scales of this type, the crystallites may be modelled as gyroid. In a finely written review, Poladian *et al.* (2009) present a thorough analysis of the possible (available) 3D pc forms and their concomitants.

Figure 21 shows a fractured *Parides sesostris* scale with what looks like a 3D pc slab (as opposed to the crystallites shown in Figs. 19 and 20). This scale also has the ridges prolonged into a series of raised ''footings'', referred to as a ''honeycomb'' in Poladian *et al.* (2009). According to these authors, the honeycomb structure may collimate the incoming light and/or randomize the angles of incoming and reflected light (in effect ''frosting'' the reflection), in either case controlling the iridescence by muting or suppressing it. In another Papilionid, *Teinopalpus imperialis* (Fig. 22), the honeycomb is much reduced and

FIG. 20 *Mitoura grynea*, edge of fractured scale (stripped of part of the envelope), showing several reflective crystallites. Each of these is a 3D photonic crystal that produces a structural colour. Scale bar, 1 μm.

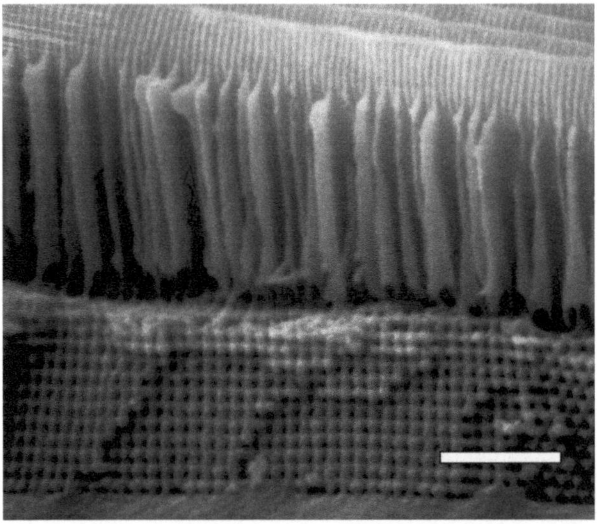

FIG. 21 *Parides sesostris*, fractured iridescent scale, longitudinal view. The ridges, which are raised on high ''footings'' that may actually suppress iridescence (Poladian *et al.*, 2009), lack lamellae, having only microribs. The scale lumen contains what may be one large 3D photonic crystal. (The mounting medium at bottom has covered the scale's lower surface.) Scale bar, 2 μm.

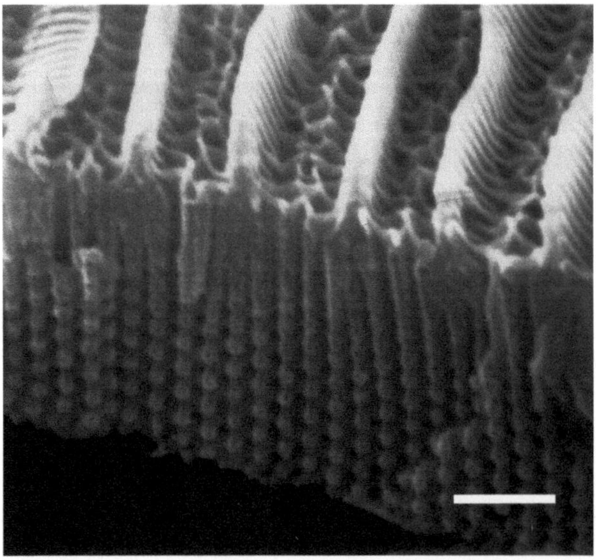

FIG. 22 *Teinopalpus imperialis*, fractured 3D photonic crystal scale. Here, the ridge footings are shorter, and upper layers of the crystal show direct continuity with the alveoli in the window regions, which latter appear to constitute a 2D photonic crystal. Scale bar, 1 μm.

has a hint of the 2D photonic crystal about it. It also seems in phase with the units of the 3D pc.

We entered this discussion by describing scales and bristles as essentially the same structure in developmental terms. The similarity carries through in the adult structures. Figure 23 shows a fractured specialized bristle from a hesperid butterfly. The surface has the usual ridges and microribs, although the crossribs and windows have been suppressed. The interior contains a cylindrical 3D pc which is responsible for the brilliant colours of this bristle. Of biomimetic interest is the fact that this is in effect a *hollow core photonic fibre*; because of their low refractive index cores ($n = 1$), such fibres conduct light faster than glass fibres and therefore their imitation is a subject of current interest.

6 Overview of macrochaete development

Having surveyed some of the many forms these scales (and bristles) may take, let us look at their development, again with special attention to that in scales. We remind the reader that a more complete discussion of the early stages is found in Ghiradella (1998). We will revisit here the question of development, but in a different context. Ghiradella and Butler (2009) review some of the following material.

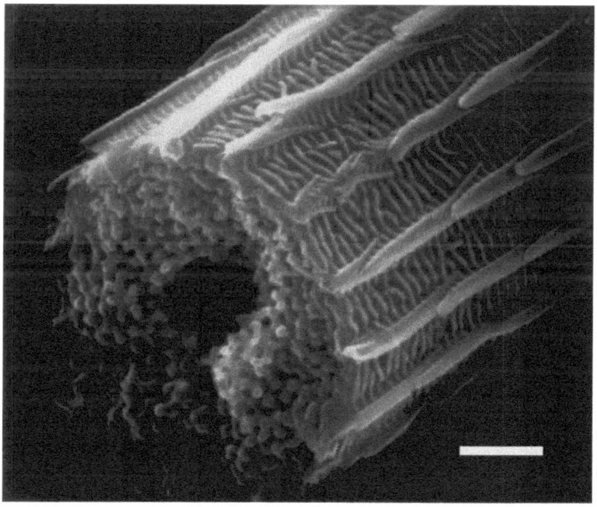

FIG. 23 *Tomares ballus ballus*, iridescent bristle. The exterior exhibits ridges with lamellae and microribs, the latter extending (with some irregularity) from ridge to ridge. The bristle interior contains a 3D photonic crystal; this is in fact a *hollow core photonic fibre*, conducting some wavelengths of light while excluding others. Scale bar, 1 μm. Reprinted from Ghiradella (2005).

The process of macrochaete building begins when an epidermal cell gets a signal to transform into a so-called *sense organ precursor* (SOP) cell. It then undergoes a division (Fig. 24) that results in two daughter cells (see Held, 2002, for a detailed discussion of control of bristle formation and differentiation). One daughter cell will divide further to produce the *trichogen* and *tormogen* cells, which in turn will form the macrochaete shaft and socket, respectively. If the macrochaete is to be a sense organ, the second daughter cell will divide to form the neuron and the sheath cell and perhaps such other accessory cells as glial cells, a ligament, etc.

Since all these structures are associated with macrochaetes, they form from processes evaginated from the SOP daughter cells. If the processes invaginate rather than evaginate, we get a homologous but very different organ, a so-called chordotonal organ, important in insect hearing and stretch reception, and possibly in some forms of infrared sensing. To say it another way, drag an innervated bristle below the cuticle surface and one has a chordotonal organ (see Held, 2002; Keil, 1997, 1998; Kernan, 2007; Lai and Orgogozo, 2004; Yack, 2004; and Yack and Dawson, 2008; Schmitz *et al.*, 2001, discuss possible chordotonal function in one type of insect thermoreceptor). The homology is reinforced by the work of Merritt (1997) who demonstrated that the mutants in the gene *cut* can transform what would otherwise be a bristle into a chordotonal organ. In this

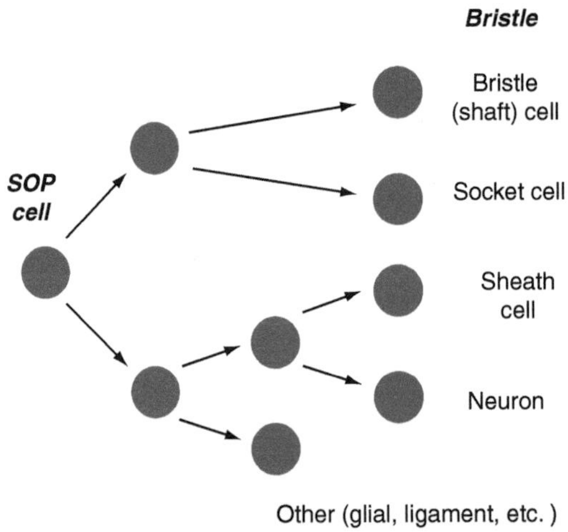

FIG. 24 Ontogeny of bristles and scales. An epidermal cell gets the signal to become a sense organ precursor cell (SOP). It then undergoes a series of stereotyped divisions that will yield a *trichogen*, or shaft cell, which will make the shaft of the scale or bristle, a *tormogen*, or socket cell that will make the socket and, if the structure is to be innervated, a neuron, a sheath cell and perhaps other accessory cells. Should the developing structure tuck inwards rather than protruding outwards, the result becomes a *chordotonal organ* characteristic of insect ears and of some proprioceptors. Modified from Ghiradella and Butler (2009).

context, we might mention again that the insect tracheoles, the ''capillaries'' of the respiratory system, each form by an invagination of the *tracheolar cell*, the last in the chain of epidermal cells surrounding that particular tracheal tube (see Fig. 30); we submit that one could consider a tracheole an invaginated bristle, an insight first expressed by Whitten (1969). In other words, there appears to be in this lineage of insect structures great variety of form derived from relatively few variations on one or a few basic themes.

Figure 25 presents a diagrammatic view of an innervated bristle. The bristle cell has made the bristle shaft; the socket cell, the socket; and the sheath cell, the sheath. The first two cells have then withdrawn, leaving a lymph space behind them; at a next moult, they will re-extend to construct a new bristle and socket. The ciliated dendrite of the neuron is attached to the socket, and the sheath maintains a second lymph space (lymph spaces are believed to allow precise control of the ions needed in transduction). The neighbouring epidermal cells are not directly involved, but if the animal grows and moults, any of them could become a SOP cell and produce its own macrochaete.

Figure 26 shows a non-innervated bristle. Shaft and socket are in place and their respective cells have drawn back into the epidermal layer. There are (usually) no other cells involved, unless the bristle or scale is also secreting something or is otherwise specialized.

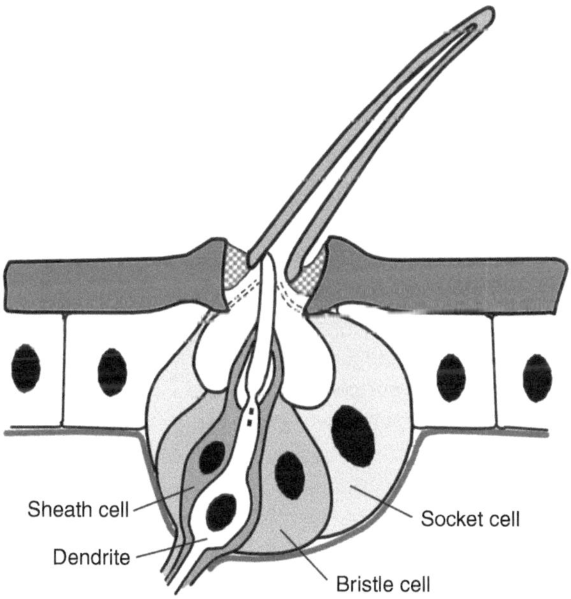

FIG. 25 Structure of an innervated bristle. Bristle, socket, and sheath cells have made their respective structures; the first two have withdrawn, but all three will be available to reconstitute their structures for any future moults. Reprinted from Ghiradella and Butler (2009).

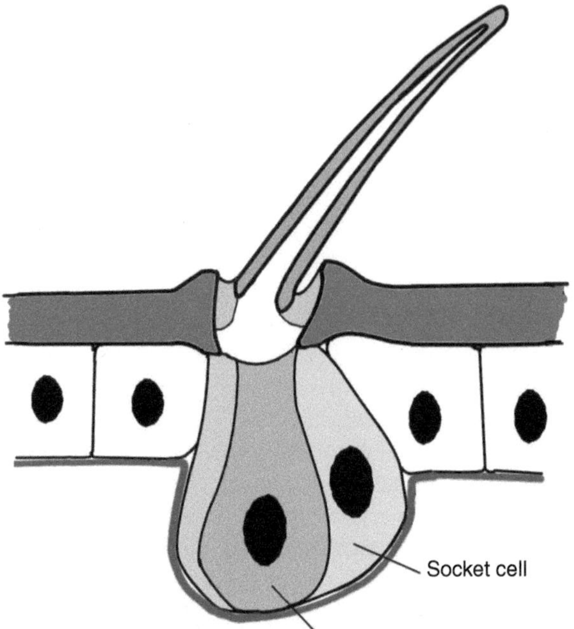

FIG. 26 A scale or non-innervated bristle. Shaft and socket cells have made their respective structures and are withdrawn as described for Fig. 25. As development proceeds, bristles remain cylindrical while scales flatten and develop their asymmetrical envelopes, ornamented on the upper surface but not on the lower one. (Figure 27 presents a diagrammatic transverse section of a scale shaft during an early stage of its development.) Reprinted from Ghiradella and Butler (2009).

As described earlier, macrochaete formation begins when the trichogen extrudes a process that projects above the epidermal cell layer. If this is to be a bristle, it remains cylindrical during its development, but if a scale, it develops a web of microtubules that apparently extend and flatten the nascent scale (Fig. 27). In either case, the periphery of the cell becomes ringed with a series of longitudinal, largely actin, bundles (Wolfrum, 1990) which are asymmetrical in size: those under the upper cell membrane are smaller than the others.

The cell starts cuticle production with the *epicuticle*, a thin initial layer that acts to contain future production and to isolate the new material from the moulting fluid and the material from the old cuticle. At this point, places along the actin bundles form localized ''regions of close contact'' (''cc regions'') (Greenstein, 1972) with the cell membrane and, at the same sites, between the cell membrane and the epicuticle. The cc regions preclude further secretion of cuticle, so that the ridges (and presumably the crossribs, although the author knows of no study of this particular point) can only form where the cc

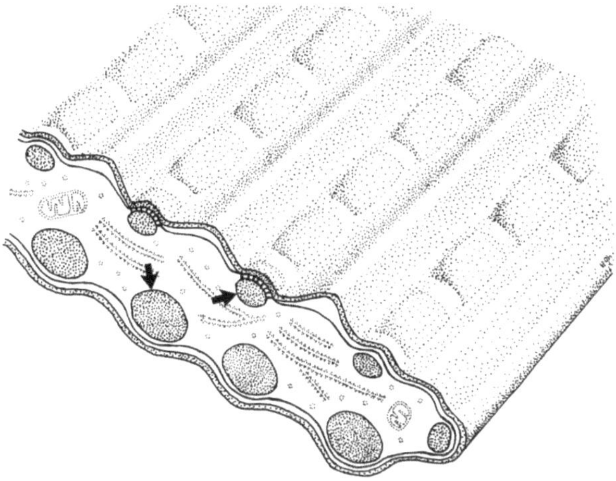

FIG. 27 A scale or bristle both start as an outward extension of the epidermal cell but in a scale, microtubules deploy so as to spread the structure in the axial plane. Actin bundles (arrows) form peripherally, just under the cell membrane; at intervals those under the developing upper surface maintain close contact (represented by the hatching) with the cell membrane. In these ''regions of close contact'', (''cc regions'') no inner layers of cuticle get laid down under the epicuticle, and here the windows will form. The actin bundles are also responsible for siting the ridges, which form between them. At these ridge sites, the cell can continue to secrete into the space under the epicuticle the precursors for the inner cuticular layers. Reprinted with permission from Ghiradella (1998).

regions are not. When the animal ecloses and the cuticle dries, the thin epicuticle will break through at these cc regions and the windows will result.

Figure 28 diagrams a standard scale at a later stage of development. The inner layers of cuticle are forming (except at the cc regions), and between the ridges the cell membrane is invaginating from both surfaces to form the trabeculae. When the animal ecloses and the cuticle dries, the lamellae settle into their final, regular forms and the fragile windows break through and open.

6.1 SPECIALIZED SCALE DEVELOPMENT

Ghiradella (1974) presents a model for the development of the ridge-iridescent scales, and Ghiradella and Radigan (1976) and Ghiradella (1989) discuss development of scales with internal structures, that is, with 3D pcs or laminae. Figure 29 shows a cross section of part of a UV-iridescent scale partway through its formation. The cell is beginning to draw back so that the inner cuticle layers can be secreted (again, except at the cc regions). The stacked lamellae are starting to appear, but they are unformed and irregular, as they will stay until the animal ecloses and the scale dries (see Ghiradella, 1974, for a more complete description).

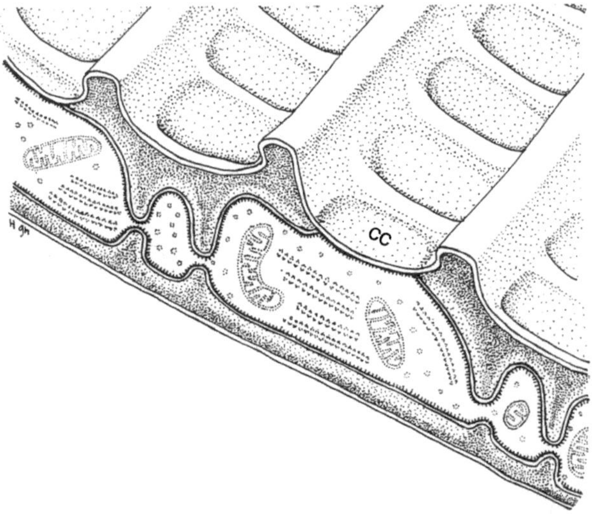

FIG. 28 Standard scale, later stage of development. The inner layers of cuticle are well on their way and the cell membrane is invaginating from top and bottom to form the trabeculae. At the regions of close contact (cc) the cuticle remains limited to the thin layer of epicuticle, which will break on eclosion to form the windows. Reprinted with permission from Ghiradella (1998).

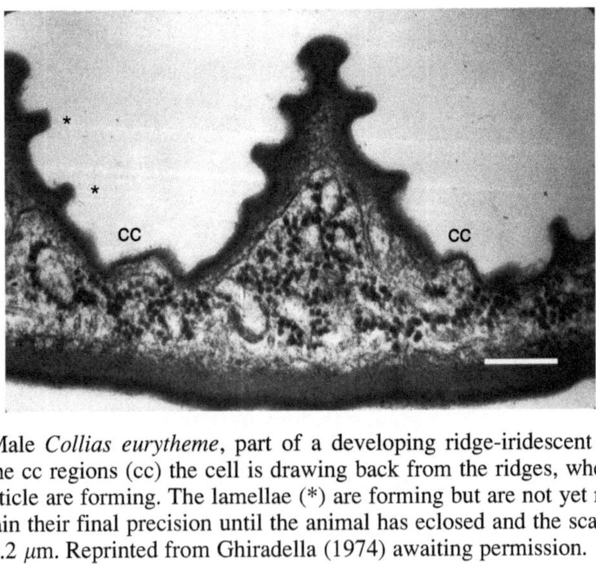

FIG. 29 Male *Collias eurytheme*, part of a developing ridge-iridescent UV + scale. Except in the cc regions (cc) the cell is drawing back from the ridges, where the inner layers of cuticle are forming. The lamellae (*) are forming but are not yet regular; they will not attain their final precision until the animal has eclosed and the scale has dried. Scale bar, 0.2 μm. Reprinted from Ghiradella (1974) awaiting permission.

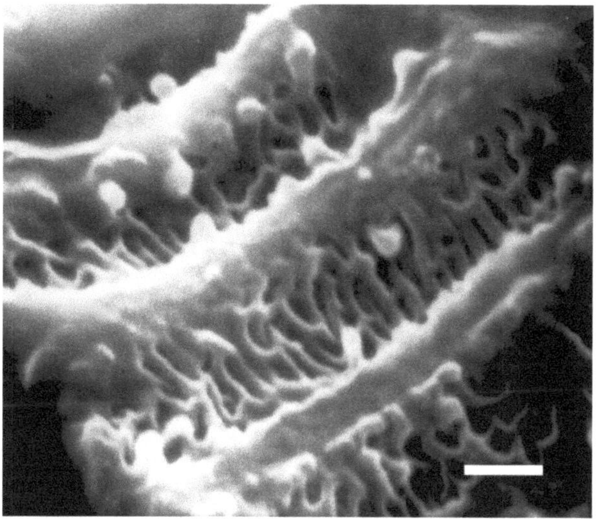

FIG. 30 *Celastrina ladon*, developing tracheole, internal view of wall. The developing tracheolar folds may in fact be homologous to the ridges and the crosswise pleats between them to the microribs. Scale bar, 0.5 μm. Reprinted with permission from Ghiradella (1998).

Figure 30 shows the interior of a fractured developing tracheole which, as noted earlier, may be thought of as an inverted scale, so that the structures we see here in the tracheole interior are equivalent to those on the exterior of a scale. The resemblance to ridges and microribs is clear, especially as these, like the lamellae on the scale in Figure 29, have not got their final form. The figures in Mill's (1998a) chapter on tracheae and tracheoles bear fair similarity in appearance to scale structures, which similarity lends support to the idea that these are developmental variants of the same basic structure, a complexly patterned cuticular tube with its attendant epidermal cell.

Given the irregularity of lamellae (or tracheolar folds) during development, we might ask what they might tell us about the mechanisms that form such regular final products. In her study of iridescent ridge development, Ghiradella (1974) saw in the micrographs of nascent ridges nothing that would suggest direct cellular involvement in lamella formation. Drawing on a suggestion by Locke (1958), she proposed that the lamellae instead might form by a physical mechanism: elastic buckling. If an elastic film is stressed, it will buckle sinusoidally (the mounting medium in Fig. 19 shows clean examples of this phenomenon). The frequency of the folds will depend on the material constants of the film (in this case, presumably the scale epicuticle) and the stresses applied. Given the deployment of the actin bundles in the scale primordium, the author suggested that were the large ones near the lower surface to contract and shorten the scale lower region, the scale would flex in its central plane and the result would be to stretch the upper surface, especially the ridges. The taller the ridges the greater the stress and, all other things being equal, the finer the folds and the more short-wave the reflected light. This might explain how ridge (lamellar) iridescent scales may be "tuned" to colours ranging from green to ultraviolet. The forms could well be uneven and irregular as the scale is developing, but upon eclosion and drying, one would expect them to "set up" and become even. . .as they do.

Timoshenko and Gere (1961) present a thorough and technical treatment of elastic buckling and related processes.

Let us now turn to the development of the scales with the 3D pcs in their lumina. Ghiradella and Radigan (1976) interpreted the crystal form of the 3D pc crystallites of *Callophrys rubi* as fcc and proposed that the formative mechanism might be emulsion polymerization: some subcellular element would form up into spheres and pack in a fcc array (a low energy configuration). On the basis of a series of Plateau constructions[2] (see Stevens, 1974), the authors proposed that surface tension might serve as the packing force—indeed all of the cell's internal structures seem to conform to this possibility of surface tension patterning. In short, the proposal was that the spheres, perhaps kept from fusing by electrostatic repulsion or an equivalent force, would be pushed together by the surface tension, resulting in the fcc arrangement. Polymerization of the continuous phase would yield the final structure.

It was somewhat sobering to discover that such a packing model was not at work here (but may exist in other systems: see Dufresne *et al.*, 2009). In a later paper, on the scales of the Lycaenid *Mitoura grynea*, Ghiradella (1989) reported that one of the cell's internal membrane systems, the SER, was involved, perhaps acting as the template for the formation of the 3D photonic crystallites. Figure 31 presents a view of such a crystallite on its way to assembly. The SER has formed a lattice and sleeves of cell membrane have woven through this to form a complementary lattice within which cuticle is secreted (extracellularly, as usual). In the same paper, Ghiradella also reported on the development of the thin-film laminae of another Lycaenid, *Celastrina ladon*. In this case, the origins are not so clear. The SER is not clearly (or at least visibly) organized, and the cell produces globules filled with some sort of fibrillar structure that may be nascent cuticle. At a later stage, however, the laminae are clearly forming within profiles of cell membrane, suggesting that all in all, formation of both internal structures, the 3D pcs and the laminae, may be by a related

[2] The Belgian physicist, J.A.F. Plateau discovered that for a 2D array of bubbles, strict rules apply. If two bubbles meet, they will partially fuse and form a partition between them. At the junction between the bubble surfaces and the end of the partition, the angles between partition and tangents to the bubble surfaces will always be 120°, as will be the angle of the radii between the two centres of curvature and the same intersection. No matter how complex the 2D system, if it is made of minimal surfaces, as bubbles usually are, all angles must be 120°. In this case, surface tension would be driving the minimization of the surfaces, but in others it may be packing or cracking (see Stevens, 1974, for a fine discussion of the principles involved).

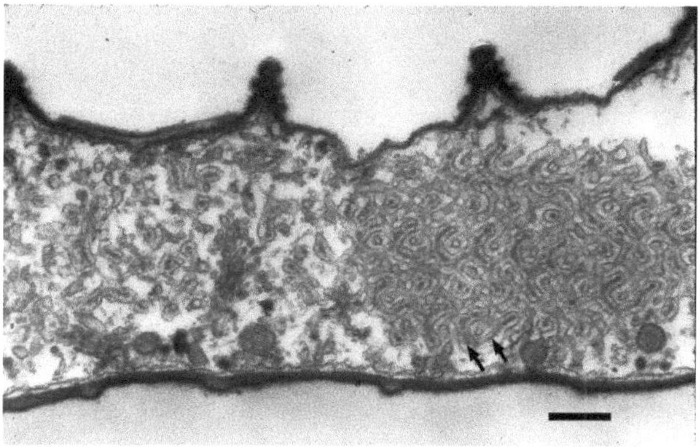

FIG. 31 *Mitoura grynea*, part of a developing reflective scale. A 3D photonic crystal is taking form at right: the smooth ER has formed a lattice through which have woven invaginated sleeves of cell membrane to form a complementary lattice. The latter is secreting the cuticle (dark material, arrows) that will form the final structure. Scale bar, 0.5 μm.

process, a view further supported by the presence in some Lycaenids of blue (laminar) scales on one side of the wing and green (3D pc) on the other.

In sum (so far), we hope that we have convinced the reader that the epidermal cell can do it all: form cuticular structures of great and specific complexity (and at very small scales—pun only partly unintended). We would like to go a little further and mention two other arthropods, one of which is relatively unknown, while the other is at least somewhat familiar. They are *Pauropus* sp., a member of the Class Pauropoda, and *Tomocerus* sp., a Collembolan.

7 Two other arthropods

Figure 32 presents a view of *Pauropus* sp. As a Pauropod, it is a minute relative of the Diplopoda (the class that includes the millipedes). Because of its small size and the fractional Reynolds number[3] at which it must be operating (this has not yet been measured), it lives in a regime in which air should be as viscous as oil and free movement should be impossible for the organism and even more so for its individual appendages. Nevertheless, the animal tools rapidly along on the

[3] Reynolds number is defined as $Re = \rho vl/\mu$, where ρ = density; v = velocity; l = length (size), usually in the direction of flow; and μ = viscosity. *Re* is essentially a measure of the ratio in a flow regime of the inertial to the viscous forces. At higher *Re*, turbulent drag becomes an issue and organisms must streamline if they are to be fast, large, and/or moving through a dense fluid. At low *Re*, turbulence is not an issue, but viscous drag is. An organism can ameliorate this drag by having surface structures that interact with the *boundary layer*, the shell of fluid that starts with the essentially static layer adhering to the organism's surface and continues outward through the additional layers that in turn surround each other and that are increasingly less reluctant to shear past each other.

FIG. 32 *Pauropus* sp., adult. This, the author's favourite arthropod, is related to the millipedes but is highly miniaturized: a large adult reaches 1.5 mm in size. As such, it is living in a physical world that biologists are only recently beginning to appreciate and understand. The numbers indicate the approximate location of the structures represented in the next three figures. Scale bar, 100 μm.

substrate and moves its antennae and their complicated branches with apparent ease. A closer look at its surface structures may yield a possible explanation.

Figure 33 shows a leg joint, together with the investiture that characterizes in one form or another the surface of the animal. Here, the structures are all hairs, hairs of two sizes and placed in precise order with respect to each other. A quick look at the scale bar shows a first question: the small hairs, at least, are so fine that it is hard to imagine how they might be formed by what we usually describe as ''cuticle formation''. There also needs explanation of the precision of the siting of the hairs.

Figure 34 shows a top view of part of the right side of the head and, at the upper right, the base of the right antenna. Besides the general hair investiture, there are several bristles of at least three different forms, all with extremely fine ornamentation that again raises the questions about their development. In the centre of the picture, where one would expect an eye, lies the *pseudoculus*, a problematical organ whose function has been tentatively ascribed on the basis of an electron microscope study (Haupt, 1973) as olfactory. This raises a second question: can an organ that is commonly an eye specialize for something so different in one lineage? Shubin *et al.* (2009) discuss some aspects of this question; we will return to it in Section 8.

The basal segments of the antenna show at least three different forms of hair investiture, each arrayed in fields whose borders are strictly delineated. And once again, the scale bar indicates that we may have to come up with some new models to explain their fabrication as well as the precision of their deployment.

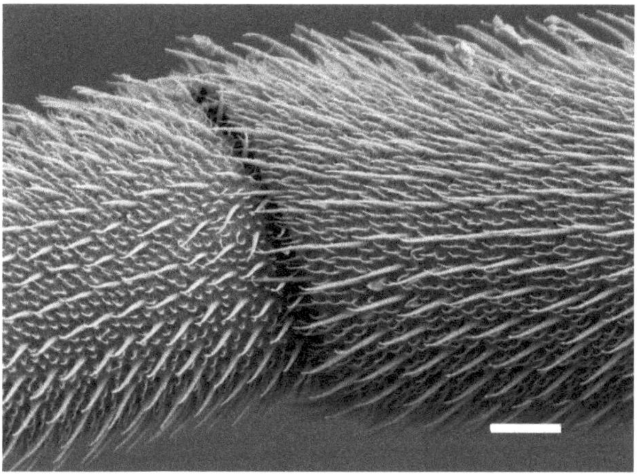

FIG. 33 *Pauropus* sp., detail of leg joint. The microchaetes here come in two sizes that are arranged in a near perfect array. The small size of the structures suggests that they may be templated on cellular filopodia. (Slightly re-touched at lower right to replace unsightly original scale bar.) Scale bar, 2 μm.

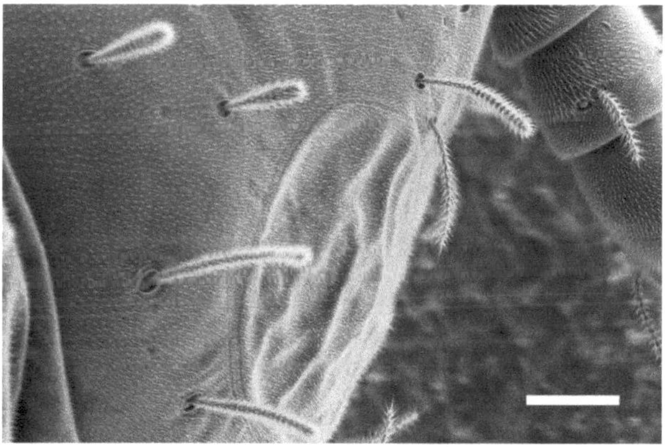

FIG. 34 *Pauropus* sp., closer view of part of head (the boundary with the thorax is at the extreme left). The "eye" region (known as the *pseudoculus*) is clearly delineated in the centre of the view, but because of diffraction issues its small size almost certainly precludes its functioning as a standard arthropod ("light pipe") eye. The surface is covered with the fine investiture we noted in Fig. 33: the bristles are similarly finely arrayed. The base of the antenna appears at the upper right: even at this small scale, the animal clearly has control of (and presumably reasons for) the precise placement of three very different patterns of surface investiture. Scale bar, 10 μm.

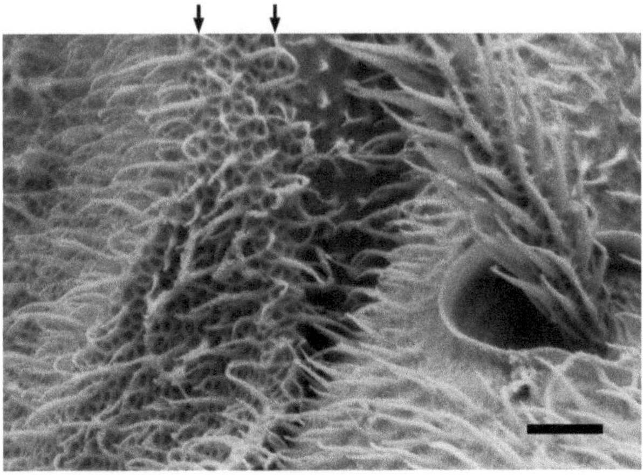

FIG. 35 *Pauropus* sp., rear edge of a left pseudoculus, which is tilted away from the observer, except for its periphery (between the arrows). The general head capsule surface (right) shows the usual complex investiture, as well as a bristle in its socket; the pseudoculus surface (seen best between the arrows) shows a pattern basically similar to those found elsewhere on the animal, but with the emphasis on a "rosette" structure formed by fine struts running from hair to hair. On the bristle, the pattern has been "stretched" into a more longitudinal form. Scale bar, 1 μm.

Figure 35 presents a view of the rear rim of a left pseudoculus. The (slightly obscured) "rosette" patterning (between the arrows) represents what we believe is the basic pattern of the hair investiture: a large hair is surrounded by several fine hairs that are joined to it and to each other by fine struts that form the rosette. The rosettes overlap to give a tesselated pattern, as shown here in the pseudoculus. In the bristle on the right, the pattern is "stretched" to become more bilateral. On the general integument (Figs. 33 and 34) the struts may be suppressed, while elsewhere (not figured) the larger hairs are, so that the appearance is simpler and flatter.

Figure 36 presents a surface view of part of our *Tomocerus* sp. The scales and bristles (which first called us to these animals-see Ghiradella and Radigan, 1974) show fine surface detailing. But it was the integument that captured our attention; it is thrown into hexagonal patterns that are characteristic of this arthropod class. Figure 37 presents a closer view of the surface pattern. Here again, there are some sturdy developmental questions that we need to answer. To the author's knowledge, there exists no developmental study dealing with this collembolan surface patterning, so we do not know what the animal actually does. But, with some simple materials and a bit of origami it is possible to come up with a fair model (Fig. 38).

In sum, the arthropod cuticle has lent itself to an astounding variety of surface structures of which we could only present the flavour here. We now need to consider some of the questions we have been raising, especially in the larger perspective of arthropod pattern formation.

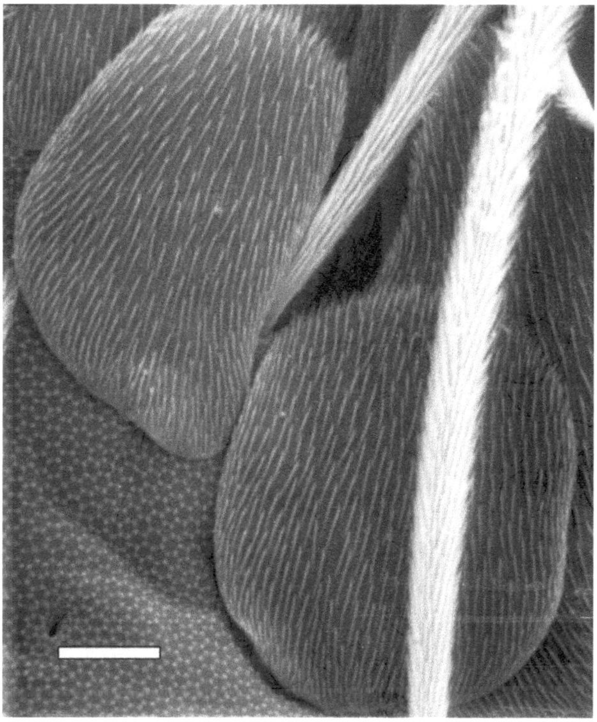

FIG. 36 *Tomocerus* sp., detail of surface. This member of the Collembola first interested us because of its scales and bristles, but these have turned out to be more or less similar to those that have already been reported in insects. The surface sculpting of the cuticle is, however, another example of what an arthropod can do. Figure 37 presents a closer view of the surface structure. Scale bar, 5 μm.

8 Discussion

Having now a general idea of the pattern-forming capabilities of the insect epidermal cell, let us step back and take a longer view to see what basic themes may emerge. In his fine review article, Locke (1998) remarks that "Insects are epidermal organisms. They are the prime example of how an infinite complexity of form may be attained by the manipulation of a layer that is only one cell thick." As various authors have remarked over the years, it is probably this two-dimensionality of body wall that has allowed, or perhaps required, the arthropods to develop a body design that consists of rolled up sheets... tubes. But any body wall is a boundary between external world and internal processes, and if it is to be one cell thick, that cell must do the job of distinguishing between out and in, exterior and interior, world and self.

Locke describes the epidermal cell as having two compartments, a basal and an apical, a distinction that is maintained tissue-wide by lateral interaction

FIG. 37 *Tomocerus* sp., end of antenna. The enlargement shows again bristles and sockets (the latter marvellously ''finished''), together with a closer view of the surface patterning. The bristles appear to have additional sculpting: for example the erect ones appear to have something resembling ''crumple zones'' near their bases. Scale bars, 20 μm (top), 2 μm (bottom).

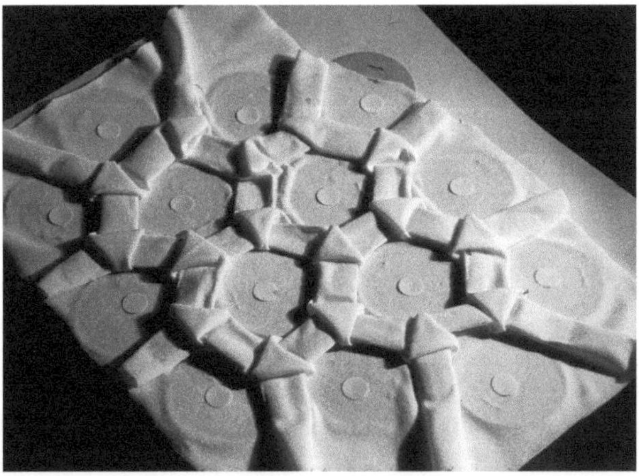

FIG. 38 The author does not know how the animal actually does it, but with some materials and a little origami it is possible quite easily to duplicate the collembolan surface patterning.

between the cell and its neighbours. The basal compartment interacts with the hemolymph, secretes some proteins, maintains the basal lamina, and endocytoses those components that are necessary to build the cuticle. The apical compartment is "microvillate" and controls a space extending to the outer epicuticle. It forms and shapes some of the structures of interest to us here. (It may also be a major player in shaping the internally derived structures, the 3D pcs and the laminae, but this is at present not as clear.)

For our discussion, we will therefore draw a somewhat modified distinction, one between (1) structures formed from templates at or on the cell membrane (the ridges, with their lamellae and microribs, the windows in their various forms, and the hairs and any other structures patterned on and/or developed from microvilli and perhaps from filopodia) and (2) structures (the 3D pcs and the laminae) whose formation is at least partly dependent on subcellular elements. Drawing on our present knowledge, we may perhaps redefine these as (1) those for whom the actin cytoskeleton is a... the?... major architectural force and (2) those that depend on the SER for their patterning.

8.1 STRUCTURES FORMATTED BY THE ACTIN CYTOSKELETON

To recap, these include the ridges, lamellae, microribs, windows, and the hairs. All these involve the deployment of the actin filaments or bundles that generally characterize epidermal cell outgrowths (Locke, 1998; Tilney and DeRosier, 2005; Tilney *et al.*, 2004). Our scenario goes as follows: early on, the cell extension must first form the outer epicuticle, the thin layer that defines the compartment within which the cell will do its patterning and which limits the extensibility of that compartment. Perhaps concurrently the actin bundles will form the cc regions, on the entire circumference in nascent bristles, but only on one side in nascent scales. These cc regions will site the ridges and probably the crossribs.

All this bespeaks control by the cell of the formation and deployment of the actin filaments and of their organization into bundles. Zhang *et al.* (2009) report on the role of a protein, RAB35, in conjunction with another protein, fascin, in forming the actin bundles involved in bristle formation in *Drosophila*. We can probably assume that this is true in all macrochaetes.

But once formed, or perhaps during their formation, the actin bundles seem to acquire complex pattern formation capabilities of their own as in, for example, deployment and patterning of the cc regions, which in turn appear to control the siting and shaping of the windows. If, as the evidence clearly suggests, the cc regions control the shapes and deployment of the windows, they may have exquisite control of that deployment. Also, if the epicuticle is in fact inextensable, it must be first laid down larger than the actual final dimensions of the macrochaete so that there will be "extra" material to buckle up into the ridges and crossribs and the microribs. This bespeaks temporal, as well as spatial control, that is, a sort of cellular "planning ahead" in the deployment of material(s). Actin filaments and/or bundles may further contribute to the final structure by shrinking the macrochaete to its final dimension(s) to allow the aforementioned buckling.

We presume that scales with "wide open" windows have larger cc regions, the extremes being those lacking crossribs altogether, as the glass scales of the morphos. Similarly, we would expect the "netted" windows of the papilionid ground scales to be formed by similarly patterned cc regions, and the alveoli of the 2D pcs by particularly minute cc regions.

We need a model for the formation of the latter. We know of no developmental study of this type of scale, but in *P. zalmoxis* one can extrapolate back from the adult structures and derive a possible developmental scenario. We can assume that secretion of the epicuticle must be the first step in any macrochaete formation. But it is not clear that these scales need or could even accommodate the more familiar actin bundles. The ridges are formed where one would expect them, but they appear to be simply narrow folds of epicuticle with no inner cuticular structures. If the cell somehow sited them and then raised a lawn of microvilli across its entire upper surface, we would have the template for the 2D pc slab.

Microvilli are supported by actin filaments (Locke, 1998) and it would not be surprising if they could establish cc regions at their apices. The epicuticle would get secreted and tacked to the microvillar apices in the minute cc regions, while the supporting structure of the 2D pc gets woven (as would peritrophic membrane) around and between the sides of the microvilli. When the animal ecloses and the cell dies back and the cuticle dries, as usual the unsupported epicuticle would break through to form what in this case are the fine alveoli.

We are at a loss to explain how this scenario could be modified to explain the "pores and plates" morphology of the androconial scales, but we still believe that this must be a variant of the process that forms the 2D pcs. One possibility might be that some of the chitin fibrils gathered in bundles between the microvilli in the nascent 2D pc scales are instead somehow spread or splayed out by mutually repellant charges or the equivalent, thereby supplying the webbing under the plates.

We need some cellular structure that can serve as a template for the very fine hairs that characterize the integument of *Pauropus* and other minute arthropods. Zhang *et al.* (2009) report that the Rab35–fascin system also raises filopodia, extremely minute cellular protrusions often found at leading edges of moving cells. Filopodia could very well serve as the templates for even the finest hairs, but we can still wonder at how such fine cellular projections can support the secretion of cuticle, especially if it is just the epicuticular layer that forms the final structure. It would be useful to have confirmation on this point, or to have a hint of what other cellular template or mechanism might form these hairs.

8.2 FORMATION OF THE LAMELLAE AND MICRORIBS

Let us briefly consider the ridges and their attendant folds. As discussed by Ghiradella (1974), we believe that the ridge lamellae may form by elastic buckling of the ridge epicuticle. By this mechanism, stresses caused during the formation of the ridges would throw them into longitudinal tension, and regular buckling would inevitably result (see Fig. 19 for an example of elastic buckling in a non-living material). The "extra" cuticle in the grooves between the lamellae would also be stressed, hence the orthogonal microribs. As mentioned earlier, by control of the cuticular material parameters and the stresses applied, such systems should easily be tunable to be reflective in any particular wavelength range. Introduction of shear into the stress will tilt the resulting lamellae, again producing substantial variation with relatively little modification of the pattern formation mechanism(s). We still have the question open of how the cell would "determine" when and how much shear (or stress, in the first place) is needed.

8.3 STRUCTURES FORMATTED BY THE SER

We now turn our attention to the 3D pcs, formed, we believe, by some sort of templating action of the SER, which is turning out to be a most versatile agent of cellular function. To list those capabilities of which the author is currently aware:

- Synthesis of lipids
- Detoxification of substances in the cell
- Storage and release of nuclear membrane components (Anderson and Hetzer, 2008)
- Production of fine spherical structures that produce a blue integumentary colour in some Odonates (Prum *et al.*, 2004)
- Reversible shuttling of Ca^{2+} into and out of the contractile machinery in voluntary muscle cells (Vogel, 2001)
- Secretion of wax from insect cell surfaces (Locke, 1985)
- Formation of the luminescent bodies ("lumisomes") in the luminescent "scales" of polynoid worms (Bassot and Nicolas, 1978, 1987)
- Acting as the template for the formation of 3D pcs within our macrochaetes (Ghiradella, 1989)

The last four functions involve some sort of close association between the SER and the cell membrane, but for such different reasons that it is clear to us that one cannot assign a particular character to a given SER in a given cell at a given time. We suspect that this is all negotiable, depending on the circumstances in which the cell finds itself.

Given the ease with which the SER can change its morphology from lattice to laminae (Almsherqi *et al.*, 2006; Snapp *et al.*, 2003), which changes may be mediated by a very local control (S. Hyde, personal communication; Powell, 2009), it is almost inconceivable that some laminar variant of the SER lattice is not also serving as a template for the laminae. The preliminary work by Ghiradella (1989) suggests otherwise, but more work is definitely needed on this particular system, especially as stretching a 3D pc in one plane should be enough to produce something very like a set of laminae, complete with spacers.

These macrochaete cells are demonstrating patterning control at yet another level, that of integration between various elements of the same scale or bristle. Three examples: (1) the restriction in such lycaenid scales as the 3D pc scales of *M. grynea* (Fig. 19) of the reflective structures to that part of the scale that is not covered by its neighbours, that is, is externally visible; (2) the deployment of distinctive high ''honeycomb'' ridges in some but not all 3D pc scales, even within a taxon; (3) the distinction between those 3D pc scales whose 3D pc reflectors are arrayed as crystallites and those which apparently make them as extended slabs. The reflective results are different in all cases, but how does the cell ''know'' and exercise the needed control in all these cases?

8.4 PAUROPUS

Such microarthropods as *Pauropus* and *Tomocerus* bring up an additional set of questions. *Pauropus* is in a lineage related to the millipedes, and it is a living example of the proposition that ''small'' does not necessarily mean ''simple''. For example, though it has a small brain with room for relatively few cells (general cell size does not change much across the taxa) it seems to have complicated behaviour, and its surface integument and antennal design bespeak pattern formation capabilities truly on a nanoscale.

The pseudoculus is where an eye would be on a larger relative, and it apparently projects to the protocerebrum (Haupt, 1973; Snodgrass, 1965). Haupt assigns to this organ an olfactory function, but we find this unlikely. The small size of the animal precludes even a fiber optic (= ommatidial) receptor design, but this does not necessarily allow us to move the organ to a different sensory domain. In their discussion of this sort of issue, Shubin *et al.* (2009) develop the concept of ''deep homology'' and review what is known of the history of photoreceptors. A main point: structures that do not appear classically homologous may nevertheless share not only a ''genetic tool kit'' but also regulatory circuits, all of which may allow the development of what appear among the taxa to be disparate features but are basically not. While physicists can predict what the optical environment might be for such a small receptor, we are a long way from understanding any possible biological response to light at such a scale. Further study may yield some interesting insights. (This author cherishes the dream that if ever a negative refractive index ''eye'' is found in a biological system, it will be here.)

We may ask *why* the animal has such an impressive investiture. Our tentative answer has to do with the boundary layer between the fluid and the organism. As noted earlier, viscous drag will be an issue for anything so small that moves or is moved through a fluid. Fine surface detailing apparently helps reduce this drag (Vogel, 1981) and at this, *Pauropus* is a master. Other minute arthropods are also complexly decorated, and probably for the same reason(s).

8.5 TOMOCERUS

Let us turn now to *Tomocerus*, our representative of the Collembola. Figure 36 presents two macrochaetes, a scale and a bristle, but other than the fineness of their ornamentation, they are not (we think) that different from insect macrochaetes. The integumental surface, however, is: a closer view (Fig. 37) shows a hexagonally packed array of what look like discs with raised ridges at their sides and triangular folds at their corners. Figure 38 presents the author's attempt to model this pattern: to do this, she tacked onto a sheet of fabric a loosely packed hexagonal array of paper cup lids and then simply moved the lids closer together so that the fabric had to bulge between them. We do not know how the animal sets up the pattern, but if it is anything like, it need only set up a hexagonal array of cc regions (the cup lids) and then shrink the underlying cellular support and let the epicuticle buckle into the final shape, a variant on the ''extra epicuticle'' model we have been proposing. If this be true, the basic patterning may be a more precise variation on that of the netted windows of the papilionid ground scales.

We originally thought of the collembolan patterning as a plastron, which might be very necessary for a small animal, especially one that lives in niches rich in surfactants and presumably therefore with water with lowered surface tension. The same would of course be true for the pauropod, but we are now also thinking for both animals in terms of boundary layer. These animals must be moving at fractional Reynolds numbers, and it is certain that they are somehow limiting viscous drag. Again, these may be fine models to study for insights into how any nanostructure may be engineered to behave as freely in its local environment as these animals do.

9 Final thoughts

In closing, let us make several points. As she thinks about these systems, the author comes increasingly to regard the eukaryotic cell as the ''real'' organism and us multicellular creatures as diverse expressions of its creative drive. Stated another way, the general lability of the developmental processes leading to these patterns and their commonality among the taxa (a theme well expressed by Shubin *et al.*, 2009) suggests that they are unreliable as taxonomic markers. A slight change in timing of some developmental element or in the precise chemical formulation of a cuticular element, or the introduction of a different degree of shear in the tension generated in an elastic buckling system (if we are right about that one) will lead to very different phenotypic outcomes.

It is almost certain that, as these processes are only specific instances of more general developmental patterns, they cannot be occupying much genetic space. For example, in studies of development in ultraviolet pierid butterfly scales, just a change in one allele was shown to determine whether or not the male would have multilayer ultraviolet reflectors on his wing scale ridges (Silberglied and Taylor, 1973). The use of microvilli as templates for such dissimilar structures as cuticular papillae, peritrophic membrane, and perhaps 2D pcs bespeaks both innovation and conservation on the part of the epidermal cell, as does the multiplicity of functions, some involving patterning, of the SER. Needless to say, the macrochaete systems alone speak volumes about possible patterning capabilities of the actin cytoskeleton.

The basic world of the very small presents real opportunities for advancement of our understanding of both the physics and biology of the world we live in. Research in this field is gaining speed: let us hope that it continues to do so, not in the least because so many of our personal body functions operate at this scale and may therefore be subject to the same or similar challenges.

And there is the field of biomimetics, the study of these and other systems in the hopes of being able to duplicate for human purposes their pattern-forming abilities. The learning goes both ways: our engineered systems sometimes give us hints of how to re-interpret the biological ones, while these latter are obviously teachers of great potential if we give their study the care it deserves. As master engineers, the arthropods are a great and talented group at whose many feet we might gainfully sit.

Acknowledgments

Over the years, the author is indebted to many who helped, either directly or indirectly in fostering this research. First and foremost, Thomas (and Maria) Eisner first introduced the subject and have supported me in it ever since. Robert Day Allen provided the place to work and learn. William Radigan and Harry L. Frisch were brilliant and able collaborators, and Bob Greenler provided insights into the field of optics. Orley Taylor and David Wright provided specimens and helpful advice. Sam Bowser of the New York State Wadsworth Center for Labs and Research and Jack Harris of Russell SageCollege are more recent collaborators. I also extend thanks to those who were not directly involved in this work but who were always there for me. In particular, I thank Doekele G. Stavenga, whose thoughtful, constructive, and meticulous review has made this a much better effort than I could have done without him. Special note is due Robert Silberglied, who was part of our early studies and whose untimely death deprived us of a talented and promising young insect biologist.

References

Almsherqi, Z. A., Kohlwein, S. D. and Deng, Y. (2006). Cubic membranes: A legend beyond the *Flatland* of cell membrane organization. *J. Cell Biol.* **173**, 839–844.

Andersen, S. O. (2009a). Cuticle. In: *Encyclopedia of Insects* (eds Resh, V. H. and Cardé, R. T.), 2nd ed. pp. 245–246. Academic Press, San Diego, CA.

Andersen, S. O. (2009b). Exoskeleton. In: *Encyclopedia of Insects* (eds Resh, V. H. and Cardé, R. T.), 2nd ed. pp. 339–342. Academic Press, San Diego, CA.
Anderson, D. J. and Hetzer, M. W. (2008). Reshaping of the endoplasmic reticulum limits the rate for nuclear envelope formation. *J. Cell Biol.* **182**, 911–924.
Bassot, J. M. and Nicolas, M. T. (1978). Similar paracrystals of endoplasmic reticulum in the photoemitters and the photoreceptors of scale-worms. *Experentia* **34**, 726–728.
Bassot, J.-M. and Nicolas, G. (1987). An optional dyadic junctional complex revealed by fast-freeze fixation in the bioluminescent system of the scale worm. *J. Cell Biol.* **105**, 2245–2256.
Binnington, K. C., Lehane, M. J. and Beaton, C. D. (1998). The peritrophic membrane. In: *Microscopic Anatomy of Invertebrates, Vol. 11B, Insecta* (ed Locke, M.), pp. 747–758. Wiley-Liss, New York.
Boppré, M. and Vane-Wright, R. I. (1989). Androconial systems in Danainae (Lepidoptera): Functional morphology of *Amauris, Danaus, Tirumala and Euploea. Zool. J. Linn. Soc.* **97**, 101–133.
Bush, J. W. M., Hu, D. L. and Prakash, M. (2008). The integument of water-walking arthropods: Form and function. *Adv. Insect. Physiol.* **34**, 117–192.
Casey, T. M. and Hegel, J. R. (1981). Caterpillar setae: Insulation for an ectotherm. *Science* **214**, 1131–1133.
Dufresne, E. R., Noh, H., Saranathan, V., Mochrie, S. G. J., Cao, H. and Prum, R. O. (2009). Self-assembly of amorphous biophotonic nanostructures by phase separation. *Soft. Matter.* **5**, 1792–1795.
Egelhaaf, A., Rick-Wagner, S. and Schneider, D. (1992). Development of the male scent organ of *Creatonotos transiens* (Lepidoptera, Arctiidae) during metamorphosis. *Zoomorphology* **111**, 125–139.
Eisner, T. and Shepherd, J. (1966). Defense mechanisms of arthropods. XIX. Inability of sundew plants to capture insects with detachable integumental outgrowths. *Ann. Entomol. Soc. Am.* **59**, 868–870.
Eisner, T., Alsop, R. and Ettershank, G. (1964). Adhesiveness of spider silk. *Science* **146**, 1058 1061.
French, A. S. and Torkkeli, P. H. (2009). Mechanoreception. In: *Encyclopedia of Insects* (eds Resh, V. H. and Cardé, R. T.), 2nd ed. pp. 610–611. Academic Press, San Diego, CA.
Ghiradella, H. (1974). Development of ultraviolet-reflecting butterfly scales: How to make an interference filter. *J. Morphol.* **142**, 395–410.
Ghiradella, H. (1989). Structure and development of iridescent butterfly scales: Lattices and laminae. *J. Morphol.* **202**, 69–88.
Ghiradella, H. (1998). Hairs, bristles and scales. In: *Microscopic Anatomy of Invertebrates, Vol. 11A, Insecta* (ed Locke, M.), pp. 257–287. Wiley-Liss, New York.
Ghiradella, H. (2005). Fine structure of basic lepidopteran scales. In: *Structural Colors in Biological Systems—Principles and Applications* (eds Kinoshita, S. and Yoshioka, S.), pp. 75–93. Osaka University Press, Osaka, Japan.
Ghiradella, H. and Butler, M. W. (2009). Many variations on a few themes: A broader look at development of iridescent scales (and feathers). *J. R. Soc. Interface* **6** (Suppl. 2), S243–S251.
Ghiradella, H. and Radigan, W. (1974). Collembolan cuticle: Wax layer and anti-wetting properties. *J. Insect Physiol.* **20**, 301–306.
Ghiradella, H. and Radigan, W. (1976). Development of butterfly scales. II. Struts, lattices, and surface tension. *J. Morphol.* **150**, 279–298.
Giraldo, M. A. and Stavenga, D. G. (2008). Wing coloration and pigment gradients in scales of pierid butterflies. *Arthropod Struct. Dev.* **37**, 118–128.

Giraldo, M. A., Yoshioka, S. and Stavenga, D. G. (2008). Far field scattering pattern of differently structured butterfly scales. *J. Comp. Physiol. A* **194**, 201–207.

Grant, G. G. and Brady, U. E. (1973). Topography of the scent scales of the male cabbage looper *Trichoplusia ni*. *J. Georgia Entomol. Soc.* **8**, 99–106.

Greenstein, M. E. (1972). The ultrastructure of developing wings in the giant silk moth *Hyalophora cecropia*. II. Scale-forming and socket-forming cells. *J. Morphol.* **136**, 23–52.

Haupt, J. (1973). Die Ultrastruktur des Pseudoculus von *Allopauropus* (Pauropoda) und die Homologie der Schläfenorgane. *Z. Morphol. Tiere* **76**, 173–191.

Heinrich, B. (1987). Thermoregulation by winter-flying endothermic moths. *J. Exp. Biol.* **127**, 313–332.

Held, L. I. Jr. (2002). Imaginal Discs: The Genetic and Cellular Logic of Pattern Formation. Cambridge University Press, Cambridge, UK.

Hölldobler, B. and Wilson, E. O. (1986). Soil-binding pilosity and camouflage in ants of the tribes Basicerotini and Stegomyrmecini (Hymenoptera, Formicidae). *Zoomorphology* **106**, 12–20.

Huxley, J. and Barnard, P. C. (1988). Wing-scales of *Pseudoleptocerus chirindensis*. Kimmins (Trichoptera: Leptoceridae). *Zool. J. Linn. Soc.* **92**, 285–312.

Keil, T. A. (1997). Functional morphology of insect mechanoreceptors. *Microsc. Res. Tech.* **39**, 506–531.

Keil, T. A. (1998). The structure of integumental mechanoreceptors. In: *Microscopic Anatomy of Invertebrates, Vol. 11B, Insecta* (ed Locke, M.), pp. 385–404. Wiley-Liss, New York.

Kernan, M. J. (2007). Mechanotransduction and auditory transduction in *Drosophila*. *Pflügers Arch. Eur. J. Physiol.* **454**, 703–720.

Kertész, K., Bálint, Z., Vértesy, Z., Márk, G. I., Lousse, V., Vigneron, J. P., Rassart, M. and Biró, L. P. (2006). Gleaming and dull surface textures from photonic-crystal-type nanostructures in the butterfly *Cyanophrys remus*. *Phys. Rev. E* **74**(021922), 1–15.

Kingsolver, J. (1988). Thermoregulation, flight, and the evolution of wing pattern in pierid butterflies: The topography of adaptive landscapes. *Am. Zool.* **28**, 899–912.

Kingsolver, J. and Moffat, R. K. (1982). Thermoregulation and the determinants of heat transfer in *Colias* butterflies. *Oecologia* **53**, 27–33.

Kinoshita, S. (2008). Structural Colors in the Realm of Nature. World Scientific, Singapore.

Kinoshita, S. and Yoshioka, S. (2005). Structural Colors in Biological Systems—Principles and Applications. Osaka University Press, Osaka, Japan.

Lai, E. C. and Orgogozo, V. (2004). A hidden program in *Drosophila* peripheral neurogenesis revealed: Fundamental principles underlying sensory organ diversity. *Dev. Biol.* **269**, 1–17.

Locke, M. (1958). The formation of tracheae and tracheoles in *Rhodnius prolixus*. *Q. J. Microsc. Sci.* **99**, 29–46.

Locke, M. (1985). A structural analysis of post embryonic development. In: *Comprehensive Insect Physiology, Biochemistry and Pharmacology* (eds Kerkut, G. A. and Gilbert, L. I.), pp. 103–165.

Locke, M. (1998). Epidermis. In: *Microscopic Anatomy of Invertebrates, Vol. 11A, Insecta* (ed Locke, M.), pp. 75–138. Wiley-Liss, New York.

Merritt, D. J. (1997). Transformation of external sensilla to chordotonal sensilla in the *cut* mutant of *Drosophila* assessed by single-cell marking in the embryo and larva. *Microsc. Res. Tech.* **39**, 492–505.

Miaoulis, I. N. and Heilman, B. D. (1998). Butterfly thin films serve as solar collectors. *Ann. Entomol. Soc. Am.* **91**, 122–127.

Michielsen, K. and Stavenga, D. G. (2008). Gyroid cuticular structures in butterfly wing scales: Biological photonic crystals. *J. R. Soc. Interface* **5**, 85–94.
Mill, P. J. (1998a). Tracheae and tracheoles. In: *Microscopic Anatomy of Invertebrates, Vol. 11A, Insecta* (ed Locke, M.), pp. 303–336. Wiley-Liss, New York.
Mill, P. J. (1998b). Gills. In: *Microscopic Anatomy of Invertebrates, Vol. 11A, Insecta* (ed Locke, M.), pp. 337–362. Wiley-Liss, New York.
Mitchell, H. K. and Peterson, N. S. (1989). Epithelial differentiation in *Drosophila* pupae. *Dev. Genet.* **10**, 42–52.
Mitchell, H. K., Roach, J. and Petersen, N. S. (1983). The morphogenesis of cell hairs on *Drosophila* wings. *Dev. Biol.* **95**, 387–398.
Mitchell, H. K., Edens, J. and Petersen, N. S. (1990). Stages of cell hair construction in *Drosophila. Dev. Genet.* **11**, 133–140.
Morris, R. B. (1975). Iridescence from diffraction structures in the wing scales of *Callophrys rubi*, the Green Hairstreak. *Proc. R. Soc. Entomol. A* **48**, 149–154.
Nachtigall, W. (1974). Insects in Flight. McGraw Hill, New York.
Neville, A. C. (1975). Biology of Arthropod Cuticle. Springer-Verlag, New York.
Neville, A. C. (1993). Biology of Fibrous Composites: Development Beyond the Cell Membrane. Cambridge University Press, Cambridge, UK.
Nijhout, H. F. (1991). The Development and Evolution of Butterfly Wing Patterns. Smithsonian Press, Washington, DC.
Palmer, C. G. (1998). Setae and microtrichia: Structures for fine-particle feeding in aquatic larvae. In: *Microscopic Anatomy of Invertebrates, Vol. 11A, Insecta* (ed Locke, M.), pp. 289–302. Wiley-Liss, New York.
Poladian, L., Wickham, S., Lee, K. and Large, M. C. J. (2009). Iridescence from photonic crystals and its suppression in butterfly scales. *J. R. Soc. Interface* **6**, (Suppl. 2), S233–S242.
Powell, K. (2009). Ahead of the curve. *Nature* **460**, 318–320.
Prum, R. O., Cole, J. A. and Torres, R. H. (2004). Blue integumentary structural colours in dragonflies (Odonata) are not produced by incoherent Tyndall scattering. *J. Exp. Biol.* **207**, 3999–4009.
Prum, R. O., Quinn, T. and Torres, R. H. (2006). Anatomically diverse butterfly scales all produce structural colours by coherent scattering. *J. Exp. Biol.* **209**, 748–765.
Schmitz, H., Schmitz, A. and Bleckmann, H. (2001). Morphology of a thermosensitive multipolar neuron in the infrared organ of *Merimna atrata* (Coleoptera, Buprestidae). *Arthropod Struct. Dev.* **30**, 99–111.
Sellier, R. (1971a). Données sur les apports de la microscopie électronique à balayage, pour l'étude ultrastructural des écailles alaires chez les Lépidoptères diurnes. *C. R. Acad. Sci. Paris Sér. D* **273**, 2097–2100.
Sellier, R. (1971b). Étude morphologique en microscope électronique à balayage de quelques types d'androconies alaires ches les Lépidoptères diurnes. *C. R. Acad. Sci. Paris Sér. D* **273**, 2550–2553.
Shubin, N., Tabin, C. and Carroll, S. (2009). Deep homology and the origins of evolutionary novelty. *Nature* **457**, 818–823.
Silberglied, R. E. and Taylor, O. R. (1973). Ultraviolet differences between the sulfur butterflies *Colias eurytheme* and *C. philodoce*, and a possible isolating mechanism. *Nature* **241**, 406–408.
Snapp, E. L., Hegde, R. S., Francolini, M., Lombardo, F., Colombo, S., Pedrazzini, E., Borgese, N. and Lippincott-Schwartz, J. (2003). Formation of stacked ER cisternae by low affinity protein interactions. *J. Cell Biol.* **163**, 257–269.
Snodgrass, R. E. (1965). A Textbook of Arthropod Anatomy. Hafner, New York.
Stavenga, D. G., Giraldo, M. and Leertouwer, H. L. (2010). Butterfly wing colors: Glass scales of *Graphium sarpedon* cause polarized iridescence and enhance blue/green pigment colouration of the wing membrane. *J. Exp. Biol.* **213**, 1731–1739.

Stevens, P. S. (1974). Patterns in Nature. Little Brown and Co., Boston.

Thorpe, W. H. and Crisp, D. J. (1949). Studies on plastron respiration. IV. Plastron respiration in the Coleoptera. *J. Exp. Biol.* **26**, 219–260.

Tilney, L. G. and DeRosier, D. J. (2005). How to make a curved *Drosophila* bristle using straight actin bundles. *Proc. Natl. Acad. Sci. USA* **102**, 18785–18792.

Tilney, L. G., Connelly, P. S., Ruggiero, L., Vranich, K. A., Guild, G. M. and DeRosier, D. (2004). The role actin filaments play in providing the characteristic curved form of *Drosophila* bristles. *Mol. Biol. Cell* **15**, 5481–5491.

Timoshenko, S. and Gere, J. (1961). Theory of Elastic Stability. McGraw-Hill, New York.

Vane-Wright, R. I. (1972). Pre-courtship activity and a new scent organ in butterflies. *Nature* **239**, 338–339.

Vogel, S. (1981). Life in Moving Fluids. Princeton University Press, Princeton.

Vogel, S. (2001). Prime Mover: A Natural History of Muscle. W.W. Norton, New York.

Vukusic, P. and Hooper, I. (2005). Directionally controlled fluorescence emission in butterflies. *Science* **310**, 1151.

Vukusic, P., Sambles, J. R. and Lawrence, C. R. (2000). Colour mixing in wing scales of a butterfly. *Nature* **404**, 457.

Wasserthal, L. T. and Wasserthal, W. (1977). Ultrastructure of a scent scale organ with pressure discharge in male *Caligo eurilochus brasiliensis* (Fldr.) (Lepidoptera: Brassolidae). *Cell Tissue Res.* **177**, 87–103.

Whitten, J. (1969). Coordinated development in the fly foot: Sequential cuticle secretion. *J. Morphol.* **127**, 73–104.

Whitten, J. (1973). Kinetosomes in insect epidermal cells and their orientation with respect to cell symmetry and intercellular patterning. *Science* **181**, 1066–1067.

Wilts, B. D., Leertouwer, H. L. and Stavenga, D. G. (2009). Imaging scatterometry and microspectrophotometry of lycaenid butterfly wing scales with perforated multilayers. *J. R. Soc. Interface* **6** (Suppl. 2), S185–S192.

Winterton, S. (2009). Scales and setae. In: *Encyclopedia of Insects* (eds Resh, V. H. and Cardé, R. T.), 2nd ed. pp. 901–904. Academic Press, San Diego, CA.

Wolfrum, U. (1990). Actin filaments: The main components of the scolopale in insect sensilla. *Cell Tissue Res.* **261**, 85–96.

Yack, J. E. (2004). Structure and function of auditory chordotonal organs in insects. *Microsc. Res. Tech.* **63**, 315–337.

Yack, J. E. and Dawson, J. W. (2008). Insect ears. In: *The Senses: A Comprehensive Reference, Vol. 3, Audition* (eds Dallos, P. and Oertel, D.), pp. 35–54. Academic Press, San Diego.

Yoshida, A. (2005). Antireflective microstructure on the wings of butterflies and moths. In: *Structural Colors in Biological Systems—Principles and Applications* (eds Kinoshita, S. and Yoshioka, S.), pp. 249–264. Osaka University Press, Osaka, Japan.

Zhang, J., Fonovic, M., Suyama, K., Bogyo, M. and Scott, M. P. (2009). Rab35 controls actin bundling by recruiting fascin as an effector protein. *Science* **325**, 1250–1254.

Structural Colours

Jean-Pol Vigneron and Priscilla Simonis

Research Center in Physics of Matter and Radiation (PMR), University of Namur (FUNDP), Namur, Belgium

1 Introduction

According to their physical mechanisms, colouring processes are traditionally classified as either 'pigmentary' or 'structural'. Pigmentary colouration is essentially obtained when part of the spectral intensity of an illuminant beam is removed as it transits through a selective absorber. Absorbers can either be dyes or pigments—'dyes' referring to light-absorbing molecules dissolved in a

ADVANCES IN INSECT PHYSIOLOGY VOL. 38
ISBN 978-0-12-381389-3
DOI: 10.1016/S0065-2806(10)38004-0

liquid matrix; 'pigments' to light-absorbing molecules distributed in a solid matrix, generally divided into separate granules. Structural colouration arises from the wavelength-dependent redirection of the incident energy in inhomogeneous transparent materials. This redirection is usually the result of a multiple reflection on a simple or an intricate pattern of interfaces and the resulting interference of the multiple outgoing waves. To be effective, the distance between interfaces should be of the order of the incident light wavelength.

This separation between pigmentary and structural colourations is a bit academic, as strong coupling between them can occur, in the more general event where the material that constitutes the multiple-scattering structure would also be selectively absorbing. Pigments, indeed, are often incorporated in a geometric structure, so that absorption and interferences can develop simultaneously (the case of the butterfly, *Troides magellanus* (Vigneron *et al.*, 2008), is a striking example). When some white light transits through a liquid coloured by the presence of a dye, nothing else occurs than a simple spectral intensity subtraction. When a pigment works on the light scattered by a coloured object, scattering and absorption has to take place simultaneously and, for efficiency, the scattering material should be appropriately structured and pigmented, and in the limit of structure sizes of the order of the wavelength, a hybrid structural–pigmentary colouration mechanism should be considered. To our knowledge, little has been done specifically to describe this hybrid mechanism on specific examples and this may be one of the next objectives in the study of natural photonics.

In the present chapter, we want to focus on the physics of structural colouration, explaining selected physical mechanisms and presenting examples of the manifestation of these effects in the living world. This will not be a bibliographical review: the reader should not expect to see this presentation as an attempt to organize the past work on the subject. It should rather be understood as a 'tutorial'. Its subject is the great diversity of optical phenomena which allow for light manipulation by living organisms. The wide diversity of organisms that will provide our examples have only been selected to express the feeling that structural colouration has naturally occurred in many evolutionary paths of animal families.

2 Iridescence from ages

Pigmentary colours are essentially associated with diffuse reflection. In fact, it is often observed that the specular reflection from the surface of a smooth, planar, selectively absorbing material is not strongly coloured: The surface of a bright blue gleaming bowl (see Fig. 1), for instance, appears blue by pigmentary diffusion. When illuminated by a much localized white-light source, the highlights formed on the varnished surface at specular directions in the images of the

FIG. 1 Illuminated bowl, showing uncoloured highlights due to specular reflection and blue coloration due to diffuse scattering by pigmented granules. (See Color Plate 18 at the back of the book.)

light spot, conserve essentially the colour of these illuminants. Pigmentary colouration implies some diffuse reflection, which generally comes from a microscopic granular structure of the pigmented materials. This, again, may hide some coupling between structure and pigments. A pigmentary colouration is then usually recognized as associated with diffuse reflection and presents roughly the same hue and luminance when perceived from various directions.

By contrast, purely structural colours are often associated with *iridescence*. An iridescent surface tends to change its colour when viewed or illuminated from different directions. Structural colours, under unpolarized light, are generally iridescent because a change of light incidence results in a change of path lengths for the light propagating in the structure. The iridescence is one of the signs by which a colour is recognized as structural. Note that this is not the only test, and that the converse is not true: a lack of iridescence is not the signature of a pigmentary mechanism. Nature has found many ways to keep using structural colours and avoid iridescence, as we will see later. Nevertheless, our knowledge of structural colours has started with the observation of iridescence, and that is really an old story.

Iridescent feathers or insect's cuticles have been incorporated in garments by very ancient civilizations (including Egyptian or pre-Columbian), which suggests that the observation of iridescence dates back from many thousands years. However, one of the first mentions of iridescence in a book can be found in '*De Rerum Natura*' (*On the Nature of Things*) by the Roman poet and philosopher *Titus Lucretius Carus*, who lived in the first century before Christ (ca. 99–55 BC). The poem in the book contains the following excerpt, which mentions the iridescence of the neck of the dove (I would rather say 'pigeon') and the iridescence of the peacock feather (translated from Latin by William Ellery Leonard; Lucretius Carus, 2008):

A colour changes, gleaming variedly,
When smote by vertical or slanting ray.
Thus in the sunlight shows the down of doves
That circle, garlanding, the nape and throat;
Now it is ruddy with a bright gold-bronze,
Now, by a strange sensation it becomes
Green-emerald blended with the coral-red.
The peacock's tail, filled with the copious light,
Changes its colour likewise, when it turns.

It is pleasurable to mention that the physical origin of the iridescence of the pigeon neck and the iridescence of the feathers from the train of the peacock have both been studied recently by Prof. Jian Zi, from Fudan University (Shanghai, China), in 2003 (peacock) (Zi *et al.*, 2003) and 2006 (pigeon) (Yin *et al.*, 2006). However, likely, Prof. Zi was inspired more by the observation of the animals than by the careful reading of Lucretius' text.

Another report of the knowledge of iridescence appears in the period named 'European Renaissance', where authors were heavily inspired by Latin literature: the French author François Rabelais (ca. 1494–April 9, 1553) insists (Rabelais, 1534):

> Pour sa robbe furent levées neuf mille six cens aulnes moins deux tiers de velours bleu comme dessus, tout porfilé d'or en figure diagonale, dont par iuste perspective issoit une couleur innomée, telle que voyez es coulz des tourterelles, qui resiouissoit merveilleusement les yeulx des spectateurs.

''For his garment, nine thousand six hundred elves of blue velvet were cut, diagonally decorated with gold, which, from an appropriate viewing point, provided an unnamed colour, such as that observed on the doves neck, a source of enjoyment for the spectator's eyes''. An iridescent colour carries strange information, as it changes so easily and, as the Renaissance writer says, 'cannot be named'.

3 Climbing the complexity hill

In the following, we will see colouring structures, from the simplest to the most complex we have understood. The classification is lead by the physical mechanism in place to produce the colouration, so that we will rationalize the presentation by first explaining the 'generic' physical device, and economize on the physical explanations in the various examples we will mention.

The simplest photonic structure is a single planar interface between media with different 'optical densities' or 'refractive index' (see Section 4). This structure is so simple that it is difficult to believe that it can have any importance for our subject. It is in fact crucial, not only because it constitutes the elementary building block of all the other structures, in which multiple reflections and transmissions define the response function, but also because this system will in certain circumstances generate total reflection that can be used to create specific visual effects. The same phenomenon can also be detrimental, for instance to terrestrial bioluminescence, and in this case, evolution has favoured the appearance of structures that corrects for internal reflection.

The optical properties of a material like chitin is described by two parameters, called 'the refractive index' and 'the absorption coefficient'. The former expresses the light deceleration experienced when entering the material: the higher the refractive index, the lower the speed. The latter expresses the rate of light energy loss during propagation.

In living organism, a 'high' refractive index is not as high as in the inorganic world. Typically, the refractive index of chitin, keratin, and other biopolymers is something like $n=1.6$. One of the highest index transparent materials from the living world could be the guanine crystal, with a value close to $n=1.8$. Occasionally, an index such as $n=2$ needs to be considered in materials containing a pigment such as melanin, essentially because there is a correlation between the refractive index and the absorption coefficient. This increased index arises from a very fundamental principle: causality. Causality implies that no response can precede its cause. Mathematically, this constraint enforces what physicists call the Kramers–Kronig relationships (Jackson, 1975), by which, close to an absorption band, the refractive index experiences spectral variations and, in particular, is larger at frequencies immediately below the absorption band. The greenish absorption of melanin leads, in melanized chitin, to dispersion (the refractive index changes with frequency) and to an increase of its refractive index for visible colours at lower frequencies. Even so, the refractive indexes remain quite moderate, compared to the indexes such as $n=3.5$ or $n=4$ encountered with inorganic semiconducting materials omnipresent in the electronic and photonic industries, or even several thousands in metals. Even these inorganic champions should be regarded as weak responders. A refractive index of 4 simply means that the light waves propagate four times slower in the dense medium than in vacuum. Such speed reduction factors are not considered large for other wave phenomena: acoustic waves, for instance, have propagation speeds of the order of 10,000 m s^{-1} in a stiff solid, but slow down to 340 m s^{-1} if it exits into the air.

The first operational filter used in living organisms for colouration is the single planar constant-thickness overlayer, used for coating a substrate. This is analogous to the oil thin film floating on a still water puddle (see Section 4). If the thickness of the layer is appropriate, light undergoes multiple reflections on both interfaces of the film, interfere to reinforce transmission or reflection, according to the film thickness and the incidence angle.

A bit more complex is the multilayer stack, where planar thin films are stacked on a substrate, producing an interferential filter (see Section 6). An important case occurs when a group of two assembled layers, with definite thicknesses and refractive indexes is repeated, as is, on top of each other, to form a thicker stack. If this repeated arrangement is infinite, it can be called a 'periodic stack', or a 'periodic multilayer', or a 'one-dimensional photonic crystal'. If it is of finite extension, containing only a finite number of repeated units, it will be called a 'one-dimensional photonic-crystal film', or a 'Bragg mirror'. In those cases, it is not difficult to get a crude idea of the colours that will be reflected, and the change of colour that occurs when the angle of incidence of the illuminant beam is changed.

Another interesting device that may be part of a colouring system is the grating. If the planar layers considered till now are mirrors—can only redirect light in 'specular' directions (i.e. like in a single surface, with a reflection angle equal to the incidence angle, and Snell's law to determine the transmitted emergence direction)—a grating can do more. It can produce the specular reflection (labelled $m = 0$, where m is called the 'order' of the diffraction), but also beams propagating in other directions: different 'diffraction orders', labelled by integer numbers, negative and positive. The important fact here is that, for orders $m \neq 0$, the direction of the emergence, after scattering by the grating, depends on the incident wavelength, that is, on the colour of the incident wave. The grating produces a decomposition of white light into spectral components which, in appropriate circumstances, produce a rainbow of colours (see Section 7).

Increasing complexity, we start finding structures that are periodic along two distinct directions, and keep homogeneous in the third one. These fibrous structures are referred to as 'two-dimensional photonic crystals' (see Section 8). Again, a distinction should be made between infinite (theoretical) structures and films of finite thickness, made from these elements and deposited on a substrate. This system is rather complex, as it can be viewed as acting simultaneously as a Bragg mirror and a grating.

Finally, the last ordered structure that will be considered is the three-dimensional photonic crystal (see also Section 8) where a localized scattering unit is repeated periodically in all three dimensions of space. This is not final touch for our story, however. Natural colouring structures are not always ordered, and disorder is an important ingredient of structural colouration photonic structures. We will discuss some of the effects of disorder in Section 9. Disorder should not be considered to be some 'flaw' for a natural photonic structure or defective, in any sense. Disorder, with all its characterization parameters, is in fact transmitted to offspring, with the structure's development information, and, when attempting to understand the optical response, should not be forgotten as it is often crucial to produce the observed visual effect. The presence of disorder in today's species indicates that it was a favourable trait for increasing populations.

4 Single planar interface

When two semi-infinite media, with different refractive indexes (incident: n_i; emergent: n_t), are separated by an infinite planar interface, the separation between reflected and transmitted waves is ruled by the so-called 'Fresnel formulas'. These are orthogonal linear polarizations. The former (s) propagates its electric field normal to the incidence plane (which contains the incident beam and the normal to the interface), and the latter (p) propagates its magnetic field normal to the incidence plane. These orientations are conserved in the reflection and transmission processes, so that reflected and transmitted waves keep the same s or p properties as the incident wave. If the refractive indexes are constant, the reflected and transmitted intensities do not vary with the incident light wavelength λ. The consequence is that, with this structure, no colouration effect occurs.

The reflectance (reflected fraction of the incident energy), for s and p polarizations, is given by Fresnel formulas:

$$R_s(\theta_i) = \left[\frac{\sin(\theta_i - \theta_t)}{\sin(\theta_i + \theta_t)}\right]^2 \tag{1}$$

$$R_p(\theta_i) = \left[\frac{\tan(\theta_i - \theta_t)}{\tan(\theta_i + \theta_t)}\right]^2 \tag{2}$$

where the angle of refraction or emergence angle θ_t is given by Snell's law as depending on the incidence angle θ_i:

$$\sin\theta_t = \frac{n_i}{n_t}\sin\theta_i \tag{3}$$

When $n_i > n_t$, an appropriate angle θ_t cannot be found for incidence angles larger than the critical angle

$$\theta_c = \arcsin\left(\frac{n_t}{n_i}\right) \tag{4}$$

This absence of any emergence angle means a total (100%) reflection, and requires a large incidence angle and an attempt of transfer energy into a weaker refractive index medium. This is shown on the right panel in Fig. 2, for an incidence refractive index $n_i = 1.56$ and transmission into vacuum. For incidence angles larger than $\theta_c = \arcsin(1/1.56) \approx 39.9°$, the reflection is 100% for both polarizations s and p.

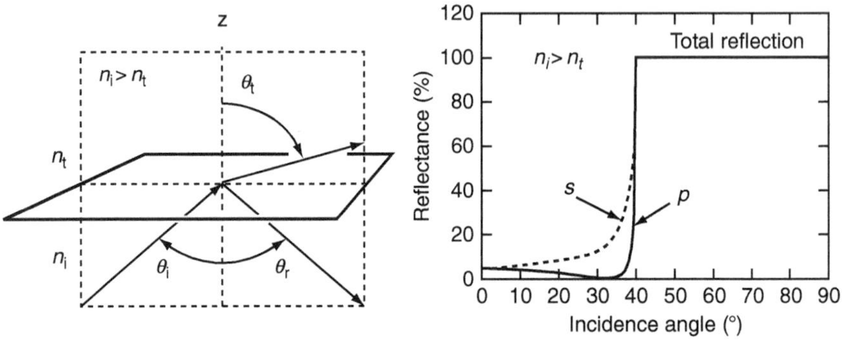

FIG. 2 Geometric parameters for the single planar interface. For a plane wave incident along a direction given by the incidence angle θ_i (measured from the normal to the interface), one observes only one reflected beam, making an 'reflection' angle θ_r equal to θ_i, and one transmitted wave in the direction indicated by the angle θ_t. Fresnel's formula gives the reflectance (intensity of the reflected beam, in units of the intensity of the incident beam), shown on the right part of the figure, for *s* (dashed) and *p* (solid) waves. In the case considered, $n_i = 1.56$ and $n_t = 1$. For incidence angles larger than about 40°, a total reflection is observed for both polarizations. The angle where the p reflectance vanishes near 32° is the Brewster angle, where a full *s* polarization is obtained.

In fact, chitin, like glass, presents a slight dispersion (Berthier *et al.*, 2003) of its refractive index and the reflectance is not exactly independent of the incident wavelength. This effect is insignificant, however, and, to our knowledge, it has not been described as a major colouration mechanism in any living organism.

Total reflection, by contrast, is more interesting and worth being examined in more detail. Actually, it plays a positive role in the colouration mechanism of some ants, and plays a negative role in bioluminescent species, forcing structures to appear, preventing these adverse effects.

4.1 EXAMPLE: THE NORTH-AFRICAN ANT *CATAGLYPHIS BOMBICINA*: PRISMATIC BRISTLES

The North-African ant *C. bombicina* provides specularly reflected light from the abdomen, the thorax and the head, in spite of the fact that the cuticle surface is neither smooth nor shiny. The cuticle is actually partly covered with setae and we should expect them to diffuse incident light in a kind of random way, at least in transverse directions. The light, however, is scattered in a much more organized way and this can be easily understood when looking at the pictures in Fig. 3. The setae, unexpectedly, show a *triangular* cross section. The basis of the equilateral triangle is flat, parallel to the cuticle surface (see middle panel of Fig. 3). The setae are transparent and their size (about 5 μm across) justifies the idea of a ray-tracing explanation. For an incidence direction close enough to the normal to the upper faces of the prismatic setae, the angle on the horizontal basis

FIG. 3 The North-African ant *Cataglyphis bombicina* is covered with setae that should diffuse the light received by the Sun, but shows actually specular reflection (left panel). The origin of this effect is the very peculiar geometry of the setae section, which appears to be triangular (middle panel, SEM imaging), with the lower basis parallel to the cuticle surface. With such a section, and an incidence angle small enough, on the lateral faces of the prism, total internal reflection can occur (right panel).

exceeds the total reflection critical angle, and the specular light escapes with a large intensity. The use of total reflection for transforming setae, from these incidences at all azimuths, into coherent mirrors is remarkable: it suffices that the cross section—usually circular—takes a somewhat modified shape to reach the appropriate function. Most photonic devices found in living organisms can serve as striking examples of gradual evolution at work.

4.2 EXAMPLE: LIGHT EXTRACTION FROM THE BIOLUMINESCENT ORGANS OF FIREFLIES

When the light encounters a flat, smooth interface between a dense incident medium and a less-dense emergence medium, total reflection severely limits light transmission. This is the main reason why solid-state sources show a very limited external efficiency, while the internal efficiency—the yield of the energy transformation that initially produces light—is near to perfect.

The limitation is more serious than usually thought. Consider, for instance, a reference system that consists of a homogeneous transparent medium terminated by a planar surface. For a point source emitting in this medium, only a fraction I_{trans}/I_{inc} of the power I_{inc} sent towards the surface will be transmitted, due to the limited transmission at small incidence angles θ_i (due to a small solid angle) and the total reflection at larger angles. We can calculate this fraction in terms of the surface transmission coefficient $T(\theta_i)$. The electromagnetic modes of such a structure can be as before classified as s- or p-polarized:

$$\Im = \frac{I_{trans}}{I_{inc}} = \int_0^{\theta_c} T(\theta_i) \sin\theta_i \, d\theta_i \tag{5}$$

where, for unpolarized light,

$$T(\theta_i) = 1 - \frac{1}{2}[R_s(\theta_i) + R_p(\theta_i)] \tag{6}$$

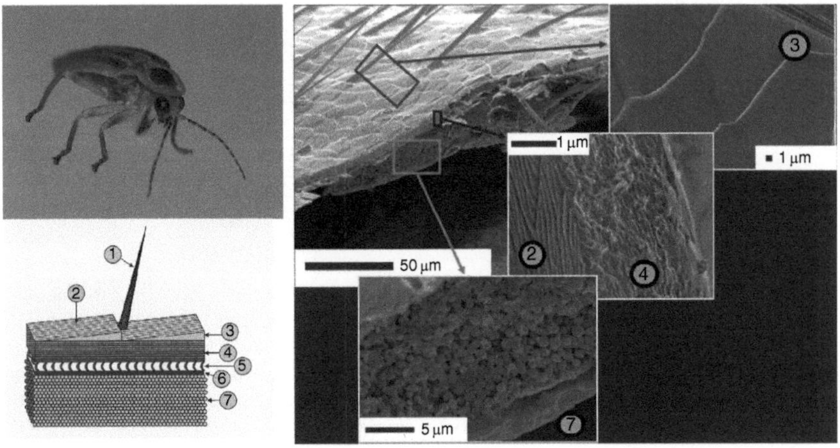

FIG. 4 Schematic view of the structure of the *Photuris* lantern. Seven substructures could be identified, some of them having an interesting impact on the light extraction. On the right: SEM views of the substructures.

This calculation is straightforward, but its result is somewhat surprising: we find that $\mathfrak{I}$ is only 20% for an incidence medium with refractive index as moderate as that of chitin, $n_{\mathrm{i}} = 1.56$ and escape into air. Semiconductor sources (LED..., OLEDs...), with refractive indexes as high as 3.5, may behave even worse. This problem keeps optical engineers in motion, in view of the economical importance of any energy saving in the domain of lightning. The same situation arises with bioluminescent fireflies which produce light inside the abdominal segments for escape into the air.

In a recent thesis, Annick Bay suggested a biomimetic approach for getting around this difficulty (Bay and Vigneron, 2009). She carefully examined each of the structures encountered by the bioluminescent light emitted by a firefly of the genus *Photuris* and looked for those who could play a role in the optimization of the light extraction. She actually found seven substructures, as summarized in Fig. 4.

Based on computations (modelling and simulations), the following results were found:

Substructure (1): The setae found on the cuticle, rigidified by longitudinal ridges, can conduct light and scatter it out, but they are not numerous and offer a small scattering cross section. They do not contribute much to the light extraction and can be neglected.

Substructure (2): The cuticle surface is divided into scales—see substructure (3)—and the scales surface show roughness arising from a grating of parallel ribs. The cross-section profile of these ribs is very smooth (protruding 100 nm) and forms a grating with step-size 250 nm. Calculation shows that this structure liberates only 1% of additional light.

Substructure (3): The cuticle of the abdominal segments is composed of scales with a misfit join at their perimeter. This is not exceptional, but here the density of joins is especially high (with small scales). The height of the protrusion due to misfit is variable, but 3 μm is a reasonable value. Computation shows that the presence of these linear abrupt scatterers improves the light transmission from 20% to 34% for s polarization of the incident wave, or 35% for incident p polarization, a very significant improvement.

Substructure (4): Just below the scales, we note a stack of about 30 inhomogeneous plates, each 80 nm thick. Each plate has a planar lower face and a wavy upper surface, mixing chitin and air. Modelling and computation shows that this structure does not improve on light extraction, again, by more than 1%.

Substructure (5): The next inner layer is seen to be essentially empty, except for a large number of fibres which bind to the neighbouring layers, substructures (4) and (6). It is difficult to provide data on the thickness of this layer because it is found to be variable over a very wide range. Many different samples provide very different values and it is plausible that this gap between solid layers is tunable by muscular contractions and this may play a role on the dynamics of the flash emission. Filled with a liquid, this structure is neutral, regarding the light extraction.

Substructure (6): The gap described in substructure (5) opens on a solid, flexible membrane, 60 nm thick, which separates this 'fibrous' gap from the chamber (7) where the bioluminescent reaction takes place. This, as well, neither contributes nor hinders the light emission.

Substructure (7): The reaction chamber is found below the above layers. It appears to be filled with granular bodies, close to spheres: the peroxysomes. Peroxisomes have been recognized as organelles by the cytologist Christian de Duve in 1967 (de Duve, 1969). One of the main functions of these organelles is to get rid of toxic peroxides. They contain a crystalline core and are known to incorporate high quantities of urate oxidase and other enzymes, which happen to have a refractive index lower than chitin. The local lowering of the refractive index in the emission region is an efficient way to get around the total internal reflection problem: modelling suggests that the efficiency shifts from 20% to 27% (incident s) or 29% (incident p) as the critical angle changes from 40° to 46°.

Finally, Annick Bay's biomimetic extractor, which gathers all substructures in a single device, can be shown to produce a very significant improvement of the light emission, reaching an external yield of 38% (incident s) or 41% (incident p), instead of 20% in the reference, flat surface system. This means that the structuring of the lantern plays a very significant role for improving the light extraction.

5 Single planar overlayer

The next photonic device, in our first step to complexity, is more properly a structural colouration device—probably the first one to have been observed and used, indeed, for instance in ceramics artworks. The structure is a simple layer of constant thickness deposited on the flat surface of a substrate. The translational invariance means that the directions of the emergence rays are simply given by the reflection and refraction rules. The intensity, however, results from the interference of multiply reflected waves exiting simultaneously from the film, in a common direction. The dominant reflected wavelength in such a situation is given by

$$\lambda = \frac{2d\sqrt{n_t^2 - n_i^2 \sin^2\theta_i}}{m + (1/2)} \tag{7}$$

where d is the film thickness, n_t is its refractive index, while n_i is the refractive index of the outer incident medium which, generally, will be air ($n_i = 1$); θ_i, as above, is the angle of incidence; m is an integer such that the resulting wavelength falls in the spectral region of interest, for instance the visible range, when we want to describe the visual effect produced by the film. The addition of a half unit (1/2) to this integer depends on the refractive index of the substrate. The reflection on an interface can lengthen the optical path by half a wavelength if the light crosses the interface with an increase of refractive index (in this case, the electric field is reflected pointing opposite to the incident field). The presence of the term (1/2) depends on the condition met at the second interface, with the substrate: if the same condition as at the first interface is met, the term (1/2) disappears, but if the ordering of indexes is different at each interface, the term (1/2) is enforced.

The interference effect, by reflection from a thin film, produces colours which were already described by Robert Boyle, a contemporary of Isaac Newton. For a chitin thin film, 600 nm thick, self-supported in air, we find the maximum of the reflected spectrum for an incident angle $\theta_i = 0$ at the following spectral locations ($m = 0, 1, 2, \ldots$): $\lambda_0 = 3744$ nm, $\lambda_1 = 1248$ nm, $\lambda_2 = 749$ nm, $\lambda_3 = 535$ nm, $\lambda_4 = 416$ nm, $\lambda_5 = 340$ nm, $\lambda_6 - 288$ nm ..., the first two lying in the infrared, the two last ones are in the ultraviolet. The visible contributions are red (λ_2), yellow-green (λ_3), purplish blue (λ_4). The line shape of the reflected band is wide and corresponds to a periodic line shape (sinusoidal for weak refractive index contrast) when presented as a function of the wave angular frequency ω. Quite a number of examples are known in the animal kingdom: the pigeon feathers and many opaque-wing wasps and bees use this simple device to produce colouration.

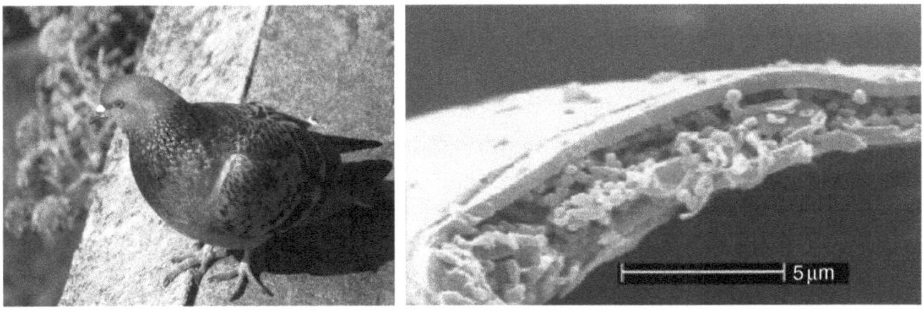

FIG. 5 An example of colouration by a simple thin film in nature: the green and violet feathers of the pigeon. The barbules are a sack filled with melanin granules (Yin *et al.*, 2006). The membrane is a transparent thin film, with a thickness remarkably constant, where light interferences take place to filter out green or violet colours, according to the membrane thickness. (See Color Plate 19 at the back of the book.)

5.1 PIGEON IRIDESCENCE

The pigeon neck displays green and violet feathers (Malloock, 1911). This colouration was studied by Yin *et al.* (2006). A bird feather has three levels of tree-like branches: the rachis, the barbs (attached to the rachis) and the barbules (attached to the barbs). In the case of pigeons, we see two types of feathers: one type green and the other type violet both observed under normal incidence. As we will see, the colouration is caused by the barbule's structure.

As Fig. 5 shows, the barbules are long, thin and flat. Scanning electron microscopy of these barbules reveals slightly curved sacks, filled with melanin granules (melanophores). The outer membrane of this sack is a hard cortex made of a transparent homogeneous material, with a remarkably constant thickness, different for each type of feather. This acts as an optical interference thin film. With a thickness of 595 nm, the 'cosine' oscillation in the reflectance spectrum, under normal incidence, produces reflected light near 743 nm (red), 530 nm (green) and 413 nm (violet) (see Eq. (7)). This combination of colours gives a desaturated green perception. For feathers closer to the wings, with a thickness of 530 nm, the reflectance spectrum peaks constructively at 472 nm (blue) and 661 nm (red), so that the colour produced is a mix of blue and red, an extra spectral hue classified as 'purple'. Iridescence can be observed on individual feathers detached from the bird's neck, but is less easy to see on the animal, as a result of a multiscale shadowing effect.

5.2 IRIDESCENCE ON THE WINGS OF A TROPICAL WASP

A second example comes from the study of the iridescent wing of a giant tropical wasp, *Megascolia procer javanensis* (Sarrazin *et al.*, 2008). In this particular case, the wing is shown to be made of rigid structure of melanized chitin, except for an overlayer, on each side of the wing. The overlayer can be shown to act as a transparent interference thin film with a thickness of 286 nm. The refractive index of the material in this layer is not precisely known, so that its analysis requires examining the reflectance spectrum in detail, for various angles of incidence. The substrate supporting this layer is better known, as a solid mix of chitin and melanin. This mix was studied by de Albuquerque *et al.* (2006), including the dispersion related to melanin absorption. This absorption is strong here, as can be seen from the opacity of the wing. An adjustment of the refractive index of the overlayer allows the reflectance spectra to be fitted quite nicely at all incidence angles, and provides a value $n = 1.76$, which is very reasonable. The reflection spectra in Fig. 6, calculated (dotted line) and measured (solid line), show that the

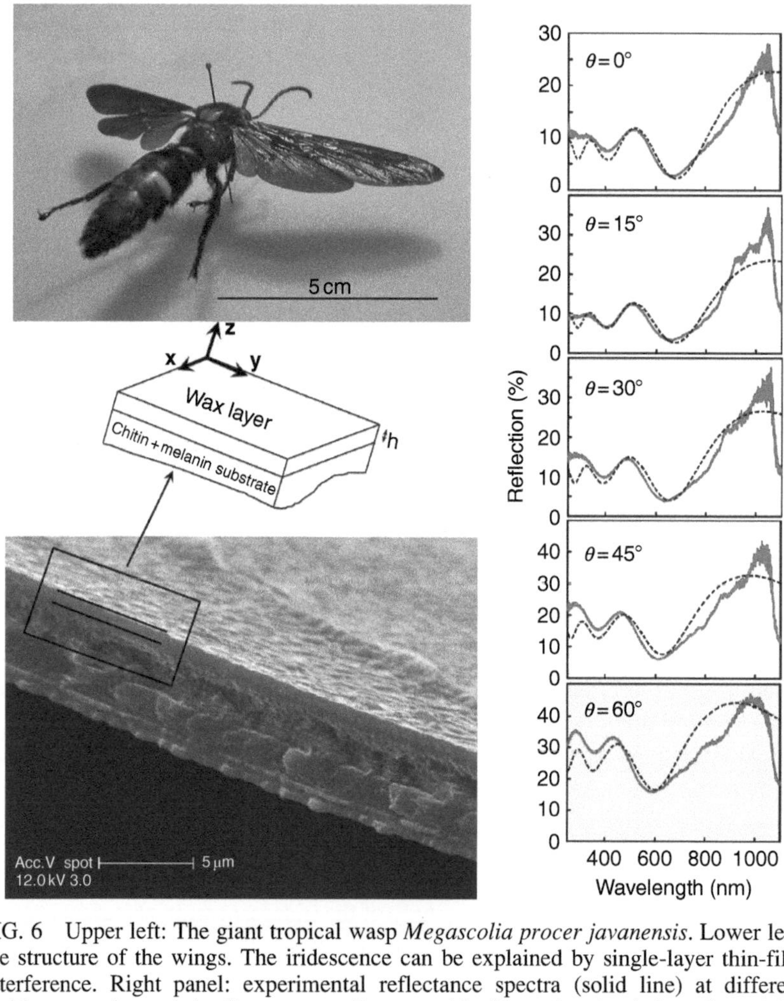

FIG. 6 Upper left: The giant tropical wasp *Megascolia procer javanensis*. Lower left: the structure of the wings. The iridescence can be explained by single-layer thin-film interference. Right panel: experimental reflectance spectra (solid line) at different incidence angles and the fit corresponding to a thin-film refractive index of $n = 1.76$. (See Color Plate 20 at the back of the book.)

interpretation of the colouration in terms of a thin-film model is quite convincing. As these spectra show, the iridescence is weak—from bluish green near-normal incidence to greenish blue at grazing incidence.

6 Planar multilayer stacks

Another structure very frequently encountered for the structural colouration of living organisms is the multilayer, where multiple layers are stacked to form a selective mirror. The multilayer stack is more a stack of interfaces than a stack of homogeneous layers: the interferences that take place in the volume of the structure arise from the superposition of multiply reflected waves at the interfaces. The multilayer structures encountered in the living world can be quite complex: in the beetle *Chrysina resplendens*, for instance, no less than 120 layers are stacked to produce a vivid metallic gold colour. The large number of layers is, for some part, needed to produce high reflection intensities to compensate the necessarily weak refractive index differences occurring at an interface with the organic materials. A second reason for staking many layers is to widen the bandwidth of reflectors: if the thicknesses of the layers repeat themselves without change, the reflection bands turn out to be quite narrow, so that the colour reflected is highly saturated. In a 'chirped' structure, the thicknesses vary gradually and the different colours can be reflected at different depths, resulting in a broadband (wide-spectrum) reflectance.

One important class of multilayers is the periodic stack, where a group of layers (generally, two layers) is stacked repeatedly to form part of a periodic multilayer. This is also called a Bragg mirror. The infinite version of this structure is a one-dimensional photonic crystal. This colouring structure is frequent with living organisms, and examples will be given below. With such a structure, it is not so difficult to predict the dominant colour that will be reflected.

Given the importance of this structure, we will somewhat detail its physical background. As illustrated in Fig 7, a one-dimensional periodic structure conserves the frequency of a monochromatic wave, but does not conserve its wave number (scaling to the inverse of the wavelength). A wave with wave number k_z will change this wave number so that its output value is a choice of any of the quantities $k_z + m(2\pi/a)$, where m is an integer and a is the period:

$$k'_z = k_z + m\frac{2\pi}{a} \tag{8}$$

This means that the electromagnetic modes (harmonic oscillations with a well-defined frequency ω) will be a superposition of those coupled waves, in the form

$$\vec{E}(\vec{\rho}, z, t) = \left[e^{ik_z z}\sum_{m=-\infty}^{+\infty} \vec{E}_m e^{im(2\pi/a)z}\right] e^{i\left(\vec{\kappa}_{\parallel}\vec{\rho} - \omega t\right)} \tag{9}$$

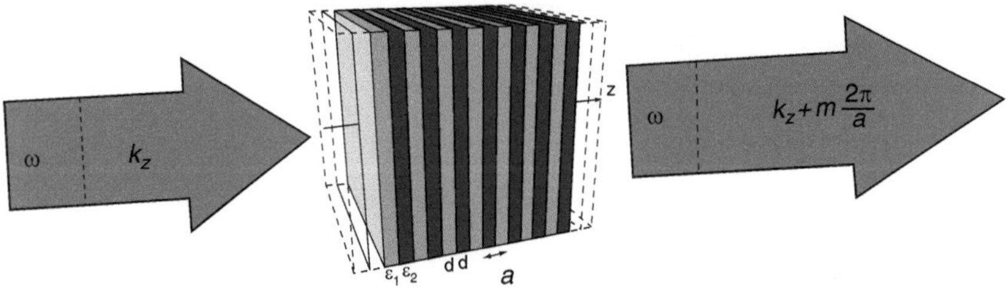

FIG. 7 A monochromatic wave (of angular frequency ω) will not keep a constant wavelength in a periodic medium such as a one-dimensional photonic crystal. Any wave with wave number k_z, propagating through a stack of period a will decay into any other waves with the same frequency but a wave vector modified by any integer multiple of the crystal structure's wave number, $2\pi/a$.

In this expression, the electric field is factorized into a Bloch wave in the coordinate z and a monochromatic travelling wave with wave vector $\overrightarrow{k}_{\|}$ in the coordinates $\overrightarrow{\rho}$ parallel to the layer interfaces.

This statement is known, in solid-state physics as the 'Bloch theorem' (Ziman, 1979), the infinite sum being a periodic function of the coordinate z, with the same period as the photonic crystal. These ideas can be used to predict the colour and the iridescence of any one-dimensional photonic crystal. Let us see first the infinite structure as a homogeneous medium, with an average refractive index $\bar{n}$. The dispersion relation of such a homogeneous structure is

$$\omega = |k_z|\frac{c}{\bar{n}} \tag{10}$$

And, in Fig. 8, it is represented by a straight (dotted) line. These modes are ordinary waves travelling in the z direction, positive or negative, according to the sign of the wave number k_z. Their propagation speed, $c/\bar{n}$ is significantly less than the speed of light in vacuum, c. If we switch on a weak periodic refractive index contrast, we start coupling waves of the same frequency, if their wave numbers k_z and k'_z are separated by $k'_z - k_z = m(2\pi/a)$. The only points where this occurs are the so-called 'Brillouin-zone boundaries' at

$$k_z = \pm\frac{\pi}{a}, \pm 2\frac{\pi}{a}, \pm 3\frac{\pi}{a}, \ldots \tag{11}$$

At those points, the dispersion relation splits because of the formation of standing waves that adopt different configurations relative to high and low refractive indexes. As Fig. 8 shows (solid line), the dispersion relation leave gaps at these zone boundaries. At the frequencies lying within the gap, no wave propagation can occur. An incident light wave with a frequency in the range of one of the narrow gaps, impinging on the surface of a semi-infinite photonic

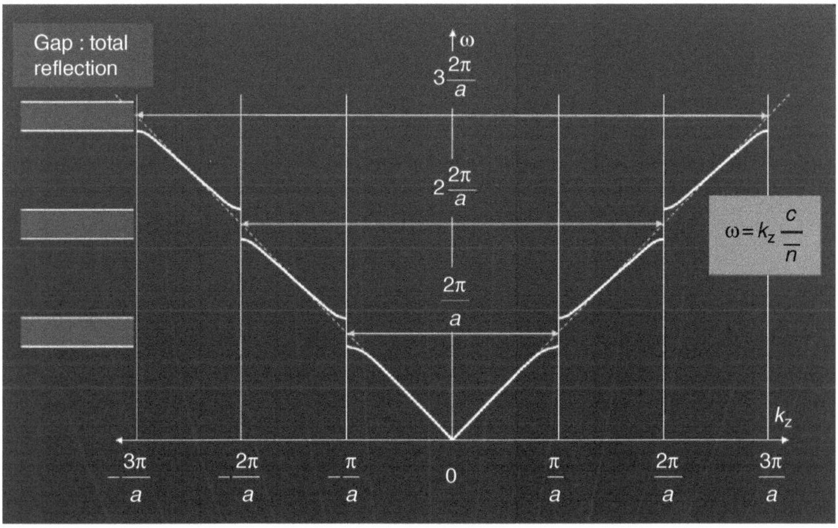

FIG. 8 Formation of the Bloch modes of a one-dimensional photonic crystal. Gaps open where a mixing occurs between waves of the same frequency, with the wave number k_z differing by only an integer multiple of the structure wave number $2\pi/a$.

crystal, will be totally reflected. We note that the total reflection occurs under the constraint of matching the gap frequency, not the usual prescription found in homogeneous media—exiting into a less-dense medium and incident in a direction such that the incidence angle is larger than the total internal reflection critical angle. On a photonic crystal surface, total reflection can occur under normal incidence, and also when air is the incidence medium. A photonic crystal produces total reflection without the constraints set by homogeneous materials.

We can easily determine the dominant reflected wavelength by locating the frequencies where the gaps occur. Under normal incidence, we simply locate the frequencies $\omega = |k_z|(c/\bar{n})$ at the border of the Brillouin zones $k_z = m(\pi/a)$. This gives $\omega = m(\pi c/a\bar{n})$ and, translated into an equivalent vacuum wavelength,

$$\lambda = \frac{2a\bar{n}}{m} \tag{12}$$

If the incidence is oblique, $\omega = \sqrt{k_y^2 + k_z^2}(c/\bar{n})$, where $k_y = (\omega/c)\sin\theta_i$. The formula giving the dominant wavelength reflected becomes

$$\lambda = \frac{2a\sqrt{\bar{n}^2 - n_i^2 \sin^2\theta_i}}{m} \tag{13}$$

FIG. 9 Iridescence of the elytra of a buprestis *Chrysochroa vittata*. When viewed under normal incidence (left) the abdomen is red. The same specimen, viewed under a grazing light emergence, becomes blue. Iridescence is the change of colour according to the viewing angle. (See Color Plate 21 at the back of the book.)

As the angle of incidence increases, the reflected wavelength undergoes a 'blue' shift (towards a shorter wavelength). This iridescence is easily seen on the pictures shown in Fig. 9, where elytra of a buprestis *Chrysochroa rajah thailandica* are seen from two different emergence angles. We see that 'normally' green areas appear blue, while red areas appear green under more grazing angles.

Many examples can be given that illustrate this behaviour. We will essentially mention two: the metallic colouration of the asiatic buprestis *Chrysochroa vittata* (Vigneron *et al.*, 2006a) and the blue colouration of the beetle *Hoplia coerulea* (Vigneron *et al.*, 2005).

6.1 CHRYSOCHROA VITTATA

The abdomen, on the ventral side, of the buprestis *C. vittata* displays a metallic red hue that turns into green when illuminated under a grazing incidence. This iridescence results from the mechanism of interference that takes place in a periodic multilayer, as explained above. The analysis of the SEM pictures (see Fig. 10) suggests that the multiple reflections takes place on a very thin air layer interspersed in the chitin plates. This layer is maintained by 'spacers', a corrugation of one of the sides of the chitin slab. In this case, the thickness of the air layer is much smaller than the thickness of the chitin slabs, so that the exact

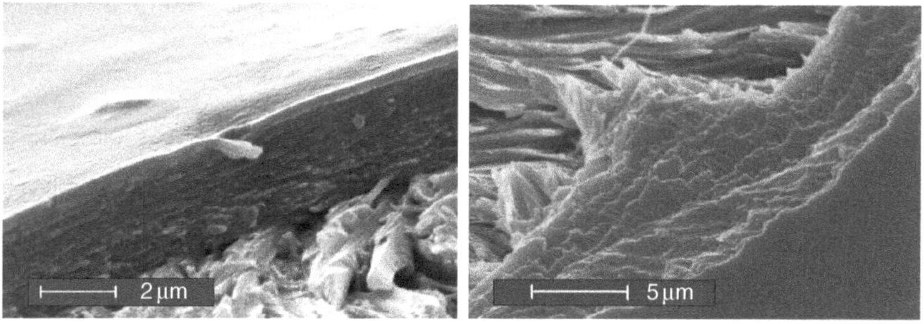

FIG. 10 Scanning electron microscope image of the layered cuticle of the buprestis *Chrysochroa vittata*. The colouring structure is made from more than 20 identical chitin slabs, separated by rough spacers that leave thin air gaps that act as reflectors. The image on the left shows a side view of the multilayer stack. SEM pictures allow to determine the multilayer period and the number of layers. The image on the right shows surface of the layers, revealing the 'spacers', which justify the hypothesis of a layer of air producing the multiple reflections.

nature of this reflecting layer does not influence significantly the average refractive index of the whole structure. Then, neither the colour nor the iridescence actually depends on the exact nature of this thin layer: the only need is that it constitutes a scattering plane, where the light can be reflected. The period, including the chitin slab and the thin junction layer is, on the average, $a = 204$ nm and the refractive index is close to that found for chitin alone, $\bar{n} = 1.54$. Then, the dominant reflected wavelength for the normal direction is easily found to be

$$\lambda = 2a\bar{n} \quad \Rightarrow \quad \lambda = 628\,\text{nm} \tag{14}$$

This wavelength is perceived as red, as standard chromaticity diagrams easily show. Under larger incidence angles, the reflected wavelength is shortened. For instance, under 45°:

$$\lambda = 2a\sqrt{\bar{n}^2 - \sin^2 45°} \quad \Rightarrow \quad \lambda = 567\,\text{nm} \tag{15}$$

which now falls in the yellow-green region. At 75°, the reflected hue turns into bluish green ($\lambda = 500$ nm).

Such iridescent material attracts interest in various areas of human activities. Franziska Schenk (Schenk and Harvey, 2008), for instance, works on the introduction of 'shifting kaleidoscopic colours' in painting and experiments with unconventional painting techniques, including the latest nature-inspired colour-shift technologies. Paint industry is also interested in iridescence: because it contains fragments of multilayer slabs that orient themselves, due to surface tension effects, some paints can change colour with the angle of viewing, as measured from the surface normal. Different parts of a car can display different hues. The same effects can be achieved, for instance, in makeups. The use of these effects is new to our imagination. Besides technical progress for the fabrication of such iridescent media, word should be given to artistic designers, to define where and when this new capability could be used.

6.2 *HOPLIA COERULEA*

Natural photonic devices are all built from a rather restricted range of materials. From the optical point of view, most transparent materials known to biology have refractive indexes close to 1.6. This limited choice does not seem to limit the availability of visual effects, as the structures tend to be complex, a necessary trade-off.

One example is provided by the 'blue beetle' *H. coerulea*, with a cuticle-bearing scales structured to filter out a spectacular blue-violet iridescence on reflection (Deparis *et al.*, 2008). The cuticle, as seen in scanning electron microscopy is shown in Fig. 11. The low-magnification image (left) shows the head of the beetle, and part of its thorax. The body is covered by small, rounded scales, or squamae, attached by a single peripheral point to the underlying cuticle. These scales are easily removed by breaking this binding.

The very flat envelope of the scales hides a structure which can be interpreted as a stack of some 20 sheets parallel to the flat surface of the scale (see central panel in Fig. 11). Each sheet is actually composed of a very thin plate of bulk chitin, bearing, on one side, a network of parallel rods with a rectangular section. The dimensions of these elements are shown in Fig. 11 (right panel). The lateral corrugation associated with the rods has a period of 170 nm, just too small to produce diffraction of light in the visible range. For those wavelengths, the rods array appears to be a homogeneous layer, and the concept of an average refractive index makes sense. The average refractive index of the whole structure was evaluated to be $\bar{n} = 1.4$ for unpolarized light near-normal incidence. As the vertical period turns out to be $120\ \text{nm} + 40\ \text{nm} = 160\ \text{nm}$, one readily find the dominant reflected wavelength,

$$\lambda = 2a\bar{n} \quad \Rightarrow \quad \lambda = 448\,\text{nm} \tag{16}$$

This wavelength is clearly the expected blue colouration. For a larger angle, such as 60°, the hue is centred on the wavelength $\lambda = 387$ nm, in the violet. The *H. coerulea* structure gives some iridescence, ranging from blue to violet, and effectively behaves as a flat multilayer structure, in spite of the lateral structuring of the rods layers. This structure was recently shown to have an optical response modifiable in presence of humidity, because water can infiltrate the voids (the structure's materials turn out to be hydrophilic) (Rassart *et al.*, 2009).

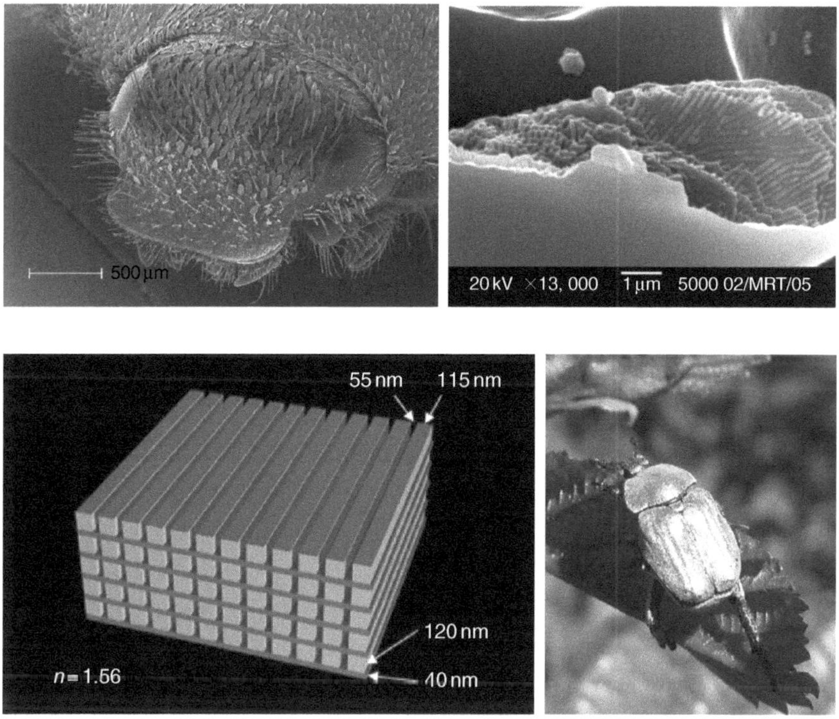

FIG. 11 *Hoplia coerulea* (lower right). The beetle's cuticle is, as seen here (higher left), covered by scales, often slightly curved out. The scales take the shape of a disc, with a diameter of about 50 µm and a thickness of 3.5 µm. The dorsal scales, seen under the optical microscope, render a blue or violet colour. The scanning electron microscope images (higher right) show the colouring structure inside the scales. An idealized model of the colouring structure, with the dimensions observed, is shown on the lower left. (See Color Plate 22 at the back of the book.)

7 Grating

Another important colouring structure is grating, a device which decomposes broadband light, for instance, in spectrophotometers, to analyse its spectral contents. The device has a surface which is periodically corrugated in some direction *along* the surface. The basic physics of the grating is just the same as that involved in a periodic multilayer stack, except for the orientation of the periodicity. If we call z the coordinate in the direction of the normal to the grating surface, and y the coordinate in the direction of the periodic corrugation (the other Cartesian coordinate x being an invariant direction), a grating just acts to modify the wave vector component in the y direction, by adding an integer multiple of the structure's wave vector $2\pi/b$, where b is the period—the spacing between lines—along y. The change of wave number in the y direction, from k_y (in) to k'_y (out) is simply

$$k'_y = k_y + m\frac{2\pi}{b} \tag{17}$$

similar to Eq. (8). This equation allows to predict the directions taken by the light, after having been scattered by the grating. Indeed, referring to Fig. 12 for the geometry, the angle of incidence θ sets the incident wave number k_y as

$$k_y = \frac{\omega}{c}\sin\theta \tag{18}$$

and the outgoing wave number k'_y determines the direction of the elastically scattered light beam. The angle of emergence ϕ is related to k'_y as

$$k'_y = \frac{\omega}{c}\sin\phi \tag{19}$$

From these considerations, one can readily see that

$$\sin\phi = \sin\theta + m\frac{\lambda}{b} \tag{20}$$

where $\lambda = (2\pi c/\omega)$; m is an integer, positive, null or negative, which indicates the 'order' of diffraction. The order $m = 0$, in reflection, means $\phi = \theta$, which can be viewed as a specular reflection from the grating surface. With orders $m = -1$

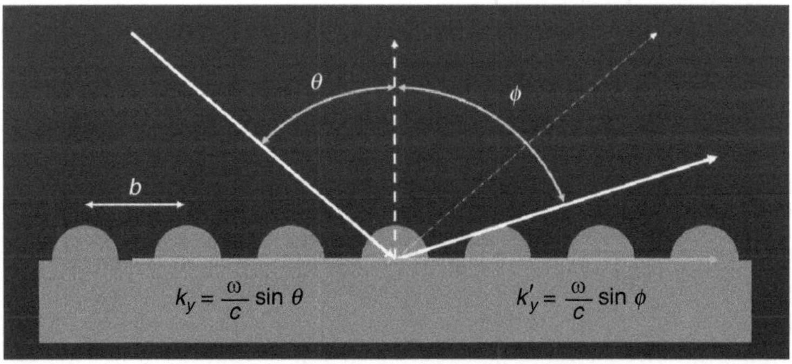

FIG. 12 Grating geometry. A grating is a substrate's surface that bears a one-dimensional periodic array of scatterers (bumps or groves) in the horizontal direction. The period of the array is noted b and has a value comparable to the light wavelength. The incidence angle θ is measured from the normal to the grating surface, as the emergence angle ϕ. Except for the diffraction order $m = 0$, which is specular, the emergence angle is not equal to the incidence angle and depends on the incident wavelength, leading to a spectral decomposition of white light and to colouration effects.

or $m = 1$, for instance, diffracted directions are not specular and the emergence angle ϕ changes with the incident wavelength λ. In this case, a broadband incident beam gets decomposed into spectral components distributed over a range of emergence angles.

The existence of diffracted beams, at an incidence angle θ, requires that the variable $x = m(\lambda/b)$ lies in the interval

$$-1 - \sin\theta < x < 1 - \sin\theta \tag{21}$$

This corresponds to a curved region of (x,θ) points (shown as a hatched area in Fig. 13) in which, for a given order m, the wavelength and incidence angle allows for a far-distance escape of constructively interfering waves. The diagram in Fig. 13 is built for a grating with spacing $b = 175$ nm. This value is interesting for different reasons. The horizontal axis labels x, which is determined from the incident wavelength λ and the diffraction order m. The zero-order diffraction is for $m = 0$, which implies $x = 0$. For all incidence angles θ, the point (x,θ) lies in the hatched region, along $x = 0$, whatever the incident wavelength, which means that a far-field travelling wave is always ready to be generated on the basis of its translational symmetry: a grating should always generate a reflected wave (however, we do not address now the question of the intensity of this wave, which depends on the grating profile). For $m = -1$, we see that the value of the parameter x depends on the wavelength and is negative. In the (human) visible range, 'blue' light components (380 nm) give an 'x' value closer to the origin $x = 0$ than 'red' (780 nm), and we see that these values are not compatible with any far-field emergence, as the corresponding (x,θ) points

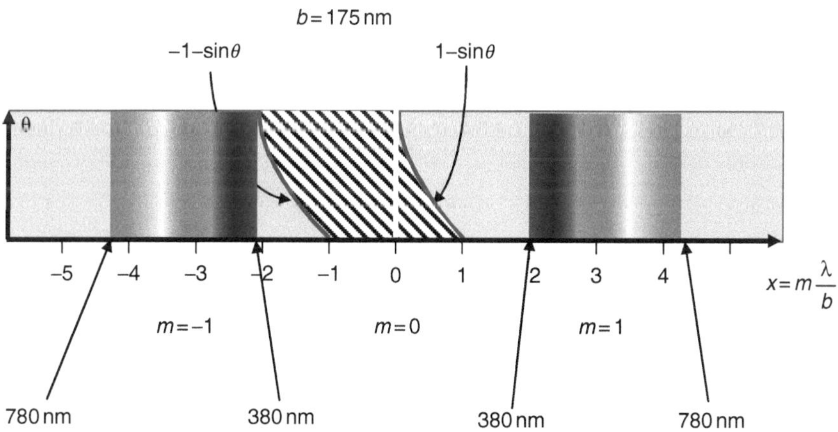

FIG. 13 The condition for diffraction by a grating can be analysed by this diagram, drawn for a given grating period (here $b = 175$ nm). Far-field emergence occurs for a given diffraction order m when the parameter x and the angle of incidence form a Cartesian point which lies in the hatched region, at the centre of the diagram. (See Color Plate 23 at the back of the book.)

all fall outside the hatched area. The only slight chance is with an extreme violet light, but only when the angle of incidence is close to 90°, that is under a grazing illumination. For all other orders, including $m = 1$, the chances of emergence are even less. The way to produce diffracted light with a grating like this one is to increase the period b, so that the parameter x gets smaller and (x,θ) points, for $m = -1$, enter the allowed emergence region (hatched). With a moderate increase, violet and blue will first enter the allowed region, at large incidences, and progressively, the whole visible range will be emerging, first at large incidence angles and then under smaller angles. For even larger periods, the 'forward' $m = 1$ order will appear, and $m = \pm 2$, in turn.

It is interesting to note that a period $b = 175$ nm is a critical period for the human colour vision range: for periods smaller than this, we have a zero-order grating, with a period too short to produce diffraction orders other than the specular order $m = 0$. For larger values of the period, diffraction occurs, first at large incidence angles, emitting violet colour.

Nature provides many examples of colouration due to gratings. One of them is a butterfly from Papua New Guinea, *Lamprolenis nitida* (Ingram *et al.*, 2008), which we will examine now. A very special grating has also been seen on another, very common butterfly from South and Central America, *Pierella luna*. This particular one provided some surprise and is worth mentioning here.

7.1 EXAMPLE: *LAMPROLENIS NITIDA*

L. nitida is a relatively rare species, which is endemic to mainland Papua New Guinea (Parsons, 1998). It inhabits forests, where it is commonly found in sunlit openings at 0–1500 m, feeding on the bamboo, *Bambusa* (Poaceae). The dorsal fore and hind wings of males and females generally appear brown in diffuse light; however, at narrow angles of observation those of the males exhibit bright iridescence.

The colouration is apparent on the different views of Fig. 14A–C, where the butterfly is pictured under different angles from the wing normal. The illumination comes from the head side of the butterfly. Red diffraction is observed under a relatively grazing incidence, while yellow and green are seen at smaller angles, closer to normal incidence. This is what is expected from a grating with period $b = 480$ nm, with a large incidence angle (75°). This 480 nm grating seen by SEM on this butterfly's scales is actually formed by the periodic arrangement of crossribs, which have here been transformed into slanted plates, as shown on the right, lower panels in Fig. 14. The intensity of the diffraction orders depends on the grating profile and can only be known in detail by solving Maxwell equations. One interesting feature of this grating is the scatterer's slant angle, which favours the intensity of the $m = -1$, and dims the specular $(m = 0)$ diffraction beam. This is what is called a 'blazed' grating, in the terminology of spectrophotometry engineering: when a grating must send a maximum of light into a specific diffraction order, a special profile of the groves is defined. This is exactly what evolution has achieved on this butterfly. It is always difficult to determine what advantage this function has brought, but we can imagine that this feature maximizes the visibility of the colouration arising from the $m = -1$ diffraction. Another advantage may be that the suppression of the energy in the specular order makes the butterfly less conspicuous from viewpoints not intended for signalling.

7.2 *PIERELLA LUNA*

This other male butterfly, *P. luna*, also produces iridescence that originates from a grating on the scales of the wing. However, as Fig. 15 shows, the ordering of the emergent colours is reversed, compared to *L. nitida*. We observe the exit of the blue end of the spectrum under a grazing emergence, while red exits closer to the wing normal.

Scanning electron microscopy, in this case, reveals that part of the scales on the iridescent sectors of the forewings are curled up, providing a 'vertical

FIG. 14 *Lamprolenis nitida*, like many butterflies, exhibits sexually dimorphic iridescence. This was found to originate from first-order blazed diffraction gratings formed by different scale nanostructures, which are angled with respect to the scale surface. (See Color Plate 24 at the back of the book.)

FIG. 15 Male and female *Pierella luna* butterflies viewed at different angles. The male is behind the female. The illumination is 45° from the normal and originates at a point opposite to the viewer (view is against the light). As we lower our angle of view, but keep the light source steady, the forewing of the male butterfly displays a rainbow of iridescence, covering nearly the entire spectrum perceivable by humans. The females lack iridescence. (See Color Plate 25 at the back of the book.)

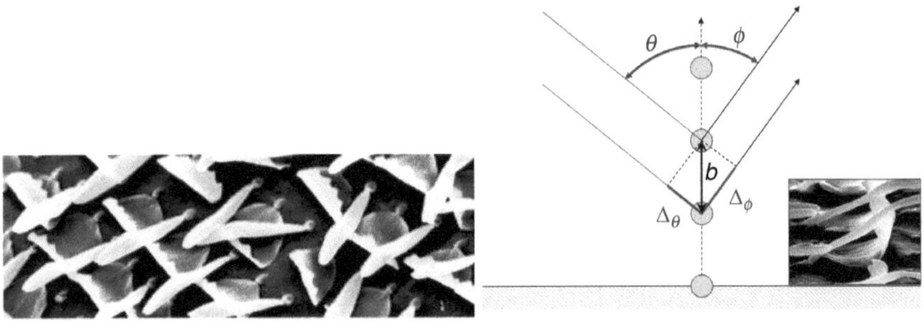

FIG. 16 The iridescent part of the wings of the male *Pierella luna* contains curled scales that function as a transmitting vertical grating. (See Color Plate 26 at the back of the book.)

grating' that functions in transmission rather than reflection. These curiously curled scales are shown in Fig. 16, together with the geometric data that allow for an understanding of how this colouration takes place.

A vertical grating works just the same as a horizontal grating, but the angles are simply not measured from the same reference origin. Here the angle of incidence and emergence are measured from the grating plane itself. The consequence, easily drawn from the diagram in Fig. 16 (middle), is that the relationship between the emergence and incidence angles is now

$$\cos\phi = -\cos\theta + m\frac{\lambda}{b} \tag{22}$$

An increase in wavelength now leads to a smaller emergence angle, with $m = 1$.

8 Photonic crystals

Photonic crystals are transparent inhomogeneous materials, with a refractive index periodic in two or three dimensions. 'One-dimensional' photonic crystals are merely the Bragg mirrors described earlier. In essence, a photonic-crystal film is at the same time a 'vertical' Bragg mirror, and a 'horizontal' grating which allows for diffraction.

The optical properties are more complex than for these two devices. However, some knowledge of 'photonic crystallography', analogous to atomic crystallography known by biologists and chemists, can help. As for the Bragg mirror and the grating, a photonic crystal is able to change the incident wave vector, but only in a very specific way. By progressing in a photonic crystal with a well-defined three-dimensional periodicity, a wave with initial wave vector $\vec{k}$ can progressively turn into a wave with wave vector $\vec{k'}$ (with $|\vec{k'}| = |\vec{k}| = \bar{n}(\omega/c)$), given by

$$\vec{k'} = \vec{k} + \vec{g} \tag{23}$$

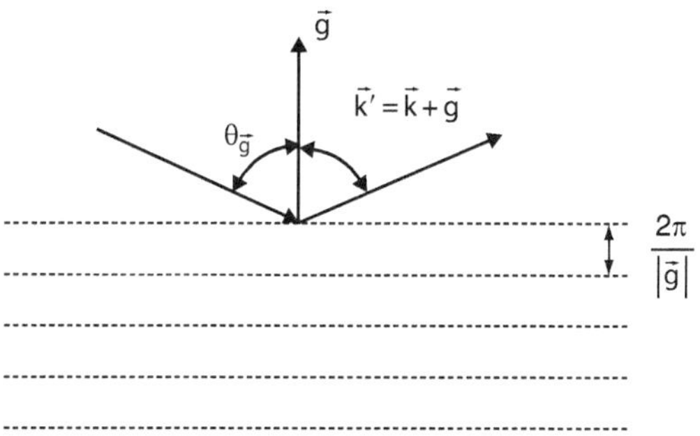

FIG. 17 Diffraction in a three-dimensional photonic crystal can be viewed as a reflection on a stack of reticular planes, just as in atomic crystallography. Each set of reticular planes in a Bravais lattice can be labelled by the shortest non-zero reciprocal lattice vector $\vec{g}$ perpendicular to these planes. The length of this reciprocal lattice vector gives the period of the stack.

The possible $\vec{g}$ vectors are discretely distributed on a regular lattice of points in space, actually related to the periodicity of the crystal. These vectors are called 'reciprocal lattice vectors' and can be calculated from the fundamental translations that leave the photonic crystal invariant (Kittel, 1995).

One property of the reciprocal lattice vectors is that they are all perpendicular to planes (called 'reticular planes') of aligned lattice nodes. A three-dimensional periodic structure can be seen as a set of criss-crossed periodic multilayers, formed by the stacks of reticular planes, one for each of the directions of the reciprocal lattice vectors $\vec{g}$ (Fig. 17). A theorem of crystallography tells us that the smallest $\vec{g}$ vector indicating a direction also tells us the period of the associated multilayer: the distance between the reticular planes normal to $\vec{g}$ is $d_{\vec{g}} = 2\pi/|\vec{g}|$. Furthermore, the emergent vector $\vec{k} + \vec{g}$ points to the direction of the wave reflected from the incident direction $\vec{k}$ on the reticular plane normal to $\vec{g}$. This means that the stack of reticular planes defines the direction of the diffracted beam, given the incident direction, and that the dominant diffracted wavelengths, with weak refractive index contrasts, is given by

$$\lambda_{\vec{g}} = \frac{2}{m}\left(\frac{2\pi}{|\vec{g}|}\right)\sqrt{\bar{n}^2 - \sin^2\theta_{\vec{g}}} \tag{24}$$

$\theta_{\vec{g}}$, in this expression, is the angle between the incident wave and the normal $\vec{g}$ to the reticular planes.

8.1 2D PHOTONIC CRYSTALS IN BIRDS: THE COMMON MAGPIE

The common magpie (Vigneron *et al.*, 2006b), *Pica pica*, gives an example of a two-dimensional or fibrous photonic crystal. This bird appears, most of the time, black and white, but under an appropriate illumination and viewing angle, green

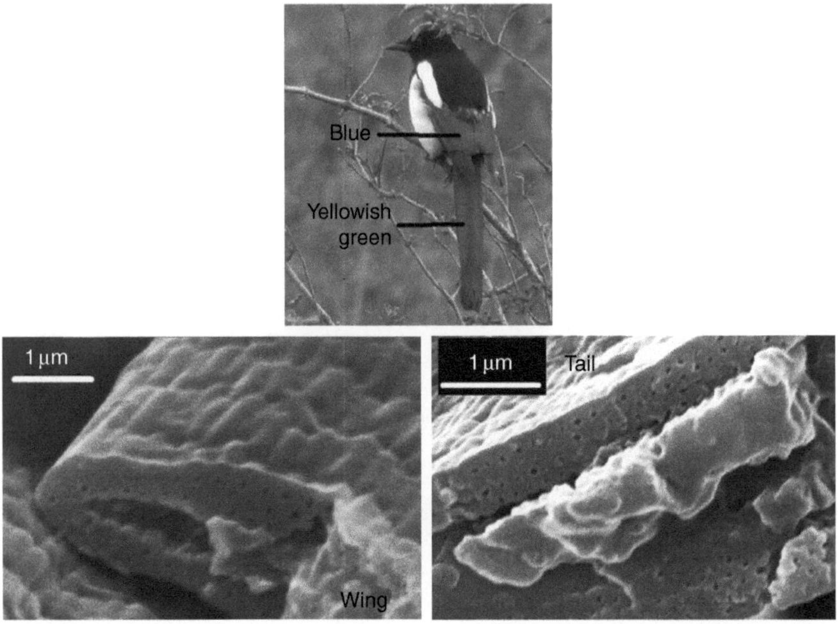

FIG. 18 The European, or common magpie (*Pica pica*) usually appears black and white, but under appropriate illumination and angle of view, shows two types of iridescence: the tail is yellowish green, with a purple termination, and the wings show deep-blue iridescence. These iridescences are caused by coherent scattering of light on a photonic crystal film formed by the cortex of the barbules (scanning electron micrographs labelled 'Tail' and 'Wings'). The magpie picture on the upper part is a grey-scale version of a picture by Thierry Tancrez (with permission).

and blue iridescence can be clearly perceived. As for the pigeon, this iridescence is caused by the structure of the cortex in the barbules. The cortex (see Fig. 18) contains melanin granules, with a hollow axis, embedded in keratin and arranged as a two-dimensional hexagonal lattice. The large reticular plane which produces the iridescence is the plane parallel to the barbules' surface and the corresponding stack spacing is $a(\sqrt{3}/2)$, where a is the distance between two neighbouring melanin granule axis. The distance between these scatterers is $a = 180$ nm for the yellow-green feathers in the tail (the calculation mentioned above leads to the yellowish green $\lambda = 2\bar{n}d/m = 592\,\text{nm}$, with $m = 1$). The blue feathers on the wing reflect a shorter wavelength, so that we would expect to find a shorter distance between the scattering vacuoles in the granules. In fact, unexpectedly, that distance is found to be $a = 270$ nm, and it should be understood that the blue colouration is caused by a reflection on the 'second gap', $m = 2$ of the 2D hexagonal photonic crystal. The dominant reflected wavelength, in this case is the blue $\lambda = 2\bar{n}d/m = 448\,\text{nm}$.

8.2 2D PHOTONIC CRYSTALS IN CTENOPHORES: *BEROË CUCUMIS*

The Ctenophore *B. cucumis* is a marine animal (Welch *et al.*, 2005), very different from those considered so far. It swims, chasing at different depth, being able to produce its faint own light by bioluminescence, and shows very rich and strong iridescence (Welch *et al.*, 2006) when illuminated from the sea surface, as it undulates to move. The iridescence of these animals has been studied by Victoria Welch, some years ago (Vigneron *et al.*, 2006b).

The origin of the colouration is found in the locomotion organs. No muscles are part of these organs, but a dense pack of cilia, producing a very precise two-dimensional structure, close to hexagonal. Figure 19 shows cross sections of the densely packed cilia, revealing an extremely coherent periodic structure. The refractive index contrasts are very weak, but the number of layers is so large that a very high light intensity is reflected. The reticular planes which selectively reflect the light are very regularly spaced, with spacing in proportion to the cilia diameter.

The main reticular plane stack produces a reflection in the red ($\sim$ 620 nm) under a normal incidence and emergence. At larger angles, the colour shifts rapidly to shorter wavelength, reaching the violet end of the human visible range (380 nm) at 60°. The iridescence is particularly rich (see the physics of the

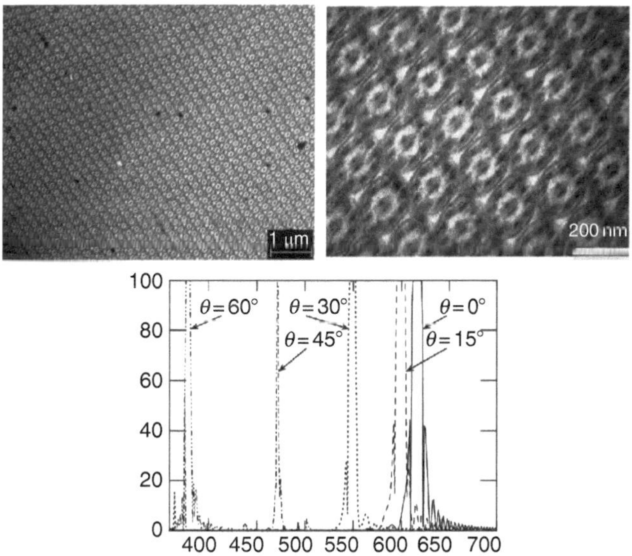

FIG. 19 The ultrastructure of packed cilia found in *Beroë cucumis*. The diameter of the cilia determines the distance between the reticular planes responsible for the selective, coloured, reflection and the iridescence. The right panel shows calculated reflectance spectra for different angles of emergence.

iridescence 'richness' in Deparis *et al.* (2006, 2008)), and this is in part due to the fact that the selective reflection occurs in water.

8.3 3D PHOTONIC CRYSTALS IN INSECTS

Chitin materials can be shaped into very complex forms and three-dimensional photonic crystals have been shown to occur in many insects: in butterflies (Berthier, 2006), first, but also in beetles like weevils or longhorns. One of the nice examples of three-dimensional structures in weevils is provided by the Brazilian 'diamond weevil', *Entimus imperialis* (see Fig. 20). This weevil bears scales, disposed in hemispheric cavities aligned on the elytra (Fig. 20, left). Illuminated with an extended white-light source, these scales reflect a variety of colours which are not due to refraction and dispersion, as in gem stones, but due to diffraction by a photonic crystal inside the scales. The photonic crystal has a face-centred cubic structure produced by the stacking of chitin slabs bearing protrusions and perforations with a triangular lattice arrangement.

In this case, several inequivalent reticular orientations can be identified, with different stack spacings, leading to different visible colours. This multi-coloured appearance is characteristic of very regular photonic crystals working in the visible, with a very large spatial coherence in the structure. Other weevils

FIG. 20 The weevil *Entimus imperialis* shows one of the most perfect three-dimensional photonic crystals in nature. The insect bears transparent scales that scatter white light as many different colours, ranging from deep blue to red. The origin of this colouration is the diffraction by the structure shown in the lower panel, occurring inside each scale. The scale itself is about 100 μm long, and contains one or two large grains of photonic crystal. (See Color Plate 27 at the back of the book.)

have structures which stabilize a desaturated colour, but then, disorder plays an important role in the final appearance. We will come back to the effects of disorder in the next section.

The crystallographic structure of such a photonic device is easy to describe, at least at the level of the Bravais lattice, defining the translational invariance. In the present case, as often with natural photonic crystals, the structure is *face-centred cubic*, with a primitive cell described by the vectors $\vec{a}_1 = (a/2)(0, 1, 1)$, $\vec{a}_2 = (a/2)(1, 0, 1)$, $\vec{a}_3 = (a/2)(1, 1, 0)$, where a is the cubic lattice parameter. In a photonic crystal, the content of the primitive cell, which corresponds to the distribution of the refractive index, is less easy to identify and to describe. Qualifications like 'opals', 'inverse opals', 'diamond', 'gyroids' have been used to describe the type of geometry, but this has still to be standardized in some rational way. Contrasting atomic crystallography, scattering centres cannot be identified and this does not help identifying a complete crystal structure.

8.4 THE LONGHORN *PROSOPOCERA LACTATOR*

From phylogeny, longhorns are not so distant from weevils, and it is interesting to note that some of them also show the presence of scales. These scales are generally elongated, with a sharp tip, rigidified by longitudinal groves on the outer part, a cortex of hard chitin. The recent discovery of a very regular photonic crystal in some of them was a total surprise (Colomer *et al.*, unpublished), as the colouration of these longhorns is not particularly bright.

Breaking one of the scales in the greenish patches of *P. lactator* reveals a three-dimensional structure made of chitin spheres connected by rigid cylindrical rods (see Fig. 21). Several crystallographic symmetries have been identified

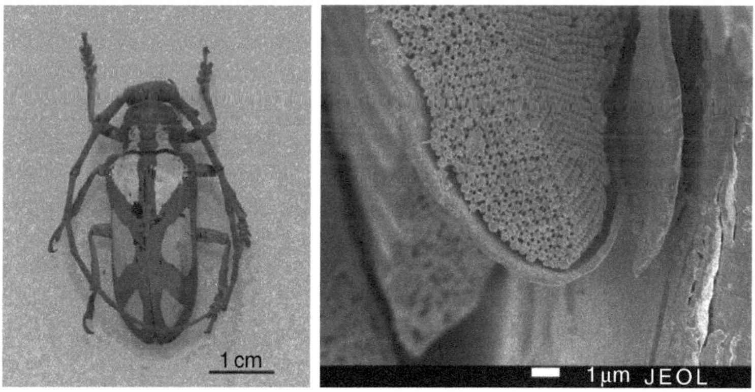

FIG. 21 *Prosopocera lactator* is a rather common longhorn of Eastern Africa, easily recognizable to the geometric pattern on the elytra and body. The light patches in this pattern are greenish, near to white, while darker zones appear light brown. The greenish areas are covered with sharp, rigid, bristles that contain a regular structure, as shown in the scanning electron image on the right.

locally. A clear determination of the structure symmetry from SEM pictures is difficult, but face-centred cubic symmetry seems to be the only structure compatible with the available images.

The colouration of these scales is much less saturated than for many other species in spite of the regularity of the photonic structure. One explanation may be thefact—visible in the scanning electron microscope in Fig. 21—that the structure is actually fractioned into grains 3–4 μm in diameter. As explained in the next section, some long-range orientation disorder can modify the visual effect to a large extent, leading to the complete destruction of the photonic-crystal iridescence or even to the loss of any colouration (Lafait *et al.*, 2009).

9 Carefully disordered structures

Disorder is not merely a negligible 'perturbation' in nature. The presence of disorder in a photonic structure actually controls visual effects, such as the lack of metallic aspect or the level of iridescence. The scales of many weevils, for instance, contain grains of photonic crystals, well-ordered inside each of them, but disordered in their orientations. The average size of the grains (say, δ) is one of the parameters that define the photonic structure, along with the parameters that define the local structure of the grains. For instance, the distance a between the scatterers inside the grains, and more precisely the associated optical path, defines their local scattering properties. In some weevils, such as *E. imperialis*, mentioned above, the grain size is of the same order of magnitude as the scale itself, so that this scale can be considered to only contain one (or just very few) large photonic monocrystal. The visual effect is a mix of several colours (see Fig. 22a) that translate into a yellowish green appearance when viewed from some distance. At the other extreme, if the grain size is of the order of the distance a between the scattering centres themselves and the structure becomes amorphous. This is the case of the white parts of the elytra in the weevil *Eupholus albofasciatus* (see Fig. 22d), where the scales contain an amorphous assembly of randomly positioned spheres, showing a maximum disorder. Between these two extremes, a weevil such as *Eupholus benetti* (Fig. 22c) shows spheres which are only locally organized (on the length scale of a grain size), suppressing iridescence as the light follows irregular paths long enough so that the memory of the initial illumination direction is lost. As analysed in Welch *et al.* (2007), this complex multiscale effect also allows for a spectral broadening, leading to the selection of a desaturated colour. The lack of iridescence is also found on the blue scales of the weevil *Eupholus schoenherri schoenherri*, shown in Fig. 22b. Note that the 'opal' structure described by Parker *et al.* (2003) some years ago, in a green weevil did not show iridescence, either. On the same *E. schoenherri schoenherri* weevil, the stripes of turquoise colour appear metallic, identifying the effect of larger grains.

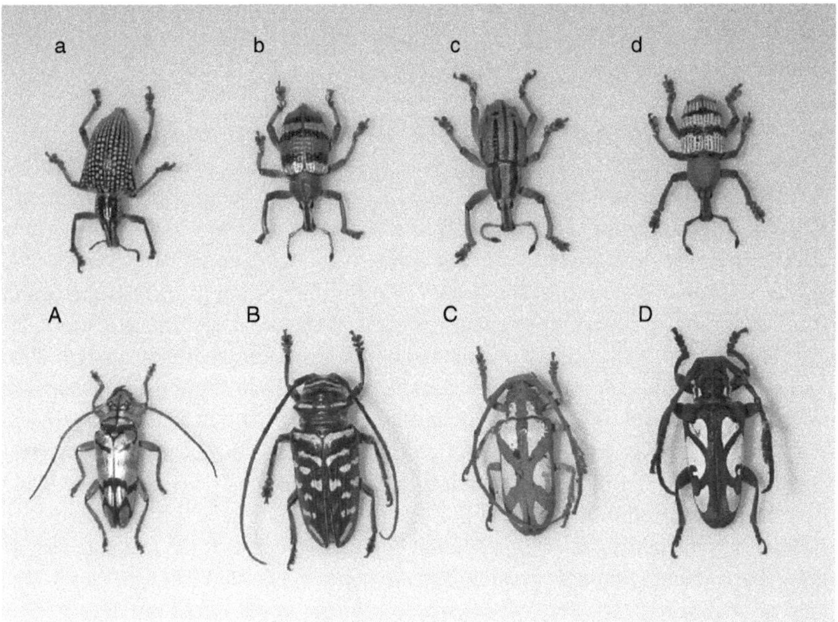

FIG. 22 Series of weevils and longhorns, where the amount of long-range disorder affects the visual effect produced by the photonic structure in the scales. In the specimens (a) and (A), the structure is iridescent and produce a metallic visual effect: the scales contain a high-coherence photonic crystal. In specimens (d) and (D), the amorphous structure produces white. In the other specimens, the variable saturation and metallic sheen results from an appropriate tuning of disorder in the photonic structure. (See Color Plate 28 at the back of the book.)

The variability is not limited to the usual parameters (hue and saturation) that define a physical colour. The visual effect also includes properties like 'metallicity', as opposed to a 'matt' appearance, related to the emergence angles distribution. Even this does not exhaust the visual aspect diversity: geometric patterns appearing on the cuticle surface further helps discrimination.

Similar effects can be seen in longhorns, where the long scales also contain photonic structures. *Tmesisternus raphaelae* (Fig. 22A) has scales that contain a structure similar to that found in *H. coerulea* (see above), and its coherence is very large, the photonic monocrystal being identified as the essential colouring device. *Calothyrza margaritifera* (Fig. 22D) has white scales which contain an amorphous random distribution of chitin spheres, while *Sternotomis virescens* (Fig. 22B) and *P. lactator* (Fig. 22C) contain grains with the right amount of disorder to reach the weak colouration and visual effect produced.

It should be emphasized that this controlled disorder is actually transmitted from one generation to the next, just as other traits are inherited. This means that this disorder affecting the coherence of the colour-producing photonic crystals should not be considered an 'imperfection', but should be regarded as part of the photonic structure design, for reaching well-defined visual effects and positive evolutionary advantages.

9.1 MORE ON WEEVILS STRUCTURES: *PACHYRRHYNCHUS CONGESTUS PAVONIUS*

A case of photonic crystal in a weevil scale has been studied in detail. *P. congestus pavonius* displays orange annular rings. The colouration is produced by scales which contain a three-dimensional photonic polycrystal. The crystallographic structure of the grains (see Fig. 23) is not so easy to perceive, from scanning electron microscope images because these images are only geometrical projection, in two dimensions, of three-dimensional objects. The easiest perception can be given by considering the three-dimensional structure as layered and built as a stack of slabs such as that shown on the right panel in Fig. 23. The slab is a thin film, profiled with a triangular lattice of cylindrical perforations and cylindrical protrusions acting as spacers. This corrugated slab is, in itself, a periodic structure, with a primitive cell containing three symmetric sites: the perforation site, the protrusion site and an empty site. The stacking of these slabs is such that a protrusion is always in contact with an empty site. This type of stacking can be shown to produce a face-centred cubic lattice.

One of the problems raised by the observation of this structure is the lack of iridescence. When the orange scales are viewed from different points of view, they always generate the same orange hue. The lack of iridescence can be traced back to a chaotic behaviour of the light propagation in the scale, where a large number of grain junction interfaces will be met (as a ball thrown in a 'chaotic'

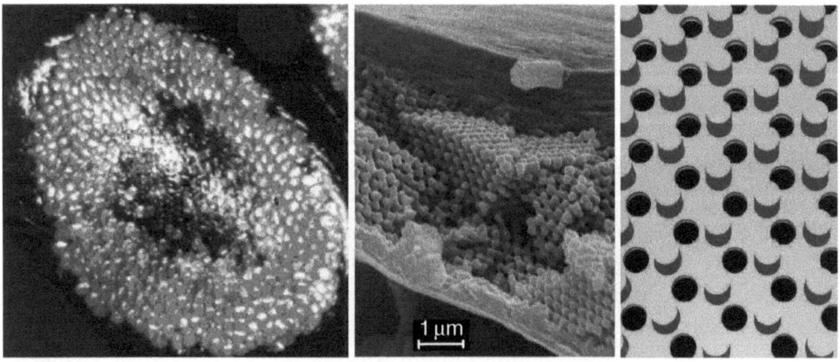

FIG. 23 The three-dimensional structure that causes the colouration of the tropical weevil *Pachyrrhynchus congestus pavonius* is a stack of corrugated layers with a triangular symmetry and three unequivalent sites A (perforation), B (protrusion) and C (flat, empty site). In the assembly of layers, the perforation lies above an empty site, which lies above a protrusion. This type of 'ABC' stacking leads to a face-centred cubic lattice. (See Color Plate 29 at the back of the book.)

billiard looses memory of its initial direction). While being multiply scattered by these interfaces, the memory of the initial incidence direction is lost, so that for all illuminations and emergence directions, the hue selection is averaged. Without any sensitivity to incidence and emergence light directions, iridescence looses all meaning.

9.2 *CYANOPHRYS REMUS* GREEN VENTRAL SIDE OF WINGS

The Brazilian butterfly *C. remus* (Kertesz *et al.*, 2006) has an important impact on the way we conceive the range of use of artificial structural colours. The ventral side of the wings of this butterfly is green, avoiding any metallic sheen or iridescence. While the individual perches, the pea-green ventral wing colouration presumably has a role in a complex survival strategy (cryptic pattern plus false-head). The green colouration is useful only when the butterfly is resting with closed wings, indicating a functional cryptic colour.

This pea-green colour is produced by light interference in a photonic polycrystal shown in Fig. 24. Actually, the local symmetry in the grain is again face-centred cubic and several colours are reflected from these grains because of the various orientations. Yellow and green grains can be identified,

FIG. 24 Details of the scales of the ventral side of the wing of *Cyanophrys remus*, seen with the scanning electron microscope (A) and a transmission electron microscope (B). The colouring structure is actually a photonic polycrystal, which produces visual effects very different from what can be expected from large monocrystals. (See Color Plate 30 at the back of the book.)

the colour variation being associated with the different orientation of the same structure. The long-range interspersed yellow and green grains produce the pea-green hue observed. However, an even more important effect occurs: the long-range disordered array of grains also acts to broaden the reflected spectrum and, at the same time, avoids iridescence. Though the grains produce a structural colour, they lead to a visual effect similar to that produced by diffusive pigments. A technological consequence is that it is perfectly conceivable to produce paints with 'structural pigments' with 'diffuse' finish.

10 Conclusion

Many optical devices that have been with us in optics and photonics have found their counterpart in living organisms. Nature knows of total reflection, thin-film filtering, gratings, photonic crystals, but also lenses, parabolic mirrors, optics fibres, solid-state light sources, fluorescent converters, broadband mirrors and much more. It can put two gratings on the same surface (Ingram *et al.*, 2008), and other things we have not yet invented.

Biology offers a wide variety of visual effects, based on the structural mechanisms of colouration. Discussing the biological function of these effects is always difficult, as a surface structure can often respond to several environment pressures. The protrusions found on the transparent wings of some moths or butterflies, for instance, may impact the quality of the membrane transparency (Deparis *et al.*, 2009), but it is also useful to avoid wetting in a humid atmosphere. Evolution and natural selection can result in extraordinarily complex and efficient structures, but their heuristic design is often the result of concurrent needs, so that a compromise is generally the rule. From the engineering point of view, we have a lot to learn there: multifunctional optimization of materials is a very difficult problem, as it is not proven that design of a structure that fulfils a single constraint can just be 'slightly perturbed' to accommodate a second one to be globally optimal. In order to properly account for several functions, complete redesign is often the best and inspiration does not easily provide the right starting point. Observing multifunctional natural structures and attempting their multiphysics reverse engineering amounts to collect new ideas for this very difficult design.

All this shows that photonic devices from living organisms should not be regarded as just the field of investigation of biologists alone. This chapter of science does include many biological concerns, but it also includes information useful for modern optical engineering. We will need light not only for lightning, but also to communicate. If we use the internet or our cellular phone, we set photons to rush through optics fibres in order to carry knowledge over long distances. Translation to and from electronic switching devices slows down the whole thing and it will soon be necessary to optically redirect photons directly, for performance purposes. Transparent homogeneous materials will likely not be smart enough for achieving such complex functions, and it is considered that inhomogeneous materials will be needed. One problem is, however, that the propagation of waves in complex inhomogeneous media has not revealed all of its secrets. Complex inhomogeneous photonic materials found at the surface of insects and other living organisms are just examples of these materials, where intricate interfaces control the optical properties. We will be able to learn this new physics by studying the mechanisms of life colouring—at least as long as these material structures do not get extinct, along with the species that may reveal them.

References

Bay, A. and Vigneron, J. P. (2009). Light extraction from the bioluminescent organs of fireflies. *Proc. SPIE* **7401**, 7401081.

Berthier, S. (2006). *Iridescences: The Physical Colors of Insects*. Springer, Berlin.

Berthier, S., Charron, E. and Silva, A. D. (2003). Determination of the cuticle index of the scales of the iridescent butterfly *Morpho menelaus*. *Opt. Commun.* **228**, 349.

Colomer, J. F., Vigneron, J. P., *et al.*, unpublished.

de Albuquerque, J., Giacomantonio, C., White, A. and Meredith, P. (2006). Study of optical properties of electropolymerized melanin films by photopyroelectric spectroscopy. *Eur. Biophys. J.* **35**, 190.

de Duve, C. (1969). The peroxisome: a new cytoplasmic organelle. *Proc. R. Soc. Lond. B Biol. Sci.* **173**(30), 71–83.

Deparis, O., Vandenbem, C., Rassart, M., Welch, V. L. and Vigneron, J.-P. (2006). Color-selecting reflectors inspired from biological periodic multilayer structures. *Opt. Express* **14**, 3547.

Deparis, O., Rassart, M., Vandenbem, C., Welch, V., Vigneron, J. P. and Lucas, S. (2008). Structurally tuned iridescent surfaces inspired by nature. *New J. Phys.* **10**, 013032.

Deparis, O., Khuzayim, N., Parker, A. and Vigneron, J. P. (2009). Assessment of the antireflection property of moth wings by three-dimensional transfer-matrix optical simulations. *Phys. Rev. E* **79**, 041910.

Ingram, A. L., Lousse, V., Parker, A. R. and Vigneron, J. P. (2008). Dual gratings interspersed on a single butterfly scale. *J. R. Soc. Interface* **5**, 1387–1390, doi:10.1098/rsif.2008.0227.

Jackson, J. D. (1975). *Classical Electrodynamics*. 2nd ed. Wiley, New York0-471-43132-XSec. 7.10.

Kertesz, K., Bálint, Zs., Vértesy, Z., Márk, G. I., Lousse, V., Vigneron, J. P., Rassart, M. and Biró, L. P. (2006). Gleaming and dull surface textures from photonic crystal type nanostructures in the butterfly *Cyanophrys remus*. *Phys. Rev. E* **74**, 021922.

Kittel, C. (1995). *Introduction to Solid State Physics*. 7th ed. Wiley, 978-0471111818, New York.

Lafait, J., Andraud, C., Berthier, S., Boulenguez, J., Callet, P., Dumazet, S., Rassart, M. and Vigneron, J.-P. (2009). Modeling the vivid white colour of *Calothyrza margaritifera* cuticle. *In* Proceedings of the E-MRS Symposium on "Bioinspired and biointegrated materials as new frontiers nanomaterials" Strasbourg.

Lucretius Carus, T. (2008). *On the Nature of Things*. BiblioBazaar, LLC, 978-1437524697.

Malloock, A. (1911). Note on the iridescent colours of the birds and insects. *Roy. Soc. Proc.* **A85**, 598.

Parker, A. R., Welch, V. L., Driver, D. and Martini, N. (2003). Structural colour: opal analogue discovered in a weevil. *Nature (Lond.)* **426**, 786.

Parsons, M. (1998). *The Butterflies of Papua New Guinea: Their Systematics and Biology*. Academic Press, London.

Rabelais, F. (1534). La Vie Inestimable du Grand Gargantua, Père De Pantagruel. *In* On les vend à Lyon, chez Françoys Iuste, devant nostre Dame de Confort Chapter vi, Lyon, France.

Rassart, M., Simonis, P., Bay, A., Deparis, O. and Vigneron, J. P. (2009). Scale colouration change following water absorption in the beetle *Hoplia coerulea* (Coleoptera). *Phys. Rev. E* **80**, 031910.

Sarrazin, M., Vigneron, J. P., Welch, V. and Rassart, M. (2008). Nanomorphology of the blue iridescent wings of a giant tropical wasp *Megascolia procer javanensis* (Hymenoptera). *Phys. Rev. E* **78**, 051902.

Schenk, F. and Harvey, J. (2008). Reflections on the Natural History Museum: the potential of the collections for art & design research. *In* Third International Conference on the Arts in Society 28–31 July, Birmingham, UK.

Vigneron, J. P., Colomer, J.-F., Vigneron, N. and Lousse, V. (2005). Natural layer-by-layer photonic structure in the squamae of *Hoplia coerulea* (Coleoptera). *Phys. Rev. E* **72**, 061904.

Vigneron, J. P., Rassart, M., Vandenbem, C., Lousse, V., Deparis, O., Biró, L. P., Dedouaire, D., Cornet, A. and Defrance, P. (2006a). Spectral filtering of visible light by the cuticle of metallic woodboring beetles and microfabrication of a matching bioinspired material. *Phys. Rev. E* **73**, 041905.

Vigneron, J. P., Colomer, J.-F., Rassart, M., Ingram, A. L. and Lousse, V. (2006b). Structural origin of the coloured reflections from the black-billed magpie feather. *Phys. Rev. E* **73**, 021914.

Vigneron, J. P., Kertész, K., Vértesy, Z., Rassart, M., Lousse, V., Bálint, Zs. and Biró, L. P. (2008). Correlated diffraction and fluorescence in the backscattering iridescence of the male butterfly *Troides magellanus* (Papilionidae). *Phys. Rev. E* **78**, 021903.

Welch, V. L., Vigneron, J. P. and Parker, A. R. (2005). The cause of colouration in the ctenophore *Beroë cucumis*. *Curr. Biol.* **15**(Suppl. 24), R985–R986.

Welch, V., Vigneron, J. P., Parker, A. and Lousse, V. (2006). Optical properties of the iridescent organ of the comb-jellyfish *Beroe cucumis* (Ctenophora). *Phys. Rev. E* **73**, 041916.

Welch, V., Lousse, V., Deparis, O., Parker, A. and Vigneron, J. P. (2007). Orange reflection from a three-dimensional photonic crystal in the scales of the weevil *Pachyrrhynchus congestus pavonius* (Curculionidae). *Phys. Rev. E* **75**, 041919.

Yin, H., Shi, L., Sha, J., Li, Y., Qin, Y., Dong, B., Meyer, S., Liu, X., Zhao, L. and Zi, J. (2006). Iridescence in the neck feathers of domestic pigeons. *Phys. Rev. E* **74**, 051916.

Zi, J., Yu, X., Li, Y., Hu, X., Xu, C., Wang, X., Liu, X. and Fu, R. (2003). Colouration strategies in peacock feathers. *Proc. Natl. Acad. Sci. USA* **100**, 12576.

Ziman, J. M. (1979). *Principles of the Theory of Solids*. 2nd ed. Cambridge University Press, Cambridge978-0521297332.

Molecular and Physiological Basis of Colour Pattern Formation

H. Frederik Nijhout

Department of Biology, Duke University, Durham, North Carolina, USA

1 Introduction

During the past decade the colour patterns of insects have emerged as model systems for the study of the interplay between development and evolution. Two features make colour patterns attractive in this regard. The first is that they occur on the surface of the animal and are essentially two-dimensional structures;

ADVANCES IN INSECT PHYSIOLOGY VOL. 38
ISBN 978-0-12-381389-3
DOI: 10.1016/S0065-2806(10)38002-7

there are no essential parts that are hidden or inaccessible underneath the visible ones, which make the patterning process easily accessible to visualization and manipulation. The second one is that colour patterns are made up of visible pigments whose biosynthetic pathways are well understood, which makes the pathway from gene expression to manifestation of the final phenotype exceptionally short, compared to that of other morphological features. This gives hope that in this system it will be possible to eventually elucidate the entire causal chain from genotype to complex phenotype. Because colour patterns are essentially a means of communication (such as mimicry, aposematism, camouflage, sexual signalling) with other animals, they are among the most visually obvious evolutionary adaptations in the natural world. The evolution of colour patterns can be viewed as two separate questions: the evolution of pigment synthesis pathways and the evolution of patterned synthesis of those pigments. Thus, pattern evolution can be readily associated with the evolution of specific and easy to understand developmental mechanisms. In this review, I focus on recent developments in our understanding of the molecular and the developmental–physiological mechanism of colour pattern formation and place these in the context of antecedent work.

2 Mechanisms of colour pattern formation

The fundamental problem of developmental biology is how a pattern arises in an initially homogeneous, undifferentiated, field of cells. This is often described as the process of determination by which the developmental fate of a subset of cells in the population becomes restricted. This is followed by the process of differentiation by which those cells develop properties different from what they had before, and different from those of the other cells in the field. In colour pattern formation, this question can be framed in terms of the process by which cells in one location come to synthesize one pigment, while cells in an adjacent location synthesize a different pigment.

Insect colour patterns are particularly suitable for the study of the mechanisms of pattern formation because they are essentially two-dimensional. This greatly simplifies analysis and interpretation because the entire pattern can be seen at once, without the need for dissection or histology. The two-dimensionality arises from the fact that colour is a feature of the integument, which is a two-layered structure composed of a monolayer of epidermal cells, which secrete the non-living overlying cuticle. Pigments are synthesized by the epidermal cells and are incorporated into the overlying cuticle (but see Section 9.1). Although epidermal cells can divide, they do not migrate nor change position relative to each other. Thus, colour pattern formation occurs primarily through processes of cell-to-cell communication in a static or growing sheet of cells.

2.1 REACTION–DIFFUSION AND GRADIENT–THRESHOLD MECHANISMS

Turing (1952) described an amazingly simple mechanism by which complex spatial patterns can arise in a sheet of cells. Turing showed that if there are two chemicals that affect each other's synthesis and degradation, and that also diffuse from areas of high concentration to areas of low concentration, it is possible to obtain a stable steady state with a standing-wave-like pattern of concentration of the two chemicals. In the absence of diffusion, Turing systems have a homogeneous steady state in which the concentrations and reaction rates are the same, and stable, throughout the field. But if the two chemicals are allowed to diffuse, the steady state becomes unstable and even a minor perturbation causes the system to move to a new steady state that is spatially inhomogeneous (i.e. both the concentration and the synthesis and degradation rates of the two chemicals will be different, but stable, at different locations). The conditions under which such reaction–diffusion systems can produce spatial patterns have been detailed by Edelstein-Keshet (2005), Meinhardt (1982) and Murray (2003). The exact pattern that is produced depends on the particulars of the reaction's mechanism, the size and shape of the field on which the reaction–diffusion takes place and the pattern of initiating stimuli. Turing (1952) proposed that the initiators of the diffusion-driven instability could be simple random events, such as thermal noise. Such initiators can lead to evenly spaced but random patterns, such as leopard spots, zebra stripes, finger prints and insect bristle patterns. The colour patterns of some tropical butterfly fish appear to be produced by such a mechanism (Kondo, 2002; Kondo and Asai, 1995). Meinhardt (1982) has explored in detail the ability of reaction diffusion mechanism to produce the diverse colour patterns formed on sea shells.

Because Turing patterns can be initiated by random perturbations, random colour patterns are the best *a priori* candidates for Turing mechanisms. In insects, colour patterns that are essentially random, such as the ripple patterns on butterfly and moth wings (Fig. 1) and the patterns on the forewings of many grasshoppers (e.g., *Schistocerca*, Fig. 1), may well be produced by Turing mechanisms. Because these patterns are random, no two individual have the identical position of bars and stripes, nor are they identical on left and right sides of an individual animal, and these patterns, like fingerprints, can be used to identify individuals.

Non-random colour patterns can also be produced by Turing-style reaction–diffusion mechanisms, but require special initiation and boundary conditions. Perhaps the most orderly of all colour patterns among the insects are those on the wings of the Lepidoptera (Nijhout, 1991). Here a system of bands and spots admits to a common groundplan that described homology relationships among colour pattern elements across a broad and diverse range of taxa. In this system, pattern formation occurs in three steps. First, a process that establishes a prepattern of sources and sinks of morphogens; second, a process by which interactions of molecular signals from those sources produce a system of

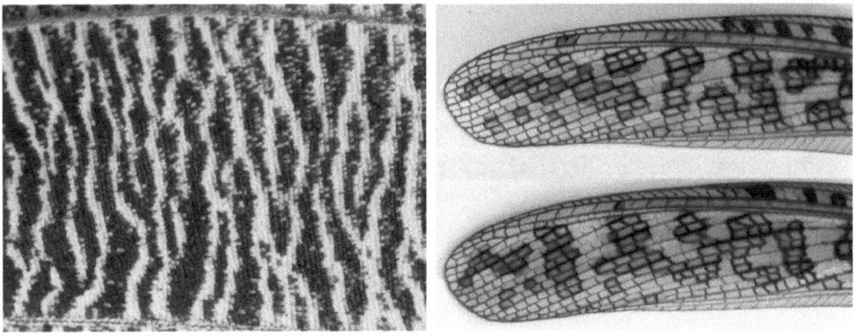

FIG. 1 Ripple patterns on the wing of *Caligo memnon* (left panel). Ripples are random colour patterns that are common in the Satyridae and related Lepidoptera. Here the ripples also illustrate stochastic pigment expression, in individual scales. Stochastic colour patterns on the wings of *Schistocerca americana* (right panel). Although there is an overall tendency towards stripes and blotches, the details of the pattern are random. (See Color Plate 31 at the back of the book.)

gradients; and third, a process whereby threshold on those gradients specify local pigment synthesis. Nijhout (1990) has shown that a Turing mechanism that relies only on the wing veins as a source that destabilizes the homogeneous steady state can produce the pattern of sources and sinks required to generate almost the entire diversity of colour patterns found in over 10,000 species of butterflies. Of course, the excellence of fit of the result does not prove the modelled mechanism is correct. The main features of lepidopteran colour patterns are produced by what appear to be gradient–threshold mechanisms (French and Brakefield, 1995; Nijhout, 1990), and such mechanisms are probably the norm for other insect patterning systems.

Turing mechanisms can also explain the bold mimicry patterns in *Papilio dardanus* (Nijhout *et al.*, 2003; Sekimura *et al.*, 2000, 2007). This species can develop some 14 distinctly different colour patterns by allelic variation at a single locus (for illustrations, see Nijhout, 2003b). Mathematical modelling of pattern formation has shown that development of these patterns depends critically on signalling from the wing margin. Using fixed boundary conditions, itwas possible to simulate a subset of realistic colour patterns. Additional realistic patterns could be simulated by changing the level at which the activator was fixed in different regions of the wing margin. This theoretical work is supported by experimental and comparative studies, which have shown that in the Lepidoptera the wing margin plays an active role in pattern determination (Koch and Nijhout, 2002; Nijhout, 1990, 1991; Reed and Gilbert, 2004; and see Section 5.1).

The attractiveness of Turing-style reaction–diffusion mechanisms arises from the fact that they can generate pattern *de novo*, in an initially unpatterned homogeneous field of cells. A large literature in theoretical mathematical biology and biophysics has developed over the years that attempts to apply

Turing mechanisms as potential solutions to a broad diversity of biological pattern-formation problems in embryonic and postembryonic development.

But Turing-style reaction–diffusion mechanisms are not the only nor even the most common mechanism for pattern formation in animals. Research in developmental genetics over the past 20 years has shown that in many developmental systems pattern formation proceeds by a succession of gradient–threshold mechanisms, each building on a previous prepattern. In brief, development in an egg starts off with graded distributions of various maternal gene products that act primarily as transcriptional regulators. There is never a homogeneous pattern-less early state. Wherever they occur above a particular threshold, these maternal transcription factors activate embryonic genes most of which likewise encode transcriptional regulators. These embryonic factors diffuse and set up secondary gradients that activate further transcriptional regulators. Due to interactions among transcriptional activators and repressors, an ever changing and progressively more complex and detailed spatial pattern of gene expression unfolds. In general, new gene expression occurs wherever a particular combination of transcription factors is above (or below) some threshold. The important thing to recognize is that for every pattern there is always an antecedent pattern, and the regression of antecedent patterns is for all practical purposes infinite. This means that initially homogeneous steady states postulated by classical Turing mechanism seldom if ever exist. In insect patterning, Turing mechanisms and gradient–threshold mechanisms operate in fields that are initially patterned, and where either the boundaries or internal inhomogeneous structures stimulate dynamic changes in the concentrations of signalling molecules or transcriptional regulators. This leads to progressive changes in the spatial pattern of gene expression that eventually becomes fixed as a spatially heterogeneous pattern of pigment synthesis.

3 Abdominal pigmentation

Abdominal spotting and striping patterns are among the most distinctive and prominent colour patterns in Diptera and Hymenoptera. Probably the best known among these are the yellow-black aposematic banding patterns of wasps and bees (Fig. 2). Many Lepidoptera also have complex abdominal colour patterns (Fig. 3). Abdominal pigmentation patterns in species of *Drosophila* consist mostly of anterior bands of melanin that can extend posteriorly along the dorsal midline. In addition, some species have distinctive medial and lateral spots (Brisson *et al.*, 2009; Hatadani *et al.*, 2004; Hollocher *et al.*, 2000; Wittkopp *et al.*, 2003). This pattern diversity is somewhat reminiscent of that of the wasps, illustrated in Fig. 2; though the fly patterns are typically less well-defined, with graded rather than sharp boundaries. Some abdominal pigment patterns in insects coincide with the locations of muscle insertions, which presumably act as pattern organizers, but most patterns are associated with either the anterior or posterior segmental margins, or the abdominal midline.

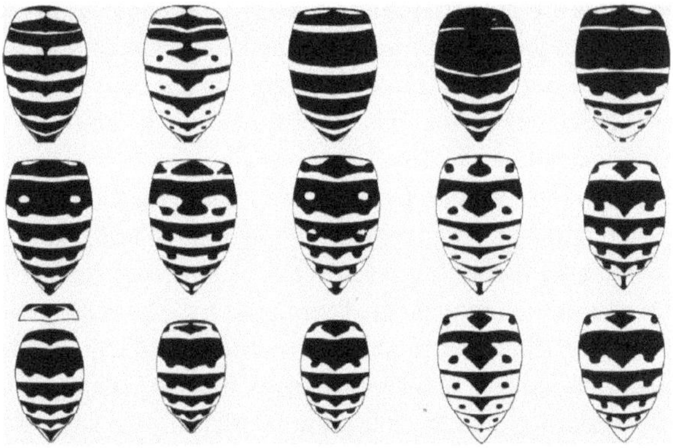

FIG. 2 Abdominal colour pattern in North American wasps of the genus *Vespula*. (From Akre *et al.*, 1981.)

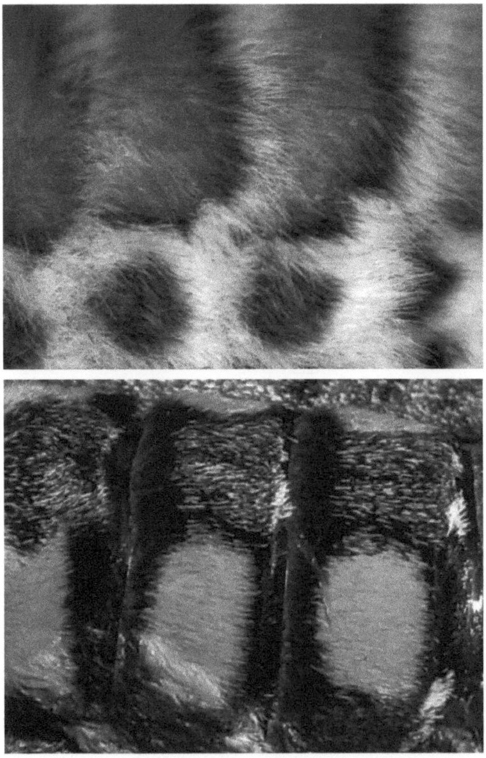

FIG. 3 Abdominal colour patterns in two moths: *Hyalophora cecropia* and *Manduca sexta*. (See Color Plate 32 at the back of the book.)

Kopp and co-workers (Kopp and Duncan, 2002; Kopp *et al.*, 1999) have elucidated a complex pattern of expression of regulatory genes during the formation of the adult abdomen in *Drosophila*, and the distribution of some of these genes mirrors the distribution of pigmentation. The adult abdominal pattern is established by a dynamically changing pattern of gene expression that accompanies the replacement of the larval epidermal cell population with histoblast-derived adult cells. Figure 4 illustrates the pattern of gene expression in one, the dorsal abdominal segment that is established stabilized after the second day of pupal development. The posterior compartment of the segment is marked by the expression of engrailed. In the pharate adult there is a band of expression of decapentaplegic (dpp) along the dorsal midline, and on either side of this midline there are two regions of wingless (wg) expression. This pattern is preceded by the expression of several genes that are expressed as bands across the width of the anterior compartment. Optomotor blind (Omb) is expressed as a broad band that straddles the compartment boundary with a sharp posterior and a graded anterior boundary of expression. Hedgehog (hh), which can act as a diffusible secreted morphogen, is expressed along the posterior half of the engrailed compartment and diffuses anteriorly (Lee *et al.*, 1992; Mohler and Vani, 1992; Tabata *et al.*, 1992). Patched (ptc), which is part of the hh receptor, is expressed as two narrow bands, one near the anterior margin and one at the posterior margin of the anterior compartment. Ptc is believed to restrict the anterior diffusion of hh (Chen and Struhl, 1996). In the anterior compartment, hh participates in the induction of wg and dpp (Ingham, 1993; Sanicola *et al.*, 1995).

These patterned regulatory genes presumably direct the development of many different features on the abdominal wall, including the more complex colour patterns. The male-specific melanic pigmentation of the terminal abdominal segment in *Drosophila* is controlled by the interaction of Abdominal-B (Abd-B) and bric-a-brac (bab), where Abd-B directly controls the expression of yellow via a specific *cis*-regulatory element in the yellow promoter (Jeong *et al.*, 2006; Kopp *et al.*, 2000).

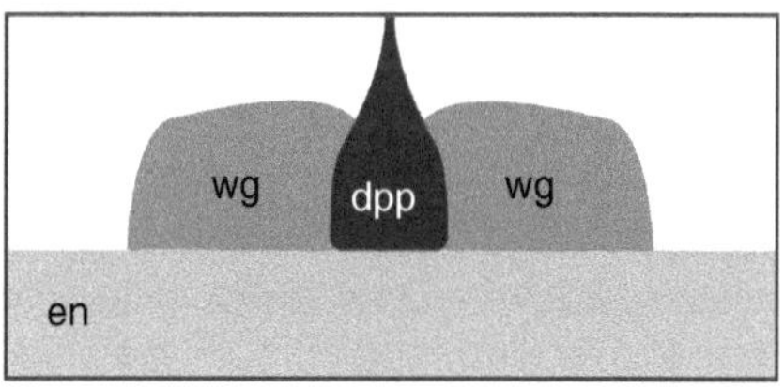

FIG. 4 Expression patterns of regulatory genes in one segment of the pharate adult abdomen of *Drosophila*, viewed from the dorsal aspect. (After Kopp and Duncan, 2002; Kopp *et al.*, 1999.)

4 Facial pigment patterns

Paper wasps in the genus *Polistes* have a highly variable facial pigmentation pattern. In *P. fuscatus* these patterns are used by an individual in a colony for individual recognition, as demonstrated by experiments in which parts of these patterns are painted over. Such altered wasps are subject to increased aggression in a colony, until the nest mates become familiar with the new pattern (Tibbetts, 2002). In *P. dominulus* the facial colour pattern is correlated with body size and also with social rank (Tibbetts and Dale, 2004). The facial pattern in this species turns out to be strongly influenced by nutrition during the larval stage: poorly fed larvae produce smaller adults that have facial patterns corresponding to individuals of a lower social rank (Tibbetts and Curtis, 2007). Nutrition is known to affect the hormonal environment during larval development (Emlen and Allen, 2003), and developmental hormones may thus be involved in regulating the development of the facial colour pattern in these wasps, just as they can affect colour pattern development in other species (see Section 9.1).

5 *Drosophila* wing patterns

Many species of Diptera have melanic wing patterns, consisting of various arrangements of patches and stripes. The best known and most spectacular examples are those of the Hawaiian *Drosophila* group (for a review of this diversity, see Edwards *et al.*, 2007). Other drosophilids have simpler patterns, consisting of a few spots or of stripes that run partially along the wing veins.

Drosophila biarmipes is a species closely related to *Drosophila melanogaster*, and differs from the latter in having, among others, a black melanin spot in the distal anterior quarter of the forewings of male flies. The forewing of *D. melanogaster* itself has a very faint overall melanization. In *D. melanogaster*, the yellow gene is expressed at low levels over the entire pupal wing, whereas in *D. biarmipes* there is a strong expression of the yellow gene in the distal anterior quadrant of the wing corresponding to the prospective location of the melanin spot.

To investigate whether the colour pattern differences between these two species were due to changes in the *cis*-regulatory region of the yellow gene or to changes in the spatial pattern of *trans*-acting transcriptional regulators, Gompel *et al.* (2005) transformed *D. melanogaster* with a green fluorescent protein (GFP)-reporter constructs that contained the non-coding sequence of the *D. biarmipes* yellow gene, and found that in the *melanogaster* wing GFP was expressed in a pattern identical to that of the *biarmipes* black spot. This experiment showed that evolutionary changes in the *cis*-regulatory region of the yellow gene was largely responsible for the evolution of the melanin spot in the *biarmipes* wing (Gompel *et al.*, 2005; Wittkopp *et al.*, 2002). Presumably, these changes involved the loss and/or gain of sites at which transcriptional regulators can bind. Phylogenetic analysis suggest that the pattern-less wing of *melanogaster* is the derived condition, so evolution of the pattern must have been due to a loss of the *melanogaster* yellow to respond to transactivators. However, the pattern of GFP expression in *melanogaster* did not precisely replicate the *biarmipes* colour pattern, which indicates that there must also be some differences in the spatial pattern of *trans*-acting transcriptional regulators. One of the regulators of yellow gene is engrailed, which represses expression of the yellow gene only in the posterior compartment of the wing, so this one regulator is clearly not sufficient to explain the spatial pattern of yellow expression. An additional level of complexity in the regulation of this pigment pattern was the finding that development of the melanin spot required not only overexpression of yellow but also down-regulation of the ebony gene (Gompel *et al.*, 2005; Wittkopp *et al.*, 2002). The protein encoded by the yellow gene is not yet known, but it is evidently required for the synthesis of melanin. The ebony (e) gene encodes the enzyme *N*-β-alanyl-dopamine (NBAD) synthetase that converts dopamine to NBAD, which is subsequently oxidized to sclerotin (Koch *et al.*, 2000; Wittkopp *et al.*, 2002; Wright, 1987). Ebony thus takes dopamine away from the melanin biosynthetic pathway and effectively acts as a suppressor of melanin synthesis. An inverse association between the expression of yellow and ebony was found in the more complex multi-spotted wing pattern of *D. guttifera* (Gompel *et al.*, 2005), which suggest that this particular polygenic control of the melanin pattern may be general for drosophilid wing patterns.

Ectopic expression of tyrosine hydroxylase (TH) in *D. melanogaster* caused the expression of an extensive and interesting ectopic melanin patterns in the wing. These patterns consisted of bands of melanin roughly parallel to the wing veins, and tapering towards the wing margin. These patterns spread gradually from the wing veins over a 3- to 5-day period after adult eclosion (True *et al.*, 1999). This ectopic colour pattern only appeared when TH was expressed ubiquitously across the wing. TH converts tyrosine to dopa, which is the first step in the synthesis of melanin (Fig. 5), and is encoded by the *Drosophila* gene *pale*. Ectopic expression of dopa-decarboxylase (DDC), the enzyme that converts dopa to dopamine (Fig. 5), did not induce this vein-dependent melanin pattern, but enhanced the effect of ectopic expression of TH. In *veinlet* and *Vein off* mutants, which have an abbreviated venation system, the pigment pattern developed only where veins were expressed and not elsewhere on the wing. The fact that the pattern gradually spread out from the wing veins suggest that the circulatory system provides the precursor, tyrosine, which gradually diffuses within the blade of the wing. Static enzymes within the wing then precipitate melanin as the tyrosine front spreads away from the veins. The melanin in this case is very pale and takes several days to develop, in contrast with natural drosophilid wing melanin patterns, which develop much more rapidly. This, and the fact that the front of melanin synthesis continues to move for several days,

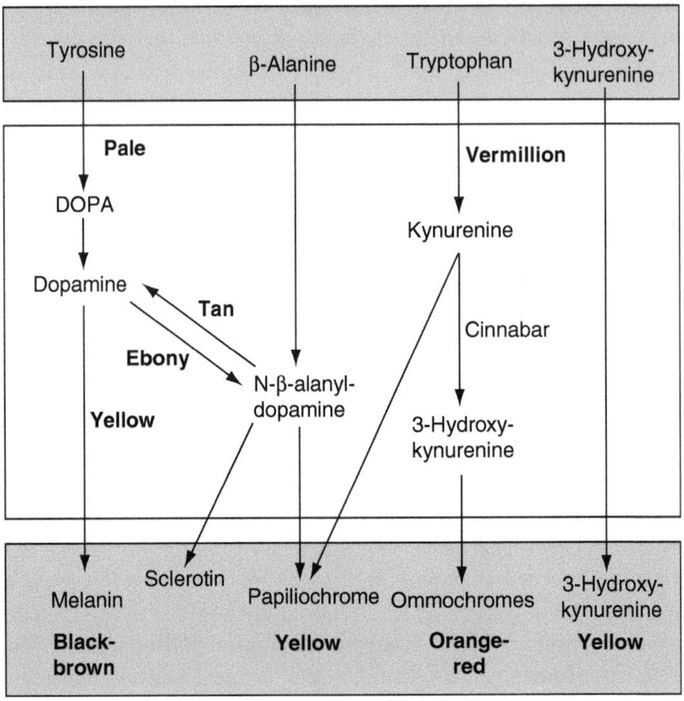

FIG. 5 Abbreviated biosynthetic pathways for the synthesis of melanin, ommochromes and papiliochrome. The locations of genes discussed in the text are shown. (After Koch *et al.*, 2000; True *et al.*, 2005; Wittkopp *et al.*, 2002; Wright, 1987.)

indicates there is not sufficient enzyme activity in the wing to convert all the tyrosine that is produced by the veins into dopa, dopamine and melanin.

These ectopic patterns bear a close resemblance to the vein-dependent colour patterns seen in many butterflies and moths (Nijhout, 1991, 2001). These lepidopteran patterns also taper towards the wing margin, and are interrupted when a vein is experimentally cut. In *Danaus plexippus* (the Monarch butterfly), the vein-dependent patterns are made up of several parallel pigment bands that look remarkably like those obtained experimentally in *Drosophila* (Fig. 6), which suggest that lateral diffusion of a substrate or a signalling molecule is responsible for these more complex pigmentation patterns as well.

5.1 VEIN AND MARGIN-DEPENDENT PATTERNS

The venation system plays an important role in pigment patterning in many insects. Among the simplest vein-dependent patterns are those in which the cuticle over all or part of the venation system is pigmented. This can give a finely lined pattern when the pigment bands are narrow, or a beautifully blotched pattern when they are wide (Fig. 7). Experiments with *Drosophila*,

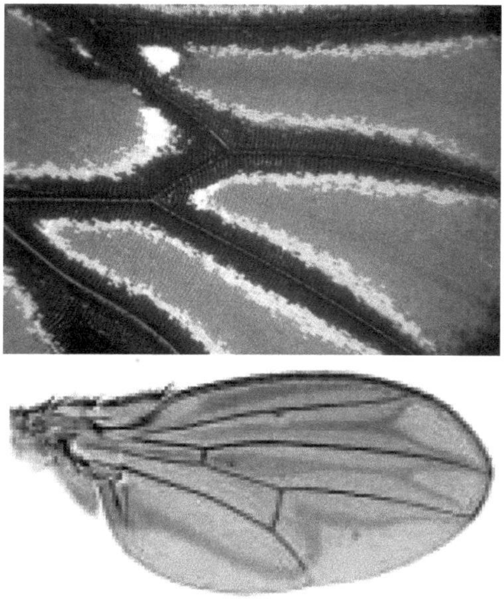

FIG. 6 Wing vein-dependent colour pattern in the Monarch butterfly, *Danaus plexippus*. Induced vein-dependent patterns in *Drosophila*. (From True *et al.*, 1999.) (See Color Plate 33 at the back of the book.)

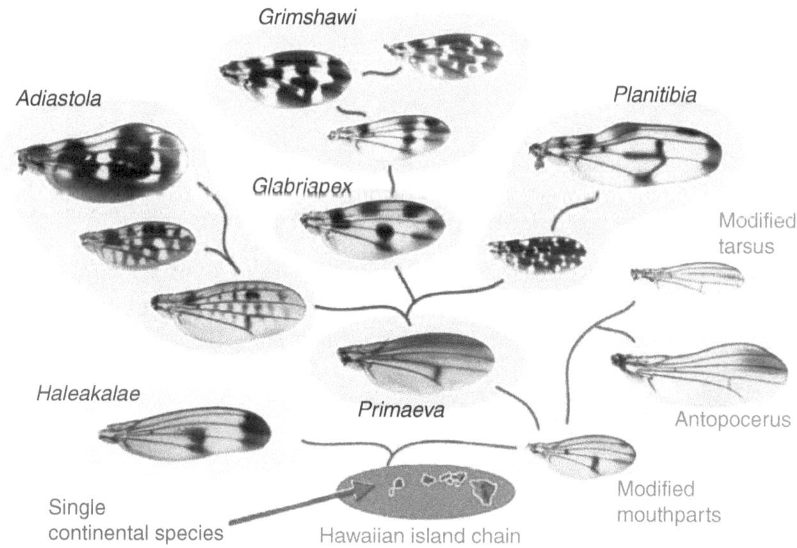

FIG. 7 Diversity of melanic wing patterns in the Hawaiian *Drosophila*. (From Edwards *et al.*, 2007.) (See Color Plate 34 at the back of the book.)

outlined in the previous section, have shown that these broader patterns can arise by diffusion of pigment substrates out of the wing veins, which are then acted on by enzymes distributed within the wing blade. Such a simple mechanism poses some advantages and some difficulties. If the enzymes are evenly distributed then they will start to work on the substrate as soon as it leaves the veins, and if the enzymes are highly active all the diffusing substrate will be converted to pigment near the wing vein, leading to narrow bands. If the enzymes experience substrate inhibition (as is likely to be the case of TH; Coleman and Neckameyer, 2004; Quinsey and Luong, 1998), then some of the substrate will escape reaction near the wing veins and be converted to pigment at some distance away, leading to broader bands (or if the inhibition is strong), to pigment bands at some distance and parallel to the veins.

In the Lepidoptera, the wing veins also play an important role in compartmentalizing the wing into regions that are semi-independent in colour pattern formation (Nijhout, 1990, 1991, 2001). Compartmentalization by wing veins leads to serial homology of pattern elements across the wing surface, and has enabled the great diversification of wing patterns in butterflies and moths. An illustration of the compartmentalizing role of wing veins is provided by the veinless mutant of *Papilio xuthus* (Fig. 8). In the absence of wing veins the colour pattern consists of a set of pigment bands parallel to the distal margin of the wing. When veins are present they restrict the propagation of signals from the wing margin so that the bands become dislocated (Koch and Nijhout, 2002). The wing veins play a similar role in compartmentalizing the colour pattern of *Heliconius* (Reed and Gilbert, 2004).

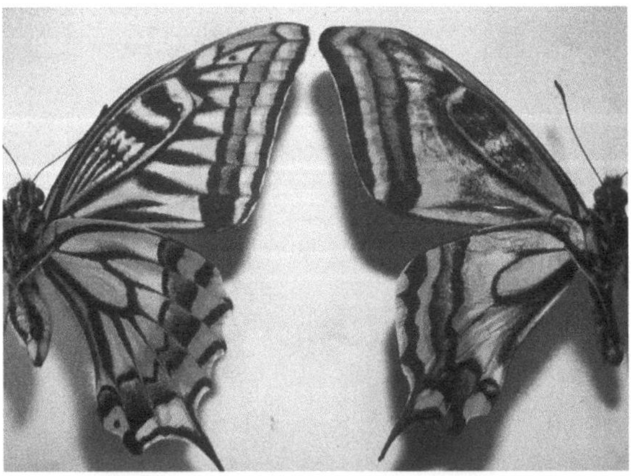

FIG. 8 Patterns in the veinless mutant of *Papilio xuthus*. In the absence of wing veins, the colour pattern forms as bands parallel to the outer wing margin, indicating the margin plays an important role in colour pattern specification. (After Koch and Nijhout, 2002.) (See Color Plate 35 at the back of the book.)

6 The eyespots of butterflies

Eyespots are among the most conspicuous colour patterns of butterflies and moths, and are also found on certain Orthoptera, Homoptera and Coleoptera. Eyespots are pigment patterns made up of concentric rings of contrasting colours. It is believed that some eyespot patterns function as devices to startle predators (Kodandaramaiah *et al.*, 2009). But many butterflies have multiple eyespots arranged as rows of ringlets whose adaptive significance is not at all obvious.

In the Lepidoptera, eyespot patterns often have a distinctive central pupil that has proven to be a key to elucidating how the pigment pattern of eyespots comes about. The centres of what will become an eyespot on the adult forewing can be easily seen on the pupal forewing cuticle as small pigmented spots, often slightly raised above the rest of the cuticle. When the cells at this centre (called the focus) are killed by cautery early in the pupal stage, the eyespot fails to develop (Fig. 9), even though the surrounding colour pattern of the wing is unaffected (Brakefield and French, 1995; French and Brakefield, 1995; Nijhout, 1980a). The focal cells can also be transplanted to a different region of the wing, where they induce a small ectopic eyespot (French and Brakefield, 1995; Nijhout, 1980a).

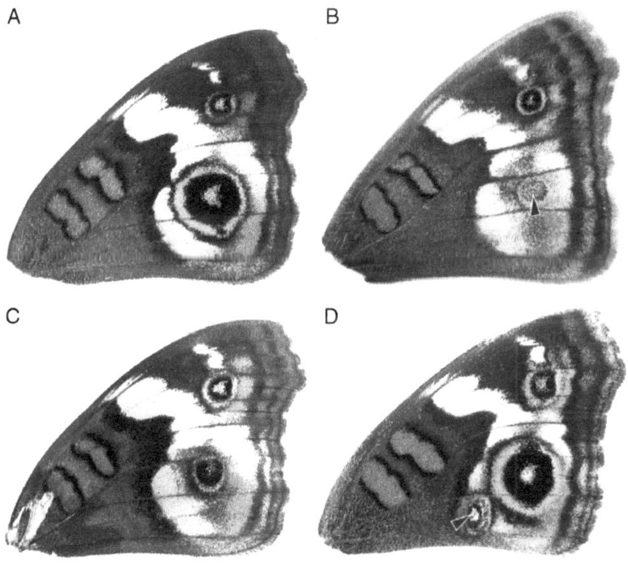

FIG. 9 Effect is cautery of the focal cells on eyespot development in *Precis coenia*: (A) normal pattern; (B) focal cell cauterized 4 h after pupation; (C) focal cells cauterized 24 h after pupation; (D) transplanted focal cells induce an ectopic eyespot. (After Nijhout, 1980a, 1991.)

French and Brakefield (1995) showed that the pigment pattern induced by a focal transplant depended on the location where the transplant was placed. Transplants in the proximal region of the wing induced a circle of orange pigmentation in the surrounding cells whereas transplants to a more distal region induced a circle of pale buff-coloured pigmentation. Similar results have been obtained by reciprocal transplants of foci between the sister species, *Precis coenia* and the Florida mangrove-feeding *Precis evarete*. Here the transplant induced eyespots that conformed to the morphology of the host tissue (*P. evarete* lacks the outer ring of the eyespot), not the donor focal cells (Nijhout, unpublished). These experiments show that the focal signal itself does not contain information about pigmentation, and the response to the focal signal is a property of the receiving cells (Brakefield *et al.*, 1996; French and Brakefield, 1995; Nijhout, 1980a).

Transplanting focal cells between genetic strains of *Bicyclus* that differ in the sizes of eyespots revealed that differences in individual eyespot size were primarily due to changes in the focal signal. By contrast, in genetic strains that differed in the size of many eyespots, this difference was found to be largely due to a change in the response sensitivity of the surrounding epidermis that receives the focal signal (Allen *et al.*, 2008; Beldade *et al.*, 2008; French and Brakefield, 1995; Monteiro *et al.*, 1994).

The foci thus behave as classical source of positional information (Wolpert, 1969, 1981, 1994). The concentric rings of the pattern show that different pigments are specified at different distances from the signalling source, but the pigment synthesized is a property of the induced cells, not of the signal. If the focal source emits a signal whose strength or concentration declines with distance, then the concentric rings of pigment could be specified by different thresholds that are sensitive to signal strength. Exactly what colours are induced, and whether there will be peripheral rings, or no rings at all, is determined by how the cells that receive the signal are genetically programmed to respond.

6.1 SIGNAL PROPAGATION

In *P. coenia*, eyespot development is completely abolished by cauterizing the focal cells 4 h after pupation (e.g., Fig. 9). But if the focal cells are cauterized at progressively later times, one obtains eyespots of small but progressively larger diameters, until about 48 h after pupation, after which cautery no longer affects normal eyespot development. Mathematical analyses show that the rate at which the diameter of the eyespot expands at different temperatures is consistent with diffusion of a small morphogen (Murray, 2003; Nijhout, 1991), and computer simulation models of gradient–threshold mechanisms suggest that diffusion gradients can account for the sizes and shapes of diverse eyespot morphologies in many species (Dilao and Sainhas, 2004; Evans and Marcus, 2006; Marcus and Evans, 2008; Monteiro *et al.*, 2001; Nijhout, 1990). In addition, reaction–diffusion mechanisms have been shown to be an adequate explanation for both detailed and global patterning on lepidopteran wings (see Section 2.1).

Some aspects of eyespot determination, however, are inconsistent with diffusion as a mechanism of signal propagation. One property of a diffusion gradient is that its level or strength decays with distance from the source. Whether the focal signal is graded is, however, an open question. The expression of genes like distalless, spalt and engrailed, which mark the disc and outer rings of eyespots in several species of butterflies (Brunetti *et al.*, 2001; and see Section 6.3), appears to be homogeneous within the field of the developing eyespot, not graded. Also, if a signal propagates by diffusion, then killing the cells at the source would not freeze further expansion of the diffusion front. Rather, the bulge of morphogen already secreted would flatten out by expanding outward and this would result in a diminished apparent rate of outward expansion of a front, and in the rapid decay of morphogen concentration to below-threshold values in regions where it was formerly above threshold. These effects do not resemble the 'freezing' of a determination front observed in cautery experiments. An alternative interpretation of the effect of cautery is that it stops the progression of eyespot determination through a trauma-induced molecular event like the induction of a heat shock response (see also Section 6.2) that halt a broad range of molecular processes and could thus appear to freeze the progress of development.

A final problem with diffusion is that no morphogen is known to have the required range. In eyespot determination in *P. coenia*, diffusion must act over a radius of about 150 cells (750 800 μm). The best-known secreted morphogens, dpp and wg, have a much shorter range (Day and Lawrence, 2000). Notch signalling, which is known to occur in the early stages of eyespot determination (Reed and Serfas, 2004), only works through cell–cell contact, and does not have the required range either, unless it is a component in a cascade (or contagion) mechanism of signal propagation. Reaction–diffusion mechanism have assumed it is possible for morphogen to diffuse from cell to cell via gap junctions (Fraser *et al.*, 1987), but such a mechanism has not yet been experimentally demonstrated in insect pigmentation.

The range of any of these mechanisms would be greatly enhanced by the use of epidermal feet. Epidermal feet (also called filopodia) were discovered by Locke and Huie (1981) and have been studied in the wing epithelium of the moth *Manduca sexta* by Nardi and Magee-Adams (1986). They are long filamentous extensions of the basal surface of epidermal cells that form a network of interconnections along the basal lamina. Epidermal feet can extend over 5–10 cell diameters, and can thus enhance the range of a contact mechanism of signal propagation. In *Drosophila*, epidermal feet mediate a long-range lateral-inhibition mechanism that uses Notch–Delta signalling (de Joussineau *et al.*, 2003).

A contagion mechanism, where one cell activates the next one in sequence, has been considered as a mechanism for eyespot patterning (Nijhout, 2001). A contagion mechanism requires a timing mechanism, rather than a signal strength sensor, to specify the various pigment rings of an eyespot. A contagion wave mechanism could produce concentric rings with different pigments by one of two mechanisms: (1) as the wave spreads outward, cells retain memory of how long they have been behind the front and adjust gene expression accordingly, or (2) the wave front activates a time-dependent cascade of gene expression so that different sets of genes are active at progressively greater times (= distances) behind the front. A contagion mechanism by itself does not help explain the apparent action-at-a-distance effect of cautery, and a heat shock response would have to be invoked here, just as in the case of signalling by diffusion.

6.2 ECTOPIC EYESPOT INDUCTION

Nijhout (1985) showed that in *P. coenia*, a small cautery on the hindwing during the first day after pupation induced a small ectopic eyespot around the site of injury (Fig. 10). Such ectopic eyespot induction was restricted to the dorsal surface of the hindwing, and did not occur on the ventral surface of the hindwing, nor on the forewing. The outer rings of the induced eyespots fuse smoothly with those of the normal eyespots that develop on the hindwing (Fig. 10), indicating that the ectopic and natural eyespots share a common patterning mechanism. When the hindwing is damaged over a larger area, for instance by surgically removing a portion of the wing disc in the larval stage (Nijhout and Grunert, 1988), the normal eyespots on the hindwing enlarges and develops as an arc-shaped pattern, open towards the cut edge of the wing (Fig. 10, middle row). These experiments suggest that damage to the hindwing induces a patterning mechanism that either interacts with, or is the same as, the one by which the normal eyespots form.

Brakefield and French (1995) made a detailed study of ectopic eyespot induction in *Bicyclus anynana* and demonstrated two important effects: (1) there is an optimal time window, between 12 and 18 h after pupation, during which damage induces ectopic eyespots on the dorsal forewing; (2) ectopic eyespot induction is easiest in the vicinity of an existing eyespot, and the frequency and size of ectopic eyespots decreases as damage is moved further away from an eyespot. As in the case of *Precis* hindwing, if the damage is done very close to an existing eyespot, the outer rings of the eyespot expands to encompass the damage site. In contrast to *Precis*, it was possible to induce ectopic eyespots on all wing surfaces in *Bicyclus*, although the ventral surface of the forewing was much less sensitive to ectopic eyespot induction than the dorsal surface. These findings illustrate that there are both spatial and temporal restrictions on the induction of ectopic eyespots, and that different species do not respond alike.

What can explain the induction of ectopic eyespots and the spatial variation in sensitivity of the wing to the induction of ectopic eyespots? The mechanism needs to account for the fact that in the *Bicyclus* forewing ectopic eyespot induction is easiest near existing eyespots and can fuse with those eyespots, and that in the

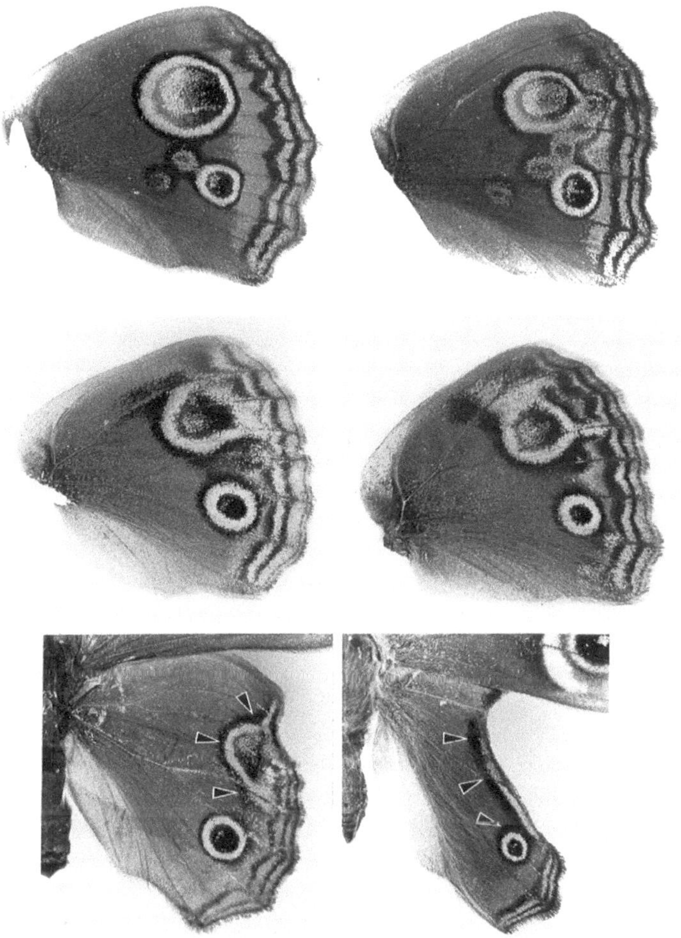

FIG. 10 Trauma-induced patterns on the hindwing of *Precis coenia*. *Top row*: induction of ectopic eyespot development by point cautery. *Middle row*: temperature-induced pattern aberrations. *Bottom row*: surgical ablation of wing margin causes eyespot (arrows) to 'open up' towards the margin, much as occurs in the temperature shock patterns. (After Nijhout, 1984, 1985, 1991; Nijhout and Grunert, 1988.)

Precis hindwing ectopic eyespots can fuse with existing eyespots and spread to the wing margin (Fig. 10). If we assume that signal propagation is by diffusion and that normal eyespot development involves a conical morphogen gradient produced at the focus, then thresholds on the gradient produce the pigment rings of the eyespots (Beldade and Brakefield, 2002; Brakefield and French, 1995; French and Brakefield, 1992; Monteiro *et al.*, 2001; Nijhout, 1985). Under such a model, cautery could produce its effects by inducing a molecule that mimics the morphogen, or it could locally alter the threshold of sensitivity of cells to the

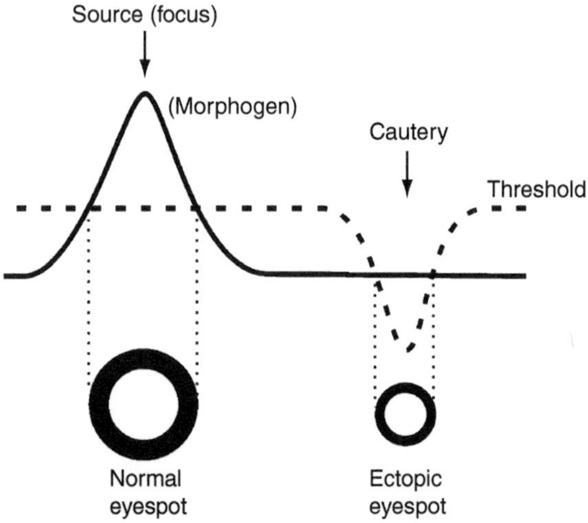

FIG. 11 The French–Brakefield model for induction of ectopic eyespots. Normal eyespots form wherever the morphogen produced by the focal cells is at and above a threshold. Cautery is assumed to reduce the threshold of sensitivity to the morphogen. (After Brakefield and French, 1995; French and Brakefield 1992.)

morphogen (Fig. 11) by destroying a chemical that is responsible for establishing the threshold (Brakefield and French, 1995). If eyespot determination does not involve a diffusion gradient but is accomplished by a contagion mechanism (see previous section), then cautery could act as a trigger or initiator of a contagion cascade. This could help explain why ectopic eyespots are always small; that is, because they are triggered relatively late in development.

6.3 GENE EXPRESSION

The developmental genetics eyespot patterns has been studied extensively, and the discovery of the cluster of genes that control the development of eyespots and their diversity of pigmentation was a key breakthrough in understanding the genetics of colour patterns in insects. The first gene discovered was *Distal-less* (Carroll *et al.*, 1994). Patterned *Dll* expression in the wing discs of *Precis* begins in about the middle of the last larval instar as a diffuse band along the distal margin of the wing imaginal disc and along the wing veins (Fig. 12A). Expression then contracts to a series of triangular wedges, each halfway between the wing veins (Fig. 12 B–D). These wedges then narrow to a line of *Dll* expression along the midline of each wing cell (Fig. 12 E and F). In the wing cells, whether eyespots will develop a rounded area of *Dll* expression expands from the tip if this line, giving it a lollypop appearance (Fig. 12 F and G). The line of *Dll* expression then disappears leaving behind spots of *Dll* expression that

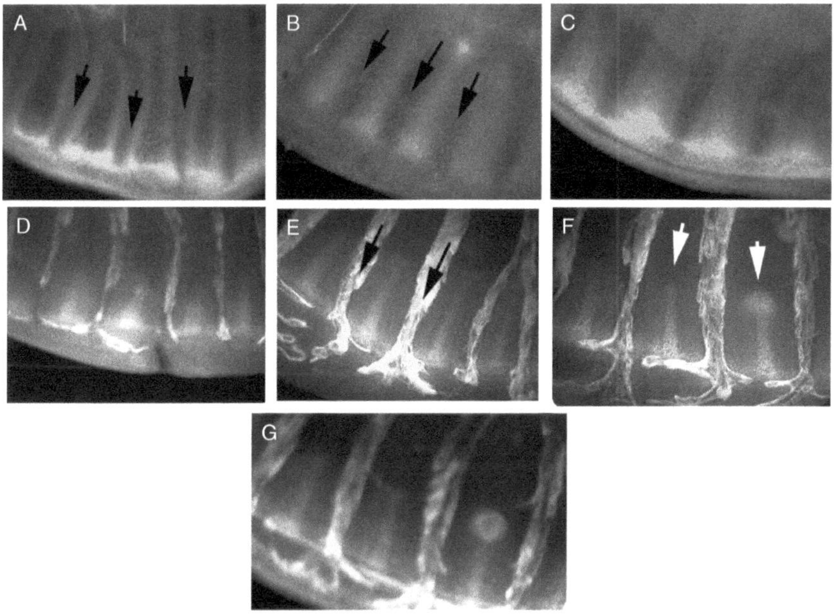

FIG. 12 Time series of the developing pattern of distal-less expression in the pupal wing of *Precis coenia*. Arrows indicate the positions of wing veins.

mark the central foci of the presumptive eyespots (Fig. 12G). This sequence of patterns is remarkably similar to that predicted by a lateral-inhibition reaction–diffusion model for pattern formation in which the activator is supplied by the venation system and diffuses into the inter-vein region (Nijhout, 1990). The model predicts an autonomous increase in activator synthesis along the midline between two veins that resolves into a single stable spot of activator synthesis. This model requires intercellular communication, and it is unlikely that *Dll* can serve that function, because it is not a secreted morphogen. However, Reed and Serfas (2004) have shown that the pattern of *Dll* expression shown in Fig. 12 is preceded by a virtually identical pattern of *Notch* expression. Thus, the *Dll* pattern could have been stimulated by *Notch* signalling. *Notch* is a *trans*-membrane receptor for short-range signals, typically using *Delta* as the ligand. Ligands for *Notch* are usually also *trans*-membrane proteins, so *Notch* signalling typically operates only between cells that are in contact with each other. *Notch* signalling over a longer range could be accomplished via epidermal feet or filopodia (de Joussineau *et al.*, 2003; Locke and Huie, 1981), which can extend 5–10 cell diameters. No other candidates for long-range signalling have as yet been elucidated, although *dpp* is patterned parallel to the central *Dll* line (Keys *et al.*, 1999; McMillan *et al.*, 2002), and may act to refine the expression domain of *Dll*. Keys *et al.* (1999) studied the roles of short-range (*hh*) and long-range (*wg*) signalling molecules in eyespot pattern formation. *hh* is expressed at two locations flanking the spot of *Dll* expression that marks the focus, and *wg* is expressed along the wing margin and two short spurs that flank the central line of *Dll* expression around the site where the focus will form, and possibly at the location of the focus as well (Keys *et al.*, 1999; McMillan *et al.*, 2002; Monteiro, 2008).

The combined spatial pattern of gene expression is illustrated in Fig. 13. This is a remarkably complex pattern whose purpose, presumably, is to produce a single spot of gene expression at the central focus of the eyespot. The exact mechanism by which *Notch*, *hh*, *wg* and *Dll* interact to produce the spatial progression of *Dll* expression is not known. Theoretical studies suggest that quantitative variations in the processes that give rise to the progression of activator expression illustrated in Fig. 12 can give rise to a surprisingly diverse array of patterns that cover some 80% of the pattern diversity observed among the butterflies (Nijhout, 1990, 1991). Thus, elucidating the manner in which this array of genes interacts, what additional genes might be involved and how variation in these interactions cause variation in the spatial pattern of their expression will be critical to develop a deep understanding of the genetic basis and diversification of colour patterns.

The focus of an eyespot is marked by the expression of *Notch* and *Dll*, *patched* (*ptc*), *cubitus interruptus* (*ci*) and *engrailed*/*invected* (*en*/*inv*) (Carroll *et al.*, 1994; Keys *et al.*, 1999; Reed and Serfas, 2004). It is not clear which, if any, of these genes is directly responsible for inducing further gene expression in the surrounding cells, but an interesting and surprisingly complex pattern of gene expression has been documented that is associated with the different pigment bands that make up an eyespot (Brunetti *et al.*, 2001). In *Bicyclus*, the eyespot on the adult wing is made up of a central spot of white scales,

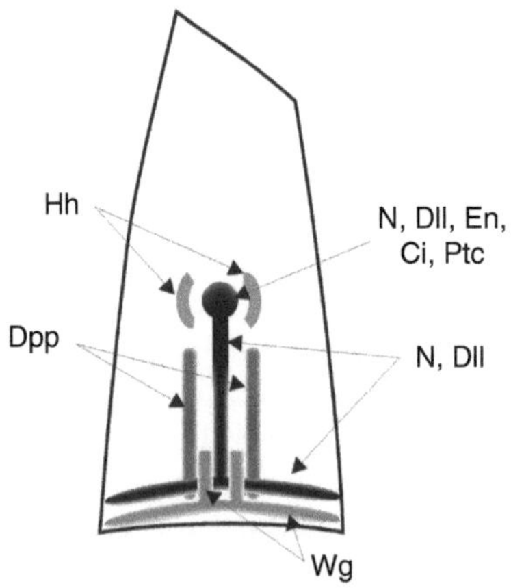

FIG. 13 Expression patterns of several regulatory genes in a single wing region. (After Carroll *et al.*, 1994; Keys *et al.*, 1999; McMillan *et al.*, 2002.)

surrounded by a disc of black scales and an outer ring of golden scales. Brunetti *et al.* (2001) found that the central spot is marked by expression of *Dll*, *en*/*inv* and *spalt* (*sal*). The cells that will make the black central disc of the eyespot express only *Dll* and *sal*, and the cells that will form the golden outer ring express only *en*/*inv*. Thus, each colour region of the eyespot is characterized by a unique permutation of expression of the three transcription factors.

Brunetti *et al.* (2001) found that this pattern of concentric ring expression develops stepwise. Initially, the cells in the area around the central focal spot begin to express *sal*. Once *sal* expression extends over the entire presumptive black region, the cells surrounding it begin to express *en*/*inv* in the presumptive region of the golden scales. *en*/*inv* expression then begins at two points at opposite sides on the periphery of the region of *sal* expression, and expands from there to form a ring around the *sal* field. Only after the *sal* and *en*/*inv* patterns are fully developed does *Dll* expression expand from the central spot to coincide with the field of *sal* expression (Fig. 14).

The final pattern of gene expression is consistent with a model that assumes a conical gradient of a morphogen, centred on the focal spot, with each of the genes expressed between a set of thresholds on this gradient, with *Dll* and *sal* encoded by a high morphogen concentration and *en*/*inv* by a lower concentration. It is unlikely, however, that the sequence in which this pattern develops reflects the expansion of the morphogenetic gradient from the central focal cells. This is because in the early stages of signalling the concentration near the focal source would be low, encoding *en*/*inv* expression, but this is not what is observed. It is possible that the positional information gradient is established first, and then a second set of signals cause the readout of this gradient: first by *sal*, followed by *en*/*inv* and last by *Dll*.

A two-step process of pattern specification, gradient setup followed by gradient interpretation, also helps to explain the diversity in gene expression pattern associated with eyespots in different species. Brunetti *et al.* (2001) examined the expression patterns of *Dll*, *sal* and *en*/*inv* during eyespot specification in four species of butterflies. They found that each species has a unique permutation by which these three transcription factors were associated with different regions of an eyespot. Even within a species (*Precis*), the forewing and hindwing eyespots differ in the pattern of expression of these genes (Fig. 14). These findings show that neither colour nor position in an eyespot is specified by a conserved set of genes. Each species has evolved a different genetic mechanism by which the colour rings of its eyespots are specified. If we assume that all species use the same morphogen to specify distance from the focal cells, then species differ in which transcription factor is activated at a particular morphogen concentration, and which pigment is specified by that particular transcription factor. It is also possible that species differ in the morphogen used, in which case every aspects of colour specification could be species-specific, and within a species, even eyespot-specific.

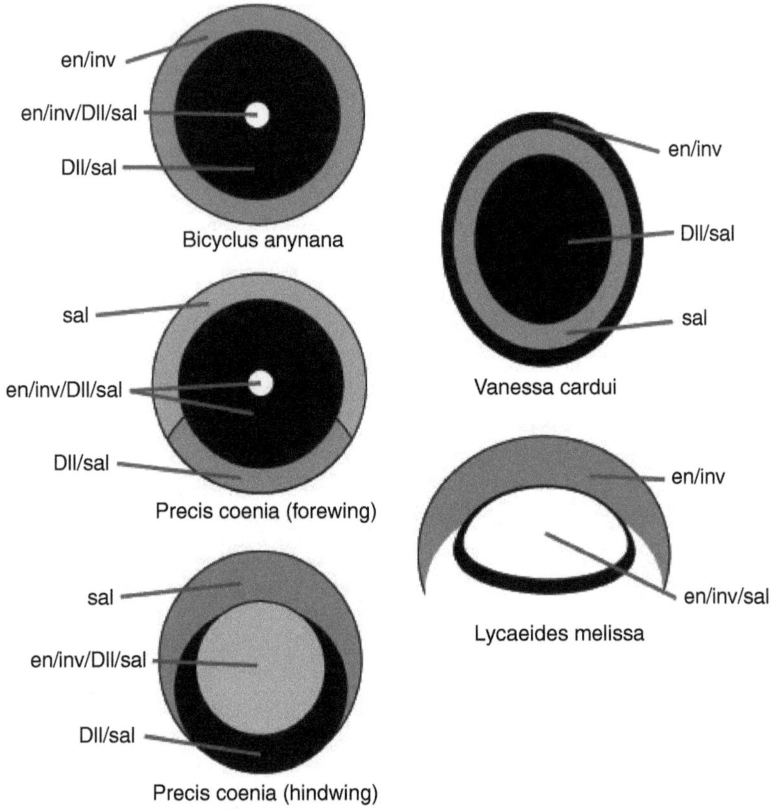

FIG. 14 The rings of eyespots in different species of butterflies are associated with different combinatorial expression patterns of a small number of regulatory genes. (After Brunetti *et al.*, 2001.) (See Color Plate 36 at the back of the book.)

These findings evidence a profound lack of conservation of these late-developing features that stands quite in contrast to the extreme conservation of genetic regulation found in early embryonic development. It may be that the evolution of late-developing features is eclectic and contingent in the 'choice' of transcriptional regulators.

6.4 NON-CIRCULAR EYESPOTS

The process of focus determination does not always terminate in a spot-like pattern, and eyespots are not always circular. Indeed, the homologues of border ocelli (the class of pattern elements to which eyespots belong) come in a broad diversity of shapes, and perfect circles are numerically less common than other shapes (Nijhout, 1990, 1991). Non-circular eyespots are formed around complexly shaped foci, as illustrated by the double-lobed eyespots of

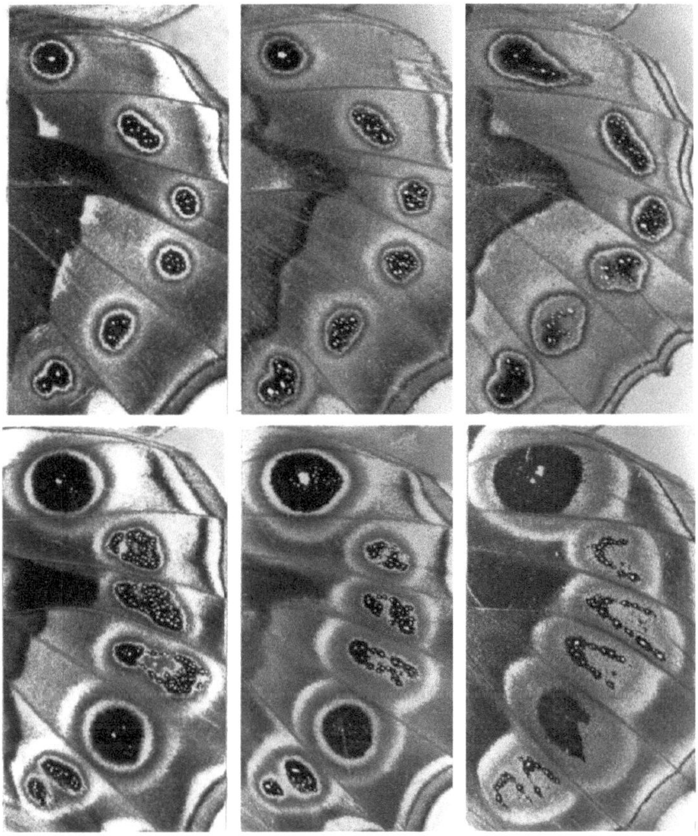

FIG. 15 Fragmented foci and individual variability of focus and eyespot expression in the genus *Lethe* (Satyridae). (After Nijhout, 1991.)

Morpho hecuba (Fig. 17). In other cases, as in *Lethe* spp. (Satyridae), foci are fragmented and cover an area that forms the centre of an irregular ocellus (Fig. 15). In *Lethe*, no two eyespots have identical foci (Fig. 15), indicating that stochastic processes play an important role on focus determination (see also Section 10).

6.5 EYESPOT MUTATIONS

Brakefield and co-workers have isolated a number of mutations that affect the eyespots of *Bicyclus* butterflies. These mutants have provided several novel insights into the genetic control of eyespot formation. Several of these mutants either appeared spontaneously in laboratory colonies or were recovered by mutagenesis. Figure 16 illustrates a number of these spectacular mutants.

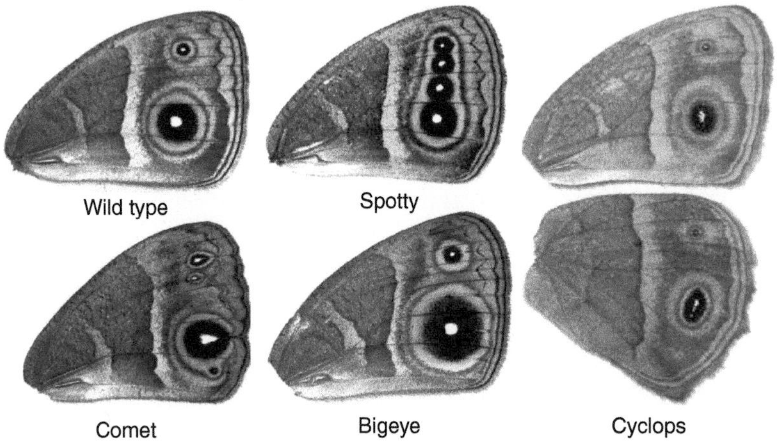

FIG. 16 Mutant eyespot pattern on the forewings of *Bicyclus* (Satyridae). The cyclops phenotype is also shown on the hindwing. (Photos courtesy of Paul Brakefield and Suzanne Saenko.) (See Color Plate 37 at the back of the book.)

Bigeye. The *Bigeye* mutation greatly enlarges all eyespots, particularly those on the ventral forewing. The *Bigeye* mutation is dominant. In this mutant the degree of *Dll* expression is not different from wild-type, so the phenotypic effect is likely due to a change in threshold of interpretation of the focal signal, rather than the signal itself (Brakefield, 1998; Brakefield *et al.*, 1996).

Spotty. The *Spotty* mutation causes extra eyespots to form on each wing surface in cells where typically no eyespots develop, although those regions typically express small dots of pigmentation. In the wild-type larval wing disc, these dots are not associated with a detectable *Dll* expression, but in the spotty mutant there is a strong *Dll* expression at those locations. It is not a threshold mutation because the size of other eyespots is not affected by *Spotty* (Brakefield *et al.*, 1996).

Goldeneye. In the *Goldeneye* mutation, the central black disc of the eyespots is reduced or absent and the golden outer ring is expanded overt the entire eyespot field (Brunetti *et al.*, 2001). This mutation presumably affects how the focal gradient is interpreted, perhaps by altering a threshold that specifies the expression of *en/inv*.

Comet. The *Comet* mutation stretches the white focus cells towards the wing margin (Brakefield, 1998, 2001). This mutation probably affects the mechanism of focal cell specification. It is possible that specification stops just before the lollypop stage (e.g., Fig. 12 F).

Cyclops. The *Cyclops* mutation is a dominant lethal. This allele causes the fusion of the foci of the middle two eyespots on the hindwing into a short bar parallel to the wing margin (Brakefield *et al.*, 1996; Saenko *et al.*, 2008).

This allele is associated with a reduction in the wing venation; specifically, the middle vein of the hindwing is truncated and does not reach the wing margin. Thus, there is no vein separating what are normally two regions of focus determination; the foci in the two adjoining cells merge to an elongated shape and can induce an abnormally large eyespot. *Dll* expression also occurs in an elongated pattern, faithfully anticipating the size and shape of the central focus of the eyespot.

Pminus. The *Pminus* mutation reduces the size of the posterior eyespot on the forewing which affects the size of the anterior eyespot (Beldade *et al*., 2005).

3 + 4. The *3 + 4* mutation is an X-ray-induced mutant that eliminates the eyespots in positions 3 and 4 on the hindwing (Monteiro *et al*., 2003). This mutation, like *Spotty* and *Pminus*, highlights the modular control over pattern development in that these genes only affect pattern in restricted regions of the wing.

In addition, it has been possible to obtain pure breeding strains for aberrant patterns by artificial selection. The *lines* phenotype extends the eyespots along the wing veins (Beldade *et al*., 2005). This genetic effect is very similar to one obtained by temperature shock in *Precis* (Fig. 10, middle row), and shows that signals from the wing veins and from foci share some patterning properties. The *position* phenotype moves all eyespots towards the wing margin (Beldade *et al*., 2005). This effect may be produced by a global change in the process of focus determination that causes the condensation of *Dll* to occur nearer the wing margin. An aberration in *Dll* specification is also indicated by the fact that some foci are elongated, and not rounded. The *spotless* phenotype eliminates eyespots on all wing surfaces (Brakefield, 1998), possibly by eliminating or foreshortening the process by which foci are established, or by raising the threshold of sensitivity to focal signals.

7 Other pattern elements

There are many pattern elements on butterfly wings other than eyespots, but the genetic control over these has not been studied explicitly. Comparative studies of gene expression have, however, revealed expression patterns that are associated with two of these. Linear intervenous pigment patterns that run along the midline of a wing cell (half way between the adjacent wing veins) pattern are associated with expression of Notch and Distalless (Reed and Serfas, 2004). These pattern elements probably arise when the progressive development of gene expression (illustrated in Fig. 12) stops at an early stage (Nijhout, 1990; Reed and Serfas, 2004). In *P. coenia*, expression of wingless marks the locations where the bands of the central symmetry system will develop (Carroll *et al*., 1994). So far, none of the other genes associated with eyespot development are known to be associated with any other pattern elements.

8 Control of pigment biosynthesis

Colour pattern determination consists of the processes that establish where on the body surface a particular pigment will be synthesized. These processes results in the spatial patterning of expression of transporters that take the precursors of pigments out of the haemolymph and into the cells that will synthesize pigments, and the spatial patterning of enzymes that will transform those precursors into the pigments. The last event in colour pattern formation is the actual biosynthesis of the pigments. Pigment biosynthesis is typically associated with a moult. Depending on the species, developmental stage and pigment, the colour of the new integument can develop shortly before or shortly after ecdysis.

Different colours on the wings of Lepidoptera are synthesized in a characteristic and stereotyped sequence. Whites are synthesized first, followed by reds and yellows, followed by black, and dark browns are synthesized last (Koch *et al.*, 1998, 2000; Nijhout, 1980b). This sequence of pigment synthesis unfolds over a period of 1 or 2 days, depending on the species. With rare exceptions, each scale on a butterfly wing synthesizes only one pigment (Koch and Kaufmann, 1995; Nijhout and Koch, 1992), so at the level of the scale-building cell there is a developmental switch that commits the scale to a single biosynthetic pathway.

Control over the precise timing of pigment synthesis can come about via several different mechanisms. It is possible to have *de novo* expression of enzymes at the time pigment synthesis starts. Alternatively, inactive proenzymes can be present throughout the integument that are activated when pigment synthesis starts. Or active enzymes can be present in the integument and the timing of pigment synthesis is controlled by transporters that import the necessary precursors. Presumably, any combination of these mechanism could be spatially patterned and operate in a given system, and different pigment pathways may utilize different mechanisms.

Patterned melanin synthesis in butterflies has been shown to be due to the patterned expression of the enzyme DDC (Koch, 1994, 1995; Koch and Kaufmann, 1995; Koch *et al.*, 1998). In *Precis*, DDC activity begins to rise about 36–48 h before adult eclosion, about 24 h before black pigment synthesis actually begins (Koch and Kaufmann, 1995). If wings are dissected out before pigment synthesis has started and incubated in a solution with an appropriate substrate (dopa or dopamine), the melanin pattern develops rapidly, just like one would develop a photograph (Koch and Kaufmann, 1995; Nijhout, 1980b). This suggests that the timing of pigment synthesis is regulated by control over the access of substrates, and not through the control of enzyme expression or activation. DDC activity is higher in presumptive black regions of the wing than in presumptive grey regions (Koch and Kaufmann, 1995), which suggest that the control of intensity of pigmentation is by means of the level of expression of DDC.

The American Tiger Swallowtail butterfly, *Papilio glaucus*, has a bold pattern of black and yellow markings. The black is melanin, and the yellow is a papiliochrome, a pigment unique to swallowtail butterflies (Ishizaki and Umebachi, 1988, 1990; Umebachi, 1985). Melanins and papiliochromes both have dopamine as a precursor (Fig. 5), and require the activity of DDC (Fig. 5). When radiolabelled tyrosine is injected 3 days before adult eclosion, the radiolabel appears only in the yellow portions of the pattern. By contrast, when radiolabelled tyrosine is injected 1 day before adult eclosion, it is incorporated only into the black portions of the pattern. Localized DDC assays show that DDC activity rises first in the presumptive yellow areas, and only about a day later in the presumptive black areas of the wing (Koch *et al.*, 1998). DDC thus contributes to different pigments at different times in development. These findings correspond to the observation that in normal development yellow is synthesized before black, but DDC by itself cannot give the required specificity. Dopamine, the product of DDC, has two fates: it can be conjugated with β-alanine by the enzyme *N*-β-alanyl-dopamine synthase (BAS), as the first step in the synthesis of papiliochromes, or it can be oxidized by diphenoloxidases to dopamine–quinone which autopolymerizes to melanin (Fig. 5). The activity of BAS rises during yellow pigment synthesis and declines soon thereafter, before melanin synthesis begins (Koch *et al.*, 2000), so it appears that the coordinated up-regulation of BAS and DDC controls papiliochrome synthesis, whereas DDC by itself yields melanin.

In *Heliconius*, the onset of pigment synthesis is controlled in part by activity of the transporters that take up pigments or their precursors from the haemolymph. The yellow wing pigment in *Heliconius* is 3-hydroxykynurenine, which is synthesized in the fat body, secreted into the haemolymph and taken up by the scale-building cells, possibly by the scarlet/white transporter (Reed *et al.*, 2008). Orange and red pigments are oxidized and reduced forms of ommochrome (xanthommatin and dihydroxanthommnatin, respectively). These are synthesized from tryptophan, which is transported into the scale-building cells from the haemolymph. The biosynthetic pathway is controlled by the genes *vermillion* (encoding the enzyme tryptophan oxidase) and *cinnabar* (encoding the enzyme kynurenine-3-hydroxylase) and the enzyme kynurinine formamidase. This pathway converts tryptophan into 3-hydroxykynurenine, which polymerizes to xanthommatin. Xanthommatin can be reduced to a brilliant red pigment dihydroxanthommatin. Red dihydroxanthommatin is readily re-oxidized to orange xanthommatin when the pigment is extracted from the wing (Nijhout, 1997), indicating that in its pure form this pigment is unstable, although *in situ* it can remain stable for a century or more, as indicated by well-coloured specimens in museum collections. *In situ*, oxidation of the pigment is inhibited by a xanthommatin-binding protein, which stabilizes the reduced state (Nijhout, 1997; Reed and Nagy, 2005; Reed *et al.*, 2008).

8.1 HORMONAL CONTROL

Pigment synthesis is one of the last events in wing development in Lepidoptera. It would be reasonable to assume that the initiation of pigment synthesis is simply the last step in a long causal cascade of events in development, where each step is the cause of the next. It appears, however, that, like many features of postembryonic development, the initiation of pigment synthesis is under central control by the brain via the secretion of ecdysone. In non-diapausing animals, ecdysone secretion begins shortly after pupation, and the increase of ecdysone causes apolysis and initiates adult development. Ecdysone continues to rise and remains high during most of the pupal stage as adult development proceeds, and then gradually declines to zero about the time adult development is completed, a short time before adult eclosion. It turns out that several late developmental events are cued by the declining titre of ecdysone (Schwartz and Truman, 1983).

The rise of DDC activity and melanin synthesis in *P. coenia* and *M. sexta* depends on the declining titre of ecdysone, and can be inhibited by exposure to ecdysone at the end of adult development (Hiruma and Riddiford, 1993; Hiruma *et al.*, 1995; Koch, 1994, 1995; Koch *et al.*, 2003; Sawada *et al.*, 2002). Likewise, the onset and duration of the expression of GTP-cyclohydrolase I, a key enzyme in the biosynthesis of pteridine pigments, has been shown to be triggered by a declining ecdysone titre during late adult development (Sawada *et al.*, 2002). Whether each step in the stereotyped sequence of pigment synthesis outlined above is triggered by a specific level of declining ecdysone concentration, or whether the decline of ecdysone below an inhibitory threshold triggers the entire sequence remains to be established.

8.2 THE LEPIDOPTERAN SCALE CASETTE

The colours on the wings of Lepidoptera are contained in the scales. Each scale appears to synthesize only a single pigment (Koch and Kaufmann, 1995; Nijhout and Koch, 1992), and the overall colour pattern is a finely tiled mosaic of scales of different colours (Figs. 1, 17 and 19). In addition to developing discrete colours, scales of different colours differ in maturation rate, as measured by the rate of cuticle synthesis and sclerotization. White scales mature most rapidly, followed by coloured scales, and black scales mature last (Koch *et al.*, 2000; Nijhout, 1980b; Reed and Nagy, 2005; Reed *et al.*, 2008). In many cases, scales of different colours also have different shapes and surface micromorphology (Burgeff and Schneider, 1979; Descimon, 1965; Gilbert *et al.*, 1988; Janssen *et al.*, 2001). In *Morpho* butterflies, the iridescent blue scales have a rounded margin whereas the melanic brown scales have a scalloped margin (Nijhout, 1991). In *M. sexta*, white scales tend to be long and narrow whereas black scales are short and wide. In *Heliconius* butterflies, scales of different colours also have characteristic shapes (Gilbert *et al.*, 1988). Mutations

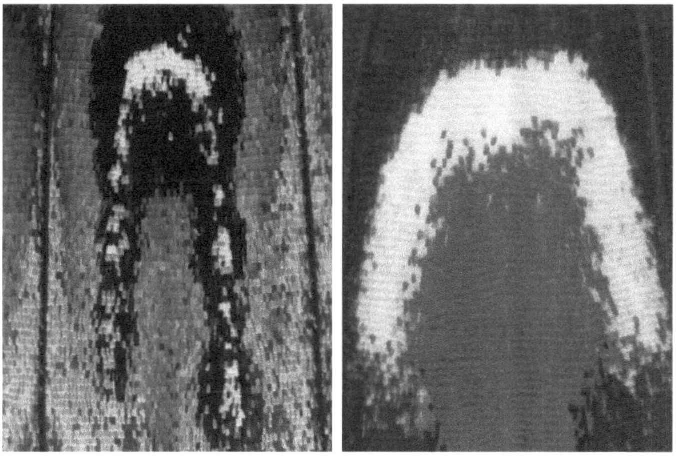

FIG. 17 The overall wing pattern is a finely tiled mosaic of monochrome scales. With rare exceptions, each scale-cell synthesize only one type of pigment. This figure shows pigmentation patterns around the eyespots of *Morpho hecuba* (Morphidae; left) and *Delias aganippe* (Pieridae; right). (See Color Plate 38 at the back of the book.)

that change the colour of scales also change their morphology and maturation rate to correspond to those of the new colour type.

Thus, scale colour, morphology and maturation rate are tightly linked in development, and develop as a unit, or cassette. The processes that determine scale colour do not just turn particular enzymatic biosynthetic pathways on or off, but they also control details of the development of each scale, such as the timing and rate of cuticle synthesis and sclerotization, the rate of growth and size and the patterns of cytoskeletal elements that give rise to the final shape and micromorphology of the scale (Ghiradella, 1984, 1985). It would be interesting to know whether the coordinate switching of such a complex and heterogeneous array of traits is managed by single regulatory genes, or whether more complex regulatory networks are involved.

9 Phenotypic plasticity and hormones

Although the overall pigmentation and colour patterns of most insects are genetically determined, the environment can alter the colouration in a variety of ways. The effect of the environment on a trait is generally referred to as phenotypic plasticity. Plasticity of pigmentation and colour pattern is widespread among insects, and can range from short-term changes associated with day–night cycles to permanent switches between alternative phenotypes. In a few cases, the colour change is a direct consequence of a change in temperature or humidity. For instance, hydration and dehydration can alter the colour of the elytra of some beetles (Hinton and Jarman, 1972; Rassart *et al.*, 2008; Vigneron *et al.*, 2007). In a great many cases, the colour change is regulated by hormones, and is associated with adaptations to particular environments and/or with transitions in developmental stages, as we will discuss in the following sections.

9.1 EPIDERMAL COLOUR CHANGES

The colour and pattern of the cuticle is generally fixed for the duration of the instar, and a change in colouration or pattern can occur only at the moult. Colour that resides in the epidermis can change during the intermoult period. Epidermal colour change has been relatively little studied. Perhaps the best known case is that of the stick insect, *Carausius morosus*, which can control the dispersion of orange and brown pigment granules in its epidermis so that the overall colour is darker at night than during the day time (Bückmann, 1977; Bückmann and Dustmann, 1962; Giersberg, 1928). This colour change is facilitated by the fact that the cuticle of walking sticks is completely transparent and colourless (Tara Maginnis, personal communication). In *Carausius*, the epidermal colour change is under the control of a circadian clock, and continues to cycle even if the animals are kept in constant darkness (Bückmann, 1977; Giersberg, 1928). Experiments by Raabe (1966) suggest that a neurosecretory hormone from the brain is involved in stimulating this daily colour rhythm. Green caterpillars likewise have a colourless cuticle and their green colour is due to pigments in the epidermis. Many green caterpillars change their epidermis to a reddish brown colour (due to the synthesis of ommochromes) at about the time they stop feeding and enter the prepupal stage (Bückmann, 1959, 1974; Nijhout, 1994). This colour shift is presumably an adaptation to the change in background colouration as the caterpillar moves from the foliage on which it feeds to the stems or leaf litter where it will pupate. In caterpillars of *Cerura vinula* and *M. sexta*, the synthesis of epidermal ommochromes is stimulated by a pulse of ecdysone that occurs at the end of larval life. This ecdysone pulse not only induces this colour change but also causes the larva to stop feeding and wander off the host plans in search for a place to pupate (Bückmann, 1959, 1974; Nijhout, 1994).

9.2 CUTICULAR COLOUR CHANGES

Phenotypic plasticity due to changes in pigmentation of the cuticle is more widespread and more profound than that involving pigmentation of the epidermis. Probably, the most widespread examples of changes in cuticular colour are those that occur in the course of normal development. For instance, the larvae of some Orthoptera and many Hemiptera typically differ in colour and pattern from the adults. The transformation is associated with the metamorphic moult and is controlled by juvenile hormone (JH). This can be experimentally demonstrated by topically applying a small drop of JH to the abdomen of a larva before the metamorphic moult. After the moult, the patch of integument at the site of application will retain the larval colour and texture, surrounded by a normal adult-patterned cuticle (Nijhout, 1994; Wigglesworth, 1959; Willis *et al.*, 1982).

In many Hemi- and Holometabola, there are more or less profound changes in colouration and pattern during larval development. In swallowtail butterflies, for instance, small early instar larvae have a black-and-white colour pattern that mimics a bird dropping, whereas later larval stages are green or boldly patterned. This shift from mimicry to camouflage has been studied in *P. xuthus* by Futahashi and Fujiwara (2008), who showed that it is under hormonal control. In this species, the transformation occurs as the larvae moult from the 4th to the 5th larval instar. Application of exogenous JH during the first day of the 4th larval instar caused retention of the 4th instar mimetic colour pattern when these larvae moulted to the 5th instar. This responsiveness to JH was lost by the end of the first day, and later treatment with JH did not prevent development of the normal 5th instar colouration. The change in patterning and pigmentation is controlled, in part, by changes in the patterned expression of TH and DDC, which control the black portions of the pattern, and the synthesis of insecticyanin and biliproteins, which control green pigmentation (Futahashi and Fujiwara, 2008).

9.3 POLYPHENISMS

The most profound cases of phenotypic plasticity in colour patterns are those associated with polyphenisms. Polyphenisms are discrete alternative phenotypes that develop in response to specific signals from the environment. Among the best known of these are the solitary and gregarious phase polyphenisms of migratory locusts, and the seasonal polyphenisms of many different insects, in which different phenotypes develop in different season of the year, cued by photoperiod, temperature or nutrition. Some of these polyphenisms are so extreme that the alternative forms were originally described as different species, even though there are no genetic differences between them. Any individual can develop into either form, depending entirely on the environment encountered during its development (Nijhout, 1994, 1999, 2003a). Many polyphenisms express both morphological and colour pattern alternatives; here I will focus on pigmentation and colour patterns.

9.3.1 *General developmental mechanism*

In all cases where the developmental mechanism underlying the expression of alternative phenotypes has been elucidated, the switch is mediated by a developmental hormone (Nijhout, 1999, 2003a). The primary hormones that play a role in polyphenic development of pigmentation are ecdysone, JH and the neurosecretory hormone corazonin. Developmental hormones act during 'hormone-sensitive periods' to direct a particular pattern of gene expression.

The general idea is that if the relevant hormone is absent or below a threshold concentration during the sensitive period, then one pattern of gene expression ensues, whereas if it is above threshold a different pattern of gene expression is stimulated. The alternative patterns of gene expression cause developmental pathways, and the resultant phenotypes, to diverge. This genetic switch has been best studied in caste determination in social insects (Evans and Wheeler, 1999, 2001; Pereboom *et al.*, 2005; Scharf *et al.*, 2003), and to a lesser degree in colour pattern switching in the Lepidoptera (Brakefield *et al.*, 1998; Koch *et al.*, 2000; Suzuki and Nijhout, 2006). Alternative environments alter either the timing or the level of hormone secretion, so that in one environment the hormone is below and in another above threshold during the hormone-sensitive period. This way, a given environment becomes associated with a particular pattern of gene expression, and a characteristic phenotype.

9.3.2 *Locust polyphenism and chromatic adaptation*

When nymphs of *Locusta migratoria* are uncrowded, they develop into adults with a green and yellow colour pattern, with relatively short wings and a solitary and sedentary behaviour. By contrast, if nymphs are crowded, they develop into adults with a bold pattern of black, yellow and orange, with longer wings, somewhat different body proportions, and with a gregarious and migratory behaviour. The control of pigmentation is complex and involves at least two hormones, the JH and the neurosecretory hormone corazonin (Pener, 1991; Tanaka, 2000a, 2004). JH is elevated in the solitary form and induces a homogeneous green pigmentation (Pener, 1991). When JH is low the pigmentation is controlled by corazonin.

Tanaka (1993) developed an unpigmented strain of *L. migratoria*, which was used to investigate the control over the brown, red and black colouration characteristic of the gregarious form. Corazonin alone proved to be able to induce a broad range of colours, depending entirely on the timing and dosage of treatment (Tanaka, 2000c). When unpigmented nymphs were injected with a high dose at the beginning of the 3rd instar, they turned completely black in the following instar, whereas nymphs injected with the same dose in the middle of the instar developed black patterns with an orange background colour, the pigmentation characteristic of normal gregarious nymphs. Injection at the end of the 3rd instar induced a red colour with few black spots. Lower doses of corazonin induced similar but less intense pigmentation patterns (Tanaka, 2000c).

The range of colours Tanaka was able to produce by timed injections of corazonin ranged from pale tan, through orange and red, to brown and black (Tanaka, 2000a,c, 2001). Adding to this, the green colour induced by JH, it is possible to produce a broad spectrum of pigmentation. *Locusta*, like some other grasshoppers, is able to adapt its pigmentation to match the background colour of their environment (Pener, 1991), and corazonin has been shown to affect body colouration in several other species of grasshoppers (Tanaka, 2000b), so it is possible that grasshoppers are able to match their body colour to their background by simple variation in the timing and intensity of corazonin secretion during the nymphal instars (Tanaka, 2000a,b,c).

9.3.3 *Butterfly seasonal forms*

Many species of butterflies have distinctive seasonal forms. These seasonal polyphenisms occur in species that have more than one generation per year, so that different generations are exposed to different seasonal environments (dry and wet seasons in the tropics, spring, summer and fall seasons in temperate zones). The alternative seasonal forms of some species differ only subtly in colouration and in the overall contrast of the pattern, while those of others look so dramatically different that they are easily mistaken for different species (Fig. 18). In general, it is the temperature or photoperiod experienced by the caterpillar that determines the colour pattern of the adult butterfly. In the tropics, high temperatures during larval life result in the development of typical dry-season forms and low temperatures in wet-season forms. In temperate zones, high temperatures and/or long-day photoperiods result in the development of summer forms, and low temperatures and/or short-day photoperiods in autumn forms. These different environments cause a change in the timing of secretion of the hormone ecdysone during the pupal stage. In normal development, ecdysone secretion begins shortly after pupation, and the rising ecdysone titre initiates adult development. In seasonally polyphonic butterflies, differences in the timing of the rise of ecdysone induce alternative adult wing pigmentation and pattern.

The European Map butterfly, *Araschnia levana*, has dramatically different adult seasonal forms in the spring and summer (Fig. 18). The orange spring form is produced when the larval stage is reared under a short-day photoperiod and

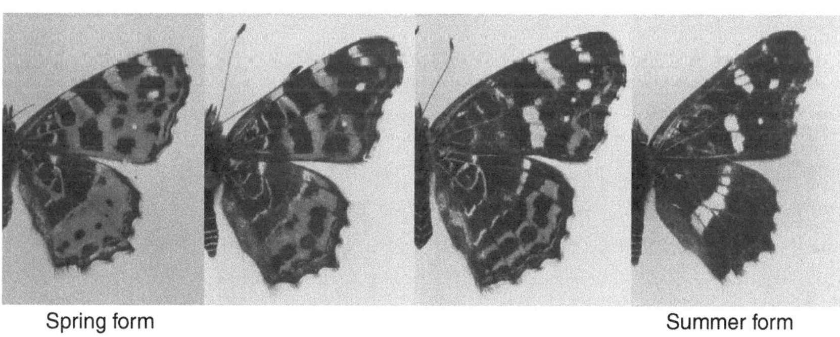

FIG. 18 Ecdysone-dependent pattern development in *Araschnia levana*. Normal spring and summer forms are illustrated on left and right, respectively. Between them are two intermediate patterns that were generated by timed injections of ecdysone into presumptive spring form pupae. (After Nijhout, 2003a, experiments by Berndt Koch.) (See Color Plate 39 at the back of the book.)

the black-and-white summer form when larvae experience long days. If pupae from larva reared under short-day conditions are injected with ecdysone during the first few days after pupation, they develop into normal summer forms, but if injection is delayed more than 10 days after pupation they develop into a normal spring form (Koch and Bückmann, 1987). Injections at intermediate times can produce adults with a whole range of patterns intermediate between the canonical spring and summer forms (Nijhout, 2003a). In the field, larvae that experience long-day conditions secrete ecdysone within a few days after pupation to initiate adult development; by contrast, under short-day conditions pupae enter an overwintering diapause and do not secrete ecdysone until several months later, long after the 10-day ecdysone-sensitive period is over. This mechanism ensures that in nature the adults develop into either the spring or summer form, without producing intermediate phenotypes.

Ecdysone also controls the seasonal forms of the North American Buckeye butterfly *Precis coenia*, though with several interesting differences. *Precis* has a summer form, with pale tan-coloured ventral hindwings, that is induced by either long day lengths or warm temperatures during the larval stage, and an autumn form, with reddish brown ventral hindwings, that is induced by short day lengths or cool temperatures (Smith, 1991). *Precis* does not have a pupal diapause and the ecdysone-sensitive period is only 20 h long, from 28 to 48 h after pupation (Rountree and Nijhout, 1995b). Under long-day condition ecdysone secretion begins about 12 h after pupation, but under short-day condition ecdysone secretion is delayed 24 h. This delay is sufficient so that ecdysone is low during the sensitive period, and this induces the autumn form. The 24-h delay of ecdysone secretion in the autumn form (Rountree and Nijhout, 1995b) results in a 24-h delay in the initiation of adult development, and a corresponding 24 h increase in the duration of the pupal stage.

An injection of ecdysone during the ecdysone-sensitive period in a short-day animal induces the development of normal long-day pigmentation (Rountree and Nijhout, 1995b). The primary difference between the two forms of *Precis* is the induction of the red-brown ommochrome pigmentation on the ventral hindwing in the autumn form, which implies that the presence of ecdysone during the sensitive period somehow causes the suppression of ommochrome synthesis in the ventral hindwing. A single gene recessive mutation that mimics the autumn form occurs in *Precis* populations at a low frequency (Rountree and Nijhout, 1995a). In a purebred strain for this mutation, the autumn form is constitutively expressed and colouration is no longer sensitive to photoperiod or temperature. Interestingly, ecdysone injection cannot induce the summer form in this genetic autumn form (Rountree and Nijhout, 1995a), indicating that the mutation is downstream of the ecdysone-sensitive step. This mutation illustrates the one mechanism by which one of the alternative forms of a polyphenism can become genetically fixed in a population, and perhaps constitutes a pathway by which new traits can evolve. Another mechanism is illustrated by the work of Brakefield *et al.* (1996), who showed that in *Bicyclus*, selection can lead to a loss of environmental sensitivity and fixation of one of the alternative phenotypes, presumably due to the fixation of polygenic variation.

The seasonal polyphenism of the tropical butterfly *B. anynana* is controlled by temperature, with a sensitive window during the last larval instar and early pupal stage. Warm temperatures induce the wet-season form, which has large eyespots in the ventral hindwing and a well-defined striping pattern. Cool temperatures induce the dry-season form, which has very small or no eyespots and a poorly defined striping pattern (Brakefield and Reitsma, 1991; Kooi and Brakefield, 1999; Windig *et al.*, 1994).

Brakefield *et al.* (1996) investigated the genetic foundation of this difference in eyespot development of the seasonal forms of *B. anynana*. The principal difference between the dry- and wet-season *Dll* expression was a low level of expression of *Dl* surrounding the focal expression. In the wet season for there was a large area of expression around the focus, roughly the size of the presumptive eyespot, but no such expression occurred around the focal expression of *Dl* in the dry-season form. The difference between the dry- and wet-season form eyespots appears to be in the degree to which this centre induces additional *Dll* expression in the surrounding cells.

Koch *et al.* (1996) used truncating selection to develop several strains of *Bicyclus* that had an altered sensitivity to temperature, so that at an intermediate temperature of 20 °C one line developed the normal dry-season pattern and another the normal wet-season pattern. The dry-season line also had a delayed adult development that was due to a delayed onset of ecdysone secretion (Koch *et al.*, 1996). Injections of ecdysone into the dry-season line during the first day after pupation shortened the pupal stage and shifted its colour pattern towards the wet-season form, inducing the formation of large eyespots and a more distinct banding (Koch *et al.*, 1996). Ecdysone injections in the wet season line had no effect on the pattern. Interestingly, the diameters of the eyespots that developed in the dry-season line were directly proportional to the dose of ecdysone injected. Not all pattern elements responded equally to ecdysone. The forewing eyespots enlarged more than the hindwing eyespots, which were almost completely insensitive to ecdysone (Brakefield *et al.*, 1998), and the banding pattern showed the strongest response of all (Koch *et al.*, 1996).

Thus, just as in *Araschnia*, the response of the colour pattern to ecdysone on *Bicyclus* is not all-or-none: each pattern element can take on intermediate appearances between those of the canonical seasonal forms, and different pattern elements respond more strongly than others to a given dose of ecdysone. It remains an open question as to the mechanism by which ecdysone could produce the alternative canonical forms in a seasonal polyphenism, and the mosaic of intermediate forms. Ecdysone is a steroid hormone that acts as a transcriptional regulator via a nuclear receptor (Riddiford *et al.*, 2000). The cascade of gene transcription that follows exposure to ecdysone has been described in *Drosophila* (Guay and Guild, 1991; White *et al.*, 1999), but none of the genes involved in this cascade are obvious candidates for a patterning function. A clue as to what is going on might be found in the fact that the intermediate patterns that can be obtained experimentally in the laboratory form a continuum between the extreme canonical forms, and this whole transformation series can be evoked by variation in the timing or dose of ecdysone during the first few days after pupation. One possibility is that the pattern is actually determined during the first few days of the pupal stage, and ecdysone simply 'freezes' the progression of pattern determination. If that is the case, then timed ecdysone injections would produce snapshots of progressive pattern determination. If this could be demonstrated to be true, then we would have an extremely useful tool with which to investigate the process of pattern formation on lepidopteran wings.

9.4 TEMPERATURE SHOCKS AND TRAUMA

More than a century ago, investigators discovered that if pupae of Lepidoptera were given a temperature shock (usually one or a few hours at a near-lethal temperature), the adults developed a highly aberrant colour pattern (Merrifield, 1890, 1891). The effect has since been documented in many species of butterflies and moths. Pattern aberrations can also be induced by chemicals. Otaki (2008) has shown that injections of heavy metals induce aberrations identical to those found with temperature shock, and Serfas and Carroll (2005) has shown that pattern aberrations can be induced by sulphated polysaccharides. For instance, in *P. coenia*, injection of heparin and chondroitin sulphates caused displacement of the banding pattern and reduction of eyespot size; dextran sulphate and fucoidan also caused a displacement of the bands but had no effect on eyespots. All these pattern aberrations, whether induced by temperature or chemicals, have several features in common: (1) not all elements of the pattern are affected, (2) eyespots are reduced in diameter, (3) eyespots fuse with parafocal elements along the midline of a wing cell, (4) overall colour of the wing is darkened and the contrast between pattern and ground is reduced, (5) the boundaries between pattern elements and between pattern and ground are less well-defined and become more stochastic (see Section 10), (6) banding patterns are shifted proximo-distally and (7) pattern at the wing margins become indistinct or are lost (see reviews in Nijhout, 1991; Otaki, 2008).

The two most interesting features of these aberrant patterns are (1) the fusion of eyespots with parafocal elements, which indicates that the two share developmental determinants and (2) the reduction in sharpness of boundaries between pattern elements and between regions of contrasting colours, which indicates that processes that establish or refine threshold are inhibited. The merger of eyespots with parafocal elements is reminiscent of some natural colour patterns found in *Neita extensa* (Satryridae), which show a clear continuity between these two pattern elements and suggest that the outer rings of eyespots and the parafocal elements may be homologous (Fig. 19). The decreased sharpness of boundaries in temperature-shocked patterns is manifested as an increased stochasticity of scale colours. This increase in stochasticity could be due to a reduction in the slope of the gradient that determines the pattern.

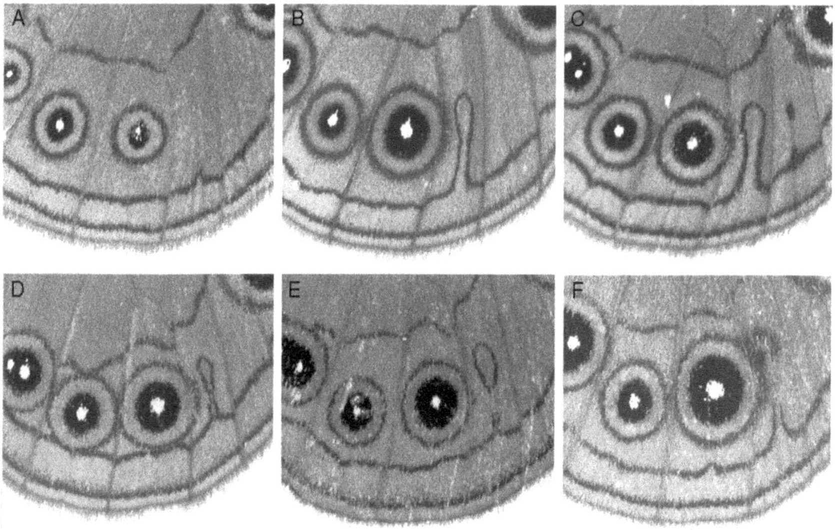

FIG. 19 Pattern variability in the parafocal elements of the satyrid, *Neita extensa*. In some specimens the parafocal element forms a loop resembling the pattern of *Dll* expression in Fig. 12 F and G. This pattern seems to merge with the outer ring of the adjoining eyespot, suggesting both are produced by a common process.

Cold shock, heat shock and chemical treatment all produce the same array of effects on the colour pattern, which suggest they may operate via a common mechanism. The most likely common factor would be the induction of heat shock proteins, which is an almost universal response to sublethal stress (Santoro, 2000). Heat shock proteins act as chaperones that bind to other proteins and can protect them from denaturation. One way in which the heat shock response could produce an effect on the colour pattern is if they temporarily inactivate proteins such as kinases or transcriptional regulators that are involved in pattern determination. Such binding would delay and disrupt ongoing development, and if not all the factors that participate in pattern determination are affected equally, then an aberrant pattern would develop. The increased stochasticity of the pattern after thermal or chemical stress could also be explained by this mechanism. If heat shock proteins bind and partially inactivate the morphogens that establish the gradients, then this would decrease the strength of the morphogenetic signal, and make it more difficult to 'interpret' by the receiving cells.

10 Stochastic processes in colour pattern formation

The concentration of transcriptional activators in a cell can be very low. For instance, if the concentration of a transcription factor is 10^{-9} M, and the volume of a cell is 100 femtolitres (fl) (10^{-13} l), then there will be only about 60 molecules in a cell. And if in a diploid cell there are only two binding sites for those transcription factors, there will be a significant stochastic or probabilistic component to the timing and rate at which a transcription factor will find a regulatory site on a promoter. As a result, gene expression is fundamentally stochastic and varies in unpredictable ways from cell to cell, and from time to time (Blake *et al.*, 2003; Cook *et al.*, 1998; Elowitz *et al.*, 2002; Raj and van Oudenaarden, 2008). Stochastic gene expression results in random variations in protein levels, temporally within a cell and spatially among cells.

Stochastic gene expression can be detected in insect colour patterns because each epidermal cell secretes the cuticle immediately overhead and any cell-level variation in the level of pigment synthetic activity or timing can result in small-scale stochastic variation in the cell-level pattern of pigmentation. Stochastic gene expression is most obvious in the wing patterns of Lepidoptera. Here the overall wing pattern is composed of several colours, but each scale expresses only one of these. Each pigment-bearing scale is the product of a single epidermal cell, and activates only a one of several potential pigment biosynthetic pathways. Exactly which of the alternative biosynthetic pathway is selected has a substantial stochastic component. This is most obvious at the boundaries between two regions of different colours: these boundaries are always 'noisy' with many scales expressing a colour that is out of place (Figs. 17 and 20). Boundaries between pigments can be sharp or gradual. No matter how sharp they are, there are always a few scales out of place. And when

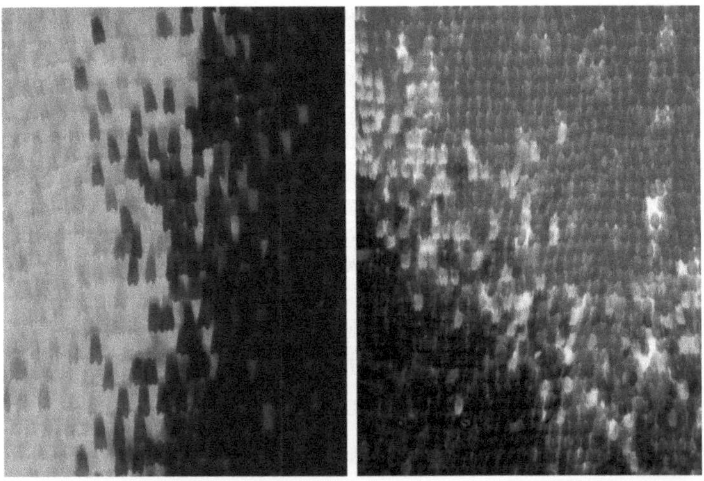

FIG. 20 Examples of stochastic gene expression leading to spatially stochastic patterns of pigment synthesis at the boundary between regions of different colours. Other examples of stochastic gene expression are shown in Figs. 1, 15 and 17. (After Nijhout, 2006.) (See Color Plate 40 at the back of the book.)

the transition between colours is gradual it is due to a graded change in the fraction of scales that are one colour or the other.

Stochastic gene expression is found in several other situations in colour pattern formation. First, in genetic polymorphisms where the alleles have incomplete dominance, the colour of heterozygotes is intermediate between those of the homozygotes and this is due to a random salt-and-pepper arrangement of scales of the two parental colours (Nijhout, 2006). Second, many species have subtle, diffuse and poorly defined patterns and those are always composed of a random salt-and-pepper pattern of scales of discrete colours. Third, the mechanism of focus determination in eyespots can be stochastic, resulting in a variable pattern of centres (Fig. 15). Finally, when the genes associated with the development of pattern first become expressed, they do not turn on simultaneously over the entire area where they will be expressed (Brunetti *et al.*, 2001; Nijhout, 2006). Rather, initial activation appears to be stochastic.

Experimentation with mathematical models of stochastic gene regulation have revealed that under some conditions graded input signals can result in threshold-like responses of gene activation (Cook *et al.*, 1998; Nijhout, 2006). A spatially graded signal across a field of cells can produce a sharp boundary of gene expression with a gene on at one side of the boundary, and off on the other. Effective Hill coefficients between 2 and 12 can be obtained, depending on the parameter values. The higher the Hill coefficient the sharper the boundary, and in the transition region around the boundary gene expression is spatially stochastic. Similarly, a temporal gradient, consisting of a gradual rise in stimulus, can cause an abrupt switch in gene activation at the threshold (Nijhout, 2006). Butterfly colour patterns provide some of the best illustrations of spatially stochastic gene expression, and can serve as a test bed for studying the mechanism and diverse consequences of stochastic gene activation.

11 Epilogue

The past 10 years have seen great developments in our understanding of the genetic, developmental and physiological mechanisms underlying colour pattern formation in insects. *Drosophila* wing patterns and butterfly eyespot patterns are now well established as model systems for understanding the developmental genetics of colour pattern formation, and the role of developmental hormones as universal mediators of environmentally induced development of alternative colour patterns is increasingly well understood. There are, however, still great gaps that need to be filled and that present exciting opportunities for original research. We still understand little about how epidermal cells select a specific pigment biosynthesis pathway, or how they switch between alternative pathways: is there a master regulatory gene for the genes that code for enzymes in each specific pigment synthesis pathway, or is each enzyme gene activated individually? What are the pattern-organizing signals that emanate from the wing veins in Lepidoptera? Although much work has been done on understanding the genetic basis of eyespot development in butterflies, this work has given little or no insight into the development of the many other elements of their colour patterns. A major open question is how the central symmetry system is determined. The central symmetry system is the dominant pattern on the wings of the majority of butterflies and moths but its development does not seem to involve any of the genes that control eyespots, which suggest that there must be other yet to be discovered clusters of regulatory genes that control the pattern on different regions of the wing. The greatest opportunities for discovery, however, will be found by expanding the kinds of species in which the developmental physiology of colour pattern formation is studied. The reason for this is that colour patterns originate late in development and continue to be actively moulded by evolution, so there are few if any of the conserved regulatory mechanisms that dominate early embryonic development. Instead, late-developing and actively evolving morphologies, such as colour patterns, capture bits and pieces of the regulatory pathways that originally evolved to control early embryonic development, and do so in diverse and unpredictable ways. Too few species and colour pattern systems have been studied to know whether there are interesting commonalities or rules that tie them together, or whether each patterning system is a one-off: a uniquely derived evolutionary and developmental novelty.

References

Akre, R., Green, A., MacDonald, J., Landolt, P. and Davis, H. (1981). The Yellowjackets of America North of Mexico. In: *Handbook No. 552*. US Department of Agriculture; pp. 120.

Allen, C., Beldade, P., Zwaan, B. and Brakefield, P. (2008). Differences in the selection response of serially repeated color pattern characters: standing variation, development, and evolution. *BMC Evol. Biol.* **8**, 94.

Beldade, P. and Brakefield, P. M. (2002). The genetics and evo-devo of butterfly wing patterns. *Nat. Rev. Genet.* **3**, 442–452.

Beldade, P., Brakefield, P. M. and Long, A. D. (2005). Generating phenotypic variation: prospects from ''evo-devo'' research on *Bicyclus anynana* wing patterns. *Evol. Dev.* **7**, 101–107.

Beldade, P., French, V. and Brakefield, P. M. (2008). Developmental and genetic mechanisms for evolutionary diversification of serial repeats: eyespot size in *Bicyclus anynana* butterflies. *J. Exp. Zool. B: Mol. Dev. Evol.* **310B**, 191–201.

Blake, W. J., Kaern, M., Cantor, C. R. and Collins, J. J. (2003). Noise in eukaryotic gene expression. *Nature* **422**, 633–637.

Brakefield, P. M. (1998). The evolution-development interface and advances with the eyespot patterns of *Bicyclus* butterflies. *Heredity* **80**, 265–272.

Brakefield, P. M. (2001). Structure of a character and the evolution of butterfly eyespot patterns. *J. Exp. Zool.* **291**, 93–104.

Brakefield, P. M. and French, V. (1995). Eyespot development on butterfly wings: the epidermal response to damage. *Dev. Biol.* **168**, 98–111.

Brakefield, P. and Reitsma, N. (1991). Phenotypic plasticity, seasonal climate and the pupation biology of *Bicyclus* butterflies (Satyridae) in Malawi. *Ecol. Entomol.* **16**, 291–303.

Brakefield, P., Gates, J., Keys, D., Kesbeke, F., Wijngaarden, P., Monteiro, A., French, V. and Carroll, S. (1996). Development, plasticity and evolution of butterfly eyespot patterns. *Nature* **384**, 236–242.

Brakefield, P., Kesbeke, F. and Koch, P. (1998). The regulation of phenotypic plasticity of eyespots in the butterfly *Bicyclus anynana*. *Am. Nat.* **152**, 853–860.

Brisson, J. A., De Toni, D. C., Duncan, I., Templeton, A. R. and Hey, J. (2009). Abdominal pigmentation variation in *Drosophila polymorpha*: geographic variation in the trait, and underlying phylogeography. *Evolution* **59**, 1046–1059.

Brunetti, C. R., Selegue, J. E., Monteiro, A., French, V., Brakefield, P. M. and Carroll, S. B. (2001). The generation and diversification of butterfly eyespot color patterns. *Curr. Biol.* **11**, 1578–1585.

Bückmann, D. (1959). Die Auslösung der Umfärbung durch das Häutungshormon bei *Cerura vinula* L. (Lepidoptera, Notodontidae). *J. Insect Physiol.* **3**, 159–166, IN5, 167–189.

Bückmann, D. (1974). Die hormonale steuerung der pigmentierung und des morphologischen farbwechsels bei insekten. *Fortschr. Zool.* **22**, 1–22.

Bückmann, D. (1977). Morphological colour change: stage independent, optically induced ommochrome synthesis in larvae of the stick insect, *Carausius morosus* Br. *J. Comp. Physiol. [B]* **115**, 185–193.

Bückmann, D. and Dustmann, J. (1962). Biochemische untersuchungen uber den morphologischen farbweschel von *Carausius morosus*. *Naturwissenschaften* **49**, 379.

Burgeff, H. and Schneider, L. (1979). Elektronmikroskopische untersuchungen zur korrelation zwichen farbe und struktur bei flügelschuppen des widderchens *Zygaena ephialtes* (Lepidoptera; Zygaenidae). *Entomol. Gen.* **4**, 135–0142.

Carroll, S., Gates, J., Keys, D., Paddock, S., Panganiban, G., Selegue, J. and Williams, J. (1994). Pattern formation and eyespot determination in butterfly wings. *Science* **265**, 109–114.

Chen, Y. and Struhl, G. (1996). Dual roles for Patched in sequestering and transducing Hedgehog. *Cell* **87**, 553–563.

Coleman, C. M. and Neckameyer, W. S. (2004). Substrate regulation of serotonin and dopamine synthesis in *Drosophila*. *Invert. Neurosci.* **5**, 85–96.

Cook, D. L., Gerber, A. N. and Tapscott, S. J. (1998). Modeling stochastic gene expression: implications for haploinsufficiency. *Proc. Natl. Acad. Sci. USA* **95**, 15641–15646.

Day, S. and Lawrence, P. (2000). Measuring dimensions: the regulation of size and shape. *Development* **127**, 2977–2987.

de Joussineau, C., Soule, J., Martin, M., Anguille, C., Montcourrier, P. and Alexandre, D. (2003). Delta-promoted filopodia mediate long-range lateral inhibition in *Drosophila*. *Nature* **426**, 555–559.

Descimon, H. (1965). Ultrastructure et pigmentation des écailles des Lépidopteres. *J. Microsc.* **4**, 130.

Dilao, R. and Sainhas, J. (2004). Modelling butterfly wing eyespot patterns. *Proc. R. Soc. Lond. B Biol. Sci.* **271**, 1565–1569.

Edelstein-Keshet, L. (2005). Mathematical Models in Biology. SIAM, Philadelphia, PA.

Edwards, K. A., Doescher, L. T., Kaneshiro, K. Y. and Yamamoto, D. (2007). A database of wing diversity in the Hawaiian *Drosophila*. *PLoS ONE* **2**, e487.

Elowitz, M. B., Levine, A. J., Siggia, E. D. and Swain, P. S. (2002). Stochastic gene expression in a single cell. *Science* **297**, 1183–1186.

Emlen, D. J. and Allen, C. E. (2003). Genotype to phenotype: physiological control of trait size and scaling in insects. *Integr. Comp. Biol.* **43**, 617–634.

Evans, T. M. and Marcus, J. M. (2006). A simulation study of the genetic regulatory hierarchy for butterfly eyespot focus determination. *Evol. Dev.* **8**, 273–283.

Evans, J. and Wheeler, D. (1999). Differential gene expression between developing queens and workers in the honey bee, *Apis mellifera*. *Proc. Natl. Acad. Sci. USA* **96**, 5575–5580.

Evans, J. and Wheeler, D. (2001). Gene expression and the evolution of insect polyphenisms. *BioEssays* **23**, 62–68.

Fraser, S., Green, C., Bode, H. and Gilula, N. (1987). Selective disruption of gap junctional communication interferes with a patterning process in hydra. *Science* **237**, 49–55.

French, V. and Brakefield, P. (1992). The development of eyespot patterns on butterfly wings: morphogen sources or sinks? *Development* **116**, 103–109.

French, V. and Brakefield, P. M. (1995). Eyespot development on butterfly wings: the focal signal. *Dev. Biol.* **168**, 112–123.

Futahashi, R. and Fujiwara, H. (2008). Juvenile hormone regulates butterfly larval pattern switches. *Science* **319**, 1061.

Ghiradella, H. (1984). Structure of iridescent lepidopteran scales: variations on several themes. *Ann. Entomol. Soc. Am.* **77**, 637–645.

Ghiradella, H. (1985). Structure and development of iridescent lepidopteran scales: the Papilionidae as a showcase family. *Ann. Entomol. Soc. Am.* **78**, 252–264.

Giersberg, H. (1928). Über den morphologischen und physiologischen Farbwechsel der Stabheuschrecke *Dixippus* (*Carausius*) *morosus*. *Z. Vergl. Physiol.* **7**, 657–695.

Gilbert, L., Forrest, H., Schultz, T. and Harvey, D. (1988). Correlations of ultrastructure and pigmentation suggest how genes control development of wing scale of *Heliconius* butterflies. *J. Res. Lepid.* **26**, 141–160.

Gompel, N., Prud'homme, B., Wittkopp, P. J., Kassner, V. A. and Carroll, S. B. (2005). Chance caught on the wing: cis-regulatory evolution and the origin of pigment patterns in *Drosophila*. *Nature* **433**, 481–487.

Guay, P. S. and Guild, G. M. (1991). The ecdysone-induced puffing cascade in *Drosophila* salivary glands: a broad-complex early gene regulates Intermolt and late gene transcription. *Genetics* **129**, 169–175.

Hatadani, L. M., Baptista, J. C. R., Souza, W. N. and Klaczko, L. B. (2004). Colour polymorphism in *Drosophila mediopunctata*: genetic (chromosomal) analysis and nonrandom association with chromosome inversions. *Heredity* **93**, 525–534.

Hinton, H. and Jarman, G. (1972). Physiological color change in the Hercules beetle. *Nature* **238**, 160–161.

Hiruma, K. and Riddiford, L. (1993). Molecular mechanisms of cuticular melanization in the tobacco hornworm, *Manduca sexta* (L.) (Lepidoptera: Sphingidae). *Int. J. Insect Morphol. Embryol.* **22**, 103–117.

Hiruma, K., Carter, M. and Riddiford, L. (1995). Characterization of the dopa decarboxylase gene of *Manduca sexta* and its suppression by 20-hydroxyecdysone. *Dev. Biol.* **169**, 195–209.

Hollocher, H., Hatcher, J. L. and Dyreson, E. G. (2000). Evolution of abdominal pigmentation differences across species in the *Drosophila dunni* subgroup. *Evolution* **54**, 2046–2056.

Ingham, P. W. (1993). Localized hedgehog activity controls spatial limits of wingless transcription in the *Drosophila* embryo. *Nature* **366**, 560–562.

Ishizaki, Y. and Umebachi, Y. (1988). Level changes of β-alanine, dopamine, and *N*-β-alanyldopamine during the pupal stage of *Papilio xuthus* (Lepidoptera: Papilionidae). *Comp. Biochem. Physiol.* **90C**, 83–87.

Ishizaki, Y. and Umebachi, Y. (1990). Further studies on dopamine and *N*-acetyldopamine during the pupal stage of *Papilio xuthus* (Lepidoptera: Papilionidae). *Comp. Biochem. Physiol.* **97B**, 563–567.

Janssen, J. M., Monteiro, A. and Brakefield, P. M. (2001). Correlations between scale structure and pigmentation in butterfly wings. *Evol. Dev.* **3**, 415–423.

Jeong, S., Rokas, A. and Carroll, S. B. (2006). Regulation of body pigmentation by the abdominal-B Hox protein and its gain and loss in *Drosophila* evolution. *Cell* **125**, 1387–1399.

Keys, D., Lewis, D., Selegue, J., Pearson, B., Goodrich, L., Johnson, R., Gates, J., Scott, M. and Carroll, S. (1999). Recruitment of a hedgehog regulatory circuit in butterfly eyespot evolution. *Science* **283**, 532–534.

Koch, P. B. (1994). Wings of the butterfly *Precis coenia* synthesize dopamine melanin by selective enzyme activity of dopadecarboxylase. *Naturwissenschaften* **81**, 36–38.

Koch, P. B. (1995). *Color pattern specific melanin synthesis is controlled by ecdysteroids via dopa decarboxylase in wings of* Precis coenia *(Lepidoptera, Nymphalidae)*. Institute of Entomology, Czechoslovak Academy of Sciences, Branisovska. pp. 161–167.

Koch, P. and Bückmann, D. (1987). Hormonal control of seasonal morphs by the timing of ecdysteroid release in *Araschnia levana* L. (Nymphalidae: Lepidoptera). *J. Insect Physiol.* **33**, 823–829.

Koch, P. B. and Kaufmann, N. (1995). Pattern specific melanin synthesis and DOPA decarboxylase activity in a butterfly wing of *Precis coenia* Hübner. *Insect Biochem. Mol. Biol.* **25**, 73–82.

Koch, P. B. and Nijhout, H. F. (2002). The role of wing veins in colour pattern development in the butterfly *Papilio xuthus* (Lepidoptera: Papilionidae). *Eur. J. Entomol.* **99**, 67–72.

Koch, P., Brakefield, P. and Kesbeke, F. (1996). Ecdysteroids control eyespot size and wing color pattern in the polyphenic butterfly *Bicyclus anynana* (Lepidoptera: Satyridae). *J. Insect Physiol.* **43**, 223–230.

Koch, P., Keys, D., Rocheleau, T., Aronstein, K., Blackburn, M., Carroll, S. and ffrench-Constant, R. (1998). Regulation of dopa decarboxylase expression during colour pattern formation in wild-type and melanic tiger swallowtail butterflies. *Development* **125**, 2303–2313.

Koch, P. B., Lorenz, U., Brakefield, P. M. and ffrench-Constant, R. H. (2000). Butterfly wing pattern mutants: developmental heterochrony and co-ordinately regulated phenotypes. *Dev. Genes. Evol.* **210**, 536–544.

Koch, B., Merk, R., Reinhardt, R. and Weber, P. (2003). Localization of ecdysone receptor protein during colour pattern formation in wings of the butterfly *Precis coenia* (Lepidoptera: Nymphalidae) and co-expression with Distal-less protein. *Dev. Genes. Evol.* **212**, 571–584.

Kodandaramaiah, U., Vallin, A. and Wiklund, C. (2009). Fixed eyespot display in a butterfly thwarts attacking birds. *Anim. Behav.* **77**, 1415–1419.

Kondo, S. (2002). The reaction-diffusion system: a mechanism for autonomous pattern formation in the animal skin. *Genes to Cells* **7**, 535–541.

Kondo, S. and Asai, R. (1995). A reaction–diffusion wave on the skin of the marine angelfish Pomacanthus. *Nature* **376**, 765–768.

Kooi, R. E. and Brakefield, P. M. (1999). The critical period for wing pattern induction in the polyphenic tropical butterfly *Bicyclus anynana* (Satyrinae). *J. Insect Physiol.* **45**, 201–212.

Kopp, A. and Duncan, I. (2002). Anteroposterior patterning in adult abdominal segments of *Drosophila*. *Dev. Biol.* **242**, 15–30.

Kopp, A., Blackman, R. and Duncan, I. (1999). Wingless, decapentaplegic and EGF receptor signaling pathways interact to specify dorso-ventral pattern in the adult abdomen of *Drosophila*. *Development* **126**, 3495–3507.

Kopp, A., Duncan, I. and Carroll, S. B. (2000). Genetic control and evolution of sexually dimorphic characters in *Drosophila*. *Nature* **408**, 553–559.

Lee, J. J., Von Kessler, D., Parks, S. and Beachy, P. A. (1992). Secretion and localized transcription suggest a role in positional signaling for products of the segmentation gene hedgehog. *Cell* **71**, 33–50.

Locke, M. and Huie, P. (1981). Epidermal feet in insect morphogenesis. *Nature* **293**, 733–735.

Marcus, J. and Evans, T. (2008). A simulation study of mutations in the genetic regulatory hierarchy for butterfly eyespot focus determination. *Biosystems* **93**, 250–255.

McMillan, W. O., Monteiro, A. and Kapan, D. D. (2002). *Development and evolution on the wing* **17**, 125–133.

Meinhardt, H. (1982). Models of Biological Pattern Formation. Academic Press, London.

Merrifield, F. (1890). Systematic temperature experiments on some Lepidoptera in all their stages. *Trans. Entomol. Soc. Lond.* **38**, 131–159.

Merrifield, F. (1891). Conspicuous effects on the markings and colouring of Lepidoptera caused by exposure of pupae to different temperature conditions. *Trans. Entomol. Soc. Lond.* **39**, 155–168.

Mohler, J. and Vani, K. (1992). Molecular organization and embryonic expression of the hedgehog gene involved in cell-cell communication in segmental patterning of *Drosophila*. *Development* **115**, 957–971.

Monteiro, A. (2008). Alternative models for the evolution of eyespots and of serial homology on lepidopteran wings. *BioEssays* **30**, 358–366.

Monteiro, A., Brakefield, P. and French, V. (1994). The evolutionary genetics and developmental basis of wing pattern variation in the butterfly *Bicyclus anynana*. *Evolution* **48**, 1147–1157.

Monteiro, A., French, V., Smit, G., Brakefield, P. and Metz, J. (2001). Butterfly eyespot patterns: evidence for specification by a morphogen diffusion gradient. *Acta Biotheor.* **49**, 77–88.

Monteiro, A., Prijs, J., Bax, M., Hakkaart, T. and Brakefield, P. M. (2003). Mutants highlight the modular control of butterfly eyespot patterns. *Evol. Dev.* **5**, 180–187.

Murray, J. D. (2003). Mathematical Biology, II. Springer, New York.

Nardi, J. B. and Magee-Adams, S. M. (1986). Formation of scale spacing patterns in a moth wing: I. Epithelial feet may mediate cell rearrangement. *Dev. Biol.* **116**, 278–290.

Nijhout, H. (1980a). Pattern formation on lepidopteran wings: determination of an eyespot. *Dev. Biol.* **80**, 267–274.

Nijhout, H. F. (1980b). Ontogeny of the color pattern on the wings of *Precis coenia* (Lepidoptera, Nymphalidae). *Dev. Biol.* **80**, 275–288.

Nijhout, H. F. (1984). Colour pattern modification by coldshock in Lepidoptera. *J. Embryol. Exp. Morphol.* **18**, 287–305.

Nijhout, H. (1985). Cautery-induced colour patterns in *Precis coenia* (Lepidoptera: Nymphalidae). *J. Embryol. Exp. Morphol.* **86**, 191–203.

Nijhout, H. F. (1990). A comprehensive model for colour pattern formation in butterflies. *Proc. R. Soc. Lond. B Biol. Sci.* **239**, 81–113.

Nijhout, H. F. (1991). *The Development and Evolution of Butterfly Wing Patterns*. Smithsonian Institution Press, Washington, DC.

Nijhout, H. F. (1994). Insect Hormones. Princeton University Press, Princeton, NJ.

Nijhout, H. F. (1997). Ommochrome pigmentation of the *linea* and *rosa* seasonal forms of *Precis coenia* (Lepidoptera: Nymphalidae). *Arch. Insect Biochem. Physiol.* **36**, 215–222.

Nijhout, H. F. (1999). Control mechanisms of polyphenic development in insects. *Bioscience* **49**, 181–192.

Nijhout, H. F. (2001). Elements of butterfly wing patterns. *J. Exp. Zool.* **291**, 213–225.

Nijhout, H. F. (2003a). Development and evolution of adaptive polyphenisms. *Evol. Dev.* **5**, 9–18.

Nijhout, H. F. (2003b). Polymorphic mimicry in *Papilio dardanus*: mosaic dominance, big effects, and origins. *Evol. Dev.* **5**, 579–592.

Nijhout, H. F. (2006). Stochastic gene expression: dominance, thresholds and boundaries. In: *The Biology of Genetic Dominance* (ed Veitia, R.), pp. 61–75. Landes Bioscience, Georgetown, TX.

Nijhout, H. F. and Grunert, L. W. (1988). Colour pattern regulation after surgery on the wing disks of *Precis coenia* (Lepidoptera: Nymphalidae). *Development* **102**, 337–385.

Nijhout, H. F. and Koch, P. B. (1992). The distribution of radiolabeled pigment precursors in the wing patterns of nymphalid butterflies. *J. Res. Lepid.* **30**.

Nijhout, H. F., Maini, P. K., Madzvamuse, A., Wathen, A. J. and Sekimura, T. (2003). Pigmentation pattern formation in butterflies: experiments and models. *C. R. Biol.* **326**, 717–727.

Otaki, J. M. (2008). Physiologically induced color-pattern changes in butterfly wings: mechanistic and evolutionary implications. *J. Insect Physiol.* **54**, 1099–1112.

Pener, M. (1991). Locust phase polymorphism and its endocrine relations. *Adv. Insect Physiol.* **23**, 1–79.

Pereboom, J. J. M., Jordan, W. C., Sumner, S., Hammond, R. L. and Bourke, A. F. G. (2005). Differential gene expression in queen–worker caste determination in bumble bees. *Proc. R. Soc. B Biol. Sci.* **272**, 1145–1152.

Quinsey, N. S. and Luong, A. Q. (1998). Mutational analysis of substrate inhibition in tyrosine hydroxylase. *J. Neurochem.* **71**, 2132–2138.

Raabe, M. (1966). Recherches sur la neurosécrétion dans la chaine nerveuse ventrale du phasme *Carausius morosus*: liaison entre l'activité des cellules B1 et la pigmentation. *C. R. Acad. Sci. (Paris)* **263**, 408–411.

Raj, A. and van Oudenaarden, A. (2008). Nature, nurture, or chance: stochastic gene expression and its consequences. *Cell* **135**, 216–226.

Rassart, M., Colomer, J.-F., Tabarrant, T. and Vigneron, J. (2008). Diffractive hygrochromic effect in the cuticle of the hercules beetle *Dynastes hercules*. *New J. Phys.* **10**, 033.

Reed, R. D. and Gilbert, L. E. (2004). Wing venation and Distal-less expression in *Heliconius* butterfly wing pattern development. *Dev. Genes Evol.* **214**, 628–634.

Reed, R. D. and Nagy, L. M. (2005). Evolutionary redeployment of a biosynthetic module: expression of eye pigment genes *vermilion*, *cinnabar*, and *white* in butterfly wing development. *Evol. Dev.* **7**, 301–311.

Reed, R. D. and Serfas, M. S. (2004). Butterfly wing pattern evolution is associated with changes in a Notch/Distal-less temporal pattern formation process. *Curr. Biol.* **14**, 1159–1166.

Reed, R. D., McMillan, W. O. and Nagy, L. M. (2008). Gene expression underlying adaptive variation in *Heliconius* wing patterns: non-modular regulation of overlapping cinnabar and vermilion prepatterns. *Proc. R. Soc. B Biol. Sci.* **275**, 37–46.

Riddiford, L., Cherbas, P. and Truman, J. (2000). Ecdysone receptors and their biological actions. *Vitam. Horm.* **60**, 1–73.

Rountree, D. and Nijhout, H. F. (1995a). Genetic control of a seasonal morph in *Precis coenia* (Lepidoptera: Nymphalidae). *J. Insect Physiol.* **41**, 1141–1145.

Rountree, D. and Nijhout, H. F. (1995b). Hormonal control of a seasonal polyphenism in *Precis coenia* (Lepidoptera: Nymphalidae). *J. Insect Physiol.* **41**, 987–992.
Saenko, S. V., French, V., Brakefield, P. M. and Beldade, Pc. (2008). Conserved developmental processes and the formation of evolutionary novelties: examples from butterfly wings. *Philos. Trans. R. Soc. B Biol. Sci.* **363**, 1549–1556.
Sanicola, M., Sekelsky, J. J., Elson, S. and Gelbart, W. M. (1995). Drawing a stripe in *Drosophila* imaginal disks: negative regulation of decapentaplegic and patched expression by engrailed. *Genetics* **139**, 745–756.
Santoro, M. G. (2000). Heat shock factors and the control of the stress response. *Biochem. Pharmacol.* **59**, 55–63.
Sawada, H., Nakagoshi, M., Reinhardt, R. K., Ziegler, I. and Koch, P. B. (2002). Hormonal control of GTP cyclohydrolase I gene expression and enzyme activity during color pattern development in wings of *Precis coenia. Insect Biochem. Mol. Biol.* **32**, 609–615.
Scharf, M., Wu-Scharf, D., Pittendrigh, B. and Bennett, G. (2003). Caste- and development-associated gene expression in a lower termite. *Genome Biol.* **4**, R62.
Schwartz, L. M. and Truman, J. W. (1983). Hormonal control of rates of metamorphic development in the tobacco hornworm *Manduca sexta. Dev. Biol.* **99**, 103–114.
Sekimura, T., Madzvamuse, A., Wathen, A. J. and Maini, P. K. (2000). A model for colour pattern formation in the butterfly wing of *Papilio dardanus. Proc. R. Soc. Lond. B Biol. Sci.* **267**, 851–859.
Sekimura, T., Madzvamuse, A. and Maini, P. K. (2007). Pigmentation pattern formation in butterfly wings: global patterns on fore- and hindwing. *Mathematical Modeling of Biological Systems, Volume I*. pp. 142–148. Springer, New York, NY.
Serfas, M. S. and Carroll, S. B. (2005). Pharmacologic approaches to butterfly wing patterning: sulfated polysaccharides mimic or antagonize cold shock and alter the interpretation of gradients of positional information. *Dev. Biol.* **287**, 416–424.
Smith, K. (1991). The effects of temperature and daylength on the rosa polyphenism in the buckeye butterfly, *Precis coenia* (Lepidoptera: Nymphalidae). *J Res. Lepid.* **30**, 237–244.
Suzuki, Y. and Nijhout, H. F. (2006). Evolution of a polyphenism by genetic accommodation. *Science* **311**, 650–652.
Tabata, T., Eaton, S. and Kornberg, T. B. (1992). The *Drosophila* hedgehog gene is expressed specifically in posterior compartment cells and is a target of engrailed regulation. *Genes Dev.* **6**, 2635–2645.
Tanaka, S. (1993). Hormonal deficiency causing albinism in *Locusta migratoria. Zool. Sci.* **10**, 467–471.
Tanaka, S. (2000a). Hormonal control of body-color polymorphism in *Locusta migratoria*: interaction between [His7]-corazonin and juvenile hormone. *J. Insect Physiol.* **46**, 1535–1544.
Tanaka, S. (2000b). Induction of darkening by corazonins in several species of Orthoptera and their possible presence in ten insect orders. *Appl. Entomol. Zool.* **35**, 509–517.
Tanaka, S. (2000c). The role of [His7]-corazonin in the control of body-color polymorphism in the migratory locust, *Locusta migratoria* (Orthoptera: Acrididae). *J. Insect Physiol.* **46**, 1169–1176.
Tanaka, S. (2001). Endocrine mechanisms controlling body-color polymorphism in locusts. *Arch. Insect Biochem. Physiol.* **47**, 139–149.
Tanaka, S. (2004). Hormonal control of body-color polyphenism in the American grasshopper, *Schistocerca americana*: a function of [His7]-corazonin. *Ann. Entomol. Soc. Am.* **97**, 302–309.

Tibbetts, E. A. (2002). Visual signals of individual identity in the wasp *Polistes fuscatus*. *Proc. R. Soc. Lond. B Biol. Sci.* **269**, 1423–1428.
Tibbetts, E. A. and Curtis, T. R. (2007). Rearing conditions influence quality signals but not individual identity signals in *Polistes* wasps. *Behav. Ecol.* **18**, 602–607.
Tibbetts, E. A. and Dale, J. (2004). A socially enforced signal of quality in a paper wasp. *Nature* **432**, 218–222.
True, J. R., Edwards, K. A., Yamamoto, D. and Carroll, S. B. (1999). *Drosophila* wing melanin patterns form by vein-dependent elaboration of enzymatic prepatterns. *Curr. Biol.* **9**, 1382–1391.
True, J. R., Yeh, S. D., Hovemann, B. T., Kemme, T., Meinertzhagen, I. A., Edwards, T. N., Liou, S. R., Han, Q. and Li, J. (2005). *Drosophila* tan encodes a novel hydrolase required in pigmentation and vision. *PLoS Genet.* **1**, e63.
Turing, A. M. (1952). The chemical basis of morphogenesis. *Philos. Trans. R. Soc. Lond. B Biol. Sci.* **237**, 37–72.
Umebachi, Y. (1985). Papiliochrome, a new pigment group of butterfly. *Zool. Sci.* **2**, 163–174.
Vigneron, J. P., Pasteels, J. M., Windsor, D. M., Vertesy, Z., Rassart, M., Seldrum, T., Dumont, J., Deparis, O., Lousse, V., Biro, L. P., Ertz, D. and Welch, V. (2007). Switchable reflector in the Panamanian tortoise beetle *Charidotella egregia* (Chrysomelidae: Cassidinae). *Phys. Rev. E Stat. Nonlin. Soft Matter Phys.* **76**, 031907.
White, K. P., Rifkin, S. A., Hurban, P. and Hogness, D. S. (1999). Microarray analysis of *Drosophila* development during metamorphosis. *Science* **286**, 2179–2184.
Wigglesworth, V. (1959). The Control of Growth and Form: A Study of the Epidermal Cell in an Insect. Cornell University Press, Ithaca, NY.
Willis, J., Rezaur, R. and Sehnal, F. (1982). Juvenoids cause some insects to form composite cuticles. *J Embryol. Exp. Morphol.* **71**, 25–40.
Windig, J., Brakefield, P., Reitsma, N. and Wilson, J. (1994). Seasonal polyphenism in the wild: survey of five wing patterns in five species of *Bicyclus* butterflies in Malawi. *Ecol. Entomol.* **19**, 285–298.
Wittkopp, P. J., True, J. R. and Carroll, S. B. (2002). Reciprocal functions of the *Drosophila* Yellow and Ebony proteins in the development and evolution of pigment patterns. *Development* **129**, 1849–1858.
Wittkopp, P. J., Williams, B. L., Selegue, J. E. and Carroll, S. B. (2003). *Drosophila* pigmentation evolution: divergent genotypes underlying convergent phenotypes. *Proc. Natl. Acad. Sci. USA* **100**, 1808–1813.
Wolpert, L. (1969). Positional information and the spatial pattern of cellular differentiation. *J. Theor. Biol.* **25**, 1–47.
Wolpert, L. (1981). Positional information and pattern formation. *Philos. Trans. R. Soc. Lond. B Biol. Sci.* **295**, 441–450.
Wolpert, L. (1994). Positional information and pattern formation in development. *Dev. Genet.* **15**, 485–490.
Wright, T. R. F. (1987). The genetics of biogenic amine metabolism, sclerotization, and melanization in *Drosophila melanogaster*. *Adv. Genet.* **24**, 127–222.

Insect Colours and Visual Appearance in the Eyes of Their Predators

Marc Théry*,[1] and Doris Gomez*,[1]

*Département d'Ecologie et de Gestion de la Biodiversité, CNRS UMR 7179, Muséum National d'Histoire Naturelle, Brunoy, France

[1] Both authors contributed equally

ADVANCES IN INSECT PHYSIOLOGY VOL. 38
ISBN 978-0-12-381389-3
DOI: 10.1016/S0065-2806(10)38001-5

1 Introduction

Insects exhibit a stunning diversity of colours: consider the splendid spotted butterflies or ladybirds, iridescent odonats or beetles, yellow-and-black striped wasps or flies, brown or green walking sticks or mantids. The previous chapters of this volume have reviewed in great detail the mechanisms of production and development of colouration in insects. It is now time to study insect colours from a more functional and evolutionary perspective and to provide answers to the question 'why is there such a diversity in insect colouration?'. The vast majority of insect colours and colour patterns is thought to have evolved in relation to the needs of communication with conspecifics, prey and predators. On the one hand, different camouflage strategies are selected to decrease detection probability by non-intentional receivers such as predators or prey. On the other hand, intentional receivers such as conspecifics or predators—in the case of warning colouration—select for more conspicuous colours because they are easier to detect and interpret.

Predators are essential in the evolution of insect colouration. Nearly all animal taxonomic groups comprise insectivorous species and these predators almost all rely, at least partially, on vision for hunting. Spatial vision—the ability to exploit the spatial visual information provided by the environment—appeared more than 500 million years ago (Land and Nilsson, 2002). The advantages that vision provided for orientation, navigation, foraging or reproduction explained its rapid and large evolutionary success, as attested by traces of visual structures in a highly diverse array of fossil records (Land and Nilsson, 2002). Insects, which appeared approximately 100 million years afterwards, had to face visually guided predators from their early evolution. All animals endowed with vision are able to exploit brightness (lightness or luminance) information of their environment. However, some animal species have developed colour vision, that is the ability to distinguish objects differing only by the spectral distribution of the radiant energy (see Kelber *et al.*, 2003 for discussion; Skorupski and Chittka, 2009), to exploit wavelength (chromatic) information independent of brightness. Insect predators add to their taxonomic diversity a diversity in visual performance which results in a manifold pressure on all components of insect colouration, both achromatic (brightness) and chromatic components. In this chapter, we will include all these aspects of colours shown by insects, and we will consider white, grey and black as being colours, to encompass the highest possible number of relevant studies.

It is important to consider all visual features that influence predators' perception of their insect prey. Not only insect colour and colour pattern (i.e. the spatial organization of colour patches on the insect body) are important for perception, but insect size, shape and texture, as many spatial frames in which colouration is expressed, are also important contributors to visual perception (Troscianko *et al.*, 2009). Moreover, insect visual appearance is subject to important temporal and spatial changes due to background and ambient light variability and more importantly to the movements of animals in their environment. Although the central target of this chapter will be colouration and colour pattern, we will also consider the other aforementioned aspects to provide a more complete picture of how predators view insects.

Numerous studies have tackled some issues concerning the appearance of insects to predators, either from a correlative, experimental or from a theoretical modelling approach. Most of them have focused on a few signallers or receivers (e.g. Eisner *et al.*, 1967; Schlee, 1986; Exnerová *et al.*, 2006; Jackson and Pollard, 2007; Ioannou and Krause, 2009). They have explored a variety of colour signals, except iridescent signals, the functional significance of which in relation to predators remains unstudied (Doucet and Meadows, 2009). They rarely considered a comparative approach at a large interspecific scale from the signaller side (but see Robbins, 1981; Williams, 2007; Song and Wenzel, 2008). It is thus interesting to generate a more synthetic view and to consider the strategies evolved by both predators and insects. We will first detail predators' visual performance and their use of vision in prey capture. This will provide the basis for a methodological and conceptual discussion on how to investigate insect colouration viewed by predators. We will then review the strategies evolved by insects to decrease predation risk—namely camouflage and warning colouration—and the impact of the visual interplay between insects and predators on insect population dynamics in the particular case of insect colour polymorphism. We deliberately excluded the evolution of insect Müllerian and Batesian mimicry from this chapter for two reasons. First, the same perceptual principles explain how mimetic and non-mimetic insects are viewed by predators. Second, mimicry itself—origin and maintenance throughout evolution—goes far beyond our initial scope. Similarly, we did not consider in detail cognitive aspects such as learning or memory in predator perception.

2 Predator vision

2.1 PREDATOR VISUAL PERFORMANCE

2.1.1 *Visual equipment of insect predators*

Insects are preyed upon by a large array of organisms. All vertebrate classes (amphibians, reptiles, birds, mammals and fish) and major invertebrate classes (insects and chelicerates) contain a large number of insectivorous predators. For more information, the reader can consult extensive reviews specific of insects (Briscoe and Chittka, 2001), birds (Hart, 2001b; Ödeen and Håstad, 2003), mammals (Jacobs, 1993), other animal groups (Kelber *et al.*, 2003; Warrant and Nilsson, 2006) or more up-to-date studies on groups like marsupials or bats (Cowing *et al.*, 2008; Müller *et al.*, 2009). We present here a selection of examples of predator species (Table 1). Species have been chosen for their

TABLE 1

Receptor sensitivity maxima for selected examples of insect predators

Animal group and species	Sensitivity maxima (nm)	References
Insects		
Odonata		
Dragonfly (*Sympetrum rubicundum*)	330, 430, 490, 520, 620	Meinertzhagen *et al*. (1983)
Dragonfly (*Aeshna cyanea*)	356, 420, 519	Autrum and Kolb (1968)
Dragonfly (*Libellula needhami*)	358, 501	Horridge (1969)
Dictyoptera		
Praying mantis (*Tenodera* sp.)[a]	500–520	Sontag (1971); Rossel (1979)
Orthoptera		
Field cricket (*Gryllus campestris*)	340, 439, 520	Zufall *et al*. (1989)
Hemiptera		
Backswimmer (*Notonecta undulata*)	345, 445, 560	Bennett and Ruck (1970)
Neuroptera		
Owl fly (*Libelloides macaronius*)	350, 530	Gogala (1967)
Coleoptera		
Ladybird (*Coccinella* sp.)	375, 475, 520	Lin (1993)
Tiger beetle (*Cicindela* sp.)	360–380, 510–530	Lin and Wu (1992)
Hymenoptera		
Beewolf (*Philanthus triangulum*)	344, 444, 524	Peitsch *et al*. (1992)
Diptera		
Hover fly (*Syrphus* sp.)[b]	440	Bernard and Stavenga (1979)
Chelicerates		
Jumping spider (*Menemerus confuses*)	360, 490, 520, 580	Yamashita and Tateda (1976)
Jumping spider (*Plexippus validus*)	360, 520	Blest *et al*. (1981)
Fish		
Goldfish (*Carrassius auratus*)	356, 447, 537, 623, rod = 522	Bowmaker *et al*. (1991)
Cichlid (*Aequideus pulcher*)	453, 530, 570, rod = 500	Kroger *et al*. (1999)

Amphibians		
Frog (*Rana* spp.)	43[illegible], 502, 562, rods = 430, 502	Koskelainen *et al*. (1994)
Salamander (*Ambystoma tigrinum*)	400, 444, 610, rod = 506	Perry and McNaughton (1991)
Reptiles		
Lizard (*Anolis carolinensis*)	358, 437, 495, 560, rod = 560	Kawamura and Yokoyama (1998)
Gecko (*Gekko gekko*)	364, 467, 521	Loew (1994)
Birds		
Blue tit (*Cyanistes caeruleus*)	371, 448, 503, 563, dc = 563, rod = 503	Hart *et al*. (2000)
Domestic chicken (*Gallus gallus domesticus*)	419, 455, 508, 570, dc = 570, rod = 506	Bowmaker *et al*. (1993)
Mammals		
Marsupials		
Fat-tailed dunnart (*Sminthopsis crassicaudata*)	363, 509, 533, rod = 512	Cowing *et al*. (2008)
Eutherians		
Mouse-eared bat (*Myotis* sp.)	< 365, 500–570, rod ~ 500	Wang et al. (2004); Ru *et al*. (2008); Müller *et al*. (2009)
European mole (*Talpa europea*)	< 400, 500–570, rod ~ 500	Glösmann *et al*. (2008)
Siberian dwarf hamster[b] (*Phodopus sungorus*)	~ 360, 506, rod ~ 500	Calderone and Jacobs (1999)
Mouse lemur (*Microcebus murinus*)	~ 430, ~ 560, rod ~ 500	Surridge *et al*. (2003)
Capuchin monkey (*Cebus* sp.)	433, [535–550–562], rod ~ 500	Surridge *et al*. (2003)

[a] Mantises likely have one photoreceptor type in their compound eyes and two in their ocelli.

[b] Robberflies are dipteran, thus closely related to hover flies (which are insectivorous as larvae); grasshopper mice are rodents, thus closely related to the hamster.

For maximal sensitivity peaks, all values separated by comas represent peaks for distinct photoreceptor classes. Values within brackets are possible alleles for the LWS cone in primates. ~ represents approximate values when precise values are not determined. – separates the upper and lower values of the range within which the peak is likely located. The abbreviation dc stands for double cones: these cones are responsible for brightness detection and equipped with LWS photopigments.

insectivorous diet or for their phylogenetic proximity to insectivorous species. To a certain extent, they represent what is known to date about the variability in visual equipment of their group. For instance, the majority of diurnal birds have five different types of cones, four types of single cones and one type of double cones. They separate into two types of vision, a sensitivity shifted towards UV represented by the blue tit and the sensitivity shifted towards violet wavelengths represented by the domestic chicken (Hart, 2001b; Ödeen and Håstad, 2003). Conversely, dragonflies show a high variability in photopigments, with a number of photopigments ranging from two to five (Table 1). Although old studies have underestimated the number of photopigments due to incomplete sampling, a large part of the variation may be relevant.

Overall, Table 1 shows that predators' variation in visual equipment likely translates into substantial differences in visual performance. In addition, most species, even nocturnal, are sensitive to ultraviolet radiation, a wavelength range likely involved in prey detection. These observations lead to interesting conclusions. First, it is crucial to consider predators' vision and not a falsely convenient human perspective when studying colouration. Although implemented long ago in insect–plant communication studies, this recommendation needed to be re-activated (Bennett *et al.*, 1994; Cuthill *et al.*, 1999) to be followed in studies focused on animal communication. For example, investigation of both predator reaction and insect colouration should use UV radiation, an aspect that can shed a new light on apparently closed cases (Majerus *et al.*, 2000). Predator species listed fulfil the minimal requirements for colour vision ability since they possess at least two different types of photoreceptors. Although this possession is by no means a proof of the ability of colour vision (Kelber *et al.*, 2003; Skorupski and Chittka, 2009), it underlines the importance of considering not only brightness but also the chromatic aspects of insect colouration when studying predators' behaviour. Major work done on the evolution of colour polymorphism in insects has excluded (for technical reasons) the chromatic aspects of colouration (Bond and Kamil, 1998, 2002, 2006), thereby giving at least a partial view of the explored issue. Including both chromatic and achromatic components of colouration is highly desirable and has been actually done in recent studies (e.g. Stobbe and Schaefer, 2008; Stobbe *et al.*, 2009).

2.1.2 *Variation in visual equipment and its relation to insect capture efficiency*

One of the best documented examples relating to visual performance and behaviour concerns the effect of visual phenotype on foraging efficiency in primates. In primates, the gene coding for S opsin is on an autosome while the gene coding for L opsin is on the X chromosome. In Old World monkeys like humans, the L gene has undergone a duplication leading to a generalized trichromacy in all individuals. In New World monkeys, the L gene presents several alleles resulting in dichromacy for males and homozygote females and trichromacy in heterozygote females (e.g. Capuchin monkey, Table 1). Variation in visual equipment translates into variation in ability to detect insects (e.g. Melin *et al.*, 2007). Studies on both human (Morgan *et al.*, 1992) and nonhuman primates (Saito *et al.*, 2005) have shown that dichromacy may be beneficial for breaking colour camouflage. Morgan *et al.* (1992) explained that colour camouflage exists when the organization of chromatic signals does not coincide, interferes with that of other characteristics like shape or texture distinctive elements. Chromatic signals mask luminance (brightness) signals; they are distractive, preventing rapid recognition of important elements. While all vision phenotypes perform equally well in bright light, dichromat white-faced capuchins make more insect capture attempts than trichromats in restricted illumination (Melin *et al.*, 2007). Dichromats are better at detecting surface-dwelling cryptic insects while trichromats are better at detecting embedded insects (Melin *et al.*, 2007). Osorio *et al.* (2004) explain why the difference of foraging ability is particularly salient in low light conditions. The yellow-blue colour channel (the only channel available in dichromatic monkeys) fails at low light intensities. While in bright light, the improved chromatic vision of dichromatic monkeys impairs their ability to find cryptic insects, they are left colour-blind in low light intensity, which improves their ability to detect cryptic prey. Caine and Mundy (2000) tested foraging efficiency in marmosets (*Callithrix geoffroyi*) by presenting them with artificially coloured green (unripe) and orange (ripe) fruit balls. They revealed that trichromats are more efficient than dichromats at detecting orange targets, but only at a long distance and not at a short distance, where both types of visual system had similar performance.

Differences in performance can be seen not only between animals differing in their number of visual pigments, but also between animals with the same number of visual pigments. Birds are a good example. Despite their diversity in ecology and species, birds have two main types of photoreceptor sets: most passerines (e.g., blue tit) and parrots have a sensitivity shifted towards UV while other species (e.g. quails or chickens) have a sensitivity shifted towards violet wavelengths (Ödeen and Håstad, 2003). Whether this differential in UV sensitivity translates into significant differences in performance at detecting objects in the environment is still controversial (Håstad *et al.*, 2005; Schaefer *et al.*, 2007). The increased sensitivity to UV may be offset by the rarity of the photoreceptors and the noise associated to their response to light (Vorobyev *et al.*, 1998).

2.1.3 *Retinal organization and neural processing influence predators' ability to detect prey*

Apart from the equipment in photoreceptors, the anatomical structure of the eye is crucial for visual performance. In insects, visual acuity, defined as the finest grating that can be resolved by the eye, is inversely proportional to the angle between two adjacent ommatidia, that is, compound eye units (Land, 1997).

The smaller the inter-ommatidial angle, the better the acuity and the greater the distance at which environmental structures can be used for visual guidance. Yet, packing ommatidia on a retina limited in size comes at a cost: ommatidia with smaller diameter have a lower capture rate, which reduces contrast resolution and degrades image quality. Most insects with the lowest inter-ommaditial angle ($< 1°$) are predators: dragonflies, mantids and sphecid wasps (Land, 1997).

In these predatory insect species, the highest acuity is obtained from an area involved in prey tacking, a dorsal band called 'fovea' in which ommatidia are much larger, with ommatidia axes nearly parallel and small inter-ommatidial angle (Olberg *et al.*, 2007). This area with maximal spatial resolution also has a high temporal resolution, which facilitates motion detection (Kral and Prete, 2004). Predatory insects such as dragonflies adjust their position relative to their prey to keep them in the fovea (Olberg *et al.*, 2007), a behaviour that optimizes prey detection. Such a fovea is also present in mantids and robberflies (Land, 1997), but not in mantispids, which have similar prey hunting tactics as mantids (Kral *et al.*, 2000). Despite similar capture behaviour, mantispids are less efficient than mantids at capturing prey (Kral *et al.*, 2000), which is likely due to their lack of fovea.

Another example of retinal structure can be seen in jumping spiders. These insectivorous species have single-lens camera-type eyes. The three pairs of secondary eyes are monochromatic, specialized in motion detection and have a poor spatial acuity. Conversely, the pair of principal eyes (frontal anterio-median eyes) have a higher visual performance and colour vision (Blest *et al.*, 1981). Principal eyes contain an everse retina with the rhabdoms—the light absorbing parts of the ommatidia—projecting towards the lens and no reflecting tapetum behind the retina (Blest, 1985; Norgaard *et al.*, 2008). Some species, namely Salticoidae and a few Spartaeines like *Portia*, have evolved principal eyes with high spatial acuity that exceeds by tenfold that of the best seeing insects of similar size (Harland and Jackson, 2004). Such visual acuity has evolved twice independently within the Salticidae (Su *et al.*, 2007). Photo-receptors in the salticid anterio-median eyes are organized in four successive layers along the light path. To reach the rearmost layer, light must pass through layers IV, III and II. Only layer I has a fine, regular mosaic of receptors necessary for detailed vision. Salticoidae and a few Spartaeines like *Portia* present a highly ordered foveal retinal layer I that contains regularly packed light guiding rhabdomeres, where light beams are compartmented more efficiently (Harland and Jackson, 2004; Su *et al.*, 2007). As a consequence, these species can distinguish between conspecifics and prey at larger distance than Lyssomaninae and most Spartaeinae species which have a less ordered foveal layer I in the retina of their principal eyes (Harland *et al.*, 1999).

Not only invertebrates but also vertebrates, like birds and mammals, present a retinal area of maximal resolution. In mammals, for instance, this region accounts for less than 1% of the retinal area; it is structurally and functionally specialized for high resolution and stereoscopic vision, a property stemming from a high number of ganglion cells and a corresponding large brain area dedicated to processing information coming from this area (Rapaport and Stone, 1984). These animals adapt their movement to maintain their prey in this area of highest resolution.

Beyond the anatomical organization of receptive cells, there often exists a functional regionalization in the retina. The different types of photopigments are not evenly or randomly distributed on the retina, creating distinct regions differing by their composition in photopigments. This regionalization in photopigment composition, which is common in insects (Briscoe and Chittka, 2001) as well as in other groups such as birds (Hart, 2001a) or mammals (Szél *et al.*, 1996), is related to animal ecology and helps to make an optimal link between the visual organization of natural scenes and the organization of light receptors. In birds, while the relative proportions of the different photoreceptor classes reflect bird phylogeny, their distribution on the retina is associated to diet, feeding behaviour and habitat (Hart, 2001a). This distribution of photoreceptors may also be complemented by a variation in oil droplet pigmentation. Oil droplets are light filters associated to cones in birds; by narrowing cone spectral absorption domain, they enhance colour discrimination (Vorobyev, 2003) at the expenses of absolute sensitivity. Areas with reduced oil droplet concentration are likely associated with prey detection, as suggested for the sacred kingfisher (*Todiramphus sanctus*; Hart, 2001a). In dragonflies, the fovea is dominated by UV and blue photoreceptors (Olberg *et al.*, 2007). Be they perchers or hawkers, dragonflies approach their prey from below. They thus usually have to detect their prey against the sky. Incorporating receptors maximally sensitive to the wavelengths dominant in the blue sky is thus an efficient strategy to extract maximal visual information from the visual environment in which prey are usually seen. This regionalization is also rendered more efficient by appropriate hunting behaviours. When perching, dragonflies orientate away from the sun, with their fovea directed to the sky, a position that optimizes prey detection (Prete, 1999; Sauseng *et al.*, 2003). When pursuing their prey, they tend to maintain a constant angle between the prey and the horizon on the retina by moving their head in order to keep their target in the fovea (Olberg *et al.*, 2007).

Finally, neural processing of photoreceptor outputs is crucial to determine the level of visual performance. In dragonflies, the complex retinal structure is complemented by neural pathways that efficiently integrate the visual information provided by photoreceptors. Two types of neurons are dedicated to recognition of small moving features in the environment; their response is related to the brightness contrast between these features and the background (O'Carroll, 1993): moving bar detectors are orientation but not direction selective neurons which respond more strongly to isolated bars than to several bars. Small target moving detectors are orientation selective neurons which respond to targets subtending visual angles equivalent to just one or two facets of a compound eye. These insects are endowed with a visual acuity of approximately 1° (between 1° and 2° in Land, 1997; 0.8° in Nordström and O'Carroll, 2006), which is remarkable for insects. As a consequence, they do not react to background motion itself; in case they do, it is in a random way, as if they were responding to small features of the background itself. They keep responding even when the background is moving at the same velocity as the target (Nordström *et al.*, 2006). These neurons help to detect moving prey in a visual clutter and are built to compensate for predators' self-motion while hunting the prey.

2.1.4 *Weighing the role of vision relative to other sensory channels*

While some predators rely heavily on vision to detect and capture their prey, other predators use more intensively different sensory modalities like olfaction and may exert little influence on the evolution of the colouration of their prey. Even if the importance of visual information can be predicted from the visual equipment of the taxonomic group to which it belongs, different species of the same group may strongly differ in the relative importance they give to different sensory modalities. For example, some predatory heteropterans (Nabidae and Reduviidae) have been experimentally shown to depend more on olfaction than on vision to locate prey patches and orientate to prey at close range (Freund and Olmstead, 2000). Other predatory heteropterans (Reduviidae) have been suggested to depend more on the visual stimuli (colour and movement) than on the scent provided by their prey (Parker, 1969a,b). Similarly, experiments performed on closely related coccinellid species revealed that *Coleomegilla maculata* shows the same predation rate in dark and light environment and no preference for any prey colouration while *Coccinella septempunctata* prey on the most conspicuous prey and more in light than in dark environment (Harmon *et al.*, 1998). While visual cues are not important for the former, they appear essential to guide predator attack in the latter species. Lizards use mainly visual cues at large distance and a combination of visual and olfactory cues at short distance (Hasegawa and Taniguchi, 1994).

Most insectivorous bats maintain a good visual sensitivity under nocturnal illumination (Ellins and Masterson, 1974). A growing body of evidence shows that not only echolocation but also vision plays a role in prey detection and capture, with a trade-off between visual equipment and hearing performance (Zhao *et al.*, 2009). Some lineages (e.g. vespertilionids) have a moderate hearing and two different types of cones—one sensitive to UV and the other to green—and a potential colour vision. Other lineages (e.g. rhinolophids) developed a more sophisticated echolocation and showed a concurrent loss of UV sensitivity, with a vision only sensitive to light intensity (Zhao *et al.*, 2009). Bats also differ in their use of vision. The gleaning bat (*Macrotus californicus*) locates prey on the ground by vision (Bell, 1985). Aerial-hawking bats have been suggested to use the bright sky at dusk to backlight their prey (Pettigrew, 1980). Experiments show that they use visual cues to locate stationary insect prey above the vegetation. Gleaning bats that forage on the ground cannot use the sky to create a brightness contrast and enhance prey conspicuousness.

However, they seem to use vision to locate their prey (Bell, 1985) and prefer visual to sonar cues to locate them (Eklöf and Jones, 2003). For all the bats studied to date, visual cues seem to facilitate the initial prey detection but may not be involved in the terminal attack (Eklöf *et al.*, 2002; Eklöf and Jones, 2003). However, little is known about the exact visual cues bats could use for prey detection (Eklöf *et al.*, 2002). Prey colouration may not play an important role as suggested by the absence of preference for dark or light targets (Eklöf *et al.*, 2002) or morphs in the peppered moth *Biston betularia* (but see Whittle *et al.*, 1976; Majerus, 2008). However, function and mechanisms of vision in bats and its importance relative to other sensory modalities have aroused research interest only recently (Müller *et al.*, 2009) and should benefit from future studies both in behaviour and physiology.

2.2 ROLE OF VISUAL SIGNALS IN INSECT PREDATION

2.2.1 *Visual cues useful for prey detection and recognition*

Predation is characterized by different steps, namely, detection, identification, attack, capture and consumption. The mechanisms of prey detection and capture have been studied in detail for a few insectivorous predators, mainly in mantids and mantispids, dragonflies, spiders, tiger beetles and toads. Experimental for their large majority, these studies have identified some of the visual features—shape, colour, movement, texture—that influence prey detection and recognition and evaluated the importance of vision in capture mechanisms. In a recent review, Troscianko *et al.* (2009) detail the relationship between object visual features and receiver cognitive processes. Although visual mechanisms are mainly known from studies of mammalian vision, they also explain how animals with different vision perceive objects like camouflaged prey in the environment, suggesting that visual mechanisms are likely similar from a functional point of view.

Prey contour and shape are fundamental cues for detection and recognition. While colour, texture or size do not uniquely identify an animal, shape is what best identifies an animal (Troscianko *et al.*, 2009). Prey contour can be inferred from the detection of animal boundaries, that is the detection of sudden changes in light intensity and spectral content. Local edge segments are detected and then assembled in objects, the shape of which is then reconstructed by the brain (Troscianko *et al.*, 2009). Animal shape is recognized for the potential target to be considered a valid prey. Insect predators often prefer elongated (worm-like silhouette) targets [mantis (Yamawaki, 1998); toad (Ewert, 2004)]. From a series of experiments conducted in the praying mantis (*Mantis religiosa*), the importance of edge in prey detection and probability of attack were determined. Praying mantis presented with objects at close distance preferentially strike at edges (Poteser and Kral, 1995) and direct more attacks at squares than discs or triangles of equal surface (higher edge to area ratio in squares compared to discs Hyden and Kral, 2005).

Prey size is also crucial for detection. Larger objects are generally more detectable than smaller objects. This increased detectability can translate directly in a higher predation risk. For instance, this is the case for birds that attack larger larvae more often than small ones, as shown by Remmel and Tammaru (2009) with field experiments involving birds preying on artificial prey items. For these large predators, larger prey are not only more detectable but also more valuable in terms of energetic content.

Prey colouration is important for detection. Without entering into too much detail, we want to underline the importance of both chromatic and achromatic components of colouration in prey detection. Detection of small (subtending a small angle on the retina, small in size and/or at long distance) or moving targets usually relies on the detection of brightness contrast [fish (Schaerer and Neumeyer, 1996); insect (Giurfa *et al.*, 1997); bird (Osorio *et al.*, 1999)]. Chromatic contrast also plays a role in detection. The relative importance of chromatic and achromatic information can depend on background complexity (example of fruit detection in birds, Schaefer *et al.*, 2006) and on the relative achromatic and chromatic contrasts offered by the prey against the background (example in blue tits, Stobbe *et al.*, 2009) and on the colour content of the prey itself.

Visual cues indispensable for detection and recognition are processed while additional information is overlooked in some cases. For instance, the grasshopper mouse uses visual cues mostly for prey detection. Once the prey is detected, visual cues allowing a precise recognition of specific species seem to be overlooked in the decision to attack; this may be due to the generally limited vision of these nocturnal animals which forage in low light conditions (Langley, 1989). Another extensively documented example is that of jumping spiders which are fundamentally guided by vision. Jumping spiders feeding on both spiders and insects generally show a preference for spiders (Jackson and Li, 2004). They do not seem to analyze the entire prey shape or contour but more to concentrate on specific visual cues that rapidly inform them about the validity and profitability of the prey. Optical cues, like the location or shape of prey eyes, are sufficient for these predators to make the difference between spiders and insects or between conspecifics and prey (Harland and Jackson, 2000, 2002; Jackson, 2000; Nelson and Jackson, 2007).

Finally, different predators attend to different visual cues to detect and recognize their potential prey. Training pigeons to differentiate wasps from various species of mimetic hoverflies, Dittrich *et al.* (1993) showed that some species that appeared poor mimics to human eyes were judged to be good mimics by pigeons. While humans appreciate hoverflies' mimicry on colour pattern mainly, pigeons use a more varied set of visual features to distinguish between different types of prey (Bain *et al.*, 2007). The set of visual cues used by predators is flexible and depends on predator's past experience (Mostler, 1935; Bain *et al.*, 2007). Mostler (1935) gave the first clear evidence for that flexibility in his well-designed large-scale experimental approach to testing the theory of mimicry. He showed that inexperienced birds of several insectivorous species attacked three noxious hymenoptera (wasps, bumblebees, honeybees) as well as their hoverfly mimics. Yet, they would rapidly learn to reject the noxious species and avoid attacking their mimics if offered after an encounter with a noxious model. Hoverflies gained their protection from their colour pattern, a visual feature to which naïve and experienced birds did not give the same importance. Similarly, in Bain *et al.*'s (2007) study, pigeons rewarded wasps used colour attributes (number of stripes, stripe colour) when asked to avoid flies while pigeons rewarded flies used mostly antennal length and also colouration to distinguish them from non-rewarding wasps.

In general, any enhancement of the differences in spectral content or intensity between the insect and the background facilitates detection (Troscianko *et al.*, 2009). For example, Remmel and Tammaru (2009) have shown that bird preference for larger insect larvae is accentuated for conspicuous prey. Because this preference is independent of prey colouration, the authors suggest that detectability more than acceptability is important in determining size-dependent predation risk in this case. Similarly, a high brightness contrast elicits a higher attack probability in the praying mantis (Yamawaki, 2000). Conversely, any increase in similarity in intensity or spectral content between the insect and the background, the production of high-contrast internal detail (internal markings) more salient than the edge or false edges likely reduces detection probability. The enhancement may be more efficient for some prey–background combinations: prey configuration (shape, size and movement) being defined, toads respond more to a black prey against a white background than to the reverse (Ewert *et al.*, 1982), probably because neurons display different responses to abrupt darkening and brightnening of the visual field (Ewert, 2004).

Prey motion is also an important visual cue for detection and attack. Motion helps to segregate objects from similarly textured backgrounds (Srinivasan *et al.*, 1999) and generally facilitates detectability. In toads, motion is crucial: an object must move to be categorized as prey (Ewert, 2004). More specifically, motion is analyzed in relation to prey size and shape. Elongated objects moving in the direction of their long axis are preferred while the same objects moving in the direction of their short axis are categorized as non-prey; this preference does not change with prey speed (Ewert, 2004). In the praying mantis, the probability of strike increases when prey motion increases (Iwasaki, 1990). Similarly, lizards prey more on moving prey than on other less active but equally palatable insects (Civantos *et al.*, 2004). Grasshopper mice attack more moving than stationary prey (Langley, 1989). In chironomid larvae, immobility decreases predation risk by sticklebacks *Gasterosteus aculeatus*, but only in cryptic prey (Ioannou and Krause, 2009). Like edge detectors, motion detectors are neural structures that are sensitive to colour and brightness. Increased contrast between target and background facilitates detection of moving objects in the environment (Ewert *et al.*, 1982; Livingstone and Hubel, 1988; Ewert, 2004). Conversely, motion camouflage (detailed in a subsequent section) can efficiently reduce detection probability.

2.2.2 *Visual cues informing about prey profitability—predator preferences*

Even if the prey is recognized as a valid prey type, it may not be attacked depending on the level of profitability expected from the attack. A prey item can be unprofitable if the attack is likely not successful, if the attack and the consumption requires more energy than the energetic content expected from the prey, or if the prey is unpalatable. Several visual features, like prey size, movement, colouration, are used by predators to determine if they should attack a potential target. This decision is subject to speed–accuracy tradeoffs (Chittka and Osorio, 2007) which are crucial in shaping the evolution of both prey appearance and predator behaviour.

First, insect size can inform on attack profitability. Although larger insects are readily more detectable, the increased detectability may not translate into an increased predation risk. If adult birds prefer large prey (Remmel and Tammaru, 2009), small or inexperienced birds prefer medium size prey over large prey because they require less handling (Moreby *et al.*, 2006). Similarly, for predators of smaller size, there exists an optimal prey size above which predation becomes less profitable—higher chance of escape for the prey, longer time for prey handling or consumption. This of course depends on the type of prey and may not hold for slow-moving insects like caterpillars. For instance, grasshopper mice prefer to attack smaller prey (Langley, 1989). Optimal prey size increases with predator size in dragonflies (Olberg *et al.*, 2005). The praying mantis pays special attention to prey volume and area but prey volume is the most important for attack decision; they attack in average smaller prey than those they pay attention to in the first place (Iwasaki, 1990) and from a large range of prey size, they prefer intermediate size (Iwasaki, 1990). Similarly, odonats presented with prey of various sizes prefer small prey (Shelly and Pearson, 1978; Rashed *et al.*, 2005).

Second, prey movement can inform on prey profitability. Palatable and unpalatable prey exhibit different flight patterns, speed and movement predictability (review in Sherratt *et al.*, 2004), all cues that can be used by predators to assess potential target profitability. These aspects will be studied in more detail in a subsequent section of this review.

Third, prey colouration can also inform the predator about the level of prey profitability. If some colours are associated with high levels of profitability while others are associated with low levels of profitability, relative preferences or avoidances for specific colours or colour patterns can evolve in the predators. At the same time, these preferences may shape the evolution of colours and patterns in insects. Attending to visual cues may be economical for predators and save them from un- or poorly rewarding attacks. Specific sections of this review will be dedicated to warning colours seen by predators or to specific cognitive processes of prey recognition (image search). Here, we want to put forward some interesting aspects of colour avoidance or preference, namely the universality of the information content of prey colouration and the role of the background colouration in the construction and the expression of a preference or an avoidance for a specific colour.

Predators can prefer or avoid specific colours and patterns. This has been extensively studied, mostly in birds (references in Ruxton *et al.*, 2004), but also in other insect predators such as insects or lizards. Most insectivorous species show unlearned avoidance for red relative to other colours such as brown, green or grey, which are preferred (Sillén-Tullberg, 1985b; Roper, 1990; Moreby *et al.*, 2006; Gamberale-Stille *et al.*, 2007). Not only colours but also patterns can elicit preferences or avoidances. Black and yellow patterns do not generate any avoidance in various insectivorous species, as shown by the high predation observed on striped hymenoptera and hoverflies (Mostler, 1935). Conversely, black and yellow patterns inhibit prey attacks in naïve domestic chicks, even for their first prey eaten (Schuler and Hesse, 1985). Aversion is not shown when black or yellow is presented separately or adjacently but not in stripes (Schuler and Hesse, 1985). A similar process can be seen in mantids (*Tenodera aridifolia*) for which black or orange solid colours are less aversive than black and orange striped patterns (Bowdish and Bultman, 1993).

The informative content associated to a given colour applies to different prey types. For instance, partridge chicks preying upon different insect species—heteroptera, coleoptera, orthoptera, hymenoptera—show a general preference for yellow-green colours over black and brown (Moreby *et al.*, 2006). However, the informative content associated to a colour has by no means a universal value. First, it is prey-type dependent. For instance, birds' avoidance of red is applicable only to insects or insect-like elongated objects while it does not exist for fruit or fruit-like round objects for which red is often preferred (Gamberale-Stille and Tullberg, 2001). Similarly, dragonflies show an aversion for yellow-black striped patterns for flies but not for wasps (Kauppinen and Mappes, 2003). Second, the informative content of a colour is also predator-type dependent. Avoidance of yellow-black striped patterns is present in a large range of insectivorous bird species but it is absent in non-insectivorous birds (Dittrich *et al.*, 1993) and turns into a preference in bee-eaters (Koenig, 1950).

Aversion or preference for specific colours is maintained regardless of the colour of the background against which these colours are presented, as long as the background colour does not significantly interfere with prey detectability (Sillén-Tullberg, 1985b; Roper, 1990; Moreby *et al.*, 2006). Associating a colour or a pattern *per se* and not a contrast between prey and background to the level of prey profitability is likely an efficient strategy since it can compensate for the spatio-temporal variations in contrast between the prey and the background. It may be particularly advantageous if prey frequent different environments, or if predators feed on a variable or large range of prey. Background colour has little effect at the time when aversion/preference is expressed, but it can have a significant effect on building, reinforcing or deactivating preferences. First, familiarity with a colour, presented in the background and not in the food itself, increases the preference for this colour (Roper, 1990) or can convert a natural aversion into a preference, as shown for chicks reared in environments presenting black and yellow elements (Roper and Cook, 1989). Second, the visual environment affects the salience of colour for young birds, probably in relation to changes in relative attractiveness or memorability. For example, chicks are more easily trained to respond to orange than to blue, although they show no innate preference for either colour (Miklósi *et al.*, 2002). Third, if background colour can increase familiarity, it has little power to convey information about prey palatability, as shown by experiments on unpalatability learning (Gamberale-Stille and Guilford, 2003). For domestic chicks, learning prey unpalatability is faster when prey colour and not background colour can be used as an indicator of this unprofitability. When both cues are available, prey colour is used as the only indicator of prey unprofitability. Moreover, learning is not possible when the only cue available is the level of contrast between the prey and the background and not colours *per se* (Gamberale-Stille and Guilford, 2003). Avoidance of unpalatable prey is learnt faster for more conspicuous prey (Harvey and Greenwood, 1978).

Predators do perform generalization based on object colouration. The experience of an association prey colour—prey profitability can be generalized to different prey types or to different colours. Such cognitive flexibility likely helps predators to adapt their foraging strategy to changing prey. First, all objects of a particular colour previously associated to a level of profitability can be assumed to bear the same profitability level. Although aversion to red is restricted to specific targets, preference for brown is generalized to all brown objects when birds are fed on brown food (Roper, 1990). Second, generalization can be applied to colours, especially in the case of novelty. For instance, birds conditioned to associate a colour to a reward can be exposed to a novel colour associated to unpalatable prey. They learn to avoid this prey based on its colour and avoid any further novel colour, showing that a generalization was made based on novelty (Schlenoff, 1984). In domestic chicks, exposure to novel food is sufficient to deactivate neophobia. Deactivation is similar regardless of the number of novel colours to which animals are exposed and is obtained even with a brief exposure (Marples *et al.*, 2007). Yet, dietary conservatism (reluctance to incorporate novel food in a diet) is stronger than neophobia (reluctance to taste novel food) and reactivation is always much more easily achieved than deactivation (Marples *et al.*, 2007). This limited flexibility in terms of reaction to colours, coupled to memorability, increase foraging efficiency by fostering safer and faster decision making.

Decision of attack can be shaped by visual information borne by insect size, colour and pattern. However, the decision can be made by integrating not only information coming from vision but also from other sensory channels as well. Communication—not only between conspecifics but also between prey and predator—is fundamentally multimodal, as shown by a growing body of evidence (Partan and Marler, 2005). For instance, lizards rely on visual cues to decide which prey to pay attention to and approach; yet, unpalatable insects camouflaged into visually palatable insects are approached but not attacked, which suggests that information about palatability can be provided by visual cues but is also accessible through olfactory cues (Hasegawa and Taniguchi, 1994).

2.2.3 *Role of visual input in capture mechanisms*

2.2.3.1. Visual estimation of distance to prey To successfully strike their prey, predators have to reliably estimate distance to them for stationary targets, and visual angle, angular velocity and distance for moving targets. For predatory insects, estimating distance is more difficult than for vertebrates. Insects have their eyes fixed relative to the head, and their compound eyes are not capable of focusing mechanisms, which could provide an indicator of distance. There are still three different mechanisms for distance estimation: image size, stereopsis and motion parallax. For an object of known size, image size measured by the number of photoreceptors excited on the retina can give a direct indication of the distance to the object. This mechanism may be used to estimate distance to conspecifics (Olberg *et al.*, 2005) but is likely of restricted relevance in the case of prey of variable size. In the case of stereopsis, the distance to an object is encoded by the difference in retinal position between right and left eyes. For insects, the amount of visual information that can be extracted by binocular vision is limited by the small head and the close distance between the eyes which reduce the information provided by the angular disparity of view between the eyes. Extraction of visual information about distance is thus mainly restricted to short distances. For example, the dragonfly *Libellula* sp. is able to accurately discriminate distance up to 1 m but models have shown that stereopsis can only contribute to determinate distance up to half a meter (Olberg *et al.*, 2005). The major mechanism used for distance estimation is motion parallax. In that case, distance is encoded by the movement velocity of its retinal image caused by the self-motion of the predator. The image of a close object is displaced faster than that of a distant one. This information does not require any binocular input. Experiments have shown that this mechanism is widely used in a large variety of insects, such as tiger beetles, mantids and mantispids or dragonflies (Poteser and Kral, 1995; Toh and Okamura, 2001 and references therein; Olberg *et al.*, 2005). For instance, the praying mantis performs peering movements, that is lateral movements of the head along a line, to estimate its distance from stationary targets and the appropriate jump distance (Poteser and Kral, 1995). Peering characteristics depend on the visual environment from which to extract visual information. For instance, *M. religiosa* lives in grassland where background elements are uniform, aligned and closely spaced. The basic peering movements it performs (Kral and Devetak, 1999) have been experimentally shown to be most efficient against horizontally or vertically striped background (Poteser and Kral, 1995). In this species, peering velocity changes with background and peering amplitude increases with prey distance. Conversely, *Empusa fasciata* which lives in habitats of shrub consisting of irregularly, variably aligned elements, translates its entire body sideways and performs back–forward movements resulting in peering of higher amplitude adapted to extract visual information from a visually more complex environment (Poteser and Kral, 1995).

2.2.3.2. Visual input during prey pursuit Pursuing moving targets is also a complex visual challenge for predators. Target detection poses problems to moving predators; the image of the prey is degraded by the relative angular velocity of the moving predator and the prey (Gilbert, 1997). Different strategies have evolved to overcome this prey localization problem. After locating their prey, some vertebrates like toads close their eyes during their fast movement to approach their prey (Lock and Collett, 1979). Suppressing any visual input may be more advantageous than coping for distorted visual information.

In some cases, visual input of the position of the prey is not even needed. This is the case of the archer fish. These fish shoot water at insects flying above the water surface and they catch their dislodged prey exactly where it falls into the water. Distance to the prey is estimated from monocular cues with an extreme precision (error of 1 mm at 80 cm), with a correction for optical distortions due to air–water interface (Schuster *et al.*, 2004). Not only memory of past experience but more likely the extrapolation of laws (concerning the changes of apparent size with distance) seem to be at play for distance estimation (Schuster *et al.*, 2004). They only need information about direction speed and prey distance in order to 'calculate' the point of incidence of the target on the water. These cues can be extracted during the first 0.15 s following target strike; the animals do not perform any visual feedback afterwards. This absence of visual feedback can be easily deduced from the unchanged ballistic trajectory reaction of individual fish towards targets for which the post-strike trajectory has been experimentally changed (Rossel *et al.*, 2002).

Motion is a real challenge for prey detection and pursuit. Because photopigments have a finite integration time, the retinal image is subject to motion blur (just as in photography) when eye and surroundings move relative to each other (Land, 1997). This occurs more strongly at high angular velocities and insects chasing their prey have evolved mechanisms to compensate for image blurring. For instance, tiger beetles are tracking predators. They often track their prey by making regular stops and goes; stops help them to relocate the prey and minimize image blurring. Experiments have shown that for some visual combinations—high visual contrast between prey and background, large prey size—tiger beetles perform tracking without stops, suggesting that visual feedback and compensation for blurring may be costly and is avoided when not indispensable for successful capture (Gilbert, 1997).

Dragonflies have evolved capture tactics which are interesting to relate to the evolution of their visual performance. First, while most insects track their prey basing their own movement to the current position of their target, dragonflies intercept their prey. Be they sit-and-wait predators or hawkers, they fly in a relatively straight line which intersects the projected flight path of their target. This strategy is likely economic in terms of visual feedback. These animals have evolved a highly developed visual system. They have evolved behaviours that make an optimal use of their visual equipment. They keep their target in their fovea—region of highest visual acuity—which is optimized to detect insect prey moving against the blue sky (Sauseng *et al.*, 2003). Their highly performing retina is relayed by neurons specialized in the detection of small features in the environment such as moving prey. These neurons are primarily sensitive to brightness (luminance) contrast and they show a maximal response for a high brightness contrast between prey and background (O'Carroll, 1993), which is highly beneficial to backlight dark prey against the light sky. These neurons are built to respond to a moving prey showing no difference in velocity compared to the background (Nordström *et al.*, 2006) which is advantageous for predators that are moving when chasing their prey.

2.2.4 *Prey visual cues and predator versatility*

Predators have evolved flexible capture tactics which are adapted to prey visual ability and behaviour. Jumping spiders stand probably as one of the best examples of this predator versatility (Curio, 1976; Harland and Jackson, 2004). Jumping spiders of the genus *Portia*, which feed on spiders and insects, adopt a cryptic stalking behaviour towards salticids: they freeze when faced by their prey and walk with their palps back. Conversely, they adopt a normal stalking behaviour towards prey other than salticids, not freezing when faced nor holding their palps back (Harland and Jackson, 2000). Visual cues that determine the type of stalking adopted is firstly the presence of eyes typical of spider principal eyes (Harland and Jackson, 2000). Normal stalking is chosen when eyes are absent, reduced or square-shaped; cryptic stalking is chosen for normal or enlarged eyes or when one eye only is visible (Harland and Jackson, 2002). Spiders adopt an ambivalent behaviour for spider-mimicking beetles or ants (Harland and Jackson, 2001, 2002), suggesting that the eyes may not be the only prey feature used by spiders to decide which capture tactic to adopt. Not only the type of stalking but also the decision to engage an attack depends on the prey and on the visual cues offered by the prey. Jumping spiders of the genus *Phaeacius* engage an attack towards salticids only if they are camouflaged from their prey by a cryptic background and if their prey is not facing them (Li *et al.*, 2003). For the other types of prey, the probability of attack does not depend on background colouration or prey orientation, suggesting that prey visual performance is not sufficient to select for more complex predation behaviours.

Predators also adapt their behaviour according to prey escape ability. When they chase prey with a low escape potential like Thysanoptera or lepidopteran larvae, salticid spiders approach their prey frontally and release them after venom injection. Conversely, when they chase prey with a high escape potential like homopteran or orthopteran species, they orientate sideways, make a rapid movement masking approach and do not release their prey after venom injection. Similarly, the praying mantis adapts its approach behaviour to the amount of prey motion (Yamawaki, 2003).

3 Methods to investigate insect colouration

3.1 MEASURING COLOURATION

Most studies of insect colouration, especially those conducted before the 1990s, involve a qualitative measurement of insect colouration, mainly through scores or human vision-based colour categorization, like 'green', 'brown' or 'black' (e.g., Kettlewell, 1955a; Sandoval, 1994; Harmon *et al.*, 1998; Civantos *et al.*, 2004; Hochkirch *et al.*, 2008). This categorization did not only apply to insects, targets or fruit items in general but also to light and visual background. At first, such a categorization may seem convenient and fairly robust, especially for animal colouration. Indeed, natural pigments and structures show a limited variation around typical reflectance spectra (e.g. pigments colouring birds' feathers; Burkhardt, 1989). For instance, 'green' colours typically have a main reflectance peak in the green range, with only secondary peaks in other regions like UV. Categorization and reflectance measurements are likely congruent in this case, making categorization a valid surrogate of reflectance spectrum, as least for species more sensitive to green than to UV wavelengths. Nevertheless, colour categorization presents numerous pitfalls (list in Endler, 1990). Scoring a colour patch is a subjective assessment; because perception is based on comparison, the scoring of a colour patch can be affected by adjacent colour patches. Similarly, it is subject to fluctuations due to lighting conditions in which scoring is performed. Finally, it is based on human vision. Colour categorization reduces variation to the human range of sensitivity, excluding regions like UV to which insects and most of their predators are sensitive (Table 1). Moreover, it ignores metamerism, that is, that several wavelength distributions can produce the same visual impression. Metamerism is common for human manufactured objects (backgrounds in experiments) and artificial lights. Yet, two colours seen as metameres by humans are likely not metameres for a different visual system. Finally, making categories implies gathering colours into groups: two colours of the same category—two greens for instance—are perceived as more similar in colouration than two colours of different categories—one green and one yellow—even if both pairs generate equal difference in photoreceptor responses. Location in wavelength of boundaries between groups, and the amount of distortion of photoreceptor responses depend on the visual system. Ideally, categorization should only come as a complementary tool to help the researchers to interpret their results or to make their message more pedagogical to a general audience but not as a primary and unique colour measurement.

Given that insects and insectivorous predators (perhaps with the exception of some insectivorous primates) have visual systems that differ from ours in terms of range of sensitivity, photopigments and visual information processing, it is essential to objectively measure colouration (Endler, 1990; Bennett *et al.*, 1994; Cuthill *et al.*, 1999). This can be done by the means of spectrometry, a powerful method which can easily be adapted to the range of sensitivity of interest by using appropriate lamps, spectrometers and optic cables. Spectrometry provides spectral data which are objective and precise quantitative measurements. Reflectance spectra give the proportion of the incident light that is reflected by the colour patch (insect colour patch or background colour patch) while irradiance spectra give the light intensity available at each wavelength (see extensive and argumented practical help in Endler, 1990; Endler and Mielke, 2005). All spectral data are based on photon counts. This approach has been successfully applied to insect colouration (e.g. Stobbe and Schaefer, 2008; Théry *et al.*, 2008; Stobbe *et al.*, 2009).

Another way of measuring colouration is to use video or photography. Photography can be particularly useful to quantify the relative importance of colour patches within a pattern or to characterize the spatial relationships between different colour patches (shapes, transitions, patch homogeneity), even about temporal changes in colour aspects (Stevens *et al.*, 2007b). Compared to spectrometry, photography is often viewed as a convenient and inexpensive means to rapidly collect large quantities of data. Nevertheless, different adjustments are needed if photography is used to investigate how insect colours are seen by potential predators.

Stevens *et al.* (2007b) provide a useful help guide to accompany potential users in all needed correction steps. They list the important technical characteristics to pay attention to when choosing a camera, they explain the calibration protocol and the process to linearize camera's response to light intensity, how to correct the response of RGB channels to extract information about brightness. Finally, they detail how to transform data into camera-independent quantitative measures of colouration to compare measures from different cameras or to incorporate the data into models of animal vision. They draw attention to two main problems that are particularly relevant for studies of insect colouration. First, most cameras exclude the UV range to which humans are blind but to which most insect predators are sensitive. It is not difficult to do UV imaging (see application for the study of flower colouration, Kevan, 1972) and it may be a useful complement to classic photography (Stevens *et al.*, 2007b). Second, information delivered by the camera's sensors is biased towards specific wavebands which spacing is a closer approximation to visual systems (like that of birds) with regularly spaced sensitivity than to human's vision. Working not on the RGB responses but on linear combinations of these responses may be useful to adapt to specific visual systems (example of mapping in Stevens and Cuthill, 2006).

3.2 BUILDING ARTIFICIALLY COLOURED STIMULI

Investigating how predators see insect colours and identifying the visual cues most relevant to predators requires an experimental approach. This imposes the manipulation of specific visual aspects of the prey (colour, shape, contrast, pattern, size) and/or the background to test their separate effect on predators' reaction. This can be done by using dead versus live prey (e.g. Eklöf *et al.*, 2002) artificially painted targets or manufactured objects (e.g. Bowdish and Bultman, 1993; Gilbert, 1997; Cuthill *et al.*, 2005; Hyden and Kral, 2005; Marples *et al.*, 2007; Wennersten and Forsman, 2009) and computer images or pictures (e.g. Dittrich *et al.*, 1993; Bond and Kamil, 2002, 2006; Bain *et al.*, 2007). Background is often 'manipulated' simply by performing laboratory experiments and choosing white, grey or black background (Harland and Jackson, 2000; Yamawaki, 2003) or by constructing computer images of visual structures (Bond and Kamil, 2002, 2006).

These manipulations—for example, painting targets, taking pictures, or setting ambient light and background—are often done without controlling visual conditions created for the experiments, which may compromise the validity of the results observed. As an example, Harmon *et al.* (1998) tested predators' reaction to different combinations of natural red and green prey morphs against artificial green and red backgrounds. They assumed that a red target was cryptic on a red background, which is not a trivial assumption. Paints are based on colourants that highly differ in terms of reflectance from natural colourants. Since they largely call for metamerism in human vision, paints likely appear really different to a different visual system, especially under artificial illumination. A red target may show a visual contrast for the tested predator and not for humans. Ignoring the receivers' view of the combinations target background may lead to an incorrect interpretation of the results. Measuring both object and background colouration as well as ambient light is indispensable to conduct thorough experimental tests of the role of insect colouration on predation.

Using videos or images present specific problems. As we saw above, computer screens and cameras are built to render objects as humans perceive them in nature. As a consequence, they likely fail to render insect colours as most predators perceive them in nature. Videos or image colours should be modified to match receivers' vision if they are used for the purpose of testing vision (Fleishman *et al.*, 1998). This manipulation consists in a change of the RGB combination for each colour patch independently and require the knowledge of receiver's vision, object natural colouration and natural illuminants (Fleishman *et al.*, 1998) as well as the response of display device to light intensity (Stevens *et al.*, 2007b). Again, there are two limitations: since videos do not include UV, they should be used in conditions where UV is not relevant—receiver insensitive to UV, prey showing no reflectance in the UV, light conditions rendering inefficient the exploitation of the UV range. In addition, videos create a gradient of light intensity and should be primarily dedicated to explore vision in diurnal predators.

3.3 ANALYZING COLOURATION DATA

Once colouration is measured, data have to be analyzed to extract biologically relevant visual information about achromatic and chromatic aspects of colouration. At this stage, two main options are available for quantitative data.

3.3.1 *Characterization of spectral shape*

Relevant characteristics of spectral shape can be extracted using different methods. They have in common to require information about the range of wavelengths to which the study animal is sensitive. In addition, all these methods assume that all wavelengths are treated equally by the visual system, an assumption convenient for computation but obviously violated by all biological systems. Such calculations can be performed directly on reflectance spectra but they can advantageously be performed on radiance spectra. Radiance is the fraction of the ambient light that is reflected by the object and that reaches the eye of the receiver, that is, the product of irradiance by reflectance. Such calculations are particularly interesting in the absence of knowledge of the visual system of the species of interest (or a closely related species with similar visual system).

A first possibility is to perform a principal components analysis (Cuthill *et al.*, 1999). This kind of analysis presents severe limitations. It assumes independence of observations while reflectance at different wavelengths is correlated over large bands of wavelengths. In addition, outputs are highly sensitive to the set of colours analyzed, and decomposition between achromatic and chromatic components has to be forced to be biologically relevant (see Endler and Mielke, 2005 for discussion about the limitations).

A second possibility can be a direct computation of parameters that characterize achromatic (brightness) and chromatic (hue and chroma) components of colouration. Brightness can be computed as the average reflectance/radiance over the total range of sensitivity (Endler, 1990), hue (colour in its common sense) as the location in wavelength of the maximal reflectance/radiance or most important change in reflectance and chroma (colour purity or saturation for pigment-based colouration) as a ratio to estimate the prevalence of a certain range of wavelengths (examples in studies of bird colouration: Örnborg *et al.*, 2002; Doutrelant *et al.*, 2008). This approach has major limitations. First, the simple and 'all-purpose' formulas to obtain these parameters lead to the incorrect, although tempting, consideration that these parameters describe inherent properties of an object. Brightness, chroma and hue are perceptual terms and as so, they should be fed by a thorough understanding of perception to be relevant for a particular species. Conversely, the simplistic formulas ignore important perceptual mechanisms. For instance, computing hue as maximal reflectance or as many chroma values as peaks in a spectrum ignores the fact that chromaticity comes from the comparison between different wavebands. Similarly, brightness perception is assumed to be uniform over the range of sensitivity while it is not. These limitations should invite researchers to make effort and feed more biology and physiology into colour data.

3.3.2 *A more integrated approach: analysis using 'physiological' models*

To date, different 'physiological' models are available in the literature, among which the most popular and conceptually interesting are Chittka's colour hexagon (Chittka, 1992), Vorobyev and Osorio's discriminability threshold model (Vorobyev and Osorio, 1998) and Endler and Mielke (2005) model for analyzing colour patterns. The aim here is neither to present the models in detail nor to compare them in terms of efficiency for studying particular signals. However, it is interesting to comment on several points.

All visual information comes ultimately from the capture of photons by photoreceptors in the eye. All models rely on photon capture as the first step (Eq. (1)). Applying such equations to data collected from digital photographic equipment and corrected to be camera independent can help to analyze insect colours using models of predator vision (Stevens *et al.*, 2007b).

$$\mathrm{qc}_i^{\mathrm{object}} = \int_{300}^{700} L(\lambda)S_i(\lambda)\mathrm{d}\lambda \qquad (1)$$

where $L(\lambda)$is the spectrum of the light entering the eye and $S_i(\lambda)$ is the spectral sensitivity of the photoreceptor *i*. The number of types of photoreceptors determines the number of quantum catches values obtained. The neural response of the photoreceptor *i* is not a passive electric transfer of the photon catch but it increases non-linearly with light intensity and reaches a maximum (eq. (2), also called Michaelis–Menton equation applied to photoreceptor response). The coefficient *k* (also called von Kries coefficient) is the reciprocal value of the photon flux evoking a response that is half the maximal response of the photoreceptor (Eq. (3)). This coefficient considers the fact that a photoreceptor gets adapted to its light environment.

$$V_i^{\mathrm{object}} = V_i^{\max} \frac{\left(k\mathrm{qc}_i^{\mathrm{object}}\right)^n}{\left(k\mathrm{qc}_i^{\mathrm{object}}\right)^n + 1} \qquad (2)$$

where

$$k = \frac{1}{\mathrm{qc}_i^{\mathrm{o}}} \tag{3}$$

V as a function of $\ln(\mathrm{qc})$ is a sigmoid, where n represents the slope of the curve of response of the photoreceptor ($n = 1$ for many species). Photoreceptor's adaptation to light is a dynamic and reflexive process so that it is possible to consider that the mean photon flux of the field of view becomes $\mathrm{qc}_i^{\mathrm{o}}$.

Taken alone, a photoreceptor only records information about light intensity level (brightness) but cannot perform any wavelength discrimination. For objects, the spectrum of the radiant light can be seen as the proportion of the ambient light that is reflected by the object, that is the product of $R_{\mathrm{object}}(\lambda)$ the reflectance spectrum of the object viewed by $I(\lambda)$ the irradiance spectrum of the ambient light illuminating the object. A visual system with n types of different photoreceptors can exploit information coming from n different independent photoreceptors. It can thus be represented by a colour space with n dimensions (Wyszecki and Stiles, 1982; Kelber *et al.*, 2003; Endler and Mielke, 2005). Being the principle of photon catch, information about brightness is accessible to all visual systems, even the simplest ones containing only one photoreceptor type. It can be seen as one dimension of the receptor space. The remaining $n - 1$ dimensions constitute the chromatic space. For two, three and four photoreceptors, the chromatic space can be represented by a line, a triangle and a tetrahedron, respectively.

Chittka (1992) proposed a model for trichromatic bee vision—the colour hexagon—which relies on simple assumptions. This model takes the photoreceptor responses to build a space that can be regarded as representing colour opponent relations. The model built as a stimulus space and not a perceptual space has been widely used to study how insects view flower (e.g. Chittka, 1996) or animal (e.g. Théry and Casas, 2002) colouration and can be used for exploring colour patterns. Nevertheless, distances are not meant to be perceptual distances. In addition, some options of calculation of receptor excitations make statistical tests of differences between sets of colours more complex than in a fully developed colour space (see Endler and Mielke, 2005 for discussion).

A perceptual perspective has been chosen by Vorobyev and Osorio (1998) in their discriminability threshold model. Considering that visual performance is primarily limited by errors in photoreceptor responses, they aimed to inform about whether a given difference in colouration between two objects translates into a perception that these objects are different in colouration. As in earlier perceptual models of animal vision (model for bee vision, Backhaus and Menzel, 1987), Vorobyev and Osorio considered that colour was coded by opponency mechanisms and that discrimination was limited by noise (but see differences in Vorobyev and Osorio, 1998). Their model has been validated as successfully predicting discrimination thresholds in di, tri and tetrachromatic species in bright illumination (Vorobyev and Osorio, 1998). Modifications of receptor noise computation allowed to extend its application to conditions of restricted photon captures (dim illumination or dark objects Osorio *et al.*, 2004). In this model, the chromatic distance ΔS between two objects A and B can be expressed for a trichromatic visual system as:

$$\begin{aligned}\Delta S^2 = & \left(e_1{}^2(\Delta f_3 - \Delta f_2)^2 + e_2{}^2(\Delta f_3 - \Delta f_1)^2\right. \\ & \left.+e_3{}^2(\Delta f_1 - \Delta f_2)^2\right)/\left((e_1e_2)^2 + (e_1e_3)^2 + (e_2e_3)^2\right)\end{aligned} \quad (4)$$

The terms e_i refers to the error associated to the type of photoreceptors i and Δf_i the differences in responses of the photoreceptor i for objects A and B.

$$\Delta f_i = \ln\left(\mathrm{qc}_i^{\mathrm{A}}\right) - \ln\left(\mathrm{qc}_i^{\mathrm{B}}\right) = \ln\left(\frac{\mathrm{qc}_i^{\mathrm{A}}}{\mathrm{qc}_i^{\mathrm{B}}}\right) = \ln\left(\frac{\int_{300}^{700} R_{\mathrm{A}}(\lambda)I(\lambda)S_i(\lambda)\mathrm{d}\lambda}{\int_{300}^{700} R_{\mathrm{B}}(\lambda)I(\lambda)S_i(\lambda)\mathrm{d}\lambda}\right) \quad (5)$$

Endler and Mielke (2005) made an interesting link between eqs. (3) and (5). In eq. (3), the fact that a photoreceptor adapts continuously to the average light in its field of view results in the fact that its response is always in the linear part of the sigmoid curve, near the half maximal response. As a consequence, a good approximation of eq. (2) can be $V_i^{\mathrm{object}} \cong \mathrm{constant}\cdot\ln(\mathrm{qc}_i^{\mathrm{object}})$. It is then easy to see that the logarithm expression in eq. (5) reflects the assumptions that a photoreceptor shows a nonlinear response with light intensity, saturation and adaptation to light and that Δf_i reflects a difference in photoreceptor response between two different stimuli.

In eq. (4), photoreceptors' responses are weighed by the confidence (related to the error e) associated to each photoreceptor type and the colour space resumes to a scale of ΔS values. Equal values of ΔS correspond to equal confidence surfaces, thus becoming perceptual distances which are expressed in JNDs (just noticeable differences). Equation (4) is a signal-to-noise ratio which describes the confidence given to visual information in its totality. The value of 1 JND is often taken as the discrimination threshold: distances below 1 indicate objects similar in colouration while values above 1 indicate objects that are perceived as different in colouration. This model performs well at predicting what happens around discrimination thresholds and stands a useful tool to determine for instance which spectral tuning of photoreceptors would offer maximal discrimination of a group of natural objects (visual background features in Chiao *et al.*, 2000) that could be insect prey.

Nevertheless, this model presents two major limitations. First, it assumes that the perceived magnitude of suprathreshold difference is proportional to the minimum number of JNDs separating two colours. In fact, the model extends the assumption made for threshold stimuli to suprathreshold stimuli, a hypothesis that cannot model cognitive processes such as colour categorization. Formulated in terms of JNDs, colour categorization is the fact that two colours of the same category are perceived as more similar than two colours of different categories even if both pairs show equal dissimilarity in terms of JNDs. Colour categorization has been found in several species which belong to insectivorous groups such as birds, insects, primates or fish (references in Ham and Osorio, 2007) and is likely a general cognitive process. Second, this model is designed to compare only two colour patches at a time. Comparing sets of colours (like colour patterns) requires building multiple pairs of comparisons. Results are difficult to interpret perceptually given that chromatic distances tell nothing about the relationships between colour elements nor about how they differ and if they concern the same wavelength regions of not. Some attempts have been made to circumvent this problem (e.g. Håstad *et al.*, 2005) but they are not satisfactory simply because the model is not meant to deal with that kind of question.

Recently, Endler and Mielke (2005) proposed a simple way of analyzing colour patterns based on unconstrained colour spaces (lines, triangles, tetrahedrons), which are parsimonious in assumptions about vision mechanisms and which provide a convenient geometrical system in which spatial statistics on differences between sets of colours are possible. Going back to colour spaces allows to study not only the distance separating the colour points in the space but also their relative location within the space. This new formulation incorporates receiver's visual system as well as the visual environment in which visual signals are viewed. Even if the incorporation of receptor noise is not routinely taken into account as in Vorobyev and Osorio (1998), it can be easily calculated to give supplementary information. This representation can conveniently incorporate data on patch relative size by weighing colour points by their relative frequency in the analyzed pattern even if, as all existing models, it cannot represent the spatial organization between colour patches, an important asset for prey detection (e.g. Cuthill *et al.*, 2005; Stevens *et al.*, 2006b, 2008a, 2009c).

It is noteworthy to point two interesting conclusions. First, all the models discussed here have in common to discount achromatic mechanisms. This can be a major issue, especially when considering (see Section 1) that prey detection depends on the achromatic contrast between the prey and the background, particularly in the case of small and/or moving targets. In addition, most of the experimental investigations of predators' reaction to insect colouration have consisted in manipulating brightness and not chromatic contrast. Second, although some adaptations have been made to deal with restricted light conditions (Osorio *et al.*, 2004), these models are fundamentally adapted for vision in bright light. Dim light conditions exert specific constraints on visual systems. Photon capture by photoreceptors is complemented by neural mechanisms of temporal and/or spatial summation of photoreceptor outputs to compensate for low photon capture. Warrant (1999) proposed a model to incorporate such mechanisms and reconstruct animal sensitivity. Although this model has not been fully developed in a colour space model, it can be helpful to analyze visual signals seen by nocturnal insectivorous species. More generally, each model has its limitations (Kelber *et al.*, 2003;

Endler and Mielke, 2005) but also its valid fields of applications. Further work on vision modelling should focus on interesting but hitherto overlooked aspects of visual systems such as integration of chromatic and achromatic information, eye regionalization, colour categorization, variable or dim light conditions. It could also be interesting to build synthetic models combining photoreceptors functioning at different levels of light intensity (typically rods and cones in vertebrates) to explain transitional perceptual processes, such as colour vision based on rods and cones at dusk (Reitner *et al.*, 1991).

3.4 COLOUR INVESTIGATION: PRACTICAL RECOMMENDATIONS

To be implemented, physiological models have to be fed with data on object colouration, eventually background colouration, ambient light and receiver's visual sensitivity. Object colouration is often the elements for which most information is provided (see Endler, 1990 for information on how to acquire these data). Spectral data can be measured. Information about the location and spatial organization of colour patches within an animal or background pattern has to be collected separately. Information about the relative contribution (patch size) of colours to a pattern can be gathered separately, for instance as in Gomez and Théry (2007) or can be acquired through the spectral measurements themselves, as recommended by Endler and Mielke (2005). For instance, one can apply a grid on the animal pattern and measure one spectrum for each cell; the number of spectra collected will be a rough indication of the area occupied by each colour type on the animal.

Theoretical work on background influence on the evolution of colour signals has underlined how important it is to consider the background in its complexity (Merilaita, 2003). However, its practical incorporation in studies of colouration in relation to vision is nearly absent (few exceptions in studies about bird colouration; Endler *et al.*, 2005; Håstad *et al.*, 2005). Measuring background variability deals with 3D complexity, a technically challenging task. A good estimate of background visual complexity can be given by hyperspectral imaging, a technique which consists in taking a picture of the environment that would collect for each pixel not a small (three for typical cameras) but a high number of values which can describe an entire radiance spectrum. Without calling for this expensive technique, incorporating at least a minimal amount of background spatial heterogeneity can reveal differences in visual performances between close visual systems (Håstad *et al.*, 2005) that had apparently same performance in the case of an average background (Gomez and Théry, 2007). Models like that of Endler and Mielke (2005) offer the mathematical and statistical grounds needed for a synthetic comparison of more realistic viewing scenes. For instance, Merilaita's (2003) finding that visually complex environments lead to the evolution of less cryptic prey can be easily spatially visualized. Visual similarity between two sets of colours is represented in a colour space by the overlap, or the inclusion of the clouds of points corresponding to these sets of colours. Visually complex backgrounds are represented in colour spaces by more disperse clouds of points, in which a higher number of insect colour patterns (clouds of points) can fit in, resulting in similar levels of crypsis.

Information about ambient light should be collected in the same environmental conditions and at the time of day that the colour patches displayed by insects are normally seen by predators. If direct measurement is not possible, standard illuminants can to some extent be satisfying substitutes. Some standards exist for open areas from dusk to midday environments and are provided by the CIE (Commission Internationale de l'Eclairage).

Finally, the knowledge about visual sensitivity can be directly acquired on the species of interest through physiological and/or behavioural investigations. Although models of colour analysis vary in their requirements on this point, they have in common to need at least the spectral sensitivity curves of the different types of photoreceptors involved in vision. Curves only available from printed articles can be digitized using software like Winding (Lovy, 1996). If no curve is available, photoreceptor absorption functions can be mathematically computed based on few data (mainly peaks of absorbance of photoreceptors) and templates. Templates have been determined for insects (Stavenga *et al.*, 1993) and for vertebrates (Govardovskii *et al.*, 2000). For some groups such as birds, complementary templates for optical filters are also needed (Hart and Vorobyev, 2005). For vertebrates, it is also possible to model the visual effect induced by ocular media filters (Endler and Mielke, 2005). Even if templates may appear awkward in their mathematical formulation, they can be easily generated with a computer. One should prefer to use the information available for the species of interest whenever available.

As a conclusion, a thorough investigation of colours should be based on the method of measurement best adapted to the question asked. Although time-consuming and effort-demanding, objective and detailed measurements of the sets of visual elements important for prey detection should be performed. Special attention should be paid to signal perception if artificial prey targets are to be used. In all cases, spectrometry is a useful tool for measuring colour targets or controlling the biological validity of visual stimuli. A range of possible options is available to reconstruct insect colours as they are likely perceived by predators. Choosing a model should be guided by the type of question explored. Ideally, model choice should precede data collection in order to build a protocol best adapted to the question asked.

While models on animal colour vision are continuously attacked with the argument that they do not reflect exactly the visual system, it is important to notice that models are not meant to be perfect but to capture the essential elements. Vision models take into account the indispensable steps to explain the behavioural action spectra linked to visual detection (see assumptions in Vorobyev and Osorio, 1998). Nevertheless, they leave aside other aspects (shape and spatial organization of colour patches, behaviour) which may be important to consider for a skeptical and fruitful interpretation of the results, and even in the planification of the study.

4 Features of insect camouflage

One way to understand how insects are viewed by their predators is through the study of camouflage. Indeed, insect camouflage is often interpreted as the result of natural selection to avoid detection and recognition by predators, frequently involving body colouration. Therefore, identifying the features of camouflage could reveal how predator visual systems might be lured by insect prey. However, since the landmark books of Thayer (1909) and Cott (1940), research on natural camouflage has not progressed rapidly (Stevens and Merilaita, 2009a). This is notably due to the fact that camouflage has often been considered as obvious, not requiring formal testing. In addition, camouflage has mostly been viewed through human eyes, and not from the perspective of the appropriate receiver, the predator. It is only in the 1980s that the objective quantification of animal colouration and visual environments (Endler, 1978, 1984) promoted a more rigorous study of insect camouflage. More recently, the development of visual and computer sciences greatly facilitated the study of camouflage and induced an explosion in the number of scientists in this field, as evidenced by a recent compilation of works from biologists, visual psychologists, computer scientists and artists (Stevens and Merilaita, 2009a).

The term camouflage encompasses 'all strategies involved in concealment, including prevention of detection and recognition' (Stevens and Merilaita, 2009a). According to this recent review, there are four main tactics of camouflage: crypsis—which includes disruptive colouration, background matching, self-shadow concealment, obliterative shading, flicker-fusion camouflage, and distractive markings—masquerade, motion dazzle and motion camouflage. In this chapter, we will focus on the mostly studied form of camouflage in insects, crypsis, and review the relatively rare insect examples of masquerade and motion camouflage.

4.1 CRYPSIS

Most studies of insect camouflage have been conducted with birds as predators (Table 2) because birds are highly visually oriented predators and one of the main groups of insect predators. In addition, it is often possible to identify bird predation of insects in the wild, for example through damage on the wing of live or artificial insects.

4.1.1 *Disruptive colouration and background matching*

Among the different means of crypsis, disruptive colouration has been the most studied, and most often in insects (Table 2). Disruptive colouration 'is a set of markings that creates the appearance of false edges and boundaries and hinders the detection or recognition of an object's, or part of an object's, true outline and

TABLE 2
Experimental tests of strategies used for visual camouflage by insects

Strategy	Prey	Predator	Support	References
Crypsis				
Disruptive colouration	Peacock butterfly	Birds	No (?)	Silberglied *et al.* (1980)
Marginal pattern elements	Artificial moth	Birds	Yes	Cuthill *et al.* (2005); Schaefer and Stobbe (2006)
	Artificial prey	Great tit	Yes	Merilaita and Lind (2005)
	Artificial moth	Bird edge detection model	Yes	Stevens and Cuthill (2006)
	Artificial moth	Humans	Yes	Fraser *et al.*(2007)
Maximal disruptive contrast	Artificial moth	Birds	Yes	Cuthill *et al.* (2005)
	Artificial moth	Birds	No	Stevens *et al.* (2006b)
	Artificial butterfly	Birds	No	Stobbe and Schaefer (2008)
Coincident disruptive colouration	Artificial moth	Birds	Yes	Cuthill and Székely (2009)
Disruption of surface	Artificial moth	Birds	Yes	Stevens *et al.* (2009c)
Asymmetrical patterns	Artificial moth	Birds	Yes	Cuthill *et al.* (2006a,b)
	Artificial prey	Great tit	Yes	Merilaita and Lind (2006)
Background matching	Moth	Blue jay	Yes	Pietrewicz and Kamil (1977)
	Lepidoptera larvae	Bird with UV vision	Yes (five out of six spp.)	Church *et al.* (1998)
	Artificial prey	Great tit	Yes	Merilaita *et al.* (2001), Merilaita and Lind (2005)
	Moth	Jumping spider	Yes	Moss *et al.*(2006a)
	Artificial moth	Birds	Yes	Fraser *et al.* (2007)
	Artificial moth	Humans	Yes	Johnsson and Kjällman-Eriksson (2008)

(*continues*)

TABLE 2 (*Continued*)

Strategy	Prey	Predator	Support	References
	Beetle	Coccid	Yes	Patra *et al*. (2008)
	Chironomid larva	Three-spined stickleback	Yes	Ioannou and Krause (2009)
Distractive markings	Artificial moth	Birds	No	Stevens *et al*. (2008a)
	Artificial prey	Blue tit	Yes	Dimitrova *et al*. (2009)
Countershading	Artificial prey	Birds	Yes	Edmunds and Dewhirst (1994), Rowland *et al*. (2007, 2008)
	Artificial prey	Birds	Species dependent	Speed *et al*. (2005)
Masquerade	Caterpillars	Birds	Yes	De Ruiter (1952)
	Beetle larva	Ants	No (not visual)	Eisner *et al*. (1967)
	Chrysoptera larva	Ants	No (not visual)	Eisner *et al*. (1978)
	Assassin bugs	Spiders, gecko, centipede	Yes, from gecko	Brandt and Mahsberg (2002)
	Assassin bugs	Jumping spider	Yes	Jackson and Pollard (2007)
	Aphid	Jumping spider	Yes	Moss *et al*. (2006b)
Motion camouflage	Chironomid larva	Three-spined stickleback	Yes	Ioannou and Krause (2009)
	Mantid	Mantid, lizard	Yes	Watanabe and Yano (2009)

? Denotes a study which conclusion has been severely criticized.

shape' (Stevens and Merilaita, 2009b). Four main principles of disruptive colouration have been investigated, three of which are largely supported (Table 2). First, 'marginal pattern elements' placed on the edge of the body, which are predicted to be more disruptive than patches placed inside (Cott, 1940). Second, highly contrasting patterns, which should be more effective than less contrasting patterns ('maximum disruptive contrast', Cott, 1940). Third, continuous patterns that range over different body parts and can be used to conceal revealing body parts, such as eyes ('coincident disruptive colouration', Cott, 1940). Fourth, false edges which can be used to create an appearance of different shape and hinder recognition of the body outline ('disruption of surface', Cott, 1940). It must be noted that these different hypotheses are not mutually exclusive and can be tested simultaneously: for example, Cuthill *et al.* (2005) have tested both the principles of marginal pattern elements and maximal disruptive contrast. Another well studied mean of crypsis is background matching, 'where the appearance generally matches the colour, lightness and pattern of one (specialist) or several (compromise) background types' (Stevens and Merilaita, 2009a). Because the study of disruptive colouration has often been conducted together with the study of background matching (Merilaita and Lind, 2005; Schaefer and Stobbe, 2006; Fraser *et al.*, 2007), and since animals may use both techniques simultaneously, we will review these mechanisms in the same section.

Surprisingly, until the last 5 years, only one experimental test of the effectiveness of insect disruptive colouration against non-human predators existed (Table 2; Silberglied *et al.*, 1980). This study did not consider a specific mechanism of disruptive colouration, and to our knowledge stands as the only study of disruptive colouration conducted with living insects and predators in the wild. It was performed on a nymphalid butterfly highly palatable to birds, the banded peacock *Anartia fatima*, and apparently did not support the hypothesis of disruptive colouration. Since the butterfly frequented only one artificial clearing, the authors could capture and individually mark nearly the entire population, which allowed extremely high recapture rates and determining the minimum age of each butterfly at every capture and a minimum longevity for each individual in the population. The wing stripes of one part of the males was obliterated with black felt-tip marking pens, and equal numbers of individuals used as controls had the dye applied to the dark region basal to the wing stripe so that their appearance remained unchanged. Wing damages by birds were then monitored over 21 weeks. Unexpectedly, butterflies lacking disruptive colouration lived as long as those from the control group, a finding inconsistent with the theory of disruptive colouration. However, several objections were later made to this conclusion. First, the experimental modification might have made the butterflies look like an unpalatable species (Waldbauer and Sternburg, 1983) or directly modified their palatability (Cuthill *et al.*, 2005). Second, obliterating the wing stripe may simply have converted the pattern from one background matching type to another background matching type of equal crypsis (Endler, 1984). Third, the wing stripe has not been demonstrated to be disruptive in function, so it may have another unrelated function (Stevens *et al.*, 2006a). Clearly, more experimental support for the theory of disruptive colouration was needed, especially by taking into account the visual abilities of predators.

4.1.1.1. Marginal pattern elements A clear improvement in the investigation of disruptive colouration was made with the study of Cuthill *et al.* (2005), who tested the prediction from Thayer (1909) and Cott (1940), that disruptive colouration occurs when patches on the body's edge are more effective than patches placed randomly over the body ('marginal pattern elements'). They offered predators artificial moth-like prey consisting of dead mealworms *Tenebrio molitor* on top of coloured paper triangles pinned onto oak trees. They carefully measured the colour contrast of artificial prey with the natural background by spectrometry coupled with the modelling of predicted photon catches of a typical passerine bird, the blue tit. 'Prey' survival was monitored over 24 h, and 'predation' by birds could be separated from predation by spiders and slugs because spiders suck fluids out, leaving a hollow exoskeleton, and slugs leave slime trails. The results were clear cut: artificial moths with markings overlapping the edge showed higher survival than the other patterns, whereas a 'prey' with randomly selected markings survived better than a model with the markings that were on the edge had been brought inward so that they do not overlap the edge. Black and brown uniform models survived worst of all. This clearly supports the hypothesis of disruptive colouration. Stevens and Cuthill (2006) analyzed digital images from the second experiment of Cuthill *et al.* (2005) with an edge detection algorithm combined with the photon catches of starlings *Sturnus vulgaris*. They showed that disruptive colouration is effective because 'false' edges are detected within the body of artificial moths rather than on the body outline, therefore hindering detection of the body outline of prey by its predator. These studies, together with others mainly studying background matching (Merilaita and Lind, 2005; Schaefer and Stobbe, 2006; Fraser *et al.*, 2007), all supported the hypothesis of disruptive colouration through marginal pattern elements (Table 2).

4.1.1.2. Maximum disruptive contrast As proposed by Poulton (1890), Thayer (1909) and Cott (1940), it could be important for concealment that some colour elements contrasting in tone and different from the background should be highly conspicuous ('maximum disruptive contrast'). This question was investigated in the second experiment conducted by Cuthill *et al.* (2005), which showed that highly contrasting colours enhance the disruptive effect, again strongly supporting the theory of disruption in the field, independently of background matching (Sherratt *et al.*, 2005; Stevens *et al.*, 2006a). However, in Cuthill *et al.*'s (2005) study, all components of disruptive patterns presented colours in common to the natural background, which might not allow a proper test of the theory of maximum disruptive contrast. In order to avoid this potential bias, Stevens *et al.* (2006b) used grey and black artificial moths (rather than brown and black in Cuthill *et al.*, 2005) with the grey luminance matching, or mismatching, the background oak bark. Therefore, the targets did not match the colour of oak bark. The results confirmed the higher survival of edge disruptive patterns, and showed that disruptive patterns have a survival advantage compared to background-matching patterns which are not disruptive. As predicted by Thayer (1909) and Cott (1940), disruptive patterns still provide camouflage when the pattern elements are not matching the background, suggesting that an animal, presenting conspicuous patches involved in intraspecific communication, for example, can benefit from having them placed disruptively. However, heightened achromatic contrasts of marginal patterns did not increase the disruptive effect. Because moth models did not match bark colouration, the fact that two-tone background-matching patterns were more difficult to detect than monochrome targets supports the idea that pattern detection involves brightness contrast in birds (Osorio *et al.*, 1999; Jones and Osorio, 2004). Stobbe and Schaefer (2008) also tackled the principle of maximum disruptive contrast by using artificial models of the diurnal white admiral butterfly *Limenitis camilla* differing in the chromatic but not the achromatic contrast of their wing stripes to the background as well as to the adjacent brown colouration of the wings. The visual contrasts on oak trees were modelled using the visual sensitivities of blue tits and the prevailing ambient light. The results showed that the strength of chromatic contrast was negatively correlated with survival probability. Therefore, as in Stevens *et al.* (2006b), these experiments do not support the predictions of maximal disruptive contrast.

4.1.1.3. Coincident disruptive colouration Apart from disguising the body outline, disruption can also function to conceal other characteristic body parts, such as eyes, antennae and limbs (Cott, 1940). Concealment is obtained by using continuous patterns joining different body parts and the outline between them, and termed ‘coincident disruptive colouration’ (Cott, 1940). Patterns could be dark eye stripes that conceal the eyes, or two-tone body on wings with a two-tone central section, with dark and light sections of the wings and body coincident. The first and only experimental test of this form of disruption has been conducted by Cuthill and Székely (2009) by using artificial moths, preyed upon by wild birds, in which coincidence of colour patterns of wings and bodies was varied. They also conducted a visual search experiment on humans watching pictures of oak bark with or without artificial moths. Both experiments confirmed Cott’s (1940) principle of coincident disruptive colouration.

4.1.1.4. Disruption of surface Another way of visual disruption than ‘marginal pattern elements’ breaking up the body outline is obtained by creating ‘false edges’ away from the body outline not corresponding to any animal feature (‘disruption of surface’, Stevens and Merilaita, 2009b). While most studies of disruption have focused on marginal pattern elements, only one recent study experimentally tested the concealment effect of surface disruption (Stevens *et al.*, 2009c). This study used artificial moths of different outline and surface patterns, and of different luminance contrasts, presented to wild avian predators. It also used the avian visual model and the algorithm of edge detection used by Stevens and Cuthill (2006). High luminance of markings increased the effectiveness of disruption, and markings with highest contrast placed away from the body outline were highly effective, supporting Cott's (1940) idea of surface disruption. The avian edge model showed that surface disruption is not obtained by creating false edges away from the body outline, and that it may be related to a different visual mechanism.

4.1.1.5. Asymmetrical patterns Another challenge was to determine if symmetrical patterns, common in moths, are effective in disrupting body outlines to predators. It is indeed intuitive and demonstrated in humans that symmetry can facilitate visual search of cryptic prey, but this had not been tested in animals. All studies to date have confirmed the survival advantage of asymmetrical patterns. Again using spectrometry and artificial moths baited with mealworms, Cuthill *et al.* (2006a) tested the conspicuousness to bird predators of highly cryptic prey with or without bilateral symmetry in the matching of oak tree colouration. Following the 'survival' of targets over 5 days, they identified a small but significant fitness increase of asymmetry in insect models. The question of why asymmetry is not more common in cryptic colouration remains to be answered, and could be a challenge for developmental processes. Using the same techniques, Cuthill *et al.* (2006b) showed that symmetry of prey colouration has the same detrimental effects with regard to bird predation for both disruptive and non-disruptive background-matching patterns. Using captive great tits and artificial prey models, Merilaita and Lind (2006) confirmed the existence of a cost to cryptic prey for bilateral symmetry, even if not all bilaterally symmetric, cryptic patterns generate similar cost (one symmetric pattern in their experiment even generated no cost, but predators were very close to the prey substrate; see Cuthill *et al.*, 2006b).

4.1.1.6. Background matching and its relation with disruptive colouration Most studies conducted on background matching have supported this principle (Table 2). An early experimental study in insects has been conducted by Pietrewicz and Kamil (1977) who trained blue jays *Cyanocitta cristata* to respond to the presence or absence of bark-like *Catocala* moths in slides. Moths were less likely to be detected by birds if they were presented on their naturally colour-matching substrate, and this concealment was increased when moths were in their natural resting orientation (head down or up depending on the species) compared to a horizontal position. Prey orientation has little effect for moths placed on a non-matching substrate, showing that background matching combined with orientation are important components of crypticity for these moths.

Endler (1978) proposed that 'a color pattern is cryptic if it resembles a random sample of the background perceived by predators at the time and age, and in the microhabitat where the prey is most vulnerable to visually hunting predators'. Despite numerous studies showing that prey are more difficult to detect when they have a high degree of resemblance with their visual background, Endler's (1978) idea of camouflage through random sampling of the background had not been experimentally tested until Merilaita and Lind (2005). This study investigated the risk of detection of artificial prey by great tits *Parus major* in captivity. The 'prey' had two patterns randomly sampled from the background (difficult and easier to detect) and a disruptive pattern. Two sets of prey were used: one with pattern elements broken on the outline, and one with full pattern elements touching the outline. The results showed that prey colouration matching a random sample of the visual background was not sufficient to maximize background matching, in contradiction with Endler's (1978, 1984) suggestion. In addition, the disruptive prey and the best background-matching prey were equally cryptic. This was true for both prey sets, with whole pattern elements on the prey outline or with elements cut on the prey outline. This supports the principle of disruptive colouration, but also shows that random visual sampling of the background is neither sufficient nor necessary for efficient camouflage.

Another interesting study of background matching was conducted by Merilaita *et al.* (2001) who used great tits searching for artificial prey on two different visual backgrounds, either small or large patterned. Two prey types had the same patterns as the backgrounds, either small or large, whereas a third type had an intermediate pattern. Search time of prey was longer on the large-patterned than on the small-patterned background. Logically, the small-patterned prey was most cryptic on the small-patterned background, and the reverse was true for large-patterned prey on the large-patterned background. The compromised prey showed an intermediate crypticity on the small-patterned background, but was not significantly less cryptic than the large-patterned prey on the large-patterned background. Most interestingly, the compromised colouration had lower predation than the matching colourations, indicating that it might be the best strategy for a prey in heterogeneous habitats.

A first attempt to consider the full range of predators' visual sensitivity in a study of background matching was made by Church *et al.* (1998), who examined whether crypsis of lepidopteran larvae found on oak trees extended in the UV. The reflectance of oak leaves and of six caterpillar species was measured by UV/visible spectrometry, in the full range of wavelengths to which insectivorous birds are sensitive. For five out of the six species, crypsis of these green caterpillars extended in the UV. However, one species which appeared moderately cryptic to the human eye was found to be very conspicuous to a UV-detecting avian predator. This showed how crucial it is to take into account the visual sensitivities of predators. Stobbe *et al.* (2009) examined the underlying visual mechanisms of background matching with two artificial prey types differing in their chromatic contrast in the UV/blue range, but achromatically identical as seen in the eyes of blue tits. The backgrounds were either matching or mismatching the prey colouration. The results showed that chromaticity in the UV/blue range, as well as achromatic cues, are used by bird predators to search for prey. However, it remains to be studied if chromaticity in the medium and long wavelengths is as important as colour contrast in the short wavelengths for the visual search of avian predators.

In order to disentangle the camouflage functions of disruptive colouration and background matching, Schaefer and Stobbe (2006) designed five artificial models of the peach blossom moth, *Thyatira batis*, differing by their visual contrasts and disruptive patterns, as measured in the blue tit visual system, which were glued on differently coloured trunk backgrounds. A dead mealworm was glued underneath the wings so that less than half of the mealworm was visible to predators, and prey survival was monitored over 24 h. Interestingly, they found that chromatic contrast is more efficient than achromatic contrast to reduce predation: low chromatic contrast of cryptic moth reduced its predation risk to the level of disruptive prey. They confirmed the lower mortality of disruptively coloured forms compared to the cryptic form, and showed that disruptive forms on the body outline have higher survival in a wider range of habitats than the background matching form even though the disruptive patterns are symmetrical. Disruptive forms on the body interior provide, as background matching, efficient camouflage, but their protective value is specific to a particular background. Overall, their results confirmed the prediction of Cuthill *et al.* (2005), that disruptive colouration is more efficient on the body outline, independently of background matching.

Apart from the study of Fraser *et al.* (2007) with 'human predators', only three studies have considered other predators than birds. The first is the investigation of background matching of grass moths *Orocrambus flexuosellus* preyed upon by jumping spiders *Trite planiceps* and *Marpissa marina* (Moss *et al.*, 2006a), who showed that spiders were less efficient to detect and capture moths on a matching than on a contrasting visual background. The second involves maggots preyed upon by wild brown trout parr *Salmo trutta* on a matching or contrasting background (Johnsson and Kjällman-Eriksson, 2008). This study also found that search time for prey was increased on the colour-matching background. The third study involved predatory beetles *Jauravia* sp. feeding on coccids *Saissetia filicum* found either on fern sori where they appear cryptic, or on vegetative leaves where they appear conspicuous (Patra *et al.*, 2008). It showed that the degree of predation by beetles is higher on conspicuous than on cryptic coccids. However, it should be noted that these three studies inferred matching or contrasting backgrounds only by using human vision.

4.1.2 *Distractive markings*

In addition to disruption, prey can also avoid predation by using distractive markings (Thayer, 1909), which are highly visible patterns or patches that distract the predator attention toward them, and away from the body outline.

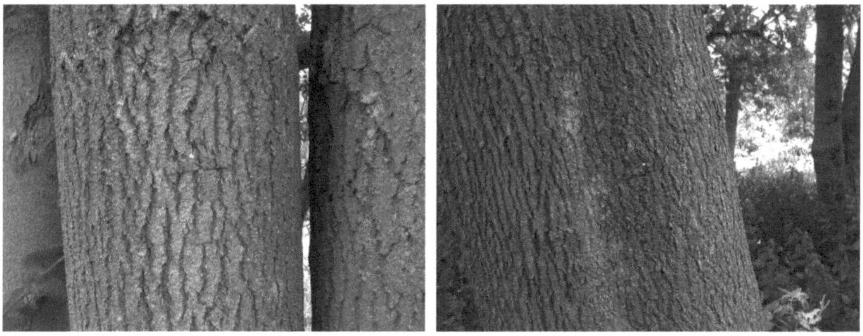

FIG. 1 Two models of artificial moth used for the study of distractive markings by Stevens *et al.* (2008a). Martin Stevens ©. (See Color Plate 41 at the back of the book.)

The first experimental test of that form of crypsis has been conducted by Stevens *et al.* (2008a) using artificial moths preyed upon by wild birds. The models used varied by the shape (irregular or circular) and the intensity of visual contrast of the markings with respect to the model (Fig. 1). A dead mealworm partially covered by the model provided an edible component to the predator. Contrary to Thayer's predictions, potentially distractive markings decreased survival of artificial moths, and circular markings were as likely detected than irregular ones. In their recent experiment, Dimitrova *et al.* (2009) used blue tits searching in a cage for artificial prey with different levels of distractive markings on different types of visual backgrounds. The backgrounds also differed by their degree of distraction. Their results confirmed Thayer's principle of distractive markings in prey improving camouflage. Interestingly, backgrounds with distractive features also increased search time of predators, independently of prey appearance. This suggests that camouflage may be favoured in highly visually contrasted habitats.

4.1.3 *Countershading*

Countershading is a means of crypsis in which animals have a darker pigmentation on the body surface mostly exposed to ambient light. It allows prey concealment through two strategies (Stevens and Merilaita, 2009a): self-shadow concealment, 'where directional light, which would lead to the creation of shadows, is cancelled out by countershading', and obliterative shading, 'where countershading leads to the obliteration of three-dimensional form' (see also Rowland, 2009). In insects, as in most animal groups, countershading has most often only been supported by indirect evidence, interpreting camouflage through human vision (for a recent review see Rowland, 2009). This was for example the case in the studies of Thayer (1909), de Ruiter (1956) and Tinbergen (1957). Few experimental tests of the adaptive value of countershading have been conducted but, because of methodology or sample size, it could not be concluded that crypsis was improved (see Rowland, 2009). More rigorous experiments were conducted by Edmunds and Dewhirst (1994), who exposed artificial prey differing by their countershading patterns to predation by wild birds and found a higher survival value of countershaded prey. However, Speed *et al.* (2005) conducted a similar experiment and found no reduction in attack of countershaded prey. None of those studies took into account the visual sensitivities of avian predators. This led Rowland *et al.* (2007) to try solving this discrepancy using similar methods, but this time including spectrometric measurements and a model of avian colour vision. In both experiments, one in which artificial prey were presented on lawns to a range of bird species, the other in which prey were presented on green boards to individual blackbirds *Turdus merula*, countershaded prey had lower levels of predation than controls. Rowland *et al.* (2008) also conducted more realistic field experiments in which they used artificial prey resembling Lepidoptera larvae presented on the upper and lower surfaces of beech tree branches to wild bird predators. When prey items were presented on the upper surface of branches, the countershaded prey had higher survival than the uniformly coloured prey. When they were presented on the underside of branches, the prey with a reversal of the orientation of countershaded colouration had highest survival. This clearly provides evidence that an increase in pigmentation on the side of the prey closest to the light source offers camouflage in the eyes of the predator. In her recent review on the function of countershading, Rowland (2009) presented the first measurements of visual contrasts of countershaded lepidopteran larvae on their food-plant visual backgrounds, showing that countershading is stronger in the achromatic channel than in the chromatic one. However, it cannot support the principle of background matching since the illuminant spectra were the same for larvae presented on the upper or underside of branches, which is a crucial parameter to consider in countershading theory. Although much is left to be done in its study and that several objections to countershading as a mean of concealment have been raised (see Rowland, 2009), our understanding of countershading has progressed rapidly in recent years, partly due to the use of more rigorous consideration of the colour of prey and of the visual sensitivity of predators.

4.2 MASQUERADE AND DECORATION

Masquerade is used when prey 'mimic inedible objects such as leaves, sticks and bird droppings, that is, objects of no inherent interest to the potential predator' (Ruxton *et al.*, 2004). Research on this camouflage tactic has been relatively limited, probably because adaptive advantages appear quite straightforward to understand (Fig. 2). Therefore, we mainly have descriptions of insects resembling inedible objects (e.g. references in Edmunds, 1990; Robinson, 1990; Canfield *et al.*, 2009), but rare experimental demonstrations

FIG. 2 An example of masquerade in the dead leaf mantid *Deroplatys desiccata* (photo Adrian Pingstone). (See Color Plate 42 at the back of the book.)

of the survival value of this tactic. An early experiment was conducted by De Ruiter (1952) who offered caterpillars of canary-shouldered thorns *Ennomos alniaria*, oak beauty *Biston strataria* and *B. histaria* to Eurasian jays *Garrulus glandarius* and chaffinches *Fringilla coelebs*. Birds were first familiarized with sticks, of the same dimension as caterpillars, taken from the trees where caterpillars were collected. Dead caterpillars were then introduced in the experimental cage, mixed with sticks. Birds had great difficulties at finding caterpillars, obviously not discriminating prey from sticks. However, birds could discriminate caterpillars from twigs of other trees. In other experiments with living stick caterpillars, prey were found and eaten as soon as they moved. Avian predators obviously could not discriminate motionless caterpillars from sticks found of the trees on which they fed. However, in this experiment, it is difficult to conclude that predators detected and misidentified prey (masquerade) rather than simply failing to detect prey (crypsis). In an ingenious experiment, Skelhorn *et al.* (2010) showed that the cognitive strategies of predators, rather than their sensory abilities, drive the evolution of masquerade. They exposed three groups of naïve domestic chicks to a hawthorn branch (the host plant of twig-like caterpillars used as prey), three groups to a hawthorn branch bound in purple cotton thread to change its visual appearance, and three groups to an empty arena. The chicks with a prior experience of twigs were slower and more cautious to attack and handle twig-like caterpillars or twigs than chicks exposed to the purple twig. Chicks with no prior experience of twigs were even faster and less cautious to peck and handle twig-like prey or twigs. Because test items were strongly contrasting on the arena substrate, these results clearly show that masquerade functions in the absence of crypsis and can provide an entirely additional benefit to it.

Another form of masquerade has been described in species that cover their body with leaf, flower parts, lichen, faecal droplets, moulted skins and other organic or non-organic material. For example, the camouflage role of the trash packet held on the back of several insect larva species has been investigated by Eisner *et al.* (1967) who exposed larvae of the thistle tortoise beetle *Cassida rubiginosa* to Allegheny mound ants *Formica exsectoides*. It was found that ants easily detected prey, but that larva used their 'faecal shield' to avoid being bitten and killed by ants, therefore rejecting the visual camouflage function. Eisner *et al.* (1978) also studied larva of the green lacewing *Chrysopa slossonae* that feeds on wooly alder aphids *Prociphilus tesselatus* and copes with ants protecting the aphid colonies by masquerading as an aphid. To do so, larvae tears away waxy tufts from aphids and loads this material on its back, closely resembling its aphid prey. Both shielded and denuded larvae were presented to ants (black carpenter ants *Camponotus pennsylvanicus*, *C. noveboracensis*, *Formica* sp.). In all cases, ants rapidly detected the larvae, but only removed denuded ones from aphid colonies. Again, this form of masquerade appears to provide protection from predators, but not through visual perception. Other evidences suggest that this trash-carrying behaviour is used as defense, not visual camouflage, against insect predators (see Eisner *et al.*, 1978).

However, in other cases, masquerading prey appears to avoid being visually identified by predators. For example, Brandt and Mahsberg (2002) studied the disguise of West African assassin bugs *Pareodocla* sp., *Acanthaspis petax* and *Acanthaspis sulcipes* with respect to their success in hunting ants (*Dorylus nigricans*, *Crematogaster* sp., *Camponotus* sp.) and to protection from predators (wall spiders, African house gecko *Hemidactylus brooki*, centipede *Scolopendra morsitans*). These bugs' nymphs have the peculiarity to cover their whole bodies with dust, sand or soil ('dust coat'), and to pile a 'backpack' of larger objects (prey corpses, other animal and vegetable matter) on their abdomens. Experiments with prey showed that better hunting success could be obtained with the dust coat impeding chemical and tactile recognition by ants. However, in experiments with predators, the main protective effect was attributed to the backpack which enhanced the concealing effect of the dust coat and confused the visually oriented predators (geckos). Jackson and Pollard (2007) also studied predation of masked or naked assassin bugs *A. petax*, but this time by three salticid spider species that also have acute vision. They confirmed results obtained by Brandt and Mahsberg (2002), showing that masquerade induced a failure of visual recognition of prey by predators. Moss *et al.* (2006b) conducted different experiments with the wooly apple aphid that covers itself with wax, *Eriosoma lanigerum*, attacked by the jumping spider *M. marina*. Their results suggest that wax is indeed hiding visual cues used for prey identification, notably the head which was sufficient for spiders to identify the aphid. Canfield *et al.* (2009) described larval decoration and morphological plasticity in larvae of Southern emerald moths *Synchlora frondaria*. They showed that larvae use anthers, pieces of petals and small leaves to cover their dorsal surfaces. Interestingly, they also showed that larvae acquire the colour of their host plants by consuming flower parts. Although both mechanisms were interpreted as obviously providing crypsis on their visual background, no test of protection from predators was conducted in this study.

4.3 MOTION CAMOUFLAGE

To our knowledge, the only study of motion camouflage involving insects in the context of predation is that of mantids *Hierodula patellifera* avoiding detection by predators (conspecific mantids and Japanese five-lined skinks *Plestiodon japonicus*) when they are exposed on vegetation moving by the wind (Watanabe and Yano, 2009). Both in the field and in the laboratory, mantids walked and swayed their body more frequently when the wind blew harder, which is suggested as adaptive in reducing the risk of predation because predators were less likely to discover prey in the swaying leaf condition.

5 Warning colourations and patterns viewed by predators

The effectiveness of warning colouration has been largely supported in a wide range of wild and domestic insect predator species (Table 3). Colours avoided by predators are often red and black-and-yellow stripes. One caveat of studies of warning colouration is the frequent absence of appropriate colour measurement and consideration of predators' visual sensitivities. To our knowledge, only Lyytinen *et al.* (2001), Schultz (2001), Gamberale-Stille (2001), Tullberg *et al.* (2005) and Prudic *et al.* (2007) used spectrometry to measure and study warning colouration. None of those has undertaken photon catch modelling.

5.1 WARNING COLOURS: LEARNT OR INNATE?

One important aspect of research on warning colouration is to determine if avoidance is learnt or innate. It has long been thought that predators learn to avoid warning colouration of prey, implying that the function of warning colouration is to facilitate avoidance learning and prey recognition (references in Schuler and Hesse, 1985). In his elegant work, Mostler (1935) underlined the importance of colour pattern in learning. Naïve insectivorous birds fed upon black and yellow hoverflies, showing no innate aversion towards striped patterns. Yet, they rapidly learnt to avoid noxious striped hymenopteran species (wasps, honeybees, bumblebees) and subsequently avoided hoverflies with striped patterns. Hoverflies gained their protection from their visual resemblance and colour patterns played an important role in avoidance learning. Since the 1950s, unlearned (in the sense of innate) avoidance of different coloured food has been demonstrated in a range of wild and poultry birds (references in Ruxton *et al.*, 2004). Using hand-reared naïve birds captured in

TABLE 3

Experimental tests of the function of warning colours and patterns

Hypothesis	Prey	Predator	Support	Avoidance	References
Warning colour	Butterflies	Birds	Yes	Novel food, red/black	Coppinger (1969, 1970)
	Butterflies	Pied flycatcher	No	No avoidance of white	Lyytinen *et al.* (1999)
	Mealworm	Domestic chick	Yes	Black/yellow stripes	Schuler and Hesse (1985)
	Mealworm	Domestic chick	Yes	Black, black-yellow stripes, red	Roper and Cook (1989)
	Mealworm	Domestic chick	Yes	Red	Roper (1990)
	Mealworm	Great tit	Yes	Black-yellow stripes	Lindström *et al.* (1999)
	Shieldbug	European blackbird	Yes	Red/black	Schlee (1986)
	Shieldbug	Blue and great tit	Yes	Red	Veselý *et al.* (2006)
	Firebug	9 passerine spp.	Yes/No	Red and black[a]	Exnerová *et al.* (2003, 2008)
	Firebug	Robin, blackcap	Yes	Red and orange	Exnerová *et al.* (2006)
	Harlequin bug	Zebra finch	Yes	Red	Sillén-Tullberg (1985a)
	Milkweed bugs	Mantid	Yes	Brightness contrast helps detection	Prudic *et al.* (2007)
	Carabid beetle	Lizard	Yes	Black with yellow spots	Hasegawa and Taniguchi (1994)
	Artificial prey	Great tit	No	No aversion to UV	Lyytinen *et al.* (2001)
	Wasps, flies	Dragonfly	Yes	Black-yellow stripes in wasp only	Kauppinen and Mappes (2003)
	Natural, artificial prey	Several dragonflies	No	Not black-yellow stripes, size	Rashed *et al.* (2005)
	Glow-worm larva	Great tit	Yes	Yellow-pink spot on jet black body	De Cock and Matthysen (2001)
	Firefly larva	Toads	Yes	Bioluminescent colouration	De Cock and Matthysen (1999, 2003)
Size and symmetry	Artificial butterfly	Domestic chick	Yes	Large, symmetric	Forsman and Merilaita (1999)
	Artificial butterfly	Domestic chick	Yes	Symmetry in colour, shape and size	Forsman and Herrström (2004)
	Artificial prey	Birds	No	No advantage of symmetry	Stevens *et al.* (2009b)

[a] Avoidance is observed in small insectivores but not in large insectivores or granivores.

The table is not exhaustive since we have not considered all studies on insect aposematism.

the USA (blue jays, red-winged blackbirds *Agelaius phoeniceus*, common grackles *Quiscalus quiscula*) fed with unfamiliar butterflies captured in Trinidad, Coppinger (1970) found a clear avoidance of novel foods and of red/black palatable butterflies (scarlet peacock *Anarthia amathea*) compared to brown and white butterflies similar in size (white peacock *Anarthia jatrophae*). Because birds were initially fed on brown food, Coppinger (1969) interpreted this result as the avoidance of novel colouration, neither learned nor innate, confirming earlier results with experienced blue jays. However, this was not a proof of the absence of unlearned avoidance of the black and red pattern. To investigate differences between the novelty and the innate avoidance, Schuler and Hesse (1985) offered naïve domestic chicks *Gallus gallus domesticus* mealworms painted either with black and yellow warningly coloured stripes or with a non-warningly coloured uniform olive green (a mix of the black and yellow paints). Chicks pecked at both prey types equally, but ate the warningly coloured ones at a much lower rate. The inhibition of attack after pecking was interpreted as resulting from the unlearned avoidance of the black and yellow colouration. To examine if the unlearned avoidance was related to particular colours and/or patterns of prey, Roper and Cook (1989) also used naïve chicks studied on their first feeding, which were offered green mealworms either painted with single colours (black, yellow or red), striped colouration (black/yellow, black/red, yellow/red) or half one colour half another (black/yellow, black/red, yellow/red). Chicks strongly avoided black prey, showed a mild aversion to black/yellow striped prey, and a mild aversion to red prey. Other colours and patterns were either neutral or preferred. It was concluded that stripes or bicolour prey were not more aversive than prey of a single colour, and that chick show specific aversions to particular colours and patterns, rather than to prey novelty or contrast. Interestingly, they showed that rearing chicks in black and yellow cages reduced or reversed their avoidance of black/yellow striped and black prey, and increased their preference for yellow and bicolour prey. Similarly, Roper (1990) showed that rearing chicks in red cages induced a preference for red prey over the usually preferred brown prey. This demonstrates that even if avoidance of particular colours and patterns are unlearned, they can be modified by experience (Mostler, 1935).

5.2 AVOIDED COLOURS

We have seen earlier that red and black-and-yellow stripes are most often colours avoided by predators (Table 3; Coppinger, 1970; Schuler and Hesse, 1985; Roper and Cook, 1989). Several other studies have confirmed these aversions, and extended our knowledge of chromatic and achromatic characteristics of warning colouration. Veselý *et al.* (2006) studied the aposematic colouration of red-black shieldbugs *Graphosoma lineatum* preyed upon by blue and great tits. They presented wild-coloured and non-aposematic brown-painted shieldbugs to birds, and found blue tits avoided both forms, whereas great tits attacked both, but the brown ones more frequently. Schlee (1986) also confirmed the anti-predatory function of the red-black shieldbug colouration with European blackbirds *T. merula*. Exnerová *et al.* (2003, 2008) confirmed that different bird species react differently to aposematic and non-aposematic prey. Overall, small insectivorous birds avoid warningly coloured prey, whereas large insectivorous birds as well as granivorous birds often attack them. Exnerová *et al.* (2006) studied predation of the chromatically polymorphic firebug *Pyrrhocoris apterus* (red in its wild form, with white, yellow and orange mutants) by great tits, robins *Erithacus rubecula* and blackcaps *Sylvia atricapilla*. They found that red and orange colouration, and not the black melanin pattern, is essential to provide protection from avian predators. White mutants were not better protected than the non-aposematic brown-painted wild form, and the reaction of birds to the yellow form was species specific. This is consistent with the study of Lyytinen *et al.* (1999) who showed that the white colouration of pierid butterflies does not have a warning function. By manipulating the colouration of common glow-worm larvae *Lampyris noctiluca*, De Cock and Matthysen (2001) showed that the colour pattern consisting of yellow-pinkish lateral spots on the jet-black background was used to learn avoiding this distasteful prey by starlings. Kauppinen and Mappes (2003) have investigated the features that intimidate dragonflies *Aeshna grandis* to attack wasps *Vespula norwegica*. In a first experiment, they painted flies either black of with yellow and black stripes, and did the same for wasps. The dragonflies showed greater aversion to wasps than to flies. Yellow-and-black striped flies were more frequently avoided than black flies, revealing the selective advantage of yellow stripes. However, yellow-and-black wasps were not more avoided than black ones, showing that some other feature(s) should make wasps intimidating to predators. In further experiments, dragonflies were offered artificial prey that were painted with either yellow-and-black stripes, solid black or solid yellow. Again, dragonflies avoided more often striped prey. This study indicates that black-and-yellow stripes alone are effective in protecting prey, even palatable ones. Using a similar approach on the field, Rashed *et al.* (2005) extended the study of Kauppinen and Mappes (2003), notably by scoring responses of several species of dragonflies, and found that these predators were not avoiding the yellow-and-black patterns present in both aposematic and mimetic prey.

By manipulating the UV reflection of artificial prey offered to great tits, Lyytinen *et al.* (2001) examined whether UV cues might function as warning signals. They found no avoidance of UV-reflecting prey and no evidence that UV cues alone can work effectively as aposematic signals. However, it should be noted that most studies of warning colours do not take into account predator visual sensitivities or control for differences in brightness or colour contrast, as it has often been the case in the more recent studies of crypsis.

Few studies have also demonstrated that non-avian predators have the ability to learn the adaptive value of firefly larvae bioluminescence. De Cock and

Matthysen (1999) presented glowing and non-glowing prey dummies to toads *Bufo bufo* captured in areas where larvae of the common glow-worm were also observed. Toads showed lower frequency of attack and longer latency to attack glowing than non-glowing prey. Toads also found glow-worm larvae distasteful; after being exposed to these larvae, they increased their attack latency to luminescent prey, but not to non-glowing prey (De Cock and Matthysen, 2003). These studies confirm the aposematic function of glow-worm larva colouration, later used by adults as a sexual signal (Lewis and Cratsley, 2008).

5.3 THE IMPORTANCE OF CONTRAST WITH THE BACKGROUND

Contrast of warning colouration with the visual background has been studied in some detail. Results are ambiguous, a number of studies finding greater effectiveness of warning colouration on a contrasting background, others not. Sillén-Tullberg (1985b) investigated the survival of wild-type red aposematic individuals and grey mutant cryptic mutants of the harlequin seed bug *Lygaeus equestris* attacked by naïve zebra finches *Taeniopygia guttata*, and showed that although the survival of the aposematic larvae was higher than that of cryptic ones, the warning signal of the red form was not decreased on the matching background. Roper and Cook (1989) and Roper (1990) also found evidence that prey colours, rather than the degree of contrast against the visual background, defines unlearned feeding preferences in chicks. Similar results were found by Lindström *et al.* (1999) with hand-reared and wild-caught great tits predating mealworms.

In contrast with the study of Sillén-Tullberg (1985a), Veselý *et al.* (2006) showed that wild-coloured shieldbugs presented on matching backgrounds were attacked less frequently than those which were presented on a white background. Similarly, Gamberale-Stille (2001) demonstrated that aposematic milkweed bug *Tropidothorax leucopterus* larvae were attacked faster by domestic chicks on contrasting than on non-contrasting background, whereas there was no significant difference on attack latency on the palatable cotton stainer *Graptostethus servus* larvae. However, one may wonder if the domestic chick is really an ideal model predator.

To our knowledge, only one study considered the importance of brightness contrast on prey detection. It was conducted by Prudic *et al.* (2007) with milkweed bugs *Oncopeltus fasciatus* preyed upon by Chinese praying mantids *T. aridifolia sinensis*. Praying mantids were used because they are though to have very limited or no colour vision. Therefore, it allowed investigating if brightness contrast (but not colour contrast) could function as a warning signal to a colour-blind predator. The palatability of milkweed bugs was manipulated by feeding them for two generations either on a diet of sunflower seeds (*Helianthus annuus*, palatable) or milkweed seeds (*Asclepias curassavica*, unpalatable). Prey were painted in either of two shades of grey and presented on a grey background. Increased brightness contrast facilitated detection of prey, predator aversion learning of unpalatable prey, and memory retention of the aversive response.

Tullberg *et al.* (2005) used human 'predators' searching on a touch screen for larva of the swallowtail *Papilio machaon* to investigate if colouration may constitute a combination between warning colouration at short distance and crypsis at longer range. Pictures of larvae were taken at different distances on shore and fen habitats, and images were manipulated to increase or reduce the warning colouration. Logically, detection times increased with distance for all colours and backgrounds. More interestingly, it was found that natural colouration is neither maximally cryptic at short range, nor maximally conspicuous at long range. Overall, the results constitute the first empirical support for the hypothesis that a colour pattern can combine warning colouration at short range with crypsis at long distance. However, human observers might not be considered as appropriate to assess camouflage or conspicuousness of prey when the natural receiver is likely different. In addition, necessary calibrations of the images or computer screens were not undertaken.

5.4 WING SPOTS AS ANTI-PREDATOR DEVICES

Wing spots, most often called eyespots, have been studied since the nineteenth century for their anti-predator function. Three principal hypotheses have been evoked to explain this function (Table 4). (1) The 'intimidation hypothesis' which considers that large circular spots located centrally on the wings mimic the eyes of the predators' own enemies, intimidating predators and allowing the prey to escape. (2) An alternative version of the intimidation hypothesis is that wing spots intimidate predators because they are highly conspicuous and contrasting with the surrounding body region, and may be avoided as novel and rare features (the 'conspicuous signal hypothesis', Stevens, 2005). (3) The 'deflection hypothesis' which states that small spots at the periphery of the wings can deflect the attack of predators to non-vital regions of the body (review by Stevens, 2005). These hypotheses have been mostly tested with birds as predators, and received mixed support (Table 4).

5.4.1 *Investigations of the intimidation hypothesis*

Apart from highly qualitative experiments, this hypothesis had not been explicitly tested before Blest (1957). In this early study, European peacock butterflies *Inachis io* (Fig. 3) were presented to yellow buntings *Emberiza citrinella*. The butterfly has one pair of large spots on each wing; usually concealed, those are exposed when the butterfly is threatened and the movement is accompanied by a hissing noise. Living butterflies were used, either with wing spots or with eye spots removed by rubbing the forewings. It was found that butterflies with eyespots were given approximately four times as many overt escape responses from avian predators as those from which the eyespots had been removed. Blest (1957) also conducted eight other experiments presenting different spots of varied shapes and complexity to chaffinches, yellow buntings and great tits.

TABLE 4

Experimental tests of hypotheses raised to study the function of wing spots

Hypothesis	Prey	Predator	Support	References
Intimidation				
	Peacock butterfly	Birds	Yes (?)	Blest (1957)
	Peacock butterfly	Blue tit	Yes (?)	Vallin *et al*. (2005)
	Peacock butterfly	Hornet	No	Wiklund (2005)
	Butterflies	Blue tit	Yes (?)	Vallin *et al*. (2006)
	Butterfly, moth	Blue tit	No	Vallin *et al*. (2007)
	Artificial moth	Birds	No	Stevens *et al*. (2007a, 2008b, 2009a)
	Peacock butterfly	Great tit	Yes (?)	Kodandaramaiah *et al*. (2009)
Conspicuous signal				
	Artificial moth	Birds	Yes	Stevens *et al*. (2007a, 2008b,c, 2009a)
Deflection				
Wing spots	Butterflies	Birds	Yes	Swynnerton (1926)
	Mealworms	Yellow bunting	Yes (?)	Blest (1957)
	Butterfly	Anole lizard, flycatcher	No	Lyytinen *et al*. (2003)
	Butterfly	Great tit, pied flycatcher	No (only naïve birds)	Lyytinen *et al*. (2004)
	Butterfly	Anole lizard	No	Vlieger and Brakefield (2007)
'False head'	Butterflies	Birds	Yes	Robbins (1981)
	Butterfly	Blue jay	Yes	Wourms and Wasserman (1985)

? Denotes studies which conclusions have been severely criticized.

FIG. 3 The European peacock butterfly *Inachis io* often used to study the hypothesis of intimidating eyespots. Joanne T. Kell ©. (See Color Plate 43 at the back of the book.)

Overall, it was found that circular shapes were more effective in releasing escape response than bars or crosses, and that a model resembling an eye with a pupil was more effective than a single circle. Blest (1957) found similar effectiveness of circular eye-looking shapes by using an apparatus in which a dead mealworm placed in the middle of an image with symmetrical patterns was presented to avian predators. Despite their interest, Blest's (1957) experiments raised a number of criticisms. First, as noted by Curio (1993, cited by Stevens, 2005) the own eyes of the human experimenter could have biased responses of birds to butterfly eyespots. Second, manipulation of the wings to remove eyespots could have altered the normal display behaviour of butterflies (Stevens, 2005). Third, Blest (1957) did not control for factors such as stimulus area, perimeter, width, the number of spot rings (Stevens *et al.*, 2008b) or capture and handling processes (Ruxton *et al.*, 2004). Finally, as pointed out by Coppinger (1969, 1970), birds may simply have avoided highly contrasting novel stimuli.

Although their sample was too small for statistical testing, further support of Blest's conclusions was obtained by Cundy and Allen (1988), who used a mechanical butterfly often eliciting flight response of birds when displaying eyespots but rarely without eyespots. As Blest, Vallin *et al.* (2005) studied the anti-predation behaviour of peacock butterflies, painting over wing spots with a marker pen and/or removing the sound-producing parts of the forewings. Confirming the anti-predator function of wing spots, they found that only 1 out of 34 butterflies with eyespots was killed by blue tits, whereas 13 out of 20 butterflies without eyespots were attacked. The hissing sound did not provide increased protection from bird predation, although it clearly has a protective effect with respect to bat predation (Møhl and Miller, 1976, cited by Vallin *et al.*, 2005).

In this study, the sound associated with the wing movement had no clear advantage for butterflies attacked by passerine birds. This was confirmed by Kodandaramaiah *et al.* (2009) who showed that motionless peacock pansy butterflies *Junonia almana* with intact wing spots have higher survival than motionless butterflies with paint-covered wing spots. Displaying the wing spots is known to be advantageous regarding bird predation, but is clearly detrimental regarding predation by European hornets *Vespa crabro* (Wiklund, 2005). By studying the differential survival of a cryptic butterfly (the comma *Polygonia c-album*), of the peacock butterfly, and of an intermediately cryptic species (the small tortoiseshell *Aglais urticae*) preyed upon by blue tits, Vallin *et al.* (2006) again confirmed the anti-predation function of wing spots. Experimenting butterflies and passerine birds of different body sizes, namely small peacocks and larger eyed hawkmoths *Smerinthus ocellatus* which have wing spots of similar sizes, preyed upon by both blue tits and great tits, Vallin *et al.* (2007) showed that wing spots of the same size do not provide the same protection from predators: the smaller peacock prey showed higher survival than the larger hawkmoth. Also, the larger bird species was less intimidated than the smallest one. Although Vallin *et al.* (2005, 2006, 2007) and Kodandaramaiah *et al.* (2009) clearly demonstrated the protective advantage of wing spots, they could not support the intimidation hypothesis through recognition of predators' eyes *per se*. The conclusion of Stevens (2005), that 'few, if any, studies have provided firm evidence that predators perceive eyespots as eyes, rather than just conspicuous or novel features', is still valid.

5.4.2 *Investigations of the conspicuous signal hypothesis*

Stevens *et al.* (2007a) explicitly tested both the 'intimidating' and the 'conspicuous signal' hypotheses. As for the study of crypsis, artificial moth-like stimuli with a dead mealworm to attract bird predators were pinned to trees and their 'survival' was monitored in the field. This time, models differing by the shape and visual contrast of wing spots were used, and their design was calibrated using spectrometry and the vision model of an insectivorous passerine. The stimuli were printed on a grey scale because of the factorial increase in possible pattern complexity of coloured eyespots. A consistent result obtained across five experiments is that significantly higher survival was obtained for models displaying high internal contrast and high contrast with the target background, irrespective of the pattern arrangement. Interestingly, circular eyespots provided models with a higher survival than less eye-like shaped ones. However, this is interpreted more as a bias originating from the visual system of avian predators, which have circular receptive fields in their retina, than to the intimidating effect of predator eyes recognized by the butterfly predator. This contrasts with the results of Blest (1957) who showed that birds were more intimidated by circular models than by bars and crosses. However, Stevens *et al.* (2007a) used achromatic circles and diamonds, which would all appear circular from the initial viewing distance in their field study, whereas bars and crosses viewed at short distance in the study of Blest (1957) and triangles used in the study of Stevens *et al.* (2007a) can clearly be separated from circles, even from a distance. This innovative study strongly supported the conspicuous signal hypothesis.

Stevens *et al.* (2008b) further tested the intimidating and conspicuous signal hypotheses by using achromatic artificial moth models with different numbers, sizes and shapes of the spots (Fig. 4A). The spots also varied by the displacement of their components toward or away from the target midline to investigate if it creates the appearance of a predator staring at its prey. Models were calibrated by spectrometry and through photon catches of the blue tit. Across three experiments, better survival was for models of large size and higher number of spots. Models with circular spots did not survive better than models with other conspicuous shapes such as bars. Models mimicking eyes staring at predators did not survive better. The same approach was used by Stevens *et al.* (2009a) who used different marking shapes (bar or circle), arrangements (eye-like and non-eye-like positions) or colours (red, yellow or blue 'iris', Fig. 4B). They found no effect of shape, arrangement, or colour on birds' aversive responses. These results unambiguously support the conspicuous signal hypothesis and show that wing spots do not necessarily mimic eyes of other animals.

When testing the conspicuous signal hypothesis, it is important to determine if conspicuous wing spots are effective if they are present on prey which is either camouflaged or conspicuous against their visual background. This question has been investigated by Stevens *et al.* (2008c) with artificial moth-like prey calibrated using spectrometry and the blue tit vision model. As in the studies of Stevens *et al.* (2007a, 2008b), wing spots were effective in decreasing predation when they were placed on conspicuous prey. In contrast, results showed that spots increased predation risk of otherwise camouflaged targets.

Although rather simplified compared to natural wing spots, the models used in the ingenious experiments conducted by Stevens and colleagues clearly improved our knowledge on the features of insect warning signals. In the future,

FIG. 4 Artificial moth models used for the study of the function of wing spots. (A) Used by Stevens *et al.* (2008b). (B) Used by Stevens *et al.* (2009a). Martin Stevens ©. (See Color Plate 44 at the back of the book.)

experiments could be refined by using more complex models on different visual backgrounds, and include the use of different vivid colourations found in nature.

5.4.3 *Investigations of the deflection hypothesis*

In an early experiment, Swynnerton (1926) marked African butterflies of the genus *Charaxes* with artificial eyes, bars, etc., on the wing margins and released them on the field. When recapturing them at intervals, he found that marked butterflies survived longer and that those showing signs of having been attacked by birds bore beak marks and wing damage near the markings, confirming the deflection hypothesis. Blest (1957) tested this hypothesis by presenting yellow buntings with dead mealworms of three different kinds: either normal, painted with a white spot bearing a black pupil, or painted with a mealworm-coloured spot. Birds pecked more often prey that displayed an eyespot than control mealworms with a mealworn-coloured spot, therefore supporting the deflection hypothesis.

Lyytinen *et al.* (2003) used three forms of the squinting bush brown butterfly *Bicyclus anynana* (the spotless dry season form, the spotted wet season form and a double mutant line showing larger spots than the wet season form) which were presented to green anoles *Anolis carolinensis* and to pied flycatchers *Ficedula hypoleuca* (Fig. 5). The presence of marginal wing spots did not increase the survival probability, nor did it influenced the location of strikes on the body or wings of butterflies. Results were similar for both birds and lizards, and do not support the deflection hypothesis. Lyytinen *et al.* (2004) found weak support for the deflection hypothesis, but only when spotted and spotless squinting bush brown butterflies were presented to naïve pied flycatchers. Again using squinting bush brown butterflies and *Anolis* lizards, but in a more standardized experimental design, Vlieger and Brakefield (2007) also found no support for the deflection

FIG. 5 The squinting bush brown butterfly *Bicyclus anynana* often used to test the hypothesis of wing spots deflection. William Piel and Antónia Monteiro ©. (See Color Plate 45 at the back of the book.)

hypothesis. As pointed out by Stevens (2005) differences in the support of this hypothesis obtained by Swynnerton and researchers using squinting bush brown butterflies is that prey had more opportunities to escape on the field than in laboratory experiments. In addition, squinting bush brown butterflies have several spots on the wings, which may direct predator attacks to different locations, reducing the opportunity to detect a positive effect. Furthermore, wing spots are small and not concealed in *B. anynana*, which may suggest that they are not used to deflect predator attacks, but may have another function. This latter explanation is supported by the fact that wing spots of *B. anynana* could be involved in female mate choice (Monteiro *et al.*, 1997; Breuker and Brakefield, 2002; Robertson and Monteiro, 2005). Nevertheless, mate attraction and predator deception may not be totally conflicting. Using phylogenetic reconstructions of wing spot evolution and comparisons of evolutionary rates, Oliver *et al.* (2009) found that dorsal characters evolved at higher rates and more often displayed sex-based differences than ventral characters, supporting the prediction that dorsal characters may be used for mate signalling and ventral characters for predator avoidance.

One variation of the deflection hypothesis is related to the fact that some butterflies display wing spots and appendages resembling antennae or legs on the ventral part of their hindwings, which are interpreted as diverting the predator attack to the less vulnerable insect end (the 'false head' hypothesis, review by Stevens, 2005). Van Someren (1922) reported that lizards preferentially attack the 'false head' of lycaenid butterflies, allowing them to escape. Robbins (1981) tested this hypothesis on about 200 lycaenid butterflies. He measured their 'predicted deceptiveness' on the basis of the number of false-head wing patterns observed, and found that species with more false-head components had more frequent wing damage than other species. This was interpreted as reflecting the higher frequency of false-head patterns to deflect predators' attacks. Wourms and Wasserman (1985) conducted two experiments in which they offered blue jays dead or live cabbage white butterflies *Pieris rapae* which were either normal or painted with six different patterns on the wings. In experiments with dead butterflies, only prey painted with wing spots showed a deflection of bird attacks towards the hind region. In experiments with live butterflies, birds redirected their handling on the head, fatal to prey, to the hind region in butterflies painted a 'false head'. As a result, live butterfly with false-head markings had higher probability of survival, supporting the deflection hypothesis. These two experiments have been criticized because butterflies are very different from species naturally showing false heads (Cordero, 2001), because there is no control on the appearance of the paint with respect to natural colouration, and because experimental markings could have altered prey behaviour and palatability (Ruxton *et al.*, 2004). By measuring the distribution of wing damage in the Burmese junglequeen butterfly *Stichophthalma louisa*, Tonner *et al.* (1993) found that symmetrical wing damages, presumably inflicted by birds when the butterfly is at rest and its underside markings visible, were obviously concentrated on the hindwing section displaying a dark spot on the underside, resembling a false head. Asymmetric damage, inflicted in flight when the false head is not visible by the predator, occurred outside the false head region. In contrast to hypotheses stating that the colouration of the anal angle of hindwings is a primary visual attraction to predators, or that the false head attracts attacks toward the posterior end of prey, Cordero (2001) argued that the head is in fact the less vulnerable part of the butterfly because the eyes of the butterfly, which are their main predator detection devices, are on the head. In that case, the false head would repel predators' attacks from the vulnerable end, and redirect them toward the true head, enhancing the opportunity of predator detection.

Evidence for the deflection hypothesis with small marginal wing spots are clearly contradictory and the use of wing damage patterns as evidence of deflected attacks has been criticized (review in Ruxton *et al.*, 2004; Stevens, 2005). However, the false-head hypothesis is rather well supported by both field and laboratory experiments.

5.5 MOTION INFORMS ABOUT PALATABILITY

Motion is another visual feature used by predator to detect prey. Although it has long been recognized that conspicuous unpalatable species tend to have slower and more predictable movements than palatable species (references in Sherratt *et al.*, 2004), different theories have been raised to explain this behaviour. Chai and Srygley (Chai and Srygley, 1990; Srygley and Chai, 1990a,b) considered that unprofitable prey should not evolve rapid movement since they have no need to escape from predators, and particularly if rapid movements are energetically costly. Slow movement may also be selected to reduce recognition errors in experienced predators (Guilford, 1986; Srygley and Chai, 1990b). Hatle and Faragher (1998) studied survivorship of the highly conspicuous chemically defended Eastern lubber grasshopper *Romalea guttata* preyed upon by northern leopard frog *Rana pipiens*. They found that lubbers moved much more slowly than undefended crickets both in the presence of a plastic or a living frog, and that slow-moving lubbers benefited of higher survivorship than fast-moving ones. Because movement triggers frog attacks, the slow movement of prey appears as an effective strategy to avoid attack by motion-oriented predators (see also Hatle *et al.*, 2002). However, as pointed out by Sherratt *et al.* (2004), the motion orientation of predators could have evolved as a secondary consequence of the fact that unprofitable prey species tend to be slow. In addition, it is difficult to understand that if slow movements can deter an attack, why profitable prey would not use such behaviour? The above theories have been tested by Sherratt *et al.* (2004) using a computer program to simulate the evolution of locomotory traits seen by human 'predators'. It was shown that unprofitable prey indeed evolve significantly slower movement than profitable prey, particularly when they are not more selected to avoid predation than profitable prey, or when it is advantageous for unprofitable prey to avoid being mistaken for profitable prey. In that sense, they confirmed the hypotheses of Guilford (1986), Chai and Srygley (1990) and Srygley and Chai (1990a,b).

Another interesting set of studies concerns the relationships between motion cues and escape tactics in butterflies. Srygley (1994), studying morphological parameters related to flight in 18 butterfly species in the tribe Heliconiini and 10 of their non-heliconiine comimics, found that particular suites of the position of centre of body mass and of wing shape were associated with evasive flight, distastefulness and mimicry. The morphology of palatable species increases flight speed and maneuverability, whereas that of unpalatable species compromise flight performance. Although unpalatable species are less maneuverable, their more regular flight pattern allows considerable energy savings (see also Srygley, 2004). Most importantly, the morphology required for evasive flight in palatable species is different from the morphology of distasteful ones, making a 'good' Batesian mimic with the morphology and flight pattern of unpalatable butterflies unlikely to evolve. Therefore, by simply observing butterfly flight patterns (Srygley, 1994) and wing motion (Srygley, 2007), predators might guess if they target a palatable or an unpalatable prey. However, it remains to be determined if birds indeed react to tiny differences in butterfly wing motion and flight pattern. The fact that the butterfly flight pattern is the first stimulus seen by birds, not its colour pattern, has even led Kassarov (2003) to suggest that birds may not constitute the principal selective constraint leading to the evolution of mimicry and aposematism in butterflies.

5.6 WARNING-PATTERN SIZE AND SYMMETRY

The first experimental investigation of the effect of warning signal size and symmetry was conducted by Forsman and Merilaita (1999) who presented domestic chicks with artificial (black paper) butterflies either non-aposematic and palatable (associated with a chick-starter crumb) or aposematic and unpalatable (with a chick-starter crumb impregnated with quinine hydrochloride). Aposematic 'prey' had either symmetric small, symmetric large or asymmetric white circles on their wings. It was found that chicks avoid the larger warning signals, particularly if the aposematic pattern is symmetric. Although the level of asymmetry used is this study seems high (33% of the wing spot diameter) this result clearly shows the importance of size and symmetry in warning colouration patterns. Forsman and Herrström (2004) refined the study of Forsman and Merilaita (1999) by using different shapes, colours and levels of asymmetry in artificial butterflies presented to naïve chicks. They confirmed the detrimental effect of size asymmetry when asymmetry in size (spot diameter) was 7.5% or higher. They also showed that warning signals with symmetric colour pattern elements were better protected than asymmetric signals with pattern elements of different colour or shape. Therefore, there is strong evidence of the protective role of conspicuous colour patterns which are symmetric in colour, shape and size. Stevens *et al.* (2009b) conducted a field study of predation of artificial butterflies with two experiments: one testing the effect of size asymmetry, between 5% and 50% (this time measured as spot area, not diameter), and the other the effect of shape asymmetry and asymmetry in position from the body midline. Results strongly differed from those obtained in laboratory experiments with chicks. Both experiments showed no survival advantage for symmetric size, position or shape. The discrepancy between these studies may be due to the fact that one study was conducted in the field, the others in captivity, to the use of wild or domestic species, or to the fact that asymmetry may be less important than other features of warning signals, like colour or size. In addition, in the field, prey items are usually encountered in a sequential manner, whereas in the lab experiments prey were presented in pairs, giving predators a choice between the symmetrical and the asymmetrical prey.

5.7 OTHER FEATURES OF APOSEMATIC COLOURATION AND MIMICRY

The question of learned, innate, or unlearned avoidance of warning colouration by predators is largely influenced by predator psychology and experience (e.g. Chittka and Osorio, 2007; Ihalainen *et al.*, 2007, 2008), which goes far beyond our initial focus on predator vision of prey colouration. In addition, visual aspects of insect aposematism are generally combined with other signal modalities (olfactory, gustatory, auditory and/or behavioural) which often enhance unlearned biases and wariness (e.g. Skelhorn and Rowe, 2005; Hauglund *et al.*, 2006; Lindström *et al.*, 2006; Skelhorn and Rowe, 2006a,b; Lindstedt *et al.*, 2008; Siddall and Marples, 2008; Skelhorn *et al.*, 2008), but fall out of the scope of our review. Furthermore, the evolution of aposematic colouration has often been related to the degree of aggregation and to body size (e.g. Hatle and Salazar, 2001; Reader and Hochuli, 2003; Beatty *et al.*, 2005; Despland and Simpson, 2005; Skelhorn and Ruxton, 2006). For a more general overview of processes involved in insect aposematism and mimicry, we also recommend the book of Ruxton *et al.* (2004).

6 Predator visual mimicry

In the preceding section, we have seen that butterfly eyespots have long been thought to mimic large predators' eyes to deter predators from attacking prey. However, in that case, it was the predator's predator that was supposed to be mimicked. Another rarely reported case is that of predator mimicry by insect prey. Both colouration and movement are important components to mimic a predator and effectively decrease predation risk. To our knowledge, the first experimental demonstration of predator mimicry is that of the snowberry fly *Rhagoletis zephyria* and the common zebra spider *Salticus scenicus*, its model and predator (Mather and Roitberg, 1987). The study compared responses of spiders to conspecifics, to house flies *Musca domestica*, to snowberry flies and to snowberry flies which wing stripes had been obliterated. The results showed that spiders treated snowberry flies as conspecifics, but that effect was not maintained in flies which had their wing stripes blackened. Most house flies were attacked, whereas few conspecifics and intact snowberry flies were. It was not likely that the higher capture rate of flies with obliterated wing stripes was due to the obliteration process. Both the wings markings and the way they were displayed to predators seemed important components of the visual mimicry effect. A comparable case of predator mimicry was demonstrated by Greene *et al.* (1987) between the tephritid fly *Zonosemata vittigera* and the 11 species of jumping spiders. To investigate if the effect of the fly wing pattern and wing-waving display was involved in the mimicry of territorial displays of spiders, five prey types were presented: normal *Zonosemata*, *Zonosemata* with *Zonosemata* wings glued on (control), *Zonosemata* with house fly wings, house flies with *Zonosemata* wings and normal house flies. The wing pattern had a strong effect on predator retreat: normal *Zonosemata* and sham-operated control flies were attacked less frequently than the three other types. Indeed these flies vigorously waved their wings at the spider approach, and spiders reacted as when they observe the territorial display of a conspecific. Interestingly, *Zonosemata* with house fly wings displayed like normal *Zonosemata*, but were always attacked. House flies with *Zonosemata* wings never showed the wing-waving display and were always attacked and killed. Therefore, the wing stripes display of *Zonosemata* appeared to mimic territorial displays of salticid spiders and protect tephritid flies. Similar experiments were conducted with other predators: nonsalticid spiders *Oxyopes salticus*, mantids *M. religiosa*, assassin bugs *Pselliopus zebra* and whiptail lizards *Cnemidophorous uniparens*. The *Zonosemata* display was ineffective against these four types of predators, showing that *Zonosemata* is a specialized mimic of salticid spiders. Because many flies have leg-like wing patterns, jumping spider mimicry may be a widespread phenomenon. Very similar effects of wing flicking, combined with a jerky motion and sudden short flights, have been described in the fruit fly *Z. vittigera* and also provide protection from sympatric jumping spiders (Whitman *et al.*, 1988). More recently, Rota and Wagner (2006) demonstrated that *Brenthia* moths mimic jumping spiders of similar size with wing markings, wing positioning, posture and movement. These moths elicit territorial display from spiders, which provides much higher rates of survival than other species which does not exhibit wing patterns and movements typical of *Brenthia* moths. A few other authors have suggested, but not demonstrated, that insects may mimic spiders (O'Brien, 1967; Santiago-Blay and Maldonado-Capriles, 1988; Zolnerowich, 1992; Floren and Otto, 2001). One case of visual mimicry has even been hypothesized between planthoppers and spiders from the early Jurassic (Shcherbakov, 2007).

7 Colour polymorphism

7.1 THE REPRESENTATIVE CASE OF THE PEPPERED MOTH

Colour polymorphism is commonly encountered in camouflaged insect species. Particularly, widespread in nocturnal moths (Kettlewell, 1973; Majerus, 1998), it is also common in other groups like grasshoppers (Dearn, 1990), walking sticks (Sandoval, 1994), water boatmen (Popham, 1941), homoptera or mantids (Evans and Schmidt, 1990). How can polymorphism be maintained through evolutionary time despite the erosive actions of natural selection or genetic drift on genetic variation? Several different mechanisms have been identified to maintain high levels of polymorphism (references in Punzalan *et al.*, 2005). What is the contribution of visual predation in this evolutionary process? Identifying the factors explaining the relative survival values of different colour morphs and assessing the exact influence of visual selective predation has been largely documented in insect species (e.g. Kettlewell, 1955b; Sandoval, 1994; Nosil, 2004). One of the most largely investigated issues is that of the melanic polymorphism, commonly found in Lepidoptera and Coleoptera (Kettlewell, 1973; Lees *et al.*, 1973; Majerus, 1998). We will detail here the case of the peppered moth (*B. betularia*), one of the most widely quoted examples of evolution in action (Fig. 6).

The melanic polymorphism of this species, as in most other species (Lees *et al.*, 1973; Brakefield, 1987), is controlled by a single gene locus with melanic allele dominant to the non-melanic 'typica' allele (Clarke and Sheppard, 1964; Lees, 1968). First recorded ca. 1848, the black form rapidly spread through industrialized northern England in half a century. It then remained for 70–80 years (or generations in this 1-year generation species) at high frequencies (above 90% and below 100%) in urban areas and low frequencies in rural areas. Coincident with the development of coal-based industrialization, similar rise of melanics was observed in several species (e.g. two spot ladybird *Adalia punctata*; Bishop *et al.*, 1978) and synchronously in Britain, Europe and North

FIG. 6 *Typica* and *carbonaria* morphs of the peppered moth *Biston betularia* on the same tree. This species has often been used to study colour polymorphism. Marteen Sanne ©. (See Color Plate 46 at the back of the book.)

America (Grant *et al.*, 1996). Differential predation on different morphs appears the principal cause of the rapid evolution of colour polymorphism in the peppered moth. Active at night, moths rest on trees during the day, where they are preyed upon by birds. The ease at which the different morphs are found by the birds depends on how well they are camouflaged. Tutt (1896) was the first to propose that typical peppered moths were better camouflaged and hidden from avian predators on lichen-covered trees in unpolluted regions while melanics were better camouflaged than typical in areas where trees have been denuded of lichens and blackened by soot. This hypothesis of differential predation by birds was later tested experimentally by Kettlewell (1955b, 1956). Kettlewell's experiments consisted in (1) quantitative rankings of camouflage effectiveness for human vision of pale and black moths placed on different backgrounds, (2) direct observations of predation by birds on moths placed on tree trunks, and (3) recapture rates of marked moths released onto trees in polluted and unpolluted woodlands. He convincingly demonstrated that the individuals most conspicuous for humans were the first to be eaten by birds and had lower recapture rates, providing strong qualitative support for the central action of differential visual predation. However, the estimates of morphs' fitness values were less reliable and strongly criticized. Predation experiments involved the release of non-living prey items in abnormally high density, in non-natural over-exposed resting sites (Bishop, 1972) and a conspicuousness ranking not based on avian vision (Majerus, 1998). All these factors may have influenced birds' hunting behaviour and relative predation rate on the different morphs. For instance, these predation experiments predated the demonstration that birds were sensitive to UV (Chen *et al.*, 1984). Lichen species similar in appearance to human eye differ in their UV reflectance; non-melanic peppered moths at rest on lichens may be more cryptic to humans than to birds (Majerus *et al.*, 2000). Again, this recent finding calls attention to the need to analyze insect colours from the predator's point of view. Yet, Kettlewell noticed that the moths ranked as most conspicuous to humans were the moths first eaten by birds, confirming a roughly congruent ranking between humans and birds (Kettlewell, 1955b, 1956). At that level, the influence of lichens on moth crypsis had likely little importance. Despite methodological limitations (one can always find limitations or flaws in classic studies), the central role of differential visual predation in driving the evolution of polymorphism in the peppered moth remains undisputable (Brakefield, 1987; Majerus, 1998) and that is what should be retained from these studies.

Although heavily criticized, Kettlewell's experiments were not renewed nor improved, underlining the difficulty for quantifying the selection exerted by visual predators on different morphs in realistic conditions. After Kettlewell's qualitative evidence for visual predation, research effort orientated towards identifying the factors accounting for the dynamics and integrating them in different models (Mani, 1990). Conversely, little research effort aimed to acquire basic knowledge on prey and predator behaviour, on the mechanisms of prey detection and on selection quantification, many issues that would bring valuable contributions to the understanding of the mechanisms of polymorphism evolution (Brakefield, 1987). Few studies have focused on morphs' general behaviour and preferences in terms of resting sites, microhabitat or pairing. For instance, Kettlewell (1955a) showed that morphs of *B. betularia* settle on backgrounds they most resembled. A similar preference was later shown in the Scalloped Hazel moth *Gonodontis bidentata* (Kettlewell and Conn, 1977). Such morph-specific background preferences were also found by Grant and Howlett (1988). An active avoidance of non-matching backgrounds has been shown in several insect species (e.g. Edmunds, 1976; Owen, 1980; Sargent, 1981). The theory of multiple niche polymorphism posits that different genotypes should evolve preferences for the microenvironment in which they enjoy the highest fitness (Levene, 1953). Liebert and Brakefield (1987) studied the survival values of female peppered moths alone and in copula. They suggested that survival depended not only on the morph of the female alone but also on the morph of the male to which they were paired, since pairing implied in this species a relatively long physical association with potential vulnerability to predation.

7.2 PREDATOR PERCEPTUAL PROCESSES AND THEIR IMPACT ON EVOLUTION MORPH FREQUENCY

If selective predation by visual predators were the only influence on colour morph frequency, it would lead to monomorphic populations (allele fixation) the appearance of which would depend upon which phenotype is most cryptic in the local population (Creed *et al.*, 1980). Several factors have been suggested to contribute to the generation and maintain of the geographical variation of melanics (review in Brakefield, 1987). Heterozygote advantage, migration or frequency-dependent selection is needed to maintain polymorphism (Creed *et al.*, 1980).

Negative frequency-dependent selection, where the relative fitness of a genotype decreases with its increasing frequency in a population, appears particularly interesting in the case of visual predation. In this context, visual predators consume different colour morphs not in direct proportion to their numbers but disproportionally more of the common ones and less of the rare ones. In other words, they switch away from once common prey when they become rare in a sigmoid functional response (Holling, 1965). With this behaviour, they can prevent rare prey types from being eliminated and constitute a powerful force able to maintain prey colour polymorphism (Punzalan *et al.*, 2005). Poulton (1884) was probably the first to recognize the importance of frequency-dependent predation and proposed it could actively maintain colour polymorphism in the geometrid moth larvae (*Cyclophora* spp.). The first experimental demonstration of this mechanism, also called apostatic selection (Clarke, 1962), was performed by Popham (1941) who found that the most abundant morphs of an aquatic insect, the water boatman (*Arctocorisa distincta*), were preyed upon faster than the others and not in direct proportion to their numbers.

Despite a high number of studies presenting evidence of frequency-dependent predation (review in Punzalan *et al.*, 2005), evidence for the cognitive proximate mechanisms of such predation patterns remain scarce. Detecting camouflaged prey is a highly demanding visual task which mobilizes a large bandwidth in a visual system constrained by a limited visual information processing. Tinbergen (1960) proposed that predators selectively focus their attention on recently and commonly encountered prey types while ignoring the alternatives. Prior experience with a certain prey-type facilitates the detection of the same prey type in subsequent encounters. Guilford and Dawkins (1987) proposed the 'search-rate' hypothesis, whereby the detection of cryptic prey is not improved by the formation of search images, but only by a lower rate of visual search for more cryptic prey. By scanning a potential patch for a longer time, predators increase their probability of detecting cryptic prey.

These hypotheses are difficult to disentangle (Guilford and Dawkins, 1987) since they both posit that detection facilitation relates to immediate past experience with a short-time memory decay, and is more important for cryptic than conspicuous prey. These hypotheses differ on one important point: the search image hypothesis supposes that the construction of the search image provokes a reduced ability of detecting any other prey type, conspicuous or other equally cryptic prey types. Such an interference is not assumed by the search-rate hypothesis (Guilford and Dawkins, 1987; Punzalan *et al.*, 2005). Only the search image process can produce frequency-dependent selection, which does not mean it actually does. The existence of colour polymorphism is by no means a proof that search image processes are operating. Other mechanisms such as migration can maintain colour polymorphism. Because the efficiency of the search image process directly depends on the level of prey crypticity and rate of prey encounter, this mechanism may be of restricted relevance in many situations, for instance, in the case of multiple backgrounds offering different crypticity levels.

However, exploring the relative importance of these perceptual processes is particularly interesting since it can provide information about the allocation of cognitive resources, how predators see their prey, what visual cues can be used for prey detection and the selective pressures driving the evolution of colour patterns. Bond and Kamil (1998, 2002, 2006) built a series of elegant experiments to assess the selective effects of visual predators on prey crypticity and phenotypic variance, in which individual blue jays searched for digital moths on computer monitors. Prey brightness (the only parameter determining colouration) evolved via a genetic algorithm by which morph frequencies in the virtual population were conditioned by moth detection probability by predators and by mutation events. Using a textured background, they showed that detection accuracy decreases with increasing dissimilarity between any two successive detected prey items, particularly not for moderately cryptic prey but for highly cryptic prey (Bond and Kamil, 2002). Results showed that birds failed to detect atypical prey among highly cryptic prey, suggesting the cognitive interference postulated by the search image hypothesis. Furthermore, this interference was higher for more cryptic prey, confirming that search image was particularly relevant in the case of highly cryptic prey. They demonstrated that frequency-dependent visual predation promotes crypticity and phenotypic variance in the prey.

7.3 ROLE OF BACKGROUND IN MORPH DETECTION

Further experiments on virtual prey explored the role of background in the visual selection operated by predators on prey brightness patterns (Fig. 7; Bond and Kamil, 2006). Background structure has two types of effects on the evolution of colour pattern in prey. First, the scale of background heterogeneity directly determines the average level of crypticity found in prey (Merilaita, 2003). Visually complex backgrounds select for less cryptic prey than do 'simple' backgrounds. Background heterogeneity determines the type of brightness patterns found in the prey. Backgrounds with small-scale spatial variation (patches lower than moth size, Fig. 7) induce the evolution of generalists. Such backgrounds are relatively homogeneous but visually complex, with no large patch of distinctive substrate, but enough complexity to enable a range of diverse forms to be equally difficult to detect. Such fine-grained backgrounds can be for instance temperate grasslands, exposed rocks and soil, leaf litter or beach gravel. Because many different morphs can produce the same level of resemblance to the background, it is impossible to predict the number of expected different prey patterns from the background structure. Conversely, coarse-grained backgrounds with large-scale spatial variation (patches larger than moth size, Fig. 7) induce the evolution of specialists. In such heterogeneous environments divided into large, disparate substrate patches, individuals occupy only one substrate type at a time. As a consequence, there can be a strong selection for a close association between morphs and backgrounds. For instance, the walking stick *Timema cristinae* (Timemidae) shows a polymorphism on colour and pattern. Striped and unstriped green morphs are closely associated with different host plant species. Experiments using lizards and birds showed that morphs were associated to the plant on which they presented a higher survival (Sandoval, 1994). Further experiments show that morphs' survival depended on the host plant species only in the presence of predators (Nosil, 2004), suggesting that the host preference is in fact a visual background preference. Predators operate a disruptive selection by reducing the survival of morphs on their non-preferred host plant. This reduced survival of 'migrants' can efficiently contribute to reproductive isolation and speciation, as shown in pea aphids, leaf beetles or butterflies (references in Nosil, 2004).

Second, background complexity also exerts additional selective effects mediated by differences in how predators search for and detect prey items. A background with high spatial variation induces slow, serial search processes where selective attention plays a central role (Bond and Kamil, 2006). When the background gets separated into larger coherent patches, it becomes beneficial for predators to focus on the currently most rewarding background

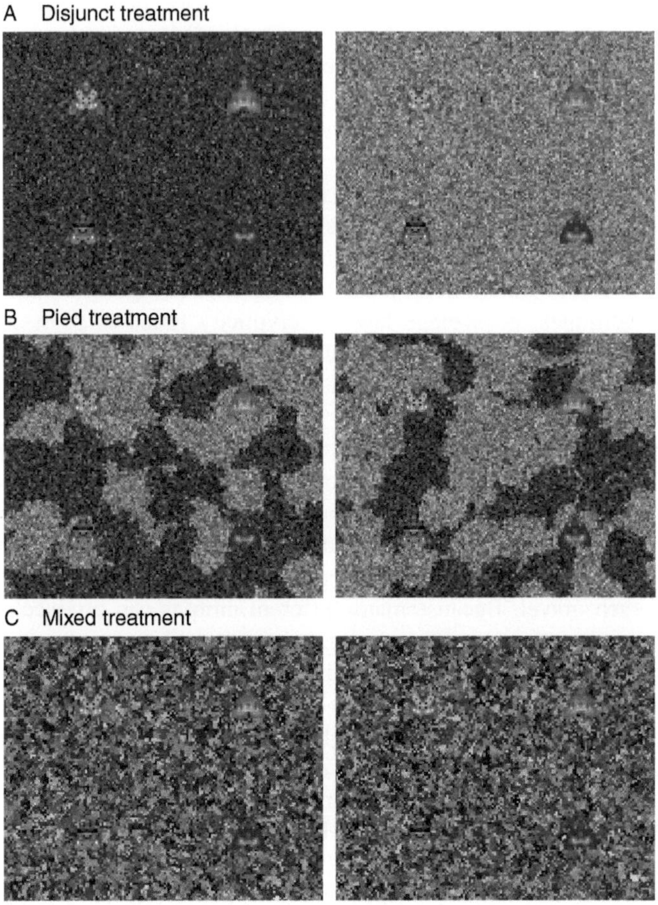

FIG. 7 Digital moths seen on disjunct (A), mottled (B) and speckled (C) backgrounds. The scale of background spatial heterogeneity is much higher than moth size (A), similar to moth size (B) or lower than moth size (C). Here, the same four moths are represented on the three backgrounds to compare their detectability. These moths evolved on the disjunct background and are the most cryptic of the individuals in their population. While on the disjunct background, moths are harder to detect on the patch they most closely resemble but can readily be located in a superficial scan, they are far more difficult to detect on backgrounds with high levels of noise at spatial frequencies comparable to moth size. After Bond and Kamil (2006). Alan Bond ©.

patch type and search entire patches rapidly in parallel. Background complexity determines predator searching behaviour and consequently the selective pressure they exert on different colour morphs.

7.4 MORPH-DEPENDENT BACKGROUND PREFERENCE, A NON-VISUAL SELECTION OF MORPH COLOURATION?

We have seen that prey detectability depends on the background against which it is seen. In some cases, there can be a strong selection for a close association between background and colouration. Some variations in colour are determined by the environment while others (and *in fine* all the limits to the variations in colour) are genetically determined. The dependence of colouration on environmental cues allows a more efficient and closer association between morph and background. The relative importance of these two levels of variation varies between species. For instance, two ground-hopper species of the genus *Tetrix* (Tetrigidae, Orthoptera) show dimorphism on colour and pattern. Contrary to pattern, colour morph frequency in nymphs responds to background colouration perceived by larvae (Hochkirch *et al.*, 2008). Many grasshopper species exhibit homochromic response to background colouration; such a phenomenon is known from burnt sites where dark individuals predominate (Rowell, 1971). The colour morph of the peppered moth larva is influenced by the background colour to which it is exposed during the previous larval instars, with an induction mediated by visual input (Noor *et al.*, 2008). Similarly, the larvae of various hawkmoth species develop different colours if reared on different trees; for these folivorous insects, the leaf reflectance and not the food itself nor the leaf texture determine the coloured developed by the larva (Grayson and Edmunds, 1989). For other species like bush crickets, the environmental cue mediating colour determination is not visual. In this species, colouration varies with the season, with green individuals in spring and brown individuals in autumn; colouration is determined by water intake, which shows a seasonal variation (Lymbery, 1992).

Selection operates not only on colouration but more generally on behavioural components linked to background exploitation. Insects are ectothermic animals for which habitat use is constrained not only by predation avoidance but also by energetic requirements. Experiments conducted on the common ground-hopper *Tetrix undulata* (Tetrigidae) showed that morphs select environments that are close to their optimal body temperature, actively avoiding high and low temperatures (Ahnesjo and Forsman, 2006). Dark individuals avoid high temperatures more actively. Besides this morph-dependent habitat preference, there is also a sex-dependent habitat preference originating from different temperature requirements in males and females. Consequently, different colour morphs (and sexes) select different habitat types due to their different thermal quality, even if at local points, the substrate seems to be chosen independently of morph type.

Colouration affects vulnerability to predation, directly through visual detectability and indirectly through thermoregulation ability and behaviour. For example, pale individuals warm up more slowly, bask more frequently for shorter periods compared to dark individuals. Because body temperature also influences escape behaviour, different morphs can show different escape potential and/or strategies. Civantos *et al.* (2004) investigated the indirect effects of morph colouration on predation risk by performing experiments in the pigmy grasshopper (*Tetrix subulata*). They tested the escape potential of black and grey morphs to predation by a lizard (*Psammodromus algirus*). In the shade, both morphs show low activity and poor escape potential, and are equally vulnerable to predators. In the sun, black individuals are more active and show a higher escape potential than grey individuals. This increased activity translates into an increased attack probability but, because of enhanced escape potential, not into an increased predation rate. Counter to intuition, the existence of these indirect effects of colouration—here on thermoregulation and behaviour—makes it possible that the evolution of colouration in the prey could be selected by predators which not primarily rely on visual cues but on cues influenced by prey colouration. In other words, a close association between colour morph and background can lay the ground for a non-visual selection of morph colouration by predators.

7.5 VISUAL DETERMINANTS OF MORPH SELECTIVE VALUE

In conclusion, we can see that the survival value of a particular colour morph results from a combination of many factors, presented here in a non-exhaustive list. Vulnerability to predation primarily depends on the colour pattern of the individual itself as well as the background on which it stands since they determine the level of detectability of the prey. Prey colour and pattern, background spatial heterogeneity and complexity play an important role. Relative colour preferences shown by predators determine the relative value of the different colour morphs in a population. For instance, various coccinellid species differ by their preference for specific morphs of the pea aphid (*Acyrthosiphon pisum*). *C. septempunctata* prey on the morphs that most contrast on the background whatever their colouration while a closely related species, *Harmonia axyridis*, shows a significant preference for red over green morph (Harmon *et al.*, 1998). The relative contribution of these predators to prey population dynamics determines which morphs are more advantaged in a population. Not only detectability but also prey motion and behaviour can also profoundly influence detectability. In insects for which mating implies a long association of partners, the phenotype of the partner also influences the survival value of a focal individual. This survival cost has been suggested for example in polymorphic moths (Liebert and Brakefield, 1987) or walking sticks (Nosil, 2004). More interestingly, predation risk also depends on the frequency of the colour morph in the population, rare morphs being relatively more protected than common ones (Popham, 1941; Bond and Kamil, 1998, 2002, 2006). Protection of rare colour morphs can be even more effective if predators are conservative in their search and choice of prey. A field test involving avian predators and prey consisting in pastry items of different colours assembled in monomorphic or polymorphic populations showed that red morphs perform worse in monomorphic than in polymorphic groups. This suggested that the initial evolution and spread of specific signals may be promoted in polymorphic groups (Wennersten and Forsman, 2009). Finally, predation risk of a given colour morph depends on its spatial proximity to other phenotypes, particularly to conspicuous ones. Kettlewell (1955b) first mentioned that some of the most cryptic peppered moths were found and eaten by birds because of their proximity to conspicuous individuals. Such a facilitation found for a polymorphic species has been recently tested in the case of different prey species showing different conspicuousness levels. Zhang and Richardson (2007) showed that trouts predating on active coloured stonefly larvae showed an enhanced predation on cryptic inactive mayfly larvae when prey individuals were in close proximity. Finally, in aposematic species, the survival value of conspicuous morphs is expected to decrease if the proportion of informed predators decreases in the population, if the proportion of mimetic non-toxic morphs of different species increases.

8 Discussion

This review has concentrated on decrypting how predators see insects' colouration and more generally insects' visual appearance and on detailing how predators use the visual information provided by insects to detect and capture them. Through examples taken from various predator–prey systems, we have revealed the evolutionary interplay between insect colouration and predator behaviour. Visual predation generates an important evolutionary pressure that can determine the evolution of specific colour patterns and behavioural strategies in insects (countershading, disruptive colouration, masquerade, warning colouration strategies). Insect colouration in its turn selects for the evolution of flexible hunting tactics in predators (versatility of predator behaviour, selective attention to prey visual features).

Insect predators are highly diverse, not only in their taxonomic position but also in the way they see insect colours and visual appearance more generally. These multiple views of common targets arise from differences not only in visual sensitivity but more widely in visual information processing—neural mechanisms of visual perception, selective attention to specific cues borne by insects, relative importance given to vision compared to other sensory channels. Such diversity generates both methodological and conceptual challenges that are progressively taken up. First, from a methodological point of view, the diversity shown by predators in their visual performance imposes the necessity to reconstruct colours as they are likely perceived by predators. Long submitted to a convenient but misleading anthropocentrism, the measurement and analysis of prey colours begin to incorporate predators' visual systems and acknowledge their importance in determining the evolution of insect colouration and colour patterns. Although this approach often requires numerous and tedious measurements of colours based on spectrometry (Endler, 1990) or digital image analysis

(Stevens *et al.*, 2007b), the objective quantitative estimates permitted by powerful techniques convey a far greater potential to reveal biologically relevant perceptual mechanisms and evolutionary processes than traditional colour scoring or human-based categorization. Some new research questions have developed thanks to this change of point of view and important technical improvements. One of the best examples is probably the role of visual cues in nocturnal predation, an area where research is still in its infancy, calling for urgent experimental and modelling studies to get a better understanding of the evolutionary pressures exerted by visual predators on the evolution of nocturnal insects colouration. Second, discrepancies in predators' view of their prey can be seen of course at high taxonomic levels but more interestingly at low taxonomic levels. Substantial differences have been observed between closely related species, for instance in visual performance, colour preferences or hunting strategies. As a consequence, investigators should acquire the knowledge about predators' behaviour for their own predator–prey system, avoiding whenever possible to take close relatives as surrogate species and to make generalization about their validity of their results. Third, predators differ in the way they perceive insect colouration. This means that the information content of colour is not universal but depends on which prey displays it and which predator views this colour or pattern. This context dependence is even reinforced by the fact that insects move in an environment showing a high temporal and spatial visual variability which maintains variability in their visual appearance to potential receivers. The non-universality of the information transmitted also arises from predators' cognitive processes, for instance unlearned preferences and avoidance for specific colours or patterns or recent experience of association between colour and profitability.

It is interesting to view the research conducted on insect colouration from a historical perspective. Most of the hypotheses about the adaptive functions of insect colours and patterns have been formulated long ago by Poulton (1890), Thayer (1909) or Cott (1940). These authors took advantage of their large naturalist culture and experience to propose conceptual advances that have proved highly relevant further on. After 1950, research on insect colouration entered an intense 'hypothesis testing' phase. Although they helped to decrypt important perceptual mechanisms (e.g. differential predation in polymorphic populations, image search hypothesis), these studies suffered from technological limitations in the measurement and analysis of insect colouration (e.g. investigations of the intimidation or deflection hypothesis for eyespots). The application of powerful techniques of colour investigation (spectrometry and digital photography) gives a new life to the study of the functional roles of insect colours and patterns at the end of the twentieth century. The impressive work of Stevens and colleagues (e.g., Cuthill *et al.*, 2005; Stevens and Cuthill, 2006; Stevens *et al.*, 2007a, 2008b, 2009b) is a good example of that 'Renaissance'. Studies relying on a thorough colour measurement independent of human vision and the incorporation of predator vision in colour analysis helped not only to test existing hypotheses but also to formulate new ones. In addition, some ingenious experimental procedures allow to progressively incorporate more realistic viewing conditions or stimuli and to simulate long-term dynamics of colour evolution, an aspect that had been hitherto impossible to tackle (e.g. computer moths with complex patterns presented on increasingly complex backgrounds in the work of Bond and Kamil, 1998, 2002, 2006). At the same time, new models are available to interpret colours from a predators' perspective (Vorobyev and Osorio, 1998; Endler and Mielke, 2005) that offer possibilities of getting closer to more realistic viewing conditions (incorporation of visual system limits, of background complexity). We should take advantage of the recent development of all these powerful tools and methods to fill some knowledge gaps that have developed all along the history of research on insect colouration and predators' behaviour.

The knowledge acquired so far about insect colouration as viewed by predators suffers from several defaults. These defaults define where to focus future research effort. First, the traditional human-based assessment of colouration is now often replaced by more objective measures of colouration and the consideration of predators' visual system, light and background spectral characteristics in colour analysis. This approach is far from being generalized (Civantos *et al.*, 2004; Hyden and Kral, 2005; Hochkirch *et al.*, 2008; Noor *et al.*, 2008) and more effort should be put in reconstruct colours as they are likely perceived by predators in the specific viewing conditions of the study.

Second, research has concentrated on a restricted number of taxonomic groups, both on the prey and the predator sides. The large number of studies dealing with birds preying on butterflies for camouflage, warning colouration or polymorphism masks a striking lack of knowledge for what is going on in other predator–prey systems. Birds are interesting models from several aspects but they are largely inappropriate to explore questions like nocturnal vision. Similarly, Lepidoptera are not representative of insect behaviour in general. This questions the external validity of the hypotheses tested on these systems. For example, would eyespots bear the same function in insect groups other than Lepidoptera or for predator other than birds? Future research effort should try to explore a larger array of prey and predator groups to question the universality of predators' perceptual mechanisms or the adaptive functions of insect colouration.

Third, research has completely left aside particular signals like iridescent signals. Yet, iridescence, that is the change of colouration with the viewing angle, is abundant in insects (reviews in Ingram and Parker, 2008; Doucet and Meadows, 2009; Seago *et al.*, 2009). Investigation of iridescent signals requires measurement tools that have been traditionally used by physics (Vukusic and Stavenga, 2009) but are rarely used by biologists, probably because of their high level of technicity. As a consequence, it is still unknown (Doucet and Meadows, 2009) whether iridescence *per se* could provide decrease predation risk thanks to the flexibility it creates in the apparent colouration. It may increase camouflage in a visually complex environment, contribute to startle predators by creating sudden changes in brightness or colour or increase the efficacy of warning colouration by providing highly detectable signals. These hypotheses have not hitherto been experimentally tested (Doucet and Meadows, 2009) but should open promising lines of research.

Fourth, experimental studies have often used simplistic artificial backgrounds and targets focusing on one predator species often placed in artificial viewing conditions (e.g. Iwasaki, 1990; Poteser and Kral, 1995; Kral and Poteser, 1997; Yamawaki, 2000; Gamberale-Stille and Tullberg, 2001; Miklósi *et al.*, 2002). Future work should try to reproduce more realistic conditions by incorporating visually more complex backgrounds and more realistic stimuli not only incorporating achromatic components (Bond and Kamil, 2002, 2006; Stevens *et al.*, 2007a, 2008b,c) but also chromatic components (Stobbe and Schaefer, 2008; Stobbe *et al.*, 2009).

Fifth, studies tend to overlook the natural behaviours shown by both the insects and their predators; this is particularly striking in the investigation of colour polymorphism, for which the peppered moth is representative of the lack of interest for behavioural and naturalistic aspects. Yet, a better understanding of the biology and ecology of both prey and predators would help to design protocols, to interpret the results obtained in artificial conditions and to formulate new hypotheses.

Sixth, the vast majority of studies adopt an experimental approach using a system with a unique prey and/or a unique predator. Such an approach is indispensable to gain valuable information about the relative importance of different visual cues in predator's hunting behaviour and to unravel the mechanisms at play in detection, recognition and capture. However, the monographic aspect of these studies restricts the potential generalization of the results obtained. It is time to go towards more integrative approaches at an interspecific level. Comparative approaches are powerful to gain understanding of the adaptive functions and the evolutionary pressures (natural or sexual selection) that determine the evolution of colouration and colour patterns. By making *a priori* hypotheses about the possible functional roles of colouration and by comparing the predictions to the realized patterns within a phylogenetic framework, it is possible to identify the communication strategies common to different species, their relationship to species' visual environment and to test the validity of hypotheses at a large scale. Encompassing the diversity shown at species level, such approaches have shed new light on the evolution of bird (Gomez and Théry, 2007) or mammal (Ortolani and Caro, 1996) colouration. Comparative analyses on insect colouration are nearly absent. A valuable study at large interspecific level, but unfortunately without phylogenetic control or consideration of the UV, was conducted by Williams (2007) who compared colour patterns in all bumblebee species worldwide and related them to possible functions like thermoregulation or communication. A phylogenetic control could have allowed him to reconstruct the history of different colour pattern characters, to individualize colour characters according to their evolutionary trajectory and to reveal the evolutionary pressures operating specifically on different components of insect colouration. More interesting is the comparative study conducted by Song and Wenzel (2008) on the evolution of colour plasticity in locusts. With a full phylogenetic control, this study established that locust basic colouration and black patterns are under different evolutionary pressures. It also showed the interest of applying the comparative method not only to fixed characters but also to plastic characters that change with environmental conditions. Although comparative methods require special effort to gather information about species biology and behaviour and phylogenetic relationships, they largely repay the effort granted. Now that a large number of hypotheses have been validated for a few species, it is time to give priority to large interspecific studies over monographic approaches in order to question the universality of predation mechanisms and insect colouration strategies.

Acknowledgements

We thank Alan Bond for producing Fig. 7 especially for this chapter. We are grateful to Martin Stevens, Joanne T. Kell, William Piel and Antónia Monteiro for providing pictures. We thank the editor Jérôme Casas for inviting us to write this review, and Lars Chittka, Alexandra Barbosa and one anonymous reviewer for their comments on the manuscript.

References

Ahnesjo, J. and Forsman, A. (2006). Differential habitat selection by pygmy grasshopper color morphs; interactive effects of temperature and predator avoidance. *Evol. Ecol.* **20**, 235–257.

Autrum, H. and Kolb, G. (1968). Spectral sensitivity of single visual cells of Aeschnidae. *Zeitschrift Fur Vergleichende Physiologie* **60**, 450–477.

Backhaus, W. and Menzel, R. (1987). Color distance derived from a receptor model of color vision in the honeybee. *Biol. Cybern.* **55**, 321–331.

Bain, R. S., Rashed, A., Cowper, V. J., Gilbert, F. S. and Sherratt, T. N. (2007). The key mimetic features of hoverflies through avian eyes. *Proc. R. Soc. Lond. B* **274**, 1949–1954.

Beatty, C. D., Bain, R. S. and Sherratt, T. N. (2005). The evolution of aggregation in profitable and unprofitable prey. *Anim. Behav.* **70**, 199–208.

Bell, G. P. (1985). The sensory basis of prey location by the California leaf-nosed bat *Macrotus californicus* (Chiroptera: Phyllostomidae). *Behav. Ecol. Sociobiol.* **16**, 343–347.

Bennett, R. R. and Ruck, P. (1970). Spectral sensitivities of dark-adapted and light-adapted *Notonecta* compound eyes. *J. Insect Physiol.* **16**, 83–88.

Bennett, A. T. D., Cuthill, I. C. and Norris, K. J. (1994). Sexual selection and the mismeasure of color. *Am. Nat.* **144**, 848–860.

Bernard, G. D. and Stavenga, D. G. (1979). Spectral sensitivities of retinular cells measured in intact, living flies by an optical method. *J. Comp. Phys.* **134**, 95–107.

Bishop, J. A. (1972). An experimental study of the cline of industrial melanism in *Biston betularia* (L) (Lepidoptera) between urban Liverpool and rural North Wales. *J. Anim. Ecol.* **41**, 209–243.

Bishop, J. A., Cook, L. M. and Muggleton, J. (1978). Response of two species of moths to industrialization in Northwest England. Part 1. Polymorphisms for melanism. *Phil. Trans. R. Soc. Lond. B* **281**, 489–515.

Blest, A. D. (1957). The function of eyespot patterns in the Lepidoptera. *Behaviour* **11**, 209–256.

Blest, A. D. (1985). The fine structure of spider photoreceptors in relation to function. In: *Neurobiology of Arachnids* (ed Barth, F. G.), pp. 79–102. Springer-Verlag, Berlin, Heidelberg, New York, Tokyo.

Blest, A. D., Hardie, R. C., McIntyre, P. and Williams, D. S. (1981). The spectral sensitivities of identified receptors and the function of retinal tiering in the principal eyes of a jumping spider. *J. Comp. Phys.* **145**, 227–239.

Bond, A. B. and Kamil, A. C. (1998). Apostatic selection by blue jays produces balanced polymorphism in virtual prey. *Nature* **395**, 594–596.

Bond, A. B. and Kamil, A. C. (2002). Visual predators select for crypticity and polymorphism in virtual prey. *Nature* **415**, 609–613.

Bond, A. B. and Kamil, A. C. (2006). Spatial heterogeneity, predator cognition, and the evolution of color polymorphism in virtual prey. *Proc. Natl. Acad. Sci. USA* **103**, 3214–3219.

Bowdish, T. I. and Bultman, T. L. (1993). Visual cues used by mantids in learning aversion to aposematically colored prey. *Am. Midl. Nat.* **129**, 215–222.

Bowmaker, J. K., Thorpe, A. and Douglas, R. H. (1991). Ultraviolet-sensitive cones in the goldfish. *Vision Res.* **31**, 349–352.

Bowmaker, J. K., Kovach, J. K., Whitmore, A. V. and Loew, E. R. (1993). Visual pigments and oil droplets in genetically manipulated and carotenoid deprived quail – a microspectrophotometric study. *Vision Res.* **33**, 571–578.

Brakefield, P. M. (1987). Industrial melanism: do we have the answers? *Trends Ecol. Evol.* **2**, 117–122.

Brandt, M. and Mahsberg, D. (2002). Bugs with a backpack: the function of nymphal camouflage in the West African assassin bugs *Paredocla* and *Acanthaspis* spp. *Anim. Behav.* **63**, 277–284.

Breuker, C. J. and Brakefield, P. M. (2002). Female choice depends on size but not symmetry of dorsal eyespots in the butterfly *Bicyclus anynana*. *Proc. R. Soc. Lond. B* **269**, 1233–1239.

Briscoe, A. D. and Chittka, L. (2001). The evolution of color vision in insects. *Annu. Rev. Entomol.* **46**, 471–510.

Burkhardt, D. (1989). UV vision: a bird's eye view of feathers. *J. Comp. Physiol. A* **164**, 787–796.

Caine, N. G. and Mundy, N. I. (2000). Demonstration of a foraging advantage for trichromatic marmosets (*Callithrix geoffroyi*) dependent on food colour. *Proc. R. Soc. Lond. B* **267**, 439–444.

Calderone, J. B. and Jacobs, G. H. (1999). Cone receptor variations and their functional consequences in two species of hamster. *Vis. Neurosci.* **16**, 53–63.

Canfield, M. R., Chang, S. and Pierce, N. E. (2009). The double cloak of invisibility: phenotypic plasticity and larval decoration in a geometrid moth, *Synchlora frondaria*, across three diet treatments. *Ecol. Entomol.* **34**, 412–414.

Chai, P. and Srygley, R. B. (1990). Predation and the flight, morphology, and temperature of neotropical rain-forest butterflies. *Am. Nat.* **135**, 748–765.

Chen, D. M., Collins, J. S. and Goldsmith, T. H. (1984). Ultraviolet receptor of bird retinas. *Science* **225**, 337–339.

Chiao, C.-C., Vorobyev, M., Cronin, T. W. and Osorio, D. (2000). Spectral tuning of dichromats to natural scenes. *Vision Res.* **40**, 3257–3271.

Chittka, L. (1992). The colour hexagon: a chromaticity diagram based on photoreceptor excitations as a generalized representation of colour opponency. *J. Comp. Physiol. A* **170**, 533–543.

Chittka, L. (1996). Optimal sets of color receptors and color opponent systems for coding of natural objects in insect vision. *J. Theor. Biol.* **181**, 179–196.

Chittka, L. and Osorio, D. (2007). Cognitive dimensions of predator responses to imperfect mimicry? *PLoS Biol.* **5**, 2754–2758.

Church, S. C., Bennett, A. T. D., Cuthill, I. C., Hunt, S., Hart, N. S. and Partridge, J. C. (1998). Does lepidopteran larval crypsis extend into the ultraviolet? *Naturwissenschaften* **85**, 189–192.

Civantos, E., Ahnesjo, J., Forsman, A., Martin, J. and Lopez, P. (2004). Indirect effects of prey coloration on predation risk: pygmy grasshoppers versus lizards. *Evol. Ecol. Res.* **6**, 201–213.

Clarke, B. C. (1962). Balanced polymorphism and the diversity of sympatric species. In: *Taxonomy and Geography: A Symposium* (ed Nichols, D.), pp. 47–70. Systematics Association, Oxford.

Clarke, C. A. and Sheppard, P. M. (1964). Genetic control of melanic form insularia of moth *Biston betularia* (L). *Nature* **202**, 215–216.

Coppinger, R. P. (1969). The effect of experience and novelty on avian feeding behavior with reference to the evolution of warning coloration in butterflies. Part I. Reactions of wild-caught adult blue jays to novel insects. *Behaviour* **35**, 45–60.

Coppinger, R. P. (1970). The effect of experience and novelty on avian feeding behavior with reference to the evolution of warning coloration in butterflies. Part II. Reactions of naive birds to novel insects. *Am. Nat.* **104**, 323–335.

Cordero, C. (2001). A different look at the false head of butterflies. *Ecol. Entomol.* **26**, 106–108.

Cott, H. B. (1940). *Adaptive Coloration in Animals*. Methuen, London.

Cowing, J. A., Arrese, C. A., Davies, W. L., Beazley, L. D. and Hunt, D. M. (2008). Cone visual pigments in two marsupial species: the fat-tailed dunnart (*Sminthopsis crassicaudata*) and the honey possum (*Tarsipes rostratus*). *Proc. R. Soc. Lond. B* **275**, 1491–1499.

Creed, E. R., Lees, D. R. and Bulmer, M. G. (1980). Pre-adult viability differences of melanic *Biston betularia* (L) (Lepidoptera). *Biol. J. Linnean Soc.* **13**, 251–262.

Cundy, J. M. and Allen, J. A. (1988). Two models for exploring the anti-predator function of eyespots. *J. Biol. Educ.* **22**, 207–210.

Curio, E. (1976). *The Ethology of Predation*. Springer Verlag, Berlin.

Cuthill, I. C. and Székely, A. (2009). Coincident disruptive coloration. *Phil. Trans. R. Soc. Lond. B* **364**, 489–496.

Cuthill, I. C., Bennett, A. T. D., Partridge, J. C. and Maier, E. J. (1999). Plumage reflectance and the objective assessment of avian sexual dichromatism. *Am. Nat.* **160**, 183–200.

Cuthill, I. C., Stevens, M., Sheppard, J., Maddocks, T., Parraga, C. A. and Troscianko, T. S. (2005). Disruptive coloration and background pattern matching. *Nature* **434**, 72–74.

Cuthill, I. C., Hiby, E. and Lloyd, E. (2006a). The predation costs of symmetrical cryptic coloration. *Proc. R. Soc. Lond. B* **273**, 1267–1271.

Cuthill, I. C., Stevens, M., Windsor, A. M. M. and Walker, H. J. (2006b). The effects of pattern symmetry on detection of disruptive and background-matching coloration. *Behav. Ecol.* **17**, 828–832.

De Cock, R. and Matthysen, E. (1999). Aposematism and bioluminescence: experimental evidence from glow-worm larvae (Coleoptera: Lampyridae). *Evol. Ecol.* **13**, 619–639.

De Cock, R. and Matthysen, E. (2001). Do glow-worm larvae (Coleoptera: Lampyridae) use warning coloration? *Ethology* **107**, 1019–1033.

De Cock, R. and Matthysen, E. (2003). Glow-worm larvae bioluminescence (Coleoptera: Lampyridae) operates as an aposematic signal upon toads (*Bufo bufo*). *Behav. Ecol.* **14**, 103–108.

De Ruiter, L. (1952). Some experiments on the camouflage of stick caterpillars. *Behaviour* **4**, 222–232.

De Ruiter, L. (1956). Countershading in caterpillars. *Arch. Néerl. Zool.* **11**, 285–341.

Dearn, J. M. (1990). Color pattern polymorphism. In: *Biology of Grasshoppers* (eds Chapman, R. F. and Joern, A.), pp. 517–549. John Wiley & Sons, New York.

Despland, E. and Simpson, S. J. (2005). Surviving the change to warning colouration: density-dependent polyphenism suggests a route for the evolution of aposematism. *Chemoecology* **15**, 69–75.

Dimitrova, M., Stobbe, N., Schaefer, H. M. and Merilaita, S. (2009). Concealed by conspicuousness: distractive prey markings and backgrounds. *Proc. R. Soc. Lond. B* **276**, 1905–1910.

Dittrich, W., Gilbert, F., Green, P., McGregor, P. and Grewcock, D. (1993). Imperfect mimicry – a pigeon's perspective. *Proc. R. Soc. Lond. B* **251**, 195–200.

Doucet, S. M. and Meadows, M. G. (2009). Iridescence: a functional perspective. *J. R. Soc. Interface* **6**, S115–S132.

Doutrelant, C., Grégoire, A., Grnac, N., Gomez, D., Lambrechts, M. M. and Perret, P. (2008). Female coloration indicates female reproductive capacity in blue tits. *J. Evol. Biol.* **21**, 226–233.

Edmunds, M. (1976). Defensive behavior of Ghanaian praying mantids with a discussion of territoriality. *Zool. J. Linn. Soc.* **58**, 1–37.

Edmunds, M. (1990). The evolution of cryptic coloration. In: *Insect Defenses – Adaptive Mechanisms and Strategies of Prey and Predators* (eds Evans, D. L. and Schmidt, J. O.), pp. 3–21. State University of New York Press, Albany, NY.

Edmunds, M. and Dewhirst, R. A. (1994). The survival value of countershading with wild birds as predators. *Biol. J. Linnean Soc.* **51**, 447–452.

Eisner, T., Van Tassell, E. and Carrel, J. E. (1967). Defensive use of a 'fecal shield' by a beetle larva. *Science* **158**, 1471–1473.

Eisner, T., Hicks, K., Eisner, M. and Robson, D. S. (1978). 'Wolf-in-sheep's-clothing' strategy of a predaceous insect larva. *Science* **199**, 790–794.

Eklöf, J. and Jones, G. (2003). Use of vision in prey detection by brown long-eared bats, *Plecotus auritus*. *Anim. Behav.* **66**, 949–953.

Eklöf, J., Svensson, A. M. and Rydell, J. (2002). Northern bats, *Eptesicus nilssonii*, use vision but not flutter-detection when searching for prey in clutter. *Oikos* **99**, 347–351.

Ellins, S. R. and Masterson, F. A. (1974). Brightness discrimination thresholds in the bat, *Eptesicus fuscus*. *Brain Behav. Evol.* **9**, 248–263.

Endler, J. A. (1978). A predator's view of animal color patterns. *Evol. Biol.* **11**, 319–364.

Endler, J. A. (1984). Progressive background matching in moths, and a quantitative measure of crypsis. *Biol. J. Linnean Soc.* **22**, 187–231.

Endler, J. A. (1990). On the measurement and classification of colour in studies of animal colour patterns. *Biol. J. Linnean Soc.* **41**, 315–352.

Endler, J. A. and Mielke, P. W. (2005). Comparing entire colour patterns as birds see them. *Biol. J. Linnean Soc.* **86**, 405–431.

Endler, J. A., Westcott, D. A., Madden, J. R. and Robson, T. (2005). Animal visual systems and the evolution of color patterns: sensory processing illuminates signal evolution. *Evolution* **59**, 1795–1818.

Evans, D. L. and Schmidt, J. O. (1990). *Insect Defenses: Adaptive Mechanisms and Strategies of Prey and Predators*. State University of New York Press, Albany, NY.

Ewert, J. P. (2004). Motion perception shapes the visual world of amphibians. In: *Complex Worlds from Simpler Nervous Systems* (ed Prete, F. R.), pp. 117–160. MIT Press, Cambridge, MA.
Ewert, J. P., Burghagen, H., Albrecht, L. and Kepper, J. (1982). Effects of background structure on the discrimination of configurational moving prey dummies by toads *Bufo bufo* (L). *J. Comp. Phys.* **147**, 179–187.
Exnerová, A., Landová, E., Štys, P., Fuchs, R., Prokopová, M. and Cehláriková, P. (2003). Reactions of passerine birds to aposematic and nonaposematic firebugs (*Pyrrhocoris apterus*; Heteroptera). *Biol. J. Linnean Soc.* **78**, 517–525.
Exnerová, A., Svádová, K., Štys, P., Barcalová, S., Landová, E., Prokopová, M., Fuchs, R. and Socha, R. (2006). Importance of colour in the reaction of passerine predators to aposematic prey: experiments with mutants of *Pyrrhocoris apterus* (Heteroptera). *Biol. J. Linnean Soc.* **88**, 143–153.
Exnerová, A., Svádová, K., Fousová, P., Fučiková, E., Ježová, D., Niederlová, A., Kopečková, M. and Štys, P. (2008). European birds and aposematic Heteroptera: review of comparative experiments. *Bull. Insectology* **61**, 163–165.
Fleishman, L. J., McClintock, W. J., D'Eath, R. B., Brainard, D. H. and Endler, J. A. (1998). Colour perception and the use of video playback experiments in animal behaviour. *Anim. Behav.* **56**, 1035–1040.
Floren, A. and Otto, S. (2001). A tropical Derbidae (Fulgoroidea, Homoptera) that mimics a predator (Salticidae, Araneae). *Ecotropica* **7**, 151–153.
Forsman, A. and Herrström, J. (2004). Asymmetry in size, shape, and color impairs the protective value of conspicuous color patterns. *Behav. Ecol.* **15**, 141–147.
Forsman, A. and Merilaita, S. (1999). Fearful symmetry: pattern size and asymmetry affects aposematic signal efficacy. *Evol. Ecol.* **13**, 131–140.
Fraser, S., Callahan, A., Klassen, D. and Sherratt, T. N. (2007). Empirical tests of the role of disruptive coloration in reducing detectability. *Proc. R. Soc. Lond. B* **274**, 1325–1331.
Freund, R. L. and Olmstead, K. L. (2000). Role of vision and antennal olfaction in habitat and prey location by three predatory heteropterans. *Environ. Entomol.* **29**, 721–732.
Gamberale-Stille, G. (2001). Benefit by contrast: an experiment with live aposematic prey. *Behav. Ecol.* **12**, 768–772.
Gamberale-Stille, G. and Guilford, T. (2003). Contrast versus colour in aposematic signals. *Anim. Behav.* **65**, 1021–1026.
Gamberale-Stille, G. and Tullberg, B. S. (2001). Fruit or aposematic insect? Context-dependent colour preferences in domestic chicks. *Proc. R. Soc. Lond. B* **268**, 2525–2529.
Gamberale-Stille, G., Hall, K. S. S. and Tullberg, B. S. (2007). Signals of profitability? Food colour preferences in migrating juvenile blackcaps differ for fruits and insects. *Evol. Ecol.* **21**, 99–108.
Gilbert, C. (1997). Visual control of cursorial prey pursuit by tiger beetles (Cicindelidae). *J. Comp. Physiol. A* **181**, 217–230.
Giurfa, M., Vorobyev, M., Brandt, R., Posner, B. and Menzel, R. (1997). Discrimination of coloured stimuli by honeybees: alternative use of achromatic and chromatic signals. *J. Comp. Physiol. A* **180**, 235–243.
Glösmann, M., Steiner, M., Peichl, L. and Ahnelt, P. K. (2008). Cone photoreceptors and potential UV vision in a subterranean insectivore, the European mole. *J. Vis.* **8**, 12.
Gogala, M. (1967). Spectral sensitivity of divided eyes of *Ascalaphus macaronius* Scop (Neuroptera Ascalaphidae). *Zeitschrift Fur Vergleichende Physiologie* **57**, 232–243.
Gomez, D. and Théry, M. (2007). Simultaneous crypsis and conspicuousness in color patterns: comparative analysis of a Neotropical rainforest bird community. *Am. Nat.* **169**, S42–S61.

Govardovskii, V. I., Fyhrquist, N., Reuter, T., Kuzmin, D. G. and Donner, K. (2000). In search of the visual pigment template. *Vis. Neurosci.* **17**, 509–528.

Grant, B. and Howlett, R. J. (1988). Background selection by the peppered moth (*Biston betularia* Linn.) – individual differences. *Biol. J. Linnean Soc.* **33**, 217–232.

Grant, B. S., Owen, D. F. and Clarke, C. A. (1996). Parallel rise and fall of melanic peppered moths in America and Britain. *J. Hered.* **87**, 351–357.

Grayson, J. and Edmunds, M. (1989). The causes of color and color-change in caterpillars of the poplar and eyed hawkmoths (*Laothoe populi* and *Smerinthus ocellata*). *Biol. J. Linnean Soc.* **37**, 263–279.

Greene, E., Orsak, L. J. and Whitman, D. W. (1987). A tephritid fly mimics the territorial displays of its jumping spider predators. *Science* **236**, 310–312.

Guilford, T. (1986). How do 'warning colours' work? Conspicuousness may reduce recognition errors in experienced predators. *Anim. Behav.* **34**, 286–288.

Guilford, T. and Dawkins, M. S. (1987). Search images not proven: a reappraisal of recent evidence. *Anim. Behav.* **35**, 1838–1845.

Ham, A. D. and Osorio, D. (2007). Colour preferences and colour vision in poultry chicks. *Proc. R. Soc. Lond. B* **274**, 1941–1948.

Harland, D. P. and Jackson, K. R. (2000). Cues by which *Portia fimbriata*, an araneophagic jumping spider, distinguishes jumping-spider prey from other prey. *J. Exp. Biol.* **203**, 3485–3494.

Harland, D. P. and Jackson, R. R. (2001). Prey classification by *Portia fimbriata*, a salticid spider that specializes at preying on other salticids: species that elicit cryptic stalking. *J. Zool.* **255**, 445–460.

Harland, D. P. and Jackson, R. R. (2002). Influence of cues from the anterior medial eyes of virtual prey on *Portia fimbriata*, an araneophagic jumping spider. *J. Exp. Biol.* **205**, 1861–1868.

Harland, D. P. and Jackson, R. R. (2004). *Portia* perceptions the umwelt of an araneophagic jumping spider. In: *Complex Worlds from Simpler Nervous Systems* (ed Prete, F. R.), pp. 4–40. MIT Press, Cambridge, MA.

Harland, D. P., Jackson, R. R. and Macnab, A. M. (1999). Distances at which jumping spiders (Araneae: Salticidae) distinguish between prey and conspecific rivals. *J. Zool.* **247**, 357–364.

Harmon, J. P., Losey, J. E. and Ives, A. R. (1998). The role of vision and color in the close proximity foraging behavior of four coccinellid species. *Oecologia* **115**, 287–292.

Hart, N. S. (2001a). Variations in cone photoreceptor abundance and the visual ecology of birds. *J. Comp. Physiol. A* **187**, 685–698.

Hart, N. S. (2001b). The visual ecology of avian photoreceptors. *Prog. Retin. Eye Res.* **20**, 675–703.

Hart, N. S. and Vorobyev, M. (2005). Modelling oil droplet absorption spectra and spectral sensitivities of bird cone photoreceptors. *J. Comp. Physiol. A* **191**, 381–392.

Hart, N. S., Partridge, J. C., Cuthill, I. C. and Bennett, A. T. D. (2000). Visual pigments, oil droplets, ocular media and cone photoreceptor distribution in two species of passerine bird: the blue tit (*Parus caeruleus* L.) and the blackbird (*Turdus merula* L.). *J. Comp. Physiol. A* **186**, 375–387.

Harvey, P. H. and Greenwood, P. J. (1978). Anti-predator defence strategies: some evolutionary problems. In: *Behavioural Ecology: An Evolutionary Approach* (eds Krebs, J. R. and Davies, N. B.), pp. 129–151. Blackwell Scientific Publications, Oxford.

Hasegawa, M. and Taniguchi, Y. (1994). Visual avoidance of a conspicuously colored carabid beetle *Dischissus mirandus* by the lizard *Eumeces okadae*. *J. Ethol.* **12**, 9–14.

Håstad, O., Victorsson, J. and Ödeen, A. (2005). Differences in color vision make passerines less conspicuous in the eyes of their predators. *Proc. Natl. Acad. Sci. USA* **102**, 6391–6394.

Hatle, J. D. and Faragher, S. G. (1998). Slow movement increases the survivorship of a chemically defended grasshopper in predatory encounters. *Oecologia* **115**, 260–267.
Hatle, J. D. and Salazar, B. A. (2001). Aposematic coloration of gregarious insects can delay predation by an ambush predator. *Environ. Entomol.* **30**, 51–54.
Hatle, J. D., Salazar, B. A. and Whitman, D. W. (2002). Survival advantage of sluggish individuals in aggregations of aposematic prey, during encounters with ambush predators. *Evol. Ecol.* **16**, 415–431.
Hauglund, K., Hagen, S. B. and Lampe, H. M. (2006). Responses of domestic chicks (*Gallus gallus domesticus*) to multimodal aposematic signals. *Behav. Ecol.* **17**, 392–398.
Hochkirch, A., Deppermann, J. and Groning, J. (2008). Phenotypic plasticity in insects: the effects of substrate color on the coloration of two ground-hopper species. *Evol. Dev.* **10**, 350–359.
Holling, C. S. (1965). The functional response of predators to prey density and its role in mimicry and population regulation. *Mem. Ent. Soc. Can.* **45**, 1–60.
Horridge, G. A. (1969). Unit studies on retina of dragonflies. *Zeitschrift Fur Vergleichende Physiologie* **62**, 1–37.
Hyden, K. and Kral, K. (2005). The role of edges in the selection of a jump target in *Mantis religiosa*. *Behav. Processes* **70**, 122–131.
Ihalainen, E., Lindström, L. and Mappes, J. (2007). Investigating Müllerian mimicry: predator learning and variation in prey defences. *J. Evol. Biol.* **20**, 780–791.
Ihalainen, E., Lindström, L., Mappes, J. and Puolakkainen, S. (2008). Can experienced birds select for Müllerian mimicry? *Behav. Ecol.* **19**, 362–368.
Ingram, A. L. and Parker, A. R. (2008). A review of the diversity and evolution of photonic structures in butterflies, incorporating the work of John Huxley (The Natural History Museum, London from 1961 to 1990). *Phil. Trans. R. Soc. Lond. B* **363**, 2465–2480.
Ioannou, C. C. and Krause, J. (2009). Interactions between background matching and motion during visual detection can explain why cryptic animals keep still. *Biol. Lett.* **5**, 191–193.
Iwasaki, T. (1990). Predatory behavior of the praying mantis, *Tenodera aridifolia*. Part 1. Effect of prey size on prey recognition. *J. Ethol.* **8**, 75–79.
Jackson, R. R. (2000). Prey preferences and visual discrimination ability of *Brettus*, *Cocalus* and *Cyrba*, araneophagic jumping spiders (Araneae: Salticidae) from Australia, Kenya and Sri Lanka. *N. Z. J. Zool.* **27**, 29–39.
Jackson, R. R. and Li, D. Q. (2004). One-encounter search-image formation by araneophagic spiders. *Anim. Cogn.* **7**, 247–254.
Jackson, R. R. and Pollard, S. D. (2007). Bugs with backpacks deter vision-guided predation by jumping spiders. *J. Zool.* **273**, 358–363.
Jacobs, G. H. (1993). The distribution and nature of colour vision among the mammals. *Biol. Rev.* **68**, 413–471.
Johnsson, J. I. and Kjällman-Eriksson, K. (2008). Cryptic prey colouration increases search time in brown trout (*Salmo trutta*): effects of learning and body size. *Behav. Ecol. Sociobiol.* **62**, 1613–1620.
Jones, C. D. and Osorio, D. (2004). Discrimination of oriented visual textures by poultry chicks. *Vision Res.* **44**, 83–89.
Kassarov, L. (2003). Are birds the primary selective force leading to evolution of mimicry and aposematism in butterflies? An opposing point of view. *Behaviour* **140**, 433–451.
Kauppinen, J. and Mappes, J. (2003). Why are wasps so intimidating: field experiments on hunting dragonflies (Odonata: *Aeshna grandis*). *Anim. Behav.* **66**, 505–511.

Kawamura, S. and Yokoyama, S. (1998). Functional characterization of visual and nonvisual pigments of American chameleon (*Anolis carolinensis*). *Vision Res.* **38**, 37–44.

Kelber, A., Vorobyev, M. and Osorio, D. (2003). Animal colour vision – behavioural tests and physiological concepts. *Biol. Rev.* **78**, 81–118.

Kettlewell, H. B. D. (1955a). Recognition of appropriate backgrounds by pale and black phases of Lepidoptera. *Nature* **175**, 943–949.

Kettlewell, H. B. D. (1955b). Selection experiments on industrial melanism in the Lepidoptera. *Heredity* **9**, 323–342.

Kettlewell, H. B. D. (1956). Further selection experiments on industrial melanism in the Lepidoptera. *Heredity* **10**, 287–301.

Kettlewell, H. B. D. (1973). *The Evolution of Melanism, the Study of a Recurring Necessity*. Clarendon Press, Oxford.

Kettlewell, H. B. D. and Conn, D. L. T. (1977). Further background-choice experiments on cryptic lepidoptera. *J. Zool.* **181**, 371–376.

Kevan, P. G. (1972). Floral colors in the high Arctic with reference to insect flower relations and pollination. *Can. J. Bot.* **50**, 2289–2316.

Kodandaramaiah, U., Vallin, A. and Wiklund, C. (2009). Fixed eyespot display in a butterfly thwarts attacking birds. *Anim. Behav.* **77**, 1415–1419.

Koenig, L. (1950). Untersuchungen über Nahrungserwerb und Beuteschema des Bienenfressers. *Zool Inf (Biol Stat Wilhelminenburg)* **2**.

Koskelainen, A., Hemila, S. and Donner, K. (1994). Spectral sensitivities of short-wavelength and long-wavelength sensitive cone mechanisms in the frog retina. *Acta Physiol. Scand.* **152**, 115–124.

Kral, K. and Devetak, D. (1999). The visual orientation strategies of *Mantis religiosa* and *Empusa fasciata* reflect differences in the structure of their visual surroundings. *J. Insect Behav.* **12**, 737–752.

Kral, K. and Poteser, M. (1997). Motion parallax as a source of distance information in locusts and mantids. *J. Insect Behav.* **10**, 145–163.

Kral, K. and Prete, F. R. (2004). In the mind of a hunter: the visual world of the praying mantis. In: *Complex Worlds from Simpler Nervous Systems* (ed Prete, F. R.), pp. 75–115. MIT Press, Cambridge, MA.

Kral, K., Vernik, M. and Devetak, D. (2000). The visually controlled prey-capture behaviour of the European mantispid *Mantispa styriaca*. *J. Exp. Biol.* **203**, 2117–2123.

Kroger, R. H. H., Bowmaker, J. K. and Wagner, H. J. (1999). Morphological changes in the retina of *Aequidens pulcher* (Cichlidae) after rearing in monochromatic light. *Vision Res.* **39**, 2441–2448.

Land, M. F. (1997). Visual acuity in insects. *Annu. Rev. Entomol.* **42**, 147–177.

Land, M. F. and Nilsson, D. E. (2002). *Animal Eyes*. Oxford University Press, Oxford.

Langley, W. M. (1989). Grasshopper mouse use of visual cues during a predatory attack. *Behav. Processes* **19**, 115–125.

Lees, D. R. (1968). Genetic control of melanic form insularia of peppered moth *Biston betularia* (L). *Nature* **220**, 1249–1250.

Lees, D. R., Creed, E. R. and Duckett, J. G. (1973). Atmospheric pollution and industrial melanism. *Heredity* **30**, 227–232.

Levene, H. (1953). Genetic equilibrium when more than one niche is available. *Am. Nat.* **87**, 331–333.

Lewis, S. M. and Cratsley, C. K. (2008). Flash signal evolution, mate choice, and predation in fireflies. *Annu. Rev. Entomol.* **53**, 293–321.

Li, D. Q., Jackson, R. R. and Lim, M. L. M. (2003). Influence of background and prey orientation on an ambushing predator's decisions. *Behaviour* **140**, 739–764.

Liebert, T. G. and Brakefield, P. M. (1987). Behavioral studies on the peppered moth *Biston betularia* and a discussion of the role of pollution and lichens in industrial melanism. *Biol. J. Linnean Soc.* **31**, 129–150.

Lin, J. T. (1993). Identification of photoreceptor locations in the compound eye of *Coccinella septempunctata* Linnaeus (Coleoptera, Coccinellidae). *J. Insect Physiol.* **39**, 555–562.

Lin, J. T. and Wu, C. Y. (1992). A comparative study on the color vision of four coleopteran insects. *Bull. Inst. Zool. Acad. Sin.* **31**, 81–88.

Lindstedt, C., Lindström, L. and Mappes, J. (2008). Hairiness and warning colours as components of antipredator defence: additive or interactive benefits? *Anim. Behav.* **75**, 1703–1713.

Lindström, L., Alatalo, R. V. and Mappes, J. (1999). Reactions of hand-reared and wild-caught predators toward warningly colored, gregarious, and conspicuous prey. *Behav. Ecol.* **10**, 317–322.

Lindström, L., Lyytinen, A., Mappes, J. and Ojala, K. (2006). Relative importance of taste and visual appearance for predator education in Müllerian mimicry. *Anim. Behav.* **72**, 323–333.

Livingstone, M. and Hubel, D. (1988). Segregation of form, color, movement, and depth – anatomy, physiology, and perception. *Science* **240**, 740–749.

Lock, A. and Collett, T. (1979). A toad's devious approach to its prey – study of some complex uses of depth vision. *J. Comp. Phys.* **131**, 179–189.

Loew, E. R. (1994). A third, ultraviolet-sensitive, visual pigment in the Today gecko (*Gekko gekko*). *Vision Res.* **34**, 1427–1431.

Lovy, D. (1996). WinDig 2.5. Freeware available from the author at dominique.lovy@-chiphy.unige.ch or by download fromhttp://www.unige.ch/sciences/chifi/epb/winding.html (verified May 2004). Department of Physical Chemistry, University of Geneva, Switzerland.

Lymbery, A. J. (1992). The environmental control of coloration in a bush cricket, *Mygalopsis marki* Bailey (Orthoptera: Tettigoniidae). *Biol. J. Linnean Soc.* **45**, 71–89.

Lyytinen, A., Alatalo, R. V., Lindström, L. and Mappes, J. (1999). Are European white butterflies aposematic? *Evol. Ecol.* **13**, 709–719.

Lyytinen, A., Alatalo, R. V., Lindström, L. and Mappes, J. (2001). Can ultraviolet cues function as aposematic signals? *Behav. Ecol.* **12**, 65–70.

Lyytinen, A., Brakefield, P. M. and Mappes, J. (2003). Significance of butterfly eyespots as an anti-predator device in ground-based and aerial attacks. *Oikos* **100**, 373–379.

Lyytinen, A., Lindström, L. and Mappes, J. (2004). Ultraviolet reflection and predation risk in diurnal and nocturnal Lepidoptera. *Behav. Ecol.* **15**, 982–987.

Majerus, M. E. N. (1998). *Melanism: Evolution in Action.* Oxford University Press, Oxford.

Majerus, M. E. N. (2008). Non-morph specific predation of peppered moths (*Biston betularia*) by bats. *Ecol. Entomol.* **33**, 679–683.

Majerus, M. E. N., Brunton, C. F. A. and Stalker, J. (2000). A bird's eye view of the peppered moth. *J. Evol. Biol.* **13**, 155–159.

Mani, G. S. (1990). Theoretical models of melanism in *Biston betularia* – a review. *Biol. J. Linnean Soc.* **39**, 355–371.

Marples, N. M., Quinlan, M., Thomas, R. J. and Kelly, D. J. (2007). Deactivation of dietary wariness through experience of novel food. *Behav. Ecol.* **18**, 803–810.

Mather, M. H. and Roitberg, B. D. (1987). A sheep in wolf's clothing: tephritid flies mimic spider predators. *Science* **236**, 308–310.

Meinertzhagen, I. A., Menzel, R. and Kahle, G. (1983). The identification of spectral receptor types in the retina and lamina of the dragonfly *Sympetrum ribicundulum*. *J. Comp. Phys.* **151**, 295–310.

Melin, A. D., Fedigan, L. M., Hiramatsu, C., Sendall, C. L. and Kawamura, S. (2007). Effects of colour vision phenotype on insect capture by a free-ranging population of white-faced capuchins, *Cebus capucinus*. *Anim. Behav.* **73**, 205–214.

Merilaita, S. (2003). Visual background complexity facilitates the evolution of camouflage. *Evolution* **57**, 1248–1254.

Merilaita, S. and Lind, J. (2005). Background-matching and disruptive coloration, and the evolution of cryptic coloration. *Proc. R. Soc. Lond. B* **272**, 665–670.

Merilaita, S. and Lind, J. (2006). Great tits (*Parus major*) searching for artificial prey: implications for cryptic coloration and symmetry. *Behav. Ecol.* **17**, 84–87.

Merilaita, S., Lyytinen, A. and Mappes, J. (2001). Selection for cryptic coloration in a visually heterogeneous habitat. *Proc. R. Soc. Lond. B* **268**, 1925–1929.

Miklósi, A., Gonda, Z., Osorio, D. and Farzin, A. (2002). The effects of the visual environment on responses to colour by domestic chicks. *J. Comp. Physiol. A* **188**, 135–140.

Monteiro, A., Brakefield, P. M. and French, V. (1997). Butterfly eyespots: the genetics and development of the color rings. *Evolution* **51**, 1207–1216.

Moreby, S. J., Aebischer, N. J. and Southway, S. (2006). Food preferences of grey partridge chicks, *Perdix perdix*, in relation to size, colour and movement of insect prey. *Anim. Behav.* **71**, 871–878.

Morgan, M. J., Adam, A. and Mollon, J. D. (1992). Dichromates detect color-camouflaged objects that are not detected by trichromates. *Proc. R. Soc. Lond. B* **248**, 291–295.

Moss, R., Jackson, R. R. and Pollard, S. D. (2006a). Hiding in the grass: background matching conceals moths (Lepidoptera: Crambidae) from detection by spider eyes (Araneae: Salticidae). *N. Z. J. Zool.* **33**, 207–214.

Moss, R., Jackson, R. R. and Pollard, S. D. (2006b). Mask of wax: secretions of wax conceal aphids from detection by spider's eyes. *N. Z. J. Zool.* **33**, 215–220.

Mostler, G. (1935). Beobachtungen zur Frage der Wespenmimikry. *Z. Morphol. Ökol. Tiere* **29**, 381–454.

Müller, B., Glösmann, M., Peichl, L., Knop, G. C., Hagemann, C. and Ammermüller, J. (2009). Bat eyes have ultraviolet-sensitive cone photoreceptors. *PLoS One* **4**, e6390.

Nelson, X. J. and Jackson, R. R. (2007). Vision-based ability of an ant-mimicking jumping spider to discriminate between models, conspecific individuals and prey. *Insect. Soc.* **54**, 1–4.

Noor, M. A. F., Parnell, R. S. and Grant, B. S. (2008). A reversible color polyphenism in American peppered moth (*Biston betularia cognataria*) catterpillars. *PLoS One* **3**, 1–5.

Nordström, K., and O'Carroll, D. C. (2006). Small object detection neurons in female hoverflies. *Proc. R. Soc. B.* **273**, 1211–1216.

Nordström, K., Barnett, P. D. and O'Carroll, D. C. (2006). Insect detection of small targets moving in visual clutter. *PLoS Biol.* **4**, 378–386.

Norgaard, T., Nilsson, D. E., Henschel, J. R., Garm, A. and Wehner, R. (2008). Vision in the nocturnal wandering spider *Leucorchestris arenicola* (Araneae: Sparassidae). *J. Exp. Biol.* **211**, 816–823.

Nosil, P. (2004). Reproductive isolation caused by visual predation on migrants between divergent environments. *Proc. R. Soc. Lond. B* **271**, 1521–1528.

O'Brien, L. B. (1967). *Caliscelis bonellii* (Latreille), a European genus of Issidae new to the United States (Homoptera: Fulgoroidea). *Pan-Pacific Entomol.* **43**, 130–133.

O'Carroll, D. (1993). Feature-detecting neurons in dragonflies. *Nature* **362**, 541–543.

Ödeen, A. and Håstad, O. (2003). Complex distribution of avian color vision systems revealed by sequencing the SWS1 opsin from DNA. *Mol. Biol. Evol.* **20**, 855–861.

Olberg, R. M., Worthington, A. H., Fox, J. L., Bessette, C. E. and Loosemore, M. P. (2005). Prey size selection and distance estimation in foraging adult dragonflies. *J. Comp. Physiol. A* **191**, 791–797.

Olberg, R. M., Seaman, R. C., Coats, M. I. and Henry, A. F. (2007). Eye movements and target fixation during dragonfly prey-interception flights. *J. Comp. Physiol. A* **193**, 685–693.

Oliver, J. C., Robertson, K. A. and Monteiro, A. (2009). Accommodating natural and sexual selection in butterfly wing pattern evolution. *Proc. R. Soc. Lond. B* **276**, 2369–2375.

Örnborg, J., Andersson, M., Griffith, S. C. and Sheldon, B. C. (2002). Seasonal changes in a ultraviolet structural colour signal in blue tits, *Parus caeruleus*. *Biol. J. Linnean Soc.* **76**, 237–245.

Ortolani, A. and Caro, T. M. (1996). The adaptive significance of colour patterns in carnivores: phylogenetic tests of classic hypotheses. In: *Carnivore Behaviour, Ecology and Evolution* (ed Gittleman, J. L.), pp. 132–188. Cornell University Press, Ithaca.

Osorio, D., Miklósi, A. and Gonda, Z. (1999). Visual ecology and perception of coloration patterns by domestic chicks. *Evol. Ecol.* **13**, 673–689.

Osorio, D., Smith, A. C., Vorobyev, M. and Buchanan-Smith, H. M. (2004). Detection of fruit and the selection of primate visual pigments for color vision. *Am. Nat.* **164**, 696–708.

Owen, D. F. (1980). *Camouflage and Mimicry*. University of Chicago Press, Chicago.

Parker, A. H. (1969a). Predatory and reproductive behaviour of *Rhinocoris bicolor* and *R. tropicus* (Hemiptera – Reduviidae). *Entomol. Exp. Appl.* **12**, 107–117.

Parker, A. H. (1969b). The predatory behaviour and life history of *Pisilus tipuliformis* Fabricius (Hemiptera: Reduviidae). *Entomol. Exp. Appl.* **8**, 1–12.

Partan, S. R. and Marler, P. (2005). Issues in the classification of multimodal communication signals. *Am. Nat.* **166**, 231–245.

Patra, B., Bera, S. and Hickey, R. J. (2008). Soral crypsis: protective mimicry of a coccid on an Indian fern. *J. Integr. Plant Biol.* **50**, 653–658.

Peitsch, D., Fietz, A., Hertel, H., Desouza, J., Ventura, D. F. and Menzel, R. (1992). The spectral input systems of Hymenopteran insects and their receptor-based color vision. *J. Comp. Physiol. A* **170**, 23–40.

Perry, R. J. and McNaughton, P. A. (1991). Response properties of cones from the retina of the tiger salamander. *J. Physiol. (London)* **433**, 561–587.

Pettigrew, J. D. (1980). Microbat vision and echolocation in an evolutionary context. In: *Animal Sonar, Processes and Performance* (eds Nachtigall, P. E. and Moore, P. W. B.), pp. 645–650. Plenum Press, New York.

Pietrewicz, A. T. and Kamil, A. C. (1977). Visual detection of cryptic prey by blue jays (*Cyanocitta cristata*). *Science* **195**, 580–582.

Popham, E. J. (1941). The variation in the colour of certain species of *Arctocorisa* (Hemiptera, Corixidae) and its significance. *Proc. Zool. Soc. Lond. A* **111**, 135–172.

Poteser, M. and Kral, K. (1995). Visual distance discrimination between stationary targets in praying-mantis – an index of the use of motion parallax. *J. Exp. Biol.* **198**, 2127–2137.

Poulton, E. B. (1884). Notes upon, or suggested by, the colours, markings, and protective attitudes of certain lepidopterous larvae and pupae, and of a hymenopterous larva. *Trans. R. Entomol. Soc. Lond.* 27–60.

Poulton, E. B. (1890). *The Colours of Animals: Their Meaning and Use. Especially Considered in the Case of Insects*. Kegan Paul, Trench Trübner & Co. Ltd, London, UK.

Prete, F. R. (1999). Prey recognition. In: *The Praying Mantids* (eds Prete, F. R., Wells, H., Wells, P. and Hurd, L. E.), pp. 141–179. Johns Hopkins University Press, Baltimore.

Prudic, K. L., Skemp, A. K. and Papaj, D. R. (2007). Aposematic coloration, luminance contrast, and the benefits of conspicuousness. *Behav. Ecol.* **18**, 41–46.

Punzalan, D., Rodd, F. H. and Hughes, K. A. (2005). Perceptual processes and the maintenance of polymorphism through frequency-dependent predation. *Evol. Ecol.* **19**, 303–320.

Rapaport, D. H. and Stone, J. (1984). The area centralis of the retina in the cat and other mammals – focal point for function and development of the visual system. *Neuroscience* **11**, 289–301.

Rashed, A., Beatty, C. D., Forbes, M. R. and Sherratt, T. N. (2005). Prey selection by dragonflies in relation to prey size and wasp-like colours and patterns. *Anim. Behav.* **70**, 1195–1202.

Reader, T. and Hochuli, D. F. (2003). Understanding gregariousness in a larval Lepidopteran: the roles of host plant, predation, and microclimate. *Ecol. Entomol.* **28**, 729–737.

Reitner, A., Sharpe, L. T. and Zrenner, E. (1991). Is color-vision possible with only rods and blue-sensitive cones? *Nature* **352**, 798–800.

Remmel, T. and Tammaru, T. (2009). Size-dependent predation risk in tree-feeding insects with different colouration strategies: a field experiment. *J. Anim. Ecol.* **78**, 973–980.

Robbins, R. K. (1981). The 'false head' hypothesis: predation and wing pattern variation of lycaenid butterflies. *Am. Nat.* **118**, 770–775.

Robertson, K. A. and Monteiro, A. (2005). Female *Bicyclus anynana* butterflies choose males on the basis of their dorsal UV-reflective eyespot pupils. *Proc. R. Soc. Lond. B* **272**, 1541–1546.

Robinson, M. H. (1990). Predator-prey interactions, informal complexity, and the origin of intelligence. In: *Insect Defenses – Adaptive Mechanisms and Strategies of Prey and Predators* (eds Evans, D. L. and Schmidt, J. O.), pp. 129–149. State University of New York Press, Albany, NY.

Roper, T. J. (1990). Responses of domestic chicks to artificially coloured insect prey: effects of previous experience and background colour. *Anim. Behav.* **39**, 466–473.

Roper, T. J. and Cook, S. E. (1989). Responses of chicks to brightly coloured insect prey. *Behaviour* **110**, 276–293.

Rossel, S. (1979). Regional differences in photoreceptor performance in the eye of the praying mantis. *J. Comp. Phys.* **131**, 95–112.

Rossel, S., Corlija, J. and Schuster, S. (2002). Predicting three-dimensional target motion: how archer fish determine where to catch their dislodged prey. *J. Exp. Biol.* **205**, 3321–3326.

Rota, J. and Wagner, D. L. (2006). Predator mimicry: metalmark moths mimic their jumping spider predators. *PLoS One* **1**, e45.

Rowell, C. H. F. (1971). The variable coloration of the Acridoid grasshoppers. *Adv. Insect Physiol.* **8**, 145–198.

Rowland, H. M. (2009). From Abbott Thayer to the present day: what have we learned about the function of countershading? *Phil. Trans. R. Soc. Lond. B* **364**, 519–527.

Rowland, H. M., Speed, M. P., Ruxton, G. D., Edmunds, M., Stevens, M. and Harvey, I. F. (2007). Countershading enhances cryptic protection: an experiment with wild birds and artificial prey. *Anim. Behav.* **74**, 1249–1258.

Rowland, H. M., Cuthill, I. C., Harvey, I. F., Speed, M. P. and Ruxton, G. D. (2008). Can't tell the caterpillars from the trees: countershading enhances survival in a woodland. *Proc. R. Soc. Lond. B* **275**, 2539–2545.

Ru, B.-H., Zhou, Y.-Y., Li, C.-J., Wang, L. and Zhao, H.-B. (2008). Short wavelength-sensitive (SWS) opsin gene in *Myotis ricketti*: RACE amplification and evolutionary analysis. *Chin. J. Zool.* **43**, 43–50.

Ruxton, G. D., Sherratt, T. N. and Speed, M. P. (2004). *Avoiding Attack – The Evolutionary Ecology of Crypsis, Warning Signals, and Mimicry*. Oxford University Press, New York.

Saito, A., Mikami, A., Kawamura, S., Ueno, Y., Hiramatsu, C., Widayati, K. A., Suryobroto, B., Teramoto, M., Mori, Y., Nagano, K., Fujita, K. Kuroshima, H., *et al.* (2005). Advantage of dichromats over trichromats in discrimination of color-camouflaged stimuli in nonhuman primates. *Am. J. Primatol.* **67**, 425–436.

Sandoval, C. P. (1994). Differential visual predation on morphs of *Timema cristinae* (Phasmatodeae, Timemidae) and its consequences for host range. *Biol. J. Linnean Soc.* **52**, 341–356.

Santiago-Blay, J. A. and Maldonado-Capriles, J. (1988). Observations on the true bugs *Emesa tenerrima*, a possible spider mimic, and *Ghilianella borincana* (Hemiptera: Reduviidae: Emesinae) from Puerto Rico. *Proc. Entomol. Soc. Wash.* **90**, 369–372.

Sargent, T. D. (1981). Antipredator adaptations of underwing moths. In: *Foraging Behavior: Ecological, Ethological and Psychological Approaches* (eds Kamil, A. C. and Sargent, T. D.), pp. 259–287. Garland STPM, New York.

Sauseng, M., Pabst, M. A. and Karl, K. (2003). The dragonfly *Libellula quadrimaculata* (Odonata: Libellulidae) makes optimal use of the dorsal fovea of the compound eyes during perching. *Eur. J. Entomol.* **100**, 475–479.

Schaefer, H. M. and Stobbe, N. (2006). Disruptive coloration provides camouflage independent of background matching. *Proc. R. Soc. Lond. B* **273**, 2427–2432.

Schaefer, H. M., Levey, D. J., Schaefer, V. and Avery, M. L. (2006). The role of chromatic and achromatic signals for fruit detection by birds. *Behav. Ecol.* **17**, 784–789.

Schaefer, H. M., Schaefer, V. and Vorobyev, M. (2007). Are fruit colors adapted to consumer vision and birds equally efficient in detecting colorful signals? *Am. Nat.* **169**, S159–S169.

Schaerer, S. and Neumeyer, C. (1996). Motion detection in goldfish investigated with the optomotor response is 'color blind'. *Vision Res.* **36**, 4025–4034.

Schlee, M. A. (1986). Avian predation on Heteroptera: experiments on the European blackbird *Turdus m. merula* L. *Ethology* **73**, 1–18.

Schlenoff, D. H. (1984). Novelty – a basis for generalization in prey selection. *Anim. Behav.* **32**, 919–921.

Schuler, W. and Hesse, E. (1985). On the function of warning coloration – a black and yellow pattern inhibits prey-attack by naive domestic chicks. *Behav. Ecol. Sociobiol.* **16**, 249–255.

Schultz, T. D. (2001). Tiger beetle defenses revisited: alternative defense strategies and colorations of two neotropical tiger beetles, *Odontocheila nicaraguensis* Bates and *Pseudoxycheila tasalis* Bates (Carabidae: Cicindelinae). *Coleopt. Bull.* **55**, 153–163.

Schuster, S., Rossel, S., Schmidtmann, A., Jager, I. and Poralla, J. (2004). Archer fish learn to compensate for complex optical distortions to determine the absolute size of their aerial prey. *Curr. Biol.* **14**, 1565–1568.

Seago, A. E., Brady, P., Vigneron, J. P. and Schultz, T. D. (2009). Gold bugs and beyond: a review of iridescence and structural colour mechanisms in beetles (Coleoptera). *J. R. Soc. Interface* **6**, S165–S184.

Shcherbakov, D. E. (2007). Mesozoic spider mimics – Cretaceous Mimarachnidae fam.n. (Homoptera: Fulgoroidea). *Russ. Entomol. J.* **16**, 259–264.

Shelly, T. E. and Pearson, D. L. (1978). Size and color discrimination of the robber fly *Efferia tricella* (Diptera Asilidae) as a predator on tiger beetles (Coleoptera Cicindelidae). *Environ. Entomol.* **7**, 790–793.

Sherratt, T. N., Rashed, A. and Beatty, C. D. (2004). The evolution of locomotory behavior in profitable and unprofitable simulated prey. *Oecologia* **138**, 143–150.

Sherratt, T. N., Rashed, A. and Beatty, C. D. (2005). Hiding in plain sight. *Trends Ecol. Evol.* **20**, 414–416.

Siddall, E. C. and Marples, N. M. (2008). Better to be bimodal: the interaction of color and odor on learning and memory. *Behav. Ecol.* **19**, 425–432.

Silberglied, R. E., Aiello, A. and Windsor, D. M. (1980). Disruptive coloration in butterflies: lack of support in *Anartia fatima*. *Science* **209**, 617–619.

Sillén-Tullberg, B. (1985a). Higher survival of an aposematic than of a cryptic form of a distasteful bug. *Oecologia* **67**, 411–415.

Sillén-Tullberg, B. (1985b). The significance of coloration per se, independent of background, for predator avoidance of aposematic prey. *Anim. Behav.* **33**, 1382–1384.

Skelhorn, J. and Rowe, C. (2005). Frequency-dependent taste-rejection by avian predation may select for defence chemical polymorphisms in aposematic prey. *Biol. Lett.* **1**, 500–503.

Skelhorn, J. and Rowe, C. (2006a). Predator avoidance learning of prey with secreted or stored defences and the evolution of insect defences. *Anim. Behav.* **72**, 827–834.

Skelhorn, J. and Rowe, C. (2006b). Prey palatability influences predator learning and memory. *Anim. Behav.* **71**, 1111–1118.

Skelhorn, J. and Ruxton, G. D. (2006). Avian predators attack aposematic prey more forcefully when they are part of an aggregation. *Biol. Lett.* **2**, 488–490.

Skelhorn, J., Griksaitis, D. and Rowe, C. (2008). Colour biases are more than a question of taste. *Anim. Behav.* **75**, 827–835.

Skelhorn, J., Rowland, H. M., Speed, M. P. and Ruxton, G. D. (2010). Masquerade: camouflage without crypsis. *Science* **327**, 51.

Skorupski, P. and Chittka, L. (2009). Is colour cognitive? *Opt. Laser Technol.* doi:10.1016/j.optlastec.2008.12.015.

Song, H. and Wenzel, J. W. (2008). Phylogeny of bird-grasshopper subfamily Cyrtacanthacridinae (Orthoptera: Acrididae) and the evolution of locust phase polyphenism. *Cladistics* **24**, 515–542.

Sontag, C. (1971). Spectral sensitivity studies on visual system of praying mantis, *Tenodera sinensis*. *J. Gen. Phys.* **57**, 93–112.

Speed, M. P., Kelly, D. J., Davidson, A. M. and Ruxton, G. D. (2005). Countershading enhances crypsis with some bird species but not others. *Behav. Ecol.* **16**, 327–334.

Srinivasan, M. V., Poteser, M. and Kral, K. (1999). Motion detection in insect orientation and navigation. *Vision Res.* **39**, 2749–2766.

Srygley, R. B. (1994). Locomotor mimicry in butterflies? The associations of positions of centers of mass among groups of mimetic, unprofitable prey. *Phil. Trans. R. Soc. Lond. B* **343**, 145–155.

Srygley, R. B. (2004). The aerodynamic costs of warning signals in palatable mimetic butterflies and their distasteful models. *Proc. R. Soc. Lond. B* **271**, 589–594.

Srygley, R. B. (2007). Evolution of the wave: aerodynamic and aposematic functions of butterfly wing motion. *Proc. R. Soc. Lond. B* **274**, 913–917.

Srygley, R. B. and Chai, P. (1990a). Flight morphology of neotropical butterflies – palatability and distribution of mass to the thorax and abdomen. *Oecologia* **84**, 491–499.

Srygley, R. B. and Chai, P. (1990b). Predation and the elevation of thoracic temperature in brightly colored butterflies. *Am. Nat.* **135**, 766–787.

Stavenga, D. G., Smits, R. P. and Hoenders, B. J. (1993). Simple exponential functions describing the absorbance bands of visual pigment spectra. *Vision Res.* **33**, 1011–1017.

Stevens, M. (2005). The role of eyespots as anti-predator mechanisms, principally demonstrated in the Lepidoptera. *Biol. Rev.* **80**, 573–588.

Stevens, M. and Cuthill, I. C. (2006). Disruptive coloration, crypsis and edge detection in early visual processing. *Proc. R. Soc. Lond. B* **273**, 2141–2147.

Stevens, M. and Merilaita, S. (2009a). Animal camouflage: current issues and new perspectives. *Phil. Trans. R. Soc. Lond. B* **364**, 423–427.

Stevens, M. and Merilaita, S. (2009b). Defining disruptive coloration and distinguishing its functions. *Phil. Trans. R. Soc. Lond. B* **364**, 481–488.

Stevens, M., Cuthill, I. C., Parraga, C. A. and Troscianko, T. (2006a). The effectiveness of disruptive coloration as a concealment strategy. *Prog. Brain Res.* **155**, 49–64.

Stevens, M., Cuthill, I. C., Windsor, A. M. M. and Walker, H. J. (2006b). Disruptive contrast in animal camouflage. *Proc. R. Soc. Lond. B* **273**, 2433–2438.

Stevens, M., Hopkins, E., Hinde, W., Adcock, A., Connolly, Y., Troscianko, T. and Cuthill, I. C. (2007a). Field experiments on the effectiveness of 'eyespots' as predator deterrents. *Anim. Behav.* **74**, 1215–1227.

Stevens, M., Parraga, C. A., Cuthill, I. C., Partridge, J. C. and Troscianko, T. S. (2007b). Using digital photography to study animal coloration. *Biol. J. Linnean Soc.* **90**, 211–237.

Stevens, M., Graham, J., Winney, I. S. and Cantor, A. (2008a). Testing Thayer's hypothesis: can camouflage work by distraction? *Biol. Lett.* **4**, 648–650.

Stevens, M., Hardman, C. J. and Stubbins, C. L. (2008b). Conspicuousness, not eye mimicry, makes 'eyespots' effective antipredator signals. *Behav. Ecol.* **19**, 525–531.

Stevens, M., Stubbins, C. L. and Hardman, C. J. (2008c). The anti-predator function of 'eyespots' on camouflaged and conspicuous prey. *Behav. Ecol. Sociobiol.* **62**, 1787–1793.

Stevens, M., Cantor, A., Graham, J. and Winney, I. S. (2009a). The function of animal 'eyespots': conspicuousness but not eye mimicry is key. *Curr. Zool.* **55**, 319–326.

Stevens, M., Castor-Perry, S. A. and Price, J. R. (2009b). The protective value of conspicuous signals is not impaired by shape, size, or position asymmetry. *Behav. Ecol.* **20**, 96–102.

Stevens, M., Winney, I. S., Cantor, A. and Graham, J. (2009c). Outline and surface disruption in animal camouflage. *Proc. R. Soc. Lond. B* **276**, 781–786.

Stobbe, N. and Schaefer, H. M. (2008). Enhancement of chromatic contrast increases predation risk for striped butterflies. *Proc. R. Soc. Lond. B* **275**, 1535–1541.

Stobbe, N., Dimitrova, M., Merilaita, S. and Schaefer, H. M. (2009). Chromaticity in the UV/blue range facilitates the search for achromatically background-matching prey in birds. *Phil. Trans. R. Soc. Lond. B* **364**, 511–517.

Su, K. F., Meier, R., Jackson, R. R., Harland, D. P. and Li, D. (2007). Convergent evolution of eye ultrastructure and divergent evolution of vision-mediated predatory behaviour in jumping spiders. *J. Evol. Biol.* **20**, 1478–1489.

Surridge, A. K., Osorio, D. and Mundy, N. I. (2003). Evolution and selection of trichromatic vision in primates. *Trends Ecol. Evol.* **18**, 198–205.

Swynnerton, C. F. M. (1926). An investigation into the defences of butterflies of the genus Charaxes. *Proc. 3rd Int. Ent. Congress, Zurich (1925)* **2**, 478–506.

Szél, A., Röhlich, P., Caffé, A. R. and vanVeen, T. (1996). Distribution of cone photoreceptors in the mammalian retina. *Microsc. Res. Techniq.* **35**, 445–462.

Thayer, G. H. (1909). *Concealing Coloration in the Animal Kingdom: an Exposition of the Laws of Disguise through Color and Pattern: Being a Summary of Abbott H. Thayer's Discoveries.* Macmillan, New York, NY.

Théry, M. and Casas, J. (2002). Predator and prey views of spider camouflage. *Nature* **415**, 133.
Théry, M., Pincebourde, S. and Feer, F. (2008). Dusk light environment optimizes visual perception of conspecifics in a crepuscular horned beetle. *Behav. Ecol.* **19**, 627–634.
Tinbergen, N. (1957). Defense by colour. *Sci. Am.* **197**, 49.
Tinbergen, L. (1960). The natural control of insects in pinewoods. Factors influencing the intensity of predation by songbirds. *Arch. Néerl. Zool.* **13**, 265–343.
Toh, Y. and Okamura, J. (2001). Behavioural responses of the tiger beetle larva to moving objects: role of binocular and monocular vision. *J. Exp. Biol.* **204**, 615–625.
Tonner, M., Novotny, V., Leps, J. and Komarek, S. (1993). False head wing pattern of the Burmese junglequeen butterfly and the deception of avian predators. *Biotropica* **25**, 474–478.
Troscianko, T., Benton, C. P., Lovell, P. G., Tolhurst, D. J. and Pizlo, Z. (2009). Camouflage and visual perception. *Phil. Trans. R. Soc. Lond. B* **364**, 449–461.
Tullberg, B. S., Merilaita, S. and Wiklund, C. (2005). Aposematism and crypsis combined as a result of distance dependence: functional versatility of the colour pattern in the swallowtail butterfly larva. *Proc. R. Soc. Lond. B* **272**, 1315–1321.
Tutt, J. W. (1896). *British Moths.* George Routledge, London.
Vallin, A., Jakobsson, S., Lind, J. and Wiklund, C. (2005). Prey survival by predator intimidation: an experimental study of peacock butterfly defence against blue tits. *Proc. R. Soc. Lond. B* **272**, 1203–1207.
Vallin, A., Jakobsson, S., Lind, J. and Wiklund, C. (2006). Crypsis versus intimidation – anti-predation defence in three closely related butterflies. *Behav. Ecol. Sociobiol.* **59**, 455–459.
Vallin, A., Jakobsson, S. and Wiklund, C. (2007). 'An eye for an eye?' – on the generality of the intimidating quality of eyespots in a butterfly and a hawkmoth. *Behav. Ecol. Sociobiol.* **61**, 1419–1424.
van Someren, V. G. L. (1922). Notes on certain colour patterns in Lycaenidae. *J. E. Afr. Uganda Nat. Hist. Soc.* **17**, 18–21.
Veselý, P., Veselá, S., Fuchs, R. and Zrzavý, J. (2006). Are gregarious red-black shieldbugs, *Graphosoma lineatum* (Hemiptera: Pentatomidae), really aposematic? An experimental approach. *Evol. Ecol. Res.* **8**, 881–890.
Vlieger, L. and Brakefield, P. M. (2007). The deflection hypothesis: eyespots on the margins of butterfly wings do not influence predation by lizards. *Biol. J. Linnean Soc.* **92**, 661–667.
Vorobyev, M. (2003). Coloured oil droplets enhance colour discrimination. *Proc. R. Soc. Lond. B* **270**, 1255–1261.
Vorobyev, M. and Osorio, D. (1998). Receptor noise as a determinant of colour thresholds. *Proc. R. Soc. Lond. B* **265**, 351–358.
Vorobyev, M., Osorio, D., Bennett, A. T. D., Marshall, N. J. and Cuthill, I. C. (1998). Tetrachromacy, oil droplets and bird plumage colours. *J. Comp. Physiol. A* **183**, 621–633.
Vukusic, P. and Stavenga, D. G. (2009). Physical methods for investigating structural colours in biological systems. *J. R. Soc. Interface* **6**, S133–S148.
Waldbauer, G. P. and Sternburg, J. G. (1983). A pitfall in using painted insects in studies of protective coloration. *Evolution* **37**, 1085–1086.
Wang, D. Y., Oakley, T., Mower, J., Shimmin, L. C., Yim, S., Honeycutt, R. L., Tsao, H. and Li, W. H. (2004). Molecular evolution of bat color vision genes. *Mol. Biol. Evol.* **21**, 295–302.
Warrant, E. J. (1999). Seeing better at night: life style, eye design and the optimum strategy of spatial and temporal summation. *Vision Res.* **39**, 1611–1630.

Warrant, E. J. and Nilsson, D.-E. (2006). *Invertebrate Vision*. Cambridge University Press, Cambridge.

Watanabe, H. and Yano, E. (2009). Behavioral response of mantid *Hierodula patellifera* to wind as an antipredator strategy. *Ann. Entomol. Soc. Am.* **102**, 517–522.

Wennersten, L. and Forsman, A. (2009). Does colour polymorphism enhance survival of prey populations? *Proc. R. Soc. Lond. B* **276**, 2187–2194.

Whitman, D. W., Orsak, L. and Greene, E. (1988). Spider mimicry in fruit flies (Diptera: Tephritidae): further experiments on the deterrence of jumping spiders (Araneae: Salticidae) by *Zonosemata vittigera* (Coquillett). *Ann. Entomol. Soc. Am.* **81**, 532–536.

Whittle, P. D. J., Clarke, C., Sheppard, P. M. and Bishop, J. A. (1976). Further studies on industrial melanic moth *Biston betularia* (L.) in northwest of British Isles. *Proc. R. Soc. Lond. B* **194**, 467–480.

Wiklund, C. (2005). Hornet predation on peacock butterflies and ecological aspects on the evolution of complex eyespots on butterfly wings. *Entomol. Fenn.* **16**, 266–272.

Williams, P. (2007). The distribution of bumblebee colour patterns worldwide: possible significance for thermoregulation, crypsis, and warning mimicry. *Biol. J. Linnean Soc.* **92**, 97–118.

Wourms, M. K. and Wasserman, F. E. (1985). Butterfly wing markings are more advantageous during handling than during the initial strike of an avian predator. *Evolution* **39**, 845–851.

Wyszecki, G. and Stiles, W. S. (1982). *Color Science, Concepts and Methods, Quantitative Data and Formulae*. Wiley, New York.

Yamashita, S. and Tateda, H. (1976). Spectral sensitivities of jumping spider eyes. *J. Comp. Phys.* **105**, 29–41.

Yamawaki, Y. (1998). Responses to non-locomotive prey models by the praying mantis, *Tenodera angustipennis* Saussure. *J. Ethol.* **16**, 23–27.

Yamawaki, Y. (2000). Effects of luminance, size, and angular velocity on the recognition of nonlocomotive prey models by the praying mantis. *J. Ethol.* **18**, 85–90.

Yamawaki, Y. (2003). Responses to worm-like-wriggling models by the praying mantis: effects of amount of motion on prey recognition. *J. Ethol.* **21**, 123–129.

Zhang, Y. X. and Richardson, J. S. (2007). Unidirectional prey-predator facilitation: apparent prey enhance predators' foraging success on cryptic prey. *Biol. Lett.* **3**, 348–351.

Zhao, H. B., Rossiter, S. J., Teeling, E. C., Li, C. J., Cotton, J. A. and Zhang, S. Y. (2009). The evolution of color vision in nocturnal mammals. *Proc. Natl. Acad. Sci. USA* **106**, 8980–8985.

Zolnerowich, G. (1992). A unique amycle nymph (Homoptera: Fulgoridae) that mimics jumping spiders (Araneae: Salticidae). *J. N. Y. Entomol. Soc.* **100**, 498–502.

Zufall, F., Schmitt, M. and Menzel, R. (1989). Spectral and polarised-light sensitivity of photoreceptors in the compound eye of the cricket (*Gryllus bimaculatus*). *J. Comp. Physiol. A* **164**, 597–608.

Index

PLATE 1 Phylogenetic diversity of sclerotization. (A) *Aglaophenia*, a common intertidal hydroid (Cnidaria). The feather-like sprigs are colonies of polyps ensheathed by a sclerotized perisarc which is a composite of chitin, polyphenols and protein. (B) *Bdelloura candida* is a commensal marine flatworm (Turbellaria, Platyhelminthes). In anticipation of egg laying, worms accumulate eggshell precursors in the vitellaria (shell glands), which are here stained with Arnow's reagent to highlight Dopa-containing proteins (i). Dopa proteins are relocated to the nascent egg-capsule (ii), and the eggs are deposited with stalks onto the book gills of *Limulus polyphemus*. (C) Sandcastle worm *Phragmatopoma californica* (Polychaeta, Annelida) uses Dopa-rich proteins for glueing together the sandgrains that comprise its tube-like home. (D) The egg capsule of a swell shark *Cephaloscyllium ventriosum* (Chondrichthyes, Chordata) contains collagens in a plywood-like arrangement cross-linked by catechols. Scale bar is 1 cm. (For b/w version, see page 78 in the volume.)

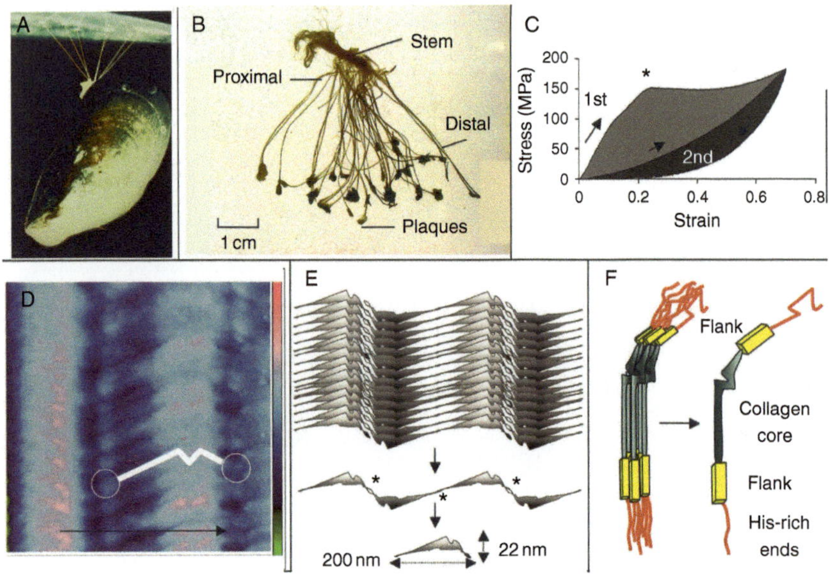

PLATE 2 Sclerotization of mussel byssus. (A) *Mytilus edulis*, the common, mussel hanging from a glass substratum by its byssus. (B) Byssus removed from a mussel and showing the different portions known as stem, proximal and distal thread and adhesive plaque. (C) Cyclic stress–strain in natural seawater of a distal byssal thread portion (strain rate = 0.2 min^{-1}) at 18 °C. If the first cycle exceeds the yield point (*), the modulus of the second cycle, which immediately follows the first, is significantly reduced. Substantial recovery of stiffness occurs after an hour or more of rest. Modulus is initial stress/strain. Engineering strain = [final length − initial length]/initial length. (D) Atomic force micrograph in tapping mode of assembled preCOLs just beneath the cuticle in a byssal thread of *M. galloprovincialis*. Arrow indicates direction of thread axis. (E) Deconstruction of preCOL assembly showing smectic register and C_2 symmetry with inversion (*), and a bent preCOL bundle. (F) A preCOL bundle of seven is further deconstructed to a single preCOL. (For b/w version, see page 80 in the volume.)

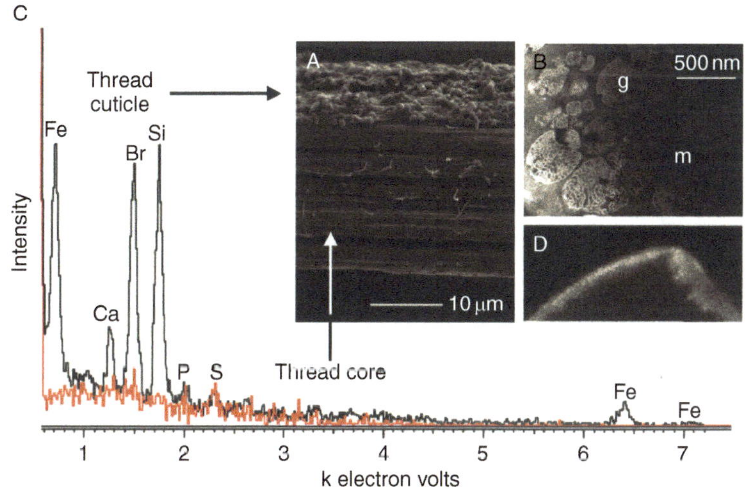

PLATE 3 Byssal cuticle architecture and chemistry. (A) Scanning electron micrograph highlighting the granular outer cuticle, and upon stripping off the cuticle the fibrous core is exposed. (B) Transmission electron micrograph of cuticle thin section stained with osmium tetroxide; G, granule; m, matrix. (C) Energy dispersive X-ray spectrum of both the cuticle and core of (D) Raman microprobe analysis of a transverse section of the cuticle and core at 460–600 nm. Scale is the same as in (A). (For b/w version, see page 85 in the volume.)

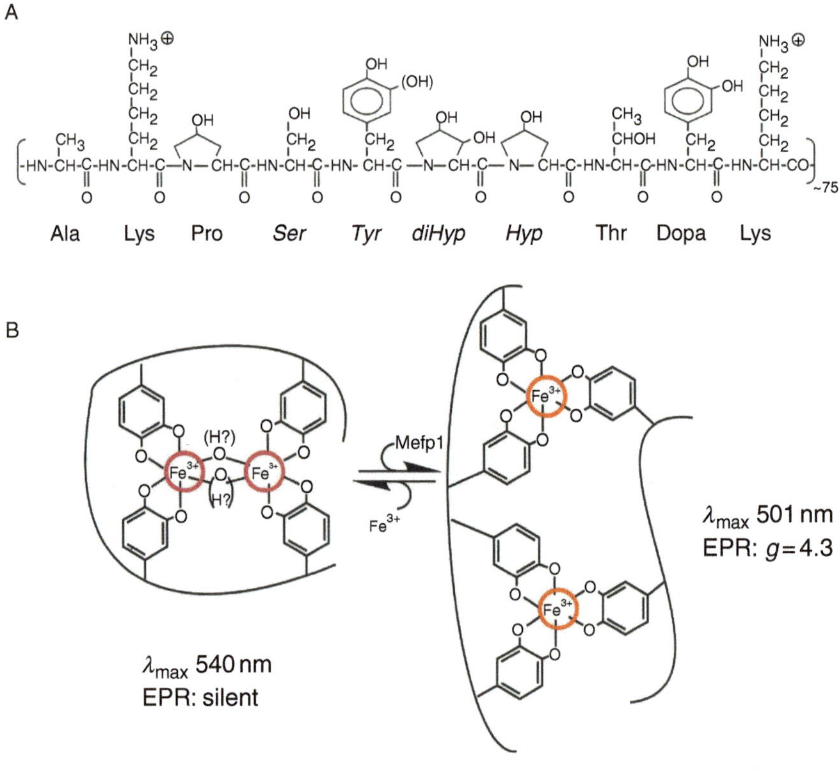

PLATE 4 Byssal cuticle protein. (A) Mfp-1, the protein constituent of cuticle, has been isolated and characterized from the mussel foot. Following a short non-repetitive N-terminal sequence, the proteins from *M. edulis* and *M. galloprovincialis* consist of 70 or more tandem repeats of a decapeptide sequence as shown. Tyrosine in the ninth position is always hydroxylated to Dopa, whereas Tyr-5 is converted less than 50% of the time. (B) At pH 7, mfp-1 is complexed to Fe^{3+} via the catecholate groups of Dopa in two modes: the red EPR active tris-catecholato-iron at high Dopa to iron ratios and the purple oxo-bridged bis-catecholato-iron species at lower ratios (Taylor *et al.*, 1996). The lines connecting the catechols represent the backbone of the same or different proteins. (For b/w version, see page 87 in the volume.)

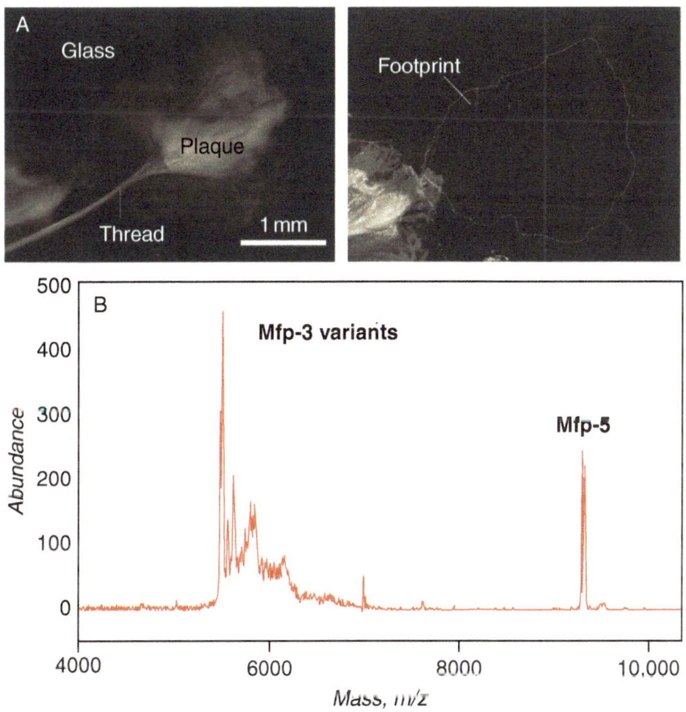

PLATE 5 (Continued)

C

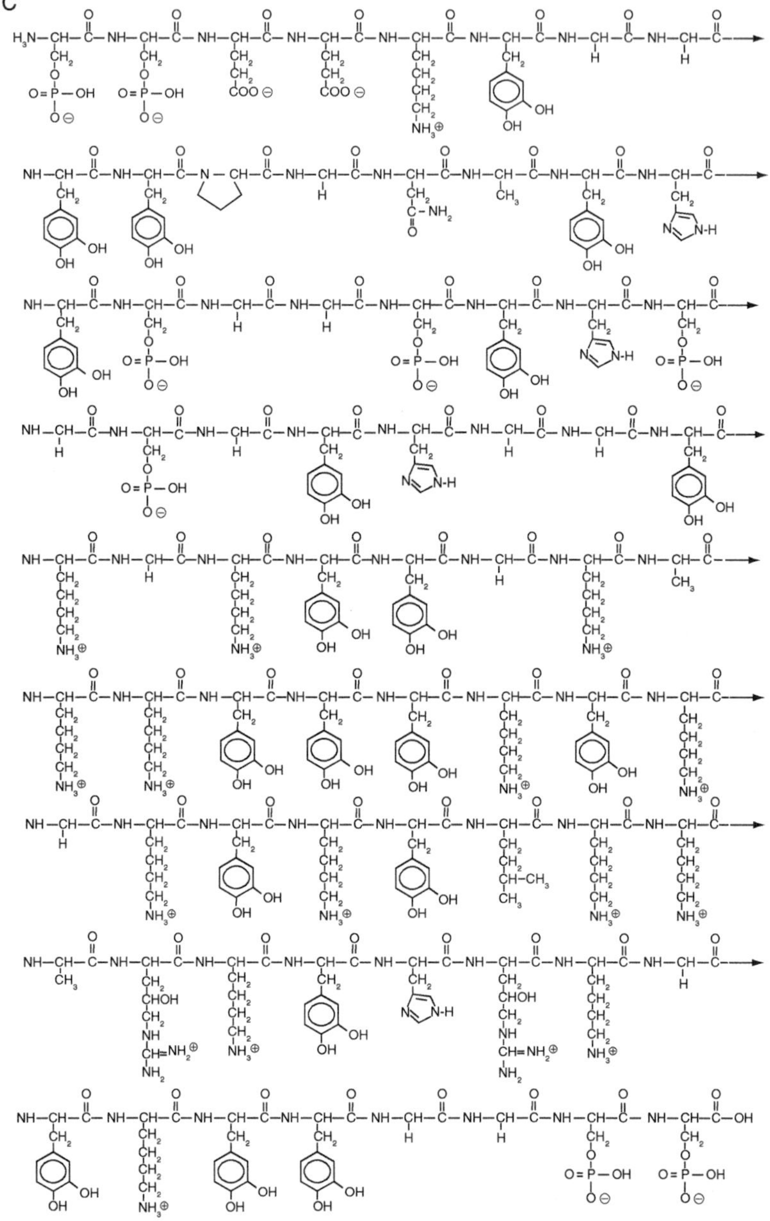

PLATE 5 Interfacial chemistry of distal plaques. (A) Plaque footprints were prepared from plaques deposited on clean glass or mica surfaces by scraping away with a clean single edged razor blade; matrix was applied to the footprint and dried before performing MALDI TOF mass spectrometry. (B) Most reproducible protein ions in the plaque footprints are mfp-3 and mfp-5. (C) Chemical sequence of mfp-5 showing the abundant Dopa and lysine residues (Waite and Qin, 2001). (For b/w version, see page 90 in the volume.)

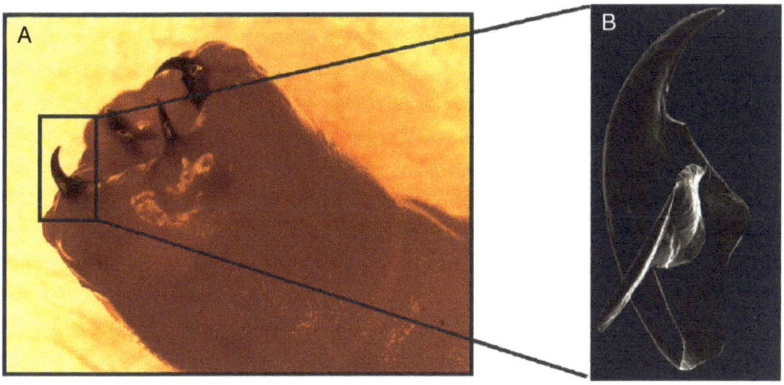

PLATE 6 *Glycera dibranchiata*. (A) Everted proboscis of the polychaete *Glycera* illustrating the presence of four sharp jaws. (B) SEM micrograph of a jaw displaying the hollow interior venom canal at lower left (Moses *et al.*, 2007). (For b/w version, see page 91 in the volume.)

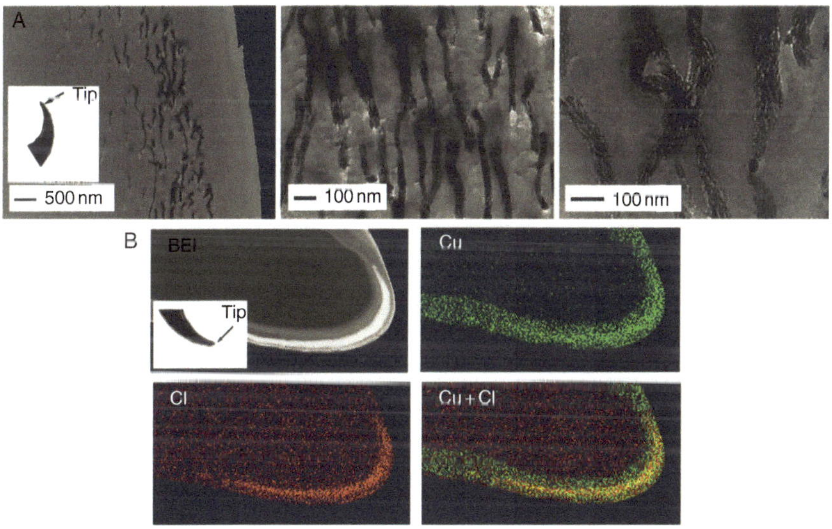

PLATE 7 Microstructural characteristics of *Glycera* jaws. (A) TEM images of longitudinal sections at various magnifications indicate the presence of elongated fibres, determined by X-ray diffraction to be Cu-based atacamite mineral (longitudinal section of tip). (B) Backscattered electron imaging (BEI) micrograph and EDS mapping of Cu and Cl, showing that the atacamite-rich region is located a few microns from the surface, while the outermost layer is composed of unmineralized Cu (Pontin *et al.*, 2007). (For b/w version, see page 93 in the volume.)

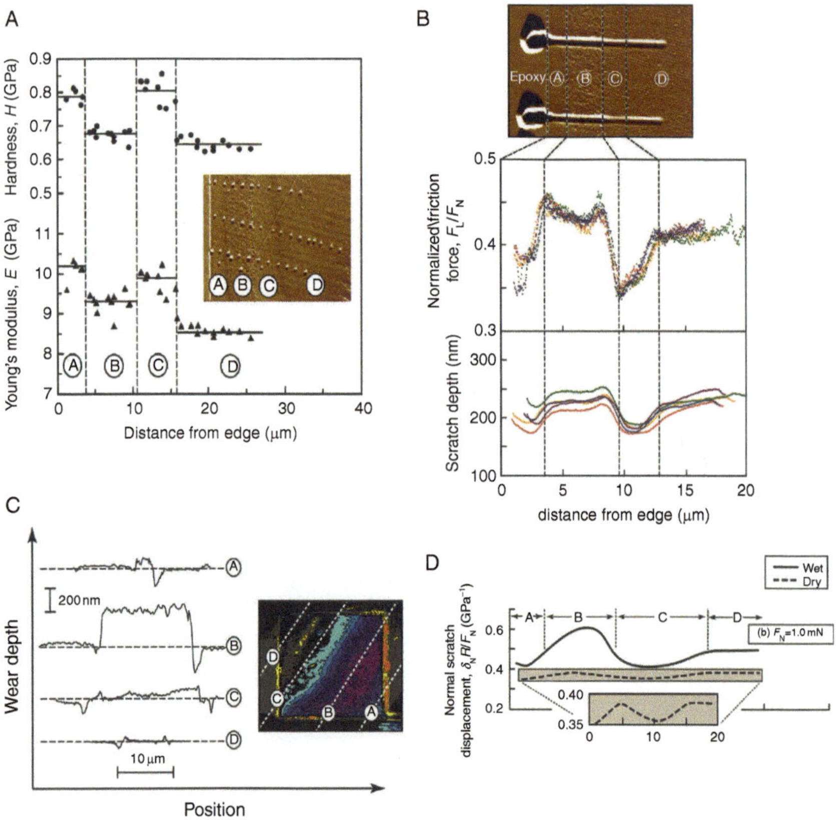

PLATE 8 Surface-probe nanomechanical investigations of *Glycera* jaws. (A) Nanoindentation profiles of the four microstructural domains present from surface to bulk, conveying that the hardest and stiffest regions are the first and third layer (those that contain unmineralized copper). (B) Nano-scratch results support the hypothesis layers one and three are most wear resistant. Across the four domains, higher scratch resistance is associated with lower normalized scratch force F_L/F_N. Layer two, which contains atacamite fibres, is the least scratch resistant. (C) Nano-wear boxes encompassing the four domains. Wear depth profiles were obtained across lines shown on the right-hand side, wear depth is larger in atacamite-rich regions. (D) Considering normalized scratch forces on hydrated specimens as compared to dry specimens, the 'hard–soft–hard–soft' theme remains, but differences in properties are magnified. The change is manifest most strikingly in the atacamite-rich second layer. Reproduced from Pontin *et al.* (2007) and Moses *et al.* (2008). (For b/w version, see page 96 in the volume.)

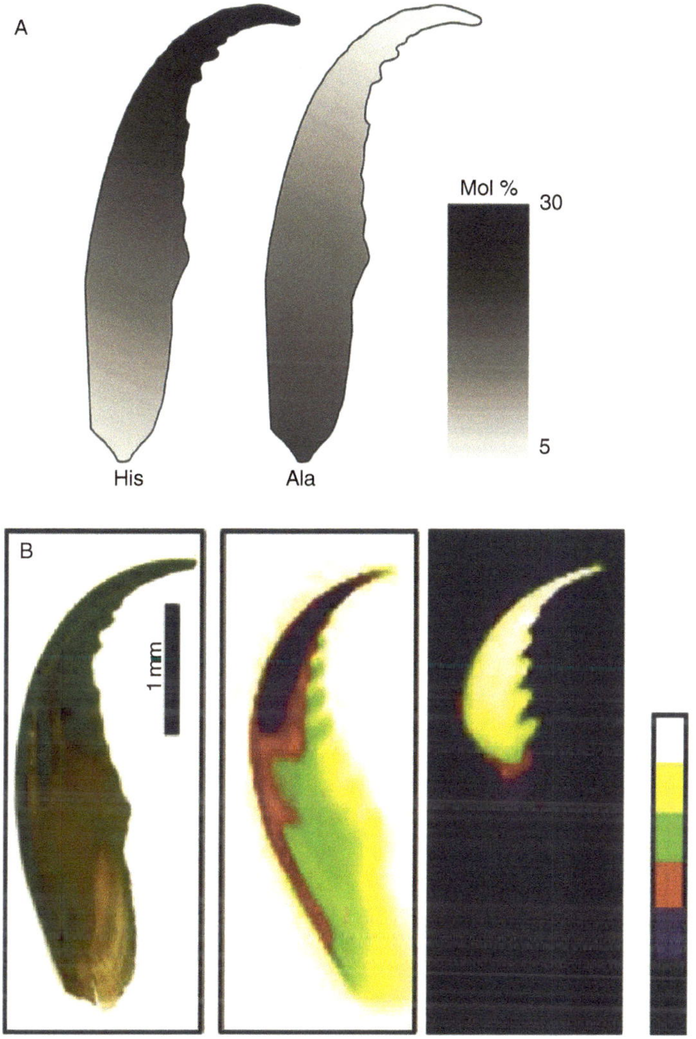

PLATE 9 Molecular gradients in *Nereis* jaws. (A) Distribution of histidine and alanine contents in biopsied samples taken from jaws at regular distal-to-proximal intervals (in relative mol% of total amino acid content post-hydrolysis). Histidine, a preferred ligand for complexes with Zn^{2+} ions is enriched towards the tip whereas alanine (the role of which is currently unknown), shows the opposite trend. (B) Zn gradient revealed by X-ray fluorescence microscopy (bright regions indicate higher Zn concentration). Zn increases towards the tip in parallel to the increase of histidine. Reproduced from Lichtenegger *et al.* (2003) and Broomell *et al.* (2007). (For b/w version, see page 99 in the volume.)

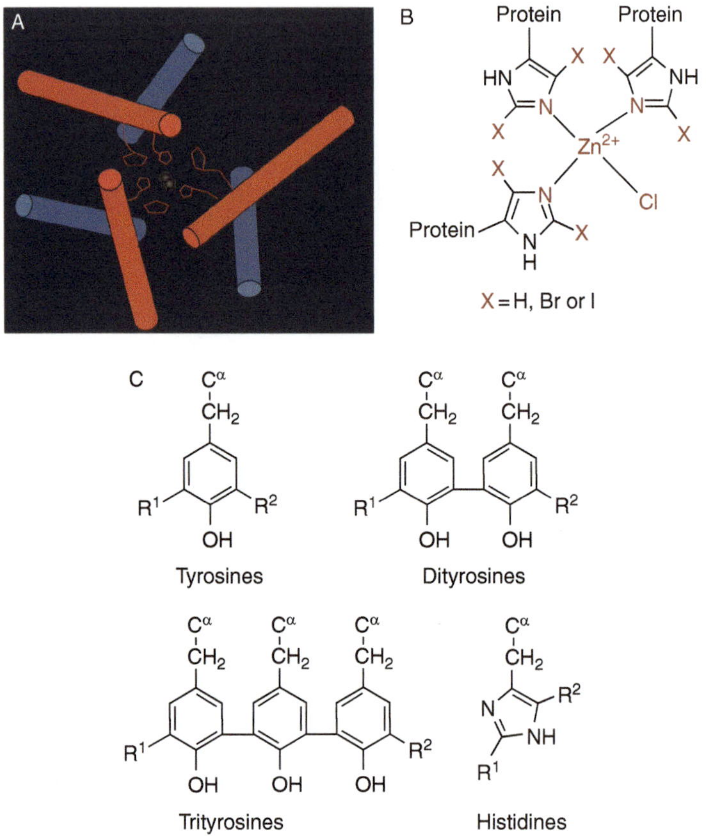

PLATE 10 Coordination complexes and halogenated amino acids in *Nereis* jaws. (A, B) Schematic of insulin-like coordination between Zn, imidazole rings and Cl. This is the best fit from the X-ray database for minerals and metalloproteins to X-ray transmission data for *Nereis* jaws. It may represent the main mechanism/complex by which *Nereis* jaw garners its robust physical properties. (C) Halogenated amino acid and cross-linking chemistry between halogenated Tyr residues. R_1 and R_2 represent iodine and/or bromine. While tyrosine cross-links have been observed, the specific role of post-translational halogenation is still unknown. Reproduced from Waite *et al.* (2004) and Birkedal *et al.* (2006). (For b/w version, see page 100 in the volume.)

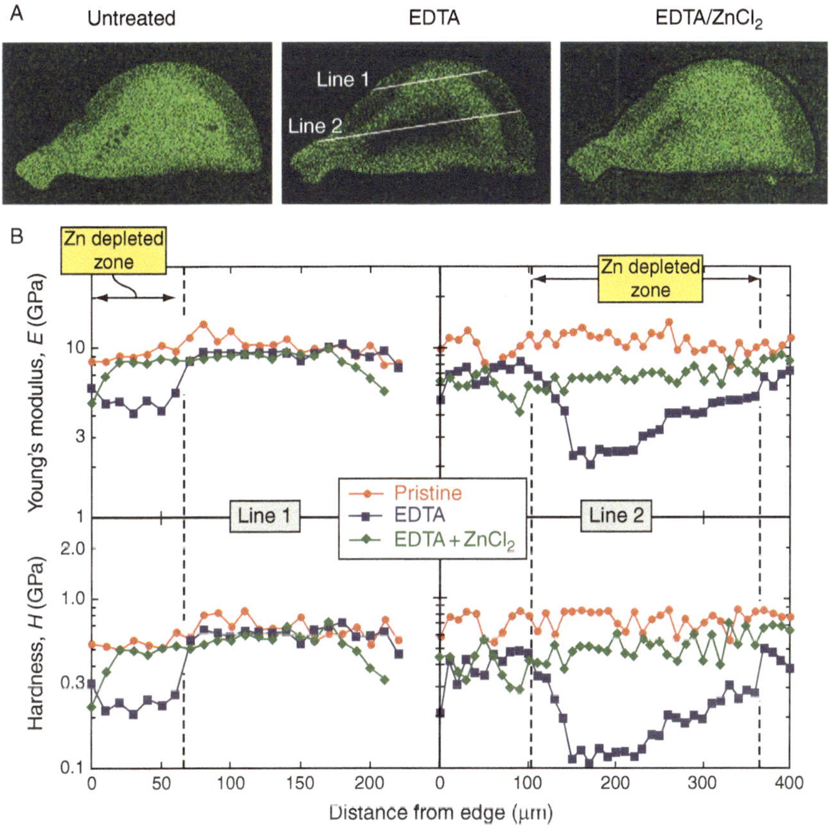

PLATE 11 Mechanical properties of *Nereis* jaws by nanoindentation. (A) Hardness and modulus profiles were assessed on (1) pristine jaw, (2) jaw with chelated Zn (by EDTA treatment), and (3) jaw reconstituted with Zn using $ZnCl_2$ solution. Lines 1 and 2 depict nanoindentation trajectories. It is evident that Zn ions may be removed from and restored into the jaw. (B) Mechanical consequences of Zn depletion and reintroduction demonstrate that the removal and reapplication of Zn^{2+} can diminish and restore the hardness and stiffness of a sclerotized tissue. Reproduced from Broomell *et al.* (2008). (For b/w version, see page 102 in the volume.)

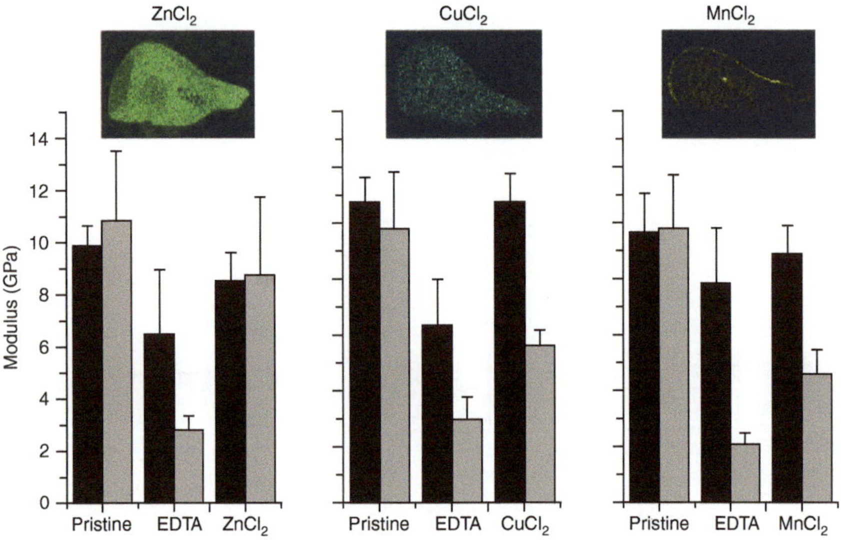

PLATE 12 Introduction of native and foreign metals in to *Nereis* jaw post-Zn chelation. Jaws were incubated in $ZnCl_2$, $CuCl_2$ and $MnCl_2$. Surrogate transition metals showed a similar recovery in material properties although Mn^{2+} has less affinity to infiltrate the structure to form complexes, providing an additional 'lever' to tune novel materials in which imidazole–metal chelation complexes are utilized. Each graph represents data that was compiled from at least two jaws tested at each stage of the treatment regimen. Black and grey bars represent results for tests in air and tests in Milli-Q water, respectively. The error bars represent standard deviations. Reproduced from Broomell *et al.* (2008). (For b/w version, see page 103 in the volume.)

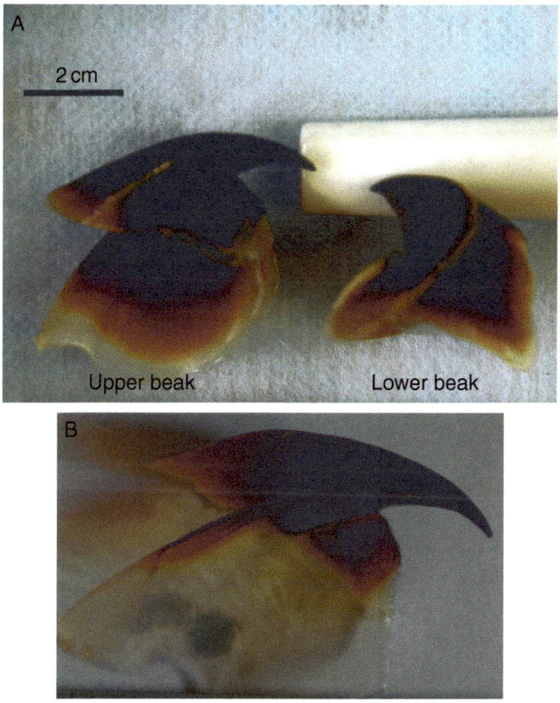

PLATE 13 *Dosidicus gigas* beaks dissected from the buccal mass. (A) Upper and lower beak clearly depicts a gradient in colour from translucent/white to opaque/black along the proximal to distal axis. (B) Upper beak after a sagittal cut along the beak with a diamond saw, illustrating the V-shape morphology with external and internal regions meeting at the sharp tip. (For b/w version, see page 104 in the volume.)

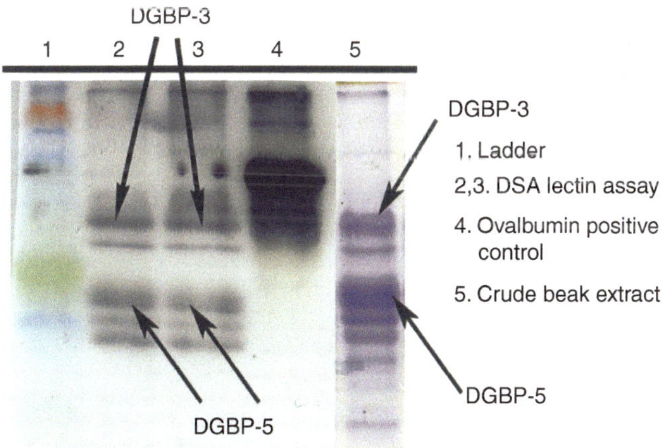

PLATE 14 Lectin-binding proteins in *Dosidicus* beak. A DSA (*Datura stramonium agglutinin*) lectin blot, known to bind glucosamine, was used to assay beak proteins for possible chitin-binding sites. A biotylated DSA conjugate was used, which was subsequently exposed to horseradish peroxidase before final detection with a chloronapthol reagent. Many beak proteins including Dgbp-3 and Dgbp-5 are decorated with glucosamine. (For b/w version, see page 107 in the volume.)

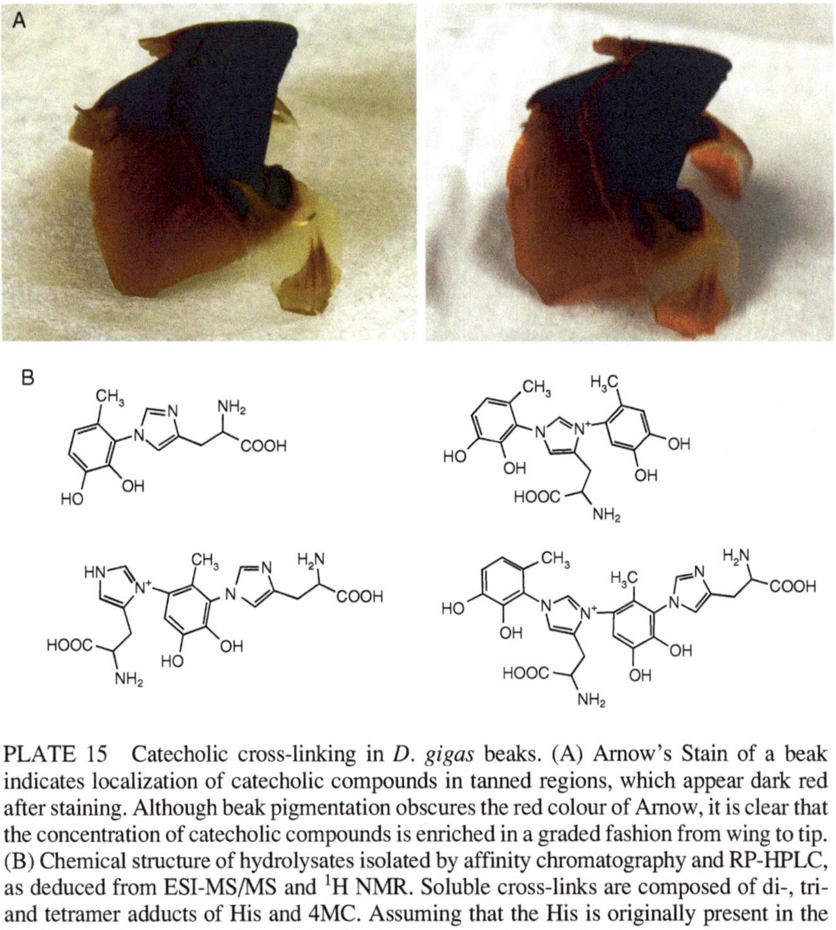

PLATE 15 Catecholic cross-linking in *D. gigas* beaks. (A) Arnow's Stain of a beak indicates localization of catecholic compounds in tanned regions, which appear dark red after staining. Although beak pigmentation obscures the red colour of Arnow, it is clear that the concentration of catecholic compounds is enriched in a graded fashion from wing to tip. (B) Chemical structure of hydrolysates isolated by affinity chromatography and RP-HPLC, as deduced from ESI-MS/MS and ^{1}H NMR. Soluble cross-links are composed of di-, tri- and tetramer adducts of His and 4MC. Assuming that the His is originally present in the backbone of a major beak protein, the trimer and tetramer structures are implicated as cross-link structures. (For b/w version, see page 108 in the volume.)

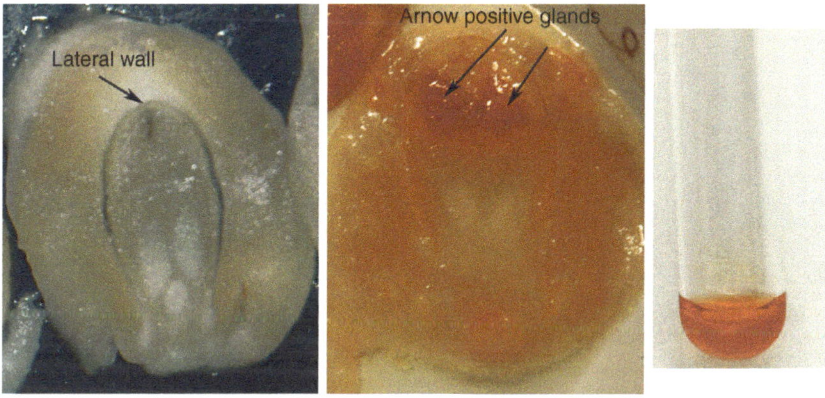

PLATE 16 The buccal mass and its catechol-rich glands. (A) Cross-section of the buccal mass, including location of the proximal region of the lateral wall (inverted U-shape). (B) Glands situated in close proximity with the lateral wall are Arnow positive suggesting that they stockpile catecholic molecules (small molecules or protein-bound catechols, i.e. DOPA). (C) Extracted compounds from the glands after late elution from gel-filtration chromatography are Arnow positive suggesting that catechols are either 'free' or are bound to small proteins/peptides. (For b/w version, see page 109 in the volume.)

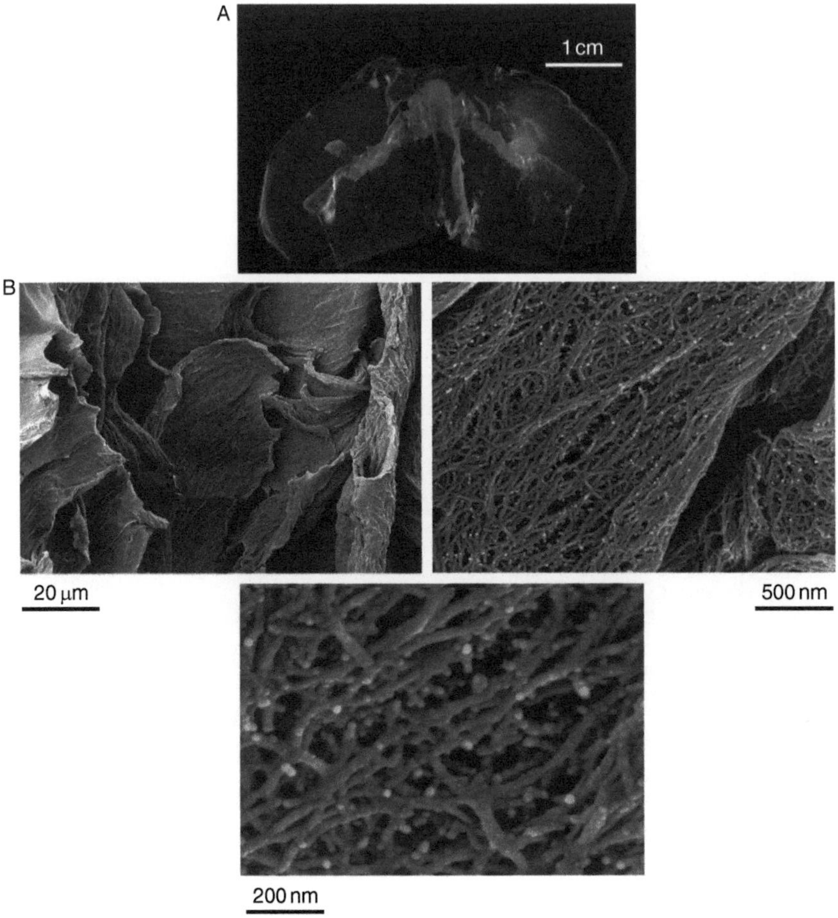

PLATE 17 *Dosidicus* beak after alkaline peroxide treatment. (A) Frontal view of a full lower beak showing that the residual shape is retained after pigment and protein removal. (B) SEM micrographs at increasing magnification of the degraded beak the after fixation at various magnifications, illustrating the residual scaffold made of leaflets containing intertangled chitin nanofibres. (For b/w version, see page 111 in the volume.)

PLATE 18 Illuminated bowl, showing uncoloured highlights due to specular reflection and blue coloration due to diffuse scattering by pigmented granules. (For b/w version, see page 183 in the volume.)

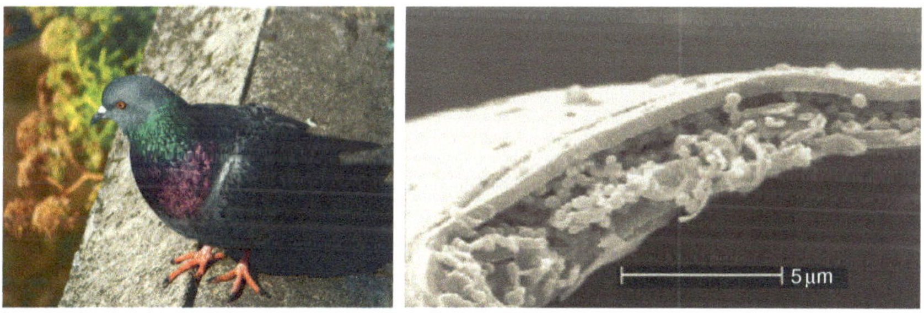

PLATE 19 An example of colouration by a simple thin film in nature: the green and violet feathers of the pigeon. The barbules are a sack filled with melanin granules (Yin *et al.*, 2006). The membrane is a transparent thin film, with a thickness remarkably constant, where light interferences take place to filter out green or violet colours, according to the membrane thickness. (For b/w version, see page 193 in the volume.)

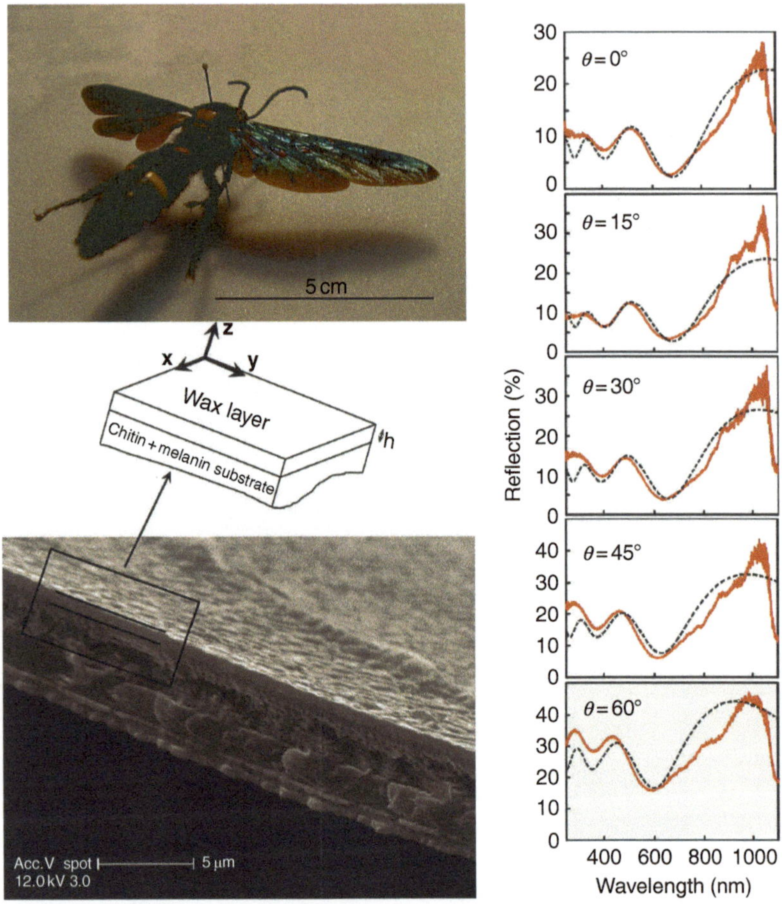

PLATE 20 Upper left: The giant tropical wasp *Megascolia procer javanensis*. Lower left: the structure of the wings. The iridescence can be explained by single-layer thin-film interference. Right panel: experimental reflectance spectra (solid line) at different incidence angles and the fit corresponding to a thin-film refractive index of $n = 1.76$. (For b/w version, see page 194 in the volume.)

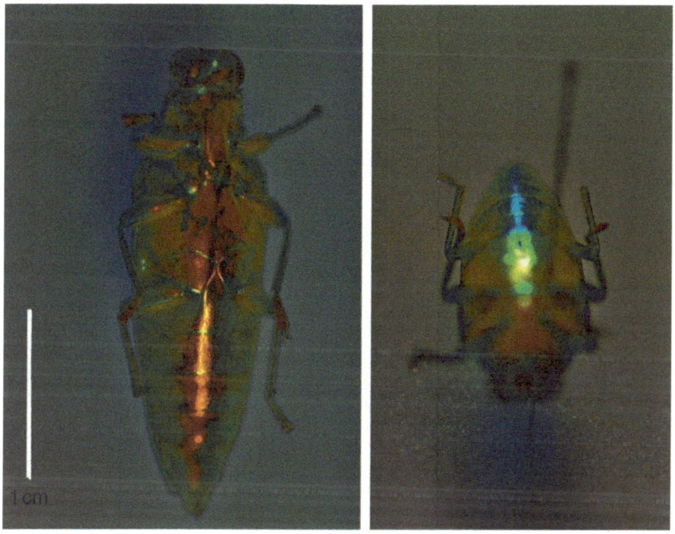

PLATE 21 Iridescence of the elytra of a buprestis *Chrysochroa vittata*. When viewed under normal incidence (left) the abdomen is red. The same specimen, viewed under a grazing light emergence, becomes blue. Iridescence is the change of colour according to the viewing angle. (For b/w version, see page 198 in the volume.)

PLATE 22 *Hoplia coerulea* (lower right). The beetle's cuticle is, as seen here (higher left), covered by scales, often slightly curved out. The scales take the shape of a disc, with a diameter of about 50 µm and a thickness of 3.5 µm. The dorsal scales, seen under the optical microscope, render a blue or violet colour. The scanning electron microscope images (higher right) show the colouring structure inside the scales. An idealized model of the colouring structure, with the dimensions observed, is shown on the lower left. (For b/w version, see page 201 in the volume.)

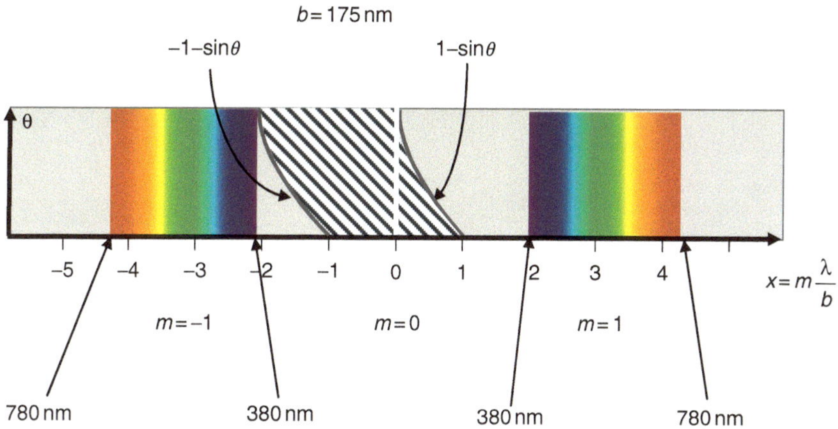

PLATE 23 The condition for diffraction by a grating can be analysed by this diagram, drawn for a given grating period (here $b = 175$ nm). Far-field emergence occurs for a given diffraction order m when the parameter x and the angle of incidence form a Cartesian point which lies in the hatched region, at the centre of the diagram. (For b/w version, see page 203 in the volume.)

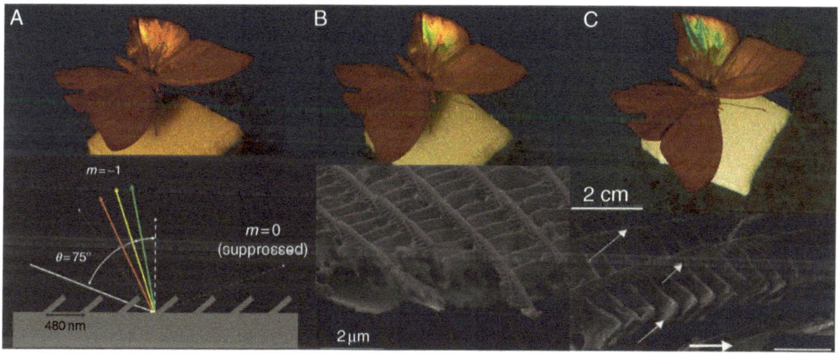

PLATE 24 *Lamprolenis nitida*, like many butterflies, exhibits sexually dimorphic iridescence. This was found to originate from first-order blazed diffraction gratings formed by different scale nanostructures, which are angled with respect to the scale surface. (For b/w version, see page 205 in the volume.)

PLATE 25 Male and female *Pierella luna* butterflies viewed at different angles. The male is behind the female. The illumination is 45° from the normal and originates at a point opposite to the viewer (view is against the light). As we lower our angle of view, but keep the light source steady, the forewing of the male butterfly displays a rainbow of iridescence, covering nearly the entire spectrum perceivable by humans. The females lack iridescence. (For b/w version, see page 205 in the volume.)

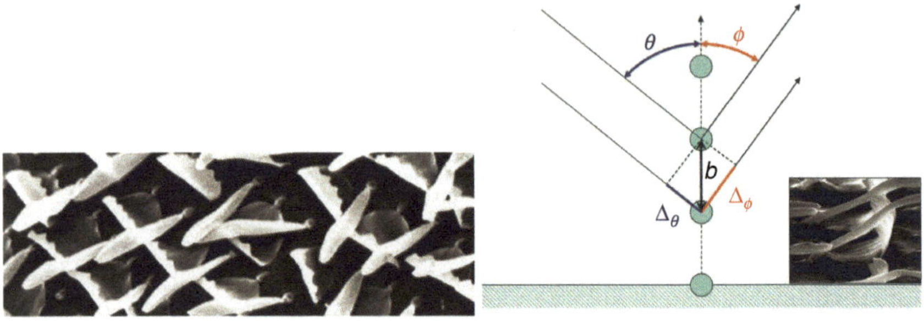

PLATE 26 The iridescent part of the wings of the male *Pierella luna* contains curled scales that function as a transmitting vertical grating. (For b/w version, see page 206 in the volume.)

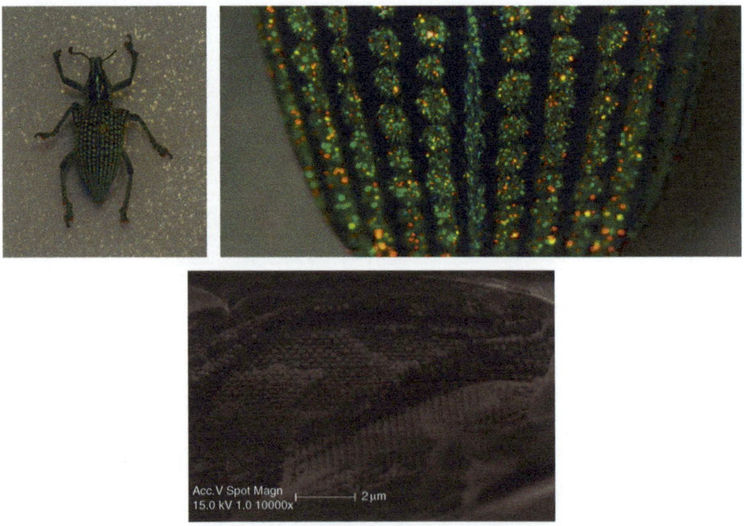

PLATE 27 The weevil *Entimus imperialis* shows one of the most perfect three-dimensional photonic crystals in nature. The insect bears transparent scales that scatter white light as many different colours, ranging from deep blue to red. The origin of this colouration is the diffraction by the structure shown in the right panel, occurring inside each scale. The scale itself is about 100 μm long, and contains one or two large grains of photonic crystal. (For b/w version, see page 210 in the volume.)

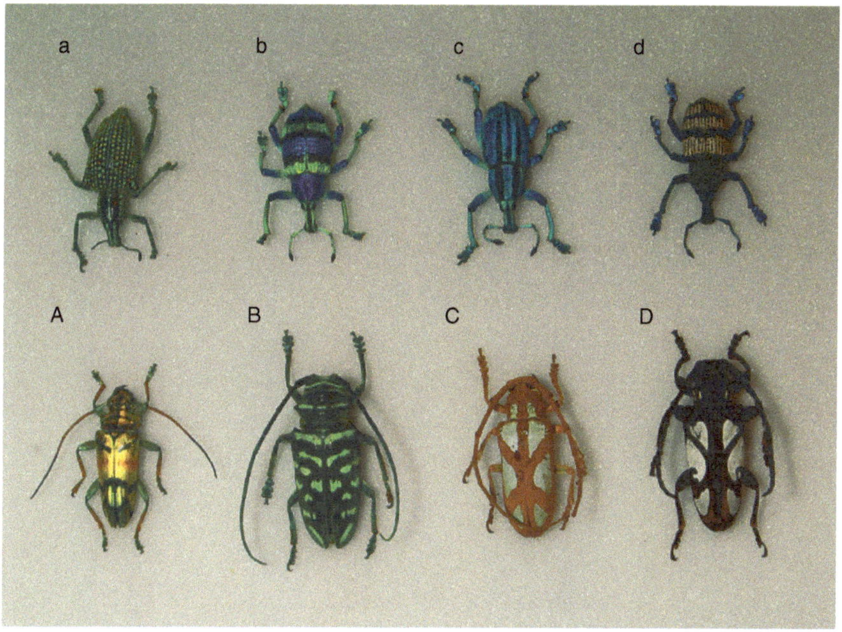

PLATE 28 Series of weevils and longhorns, where the amount of long-range disorder affects the visual effect produced by the photonic structure in the scales. In the specimens (a) and (A), the structure is iridescent and produce a metallic visual effect: the scales contain a high-coherence photonic crystal. In specimens (d) and (D), the amorphous structure produces white. In the other specimens, the variable saturation and metallic sheen results from an appropriate tuning of disorder in the photonic structure. (For b/w version, see page 213 in the volume.)

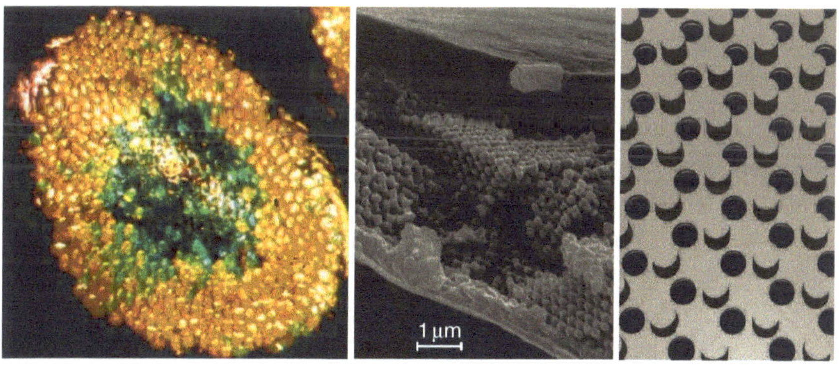

PLATE 29 The three-dimensional structure that causes the colouration of the tropical weevil *Pachyrrhynchus congestus pavonius* is a stack of corrugated layers with a triangular symmetry and three unequivalent sites A (perforation), B (protrusion) and C (flat, empty site). In the assembly of layers, the perforation lies above an empty site, which lies above a protrusion. This type of 'ABC' stacking leads to a face-centred cubic lattice. (For b/w version, see page 214 in the volume.)

PLATE 30 Details of the scales of the ventral side of the wing of *Cyanophrys remus*, seen with the scanning electron microscope (A) and a transmission electron microscope (B). The colouring structure is actually a photonic polycrystal, which produces visual effects very different from what can be expected from large monocrystals. (For b/w version, see page 215 in the volume.)

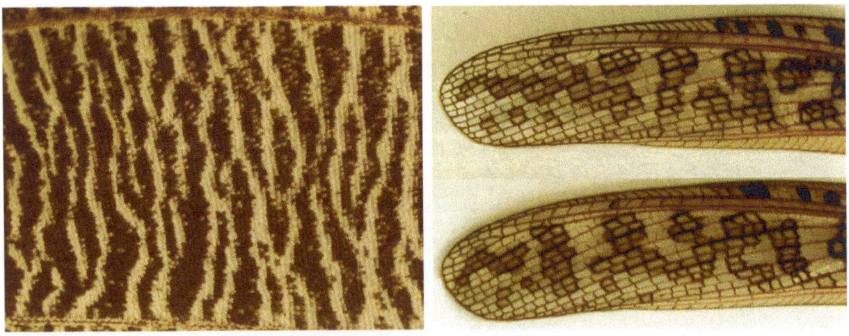

PLATE 31 Ripple patterns on the wing of *Caligo memnon* (left panel). Ripples are random colour patterns that are common in the Satyridae and related Lepidoptera. Here the ripples also illustrate stochastic pigment expression, in individual scales. Stochastic colour patterns on the wings of *Schistocerca americana* (right panel). Although there is an overall tendency towards stripes and blotches, the details of the pattern are random. (For b/w version, see page 222 in the volume.)

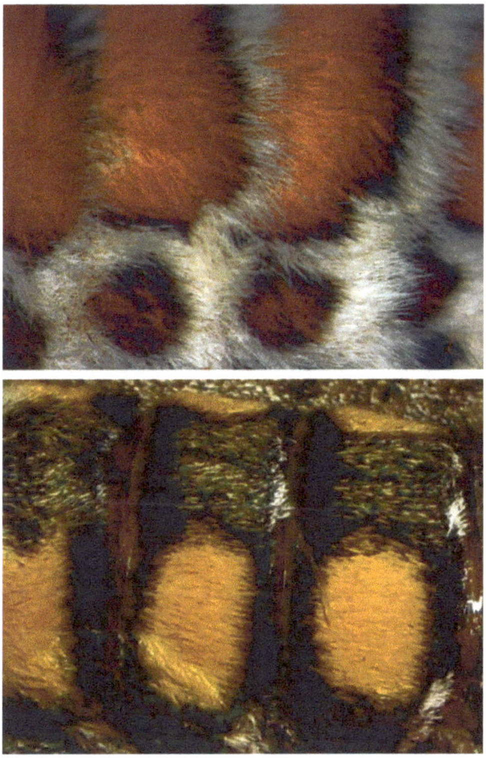

PLATE 32 Abdominal colour patterns in two moths: *Hyalophora cecropia* and *Manduca sexta*. (For b/w version, see page 224 in the volume.)

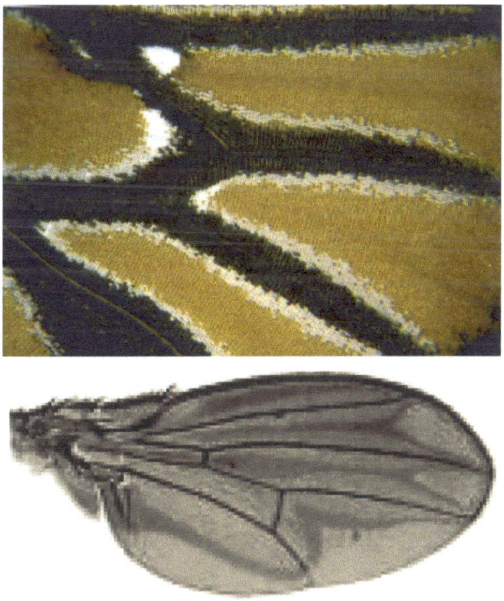

PLATE 33 Wing vein-dependent colour pattern in the Monarch butterfly, *Danaus plexippus*. Induced vein-dependent patterns in *Drosophila*. (From True *et al*., 1999.) (For b/w version, see page 229 in the volume.)

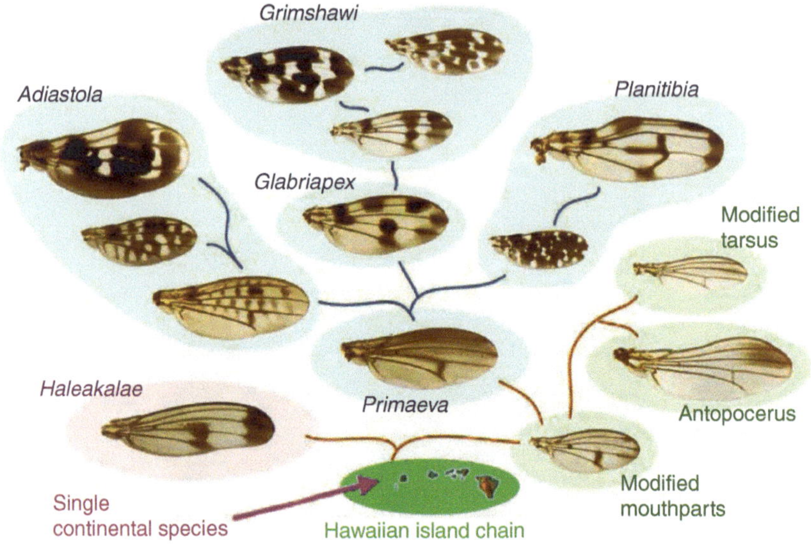

PLATE 34 Diversity of melanic wing patterns in the Hawaiian *Drosophila*. (From Edwards *et al*., 2007.) (For b/w version, see page 229 in the volume.)

PLATE 35 Patterns in the veinless mutant of *Papilio xuthus*. In the absence of wing veins, the colour pattern forms as bands parallel to the outer wing margin, indicating the margin plays an important role in colour pattern specification. (After Koch and Nijhout, 2002.) (For b/w version, see page 230 in the volume.)

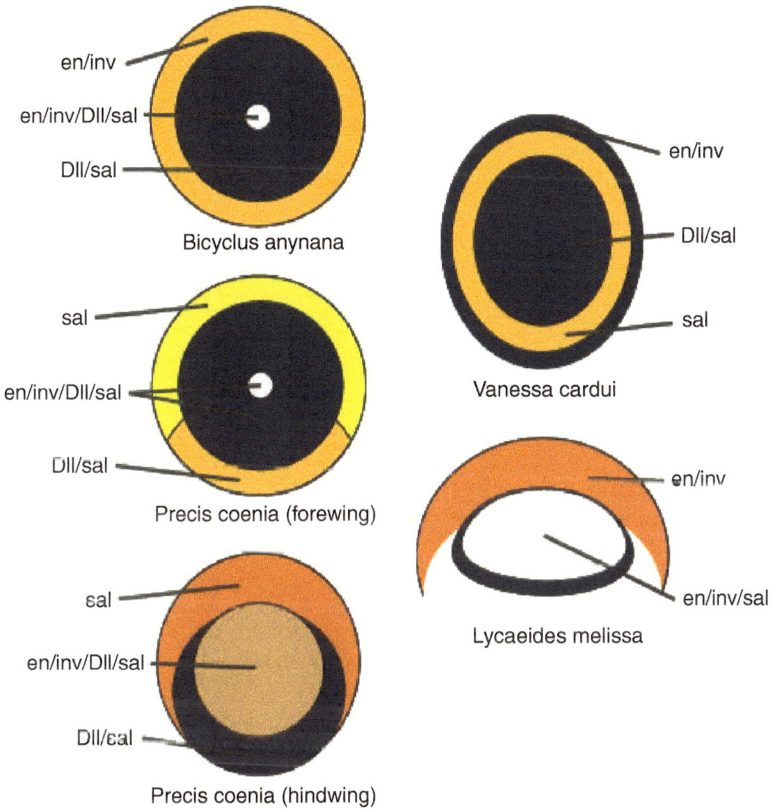

PLATE 36 The rings of eyespots in different species of butterflies are associated with different combinatorial expression patterns of a small number of regulatory genes. (After Brunetti *et al.*, 2001.) (For b/w version, see page 240 in the volume.)

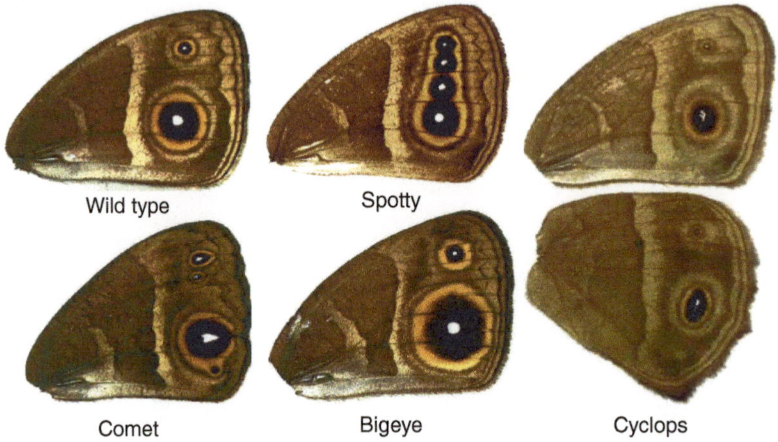

PLATE 37 Mutant eyespot pattern on the forewings of *Bicyclus* (Satyridae). The cyclops phenotype is also shown on the hindwing. (Photos courtesy of Paul Brakefield and Suzanne Saenko.) (For b/w version, see page 242 in the volume.)

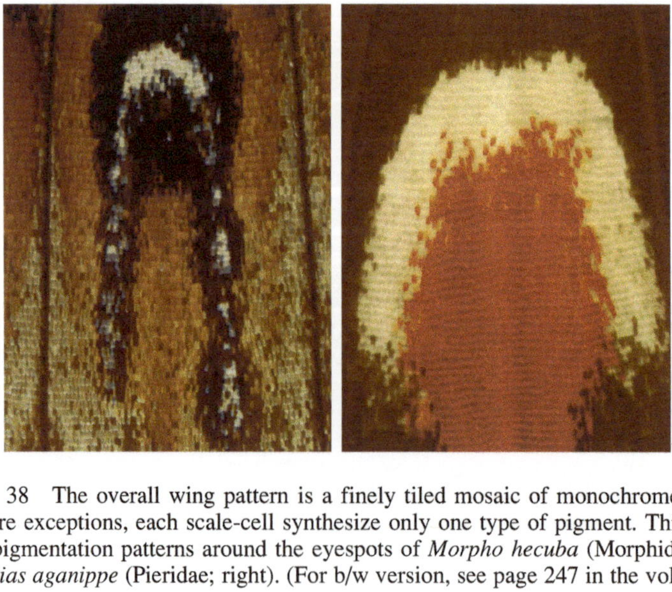

PLATE 38 The overall wing pattern is a finely tiled mosaic of monochrome scales. With rare exceptions, each scale-cell synthesize only one type of pigment. This figure shows pigmentation patterns around the eyespots of *Morpho hecuba* (Morphidae; left) and *Delias aganippe* (Pieridae; right). (For b/w version, see page 247 in the volume.)

PLATE 39 Ecdysone-dependent pattern development in *Araschnia levana*. Normal spring and summer forms are illustrated on left and right, respectively. Between them are two intermediate patterns that were generated by timed injections of ecdysone into presumptive spring form pupae. (After Nijhout, 2003a, experiments by Berndt Koch.) (For b/w version, see page 251 in the volume.)

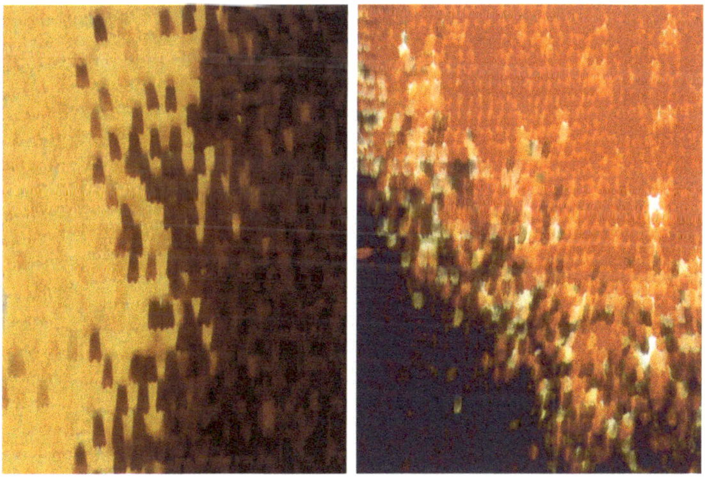

PLATE 40 Examples of stochastic gene expression leading to spatially stochastic patterns of pigment synthesis at the boundary between regions of different colours. Other examples of stochastic gene expression are shown in Figs. 1, 15 and 17. (After Nijhout, 2006.) (For b/w version, see page 256 in the volume.)

PLATE 41 Two models of artificial moth used for the study of distractive markings by Stevens *et al.* (2008a). Martin Stevens ©. (For b/w version, see page 305 in the volume.)

PLATE 42 An example of masquerade in the dead leaf mantid *Deroplatys desiccata* (photo Adrian Pingstone). (For b/w version, see page 307 in the volume.)

PLATE 43 The European peacock butterfly *Inachis io* often used to study the hypothesis of intimidating eyespots. Joanne T. Kell ©. (For b/w version, see page 316 in the volume.)

PLATE 44 Artificial moth models used for the study of the function of wing spots. (A) Used by Stevens *et al.* (2008b). (B) Used by Stevens *et al.* (2009a). Martin Stevens ©. (For b/w version, see page 318 in the volume.)

PLATE 45 The squinting bush brown butterfly *Bicyclus anynana* often used to test the hypothesis of wing spots deflection. William Piel and Antónia Monteiro ©. (For b/w version, see page 319 in the volume.)

PLATE 46 *Typica* and *carbonaria* morphs of the peppered moth *Biston betularia* on the same tree. This species has often been used to study colour polymorphism. Marteen Sanne ©. (For b/w version, see page 325 in the volume.)

FSC
www.fsc.org
MIX
Papier | Fördert
gute Waldnutzung
FSC® C083411

Kolibri 360 GmbH
Ferdinand-Jühlke-Straße 7
99095 Erfurt, Deutschland
produktsicherheit@kolibri360.de